WITHDRAWN
FAIRFIELD UNIVERSITY
LIBRARY

Methods in Enzymology

Volume 156
BIOMEMBRANES
Part P
ATP-Driven Pumps and Related Transport:
The Na,K-Pump

# METHODS IN ENZYMOLOGY

EDITORS-IN-CHIEF

John N. Abelson   Melvin I. Simon

Methods in Enzymology
Volume 156

Biomembranes

Part P
ATP-Driven Pumps and Related Transport: The Na,K-Pump

EDITED BY

Sidney Fleischer
Becca Fleischer

DEPARTMENT OF MOLECULAR BIOLOGY
VANDERBILT UNIVERSITY
NASHVILLE, TENNESSEE

Editorial Advisory Board

Yasuo Kagawa
Ronald Kaback
Martin Klingenberg
Robert L. Post

George Sachs
Antonio Scarpa
Widmar Tanner
Karl Ullrich

ACADEMIC PRESS, INC.
Harcourt Brace Jovanovich, Publishers
San Diego   New York   Berkeley   Boston
London   Sydney   Tokyo   Toronto

COPYRIGHT © 1988 BY ACADEMIC PRESS, INC.
ALL RIGHTS RESERVED.
NO PART OF THIS PUBLICATION MAY BE REPRODUCED OR
TRANSMITTED IN ANY FORM OR BY ANY MEANS, ELECTRONIC
OR MECHANICAL, INCLUDING PHOTOCOPY, RECORDING, OR
ANY INFORMATION STORAGE AND RETRIEVAL SYSTEM, WITHOUT
PERMISSION IN WRITING FROM THE PUBLISHER.

ACADEMIC PRESS, INC.
1250 Sixth Avenue
San Diego, California 92101

United Kingdom Edition published by
ACADEMIC PRESS INC. (LONDON) LTD.
24-28 Oval Road, London NW1 7DX

LIBRARY OF CONGRESS CATALOG CARD NUMBER: 54-9110

ISBN 0-12-182057-2 (alk. paper)

PRINTED IN THE UNITED STATES OF AMERICA
88 89 90 91   9 8 7 6 5 4 3 2 1

Table of Contents

CONTRIBUTORS TO VOLUME 156 ix
PREFACE xiii
VOLUMES IN SERIES xv

1. Overview: The Na,K-Pump	JENS CHRISTIAN SKOU	1

Section I. Preparation of Na^+,K^+-ATPase and Subunits

2. Purification of Na^+,K^+-ATPase: Enzyme Sources, Preparative Problems, and Preparation from Mammalian Kidney	PETER LETH JØRGENSEN	29
3. Preparation of Membrane Na^+,K^+-ATPase from Rectal Glands of *Squalus acanthias*	JENS CHRISTIAN SKOU AND MIKAEL ESMANN	43
4. Purification of Na^+,K^+-ATPase from the Supraorbital Salt Gland of the Duck	THOMAS WOODWARD SMITH	46
5. Preparation of Na^+,K^+-ATPase from Brine Shrimp	GARY L. PETERSON AND LOWELL E. HOKIN	48
6. Preparation of the $\alpha(+)$ Isozyme of the Na^+,K^+-ATPase from Mammalian Axolemma	KATHLEEN J. SWEADNER	65
7. Solubilization of Na^+,K^+-ATPase	MIKAEL ESMANN	72
8. Crystallization of Membrane-Bound Na^+,K^+-ATPase in Two Dimensions	ELISABETH SKRIVER, ARVID B. MAUNSBACH, HANS HERBERT, AND PETER L. JØRGENSEN	80
9. Preparation of Antibodies to Na^+,K^+-ATPase and Its Subunits	WILLIAM J. BALL, JR., TERENCE L. KIRLEY, AND LOIS K. LANE	87

Section II. Assay of Na^+,K^+-ATPase Activity

10. ATPase and Phosphatase Activity of Na^+,K^+-ATPase: Molar and Specific Activity, Protein Determination	MIKAEL ESMANN	105

11. Coupled Assay of Na$^+$,K$^+$-ATPase Activity — JENS G. NØRBY — 116

12. Identification and Quantitation of Na$^+$,K$^+$-ATPase by Back-Door Phosphorylation — MARILYN D. RESH — 119

Section III. Reconstitution of Na,K-Pump Activity

13. Reconstitution of Na,K-Pump Activity by Cholate Dialysis: Sidedness and Stoichiometry — STANLEY M. GOLDIN, MICHAEL FORGAC, AND GILBERT CHIN — 127

14. Reconstitution of the Na,K-Pump by Freeze–Thaw Sonication: Estimation of Coupling Ratio and Electrogenicity — LOWELL E. HOKIN AND JOHN F. DIXON — 141

15. Incorporation of $C_{12}E_8$-Solubilized Na$^+$,K$^+$-ATPase into Liposomes: Determination of Sidedness and Orientation — FLEMMING CORNELIUS — 156

Section IV. Analysis of the Pump Cycle

16. Measurement of Na$^+$ and K$^+$ Transport and Na$^+$,K$^+$-ATPase Activity in Inside-Out Vesicles from Mammalian Erythrocytes — RHODA BLOSTEIN — 171

17. Measurement of Active and Passive Na$^+$ and K$^+$ Fluxes in Reconstituted Vesicles — S. J. D. KARLISH — 179

Section V. Measurement of Ligand Binding and Distance between Ligands

18. Measurement of Binding of ATP and ADP to Na$^+$,K$^+$-ATPase — JENS G. NØRBY AND JØRGEN JENSEN — 191

19. Interaction of Cardiac Glycosides with Na$^+$,K$^+$-ATPase — EARL T. WALLICK AND ARNOLD SCHWARTZ — 201

20. Estimation of Na,K-Pump Numbers and Turnover in Intact Cells with [^{3}H]Ouabain — SETH R. HOOTMAN AND STEPHEN A. ERNST — 213

21. Measurement of Binding of Na$^+$, K$^+$, and Rb$^+$ to Na$^+$,K$^+$-ATPase by Centrifugation Methods — HIDEO MATSUI AND HARUO HOMAREDA — 229

22. Estimating Affinities for Physiological Ligands and Inhibitors by Kinetic Studies on Na$^+$,K$^+$-ATPase and Its Partial Activities — JOSEPH D. ROBINSON — 236

23. Inhibition of Translocation Reactions by Vanadate — LUIS BEAUGÉ — 251

Section VI. Measurements of Conformational States of Na$^+$,K$^+$-ATPase

24. Use of Formycin Nucleotides, Intrinsic Protein Fluorescence, and Fluorescein Isothiocyanate-Labeled Enzymes for Measurement of Conformational States of Na$^+$,K$^+$-ATPase — S. J. D. KARLISH — 271

25. Eosin as a Fluorescence Probe for Measurement of Conformational States of Na$^+$,K$^+$-ATPase — J. C. SKOU AND MIKAEL ESMANN — 278

26. Rapid Ion-Exchange Technique for Measuring Rates of Release of Occluded Ions — I. M. GLYNN, D. E. RICHARDS, AND L. A. BEAUGÉ — 281

Section VII. Modification of Na$^+$,K$^+$-ATPase

27. Proteolytic Cleavage as a Tool for Studying Structure and Conformation of Pure Membrane-Bound Na$^+$,K$^+$-ATPase — PETER L. JØRGENSEN AND ROBERT A. FARLEY — 291

28. Irreversible and Reversible Modification of SH Groups and Effect on Catalytic Activity — WILHELM SCHONER, MARION HASSELBERG, AND RALF KISON — 302

29. Photoaffinity Labeling with ATP Analogs — WILHELM SCHONER AND GEORGIOS SCHEINER-BOBIS — 312

30. Affinity Labeling of the Digitalis-Binding Site — BERNARD ROSSI AND MICHEL LAZDUNSKI — 323

31. Determination of Quaternary Structure of an Active Enzyme Using Chemical Cross-Linking with Glutaraldehyde — WILLIAM S. CRAIG — 333

32. Use of Cross-Linking Reagents for Detection of Subunit Interactions of Membrane-Bound Na$^+$,K$^+$-ATPase — WU-HSIUNG HUANG, SHAM S. KAKAR, SANKARIDRUG M. PERIYASAMY, AND AMIR ASKARI — 345

Section VIII. Magnetic Resonance Studies of Na$^+$,K$^+$-ATPase

33. Nuclear Magnetic Resonance Investigations of Na$^+$,K$^+$-ATPase — CHARLES M. GRISHAM — 353

34. Electron Spin Resonance Investigations of Na$^+$,K$^+$-ATPase — MIKAEL ESMANN — 371

Section IX. Biogenesis and Membrane Assembly

35. Molecular Cloning of Na$^+$,K$^+$-ATPase α Subunit Gene Using Antibody Probes — JAY W. SCHNEIDER, ROBERT W. MERCER, EDWARD J. BENZ, JR., AND ROBERT LEVENSON — 379

36. Preparation and Use of Monoclonal Antibodies to Na$^+$,K$^+$-ATPase — MICHAEL KASHGARIAN AND DANIEL BIEMESDERFER — 392

Section X. Microscopy of Na$^+$,K$^+$-ATPase

37. Histochemical Localization of Na$^+$,K$^+$-ATPase — HIROSHI MAYAHARA AND KAZUO OGAWA — 417

38. Analysis of Na$^+$,K$^+$-ATPase by Electron Microscopy — ARVID B. MAUNSBACH, ELISABETH SKRIVER, AND PETER L. JØRGENSEN — 430

AUTHOR INDEX 443

SUBJECT INDEX 455

Contributors to Volume 156

Article numbers are in parentheses following the names of contributors. Affiliations listed are current.

AMIR ASKARI (32), *Department of Pharmacology, Medical College of Ohio, Toledo, Ohio 43699*

WILLIAM J. BALL, JR. (9), *Department of Pharmacology and Cell Biophysics, University of Cincinnati College of Medicine, Cincinnati, Ohio 45267-0575*

LUIS BEAUGÉ (23, 26), *División de Biofísica, Instituto de Investigación Médica Mercedes y Martín Ferreyra, 5000 Córdoba, Argentina*

EDWARD J. BENZ, JR. (35), *Department of Internal Medicine and Human Genetics, Yale University School of Medicine, New Haven, Connecticut 06510*

DANIEL BIEMESDERFER (36), *Department of Cell Biology, Yale University School of Medicine, New Haven, Connecticut 06510*

RHODA BLOSTEIN (16), *Departments of Medicine and Biochemistry, McGill University, and the Montreal General Hospital Research Institute, Montreal, Quebec, Canada H3G 1A4*

GILBERT CHIN (13), *Department of Physiology and Neurology, College of Physicians and Surgeons, Columbia University, New York, New York 10032*

FLEMMING CORNELIUS (15), *Institute of Biophysics, University of Aarhus, DK-8000 Aarhus C, Denmark*

WILLIAM S. CRAIG (31), *Department of Microbiology and Biochemistry, Salk Institute Biotechnology/Industrial Associates, San Diego, California 92138*

JOHN F. DIXON (14), *Department of Pharmacology, University of Wisconsin Medical School, Madison, Wisconsin 53706*

STEPHEN A. ERNST (20), *Department of Anatomy and Cell Biology, University of Michigan, Ann Arbor, Michigan 48109*

MIKAEL ESMANN (3, 7, 10, 25, 34), *Institute of Biophysics, University of Aarhus, DK-8000 Aarhus C, Denmark*

ROBERT A. FARLEY (27), *Department of Physiology and Biophysics, University of Southern California, Los Angeles, California 90033*

MICHAEL FORGAC (13), *Department of Physiology, Tufts University School of Medicine, Boston, Massachusetts 02111*

I. M. GLYNN (26), *Physiological Laboratory, University of Cambridge, Cambridge CB2 3EG, England*

STANLEY M. GOLDIN (13), *Department of Biological Chemistry and Molecular Pharmacology, Harvard Medical School, Boston, Massachusetts 02115*

CHARLES M. GRISHAM (33), *Department of Chemistry, University of Virginia, Charlottesville, Virginia 22901*

MARION HASSELBERG (28), *Institut für Biochemie und Endokrinologie, Justus-Liebig-Universität Giessen, D-6300 Giessen, Federal Republic of Germany*

HANS HEBERT (8), *Department of Medical Biophysics, Karolinska Institutet, S-104 01 Stockholm, Sweden*

LOWELL E. HOKIN (5, 14), *Department of Pharmacology, University of Wisconsin Medical School, Madison, Wisconsin 53706*

HARUO HOMAREDA (21), *Department of Biochemistry, Kyorin University School of Medicine, Mitaka, Tokyo 181, Japan*

SETH R. HOOTMAN (20), *Department of Physiology, Michigan State University, East Lansing, Michigan 48824*

WU-HSIUNG HUANG (32), *Department of Pharmacology, Medical College of Ohio, Toledo, Ohio 43699*

ix

JØRGEN JENSEN (18), *Institute of Physiology, University of Aarhus, DK-8000 Aarhus C, Denmark*

PETER L. JØRGENSEN (2, 8, 27, 38), *Institute of Physiology, University of Aarhus, DK-8000 Aarhus C, Denmark*

SHAM S. KAKAR (32), *Department of Pharmacology, Medical College of Ohio, Toledo, Ohio 43699*

S. J. D. KARLISH (17, 24), *Department of Biochemistry, The Weizmann Institute of Science, Rehovot 76100, Israel*

MICHAEL KASHGARIAN (36), *Department of Pathology, Yale University School of Medicine, New Haven, Connecticut 06510*

TERENCE L. KIRLEY (9), *Department of Pharmacology and Cell Biophysics, University of Cincinnati College of Medicine, Cincinnati, Ohio 45267-0575*

RALF KISON (28), *Institut für Biochemie und Endokrinologie, Justus-Liebig-Universität Giessen, D-6300 Giessen, Federal Republic of Germany*

LOIS K. LANE (9), *Department of Pharmacology and Cell Biophysics, University of Cincinnati College of Medicine, Cincinnati, Ohio 45267-0575*

MICHEL LAZDUNSKI (30), *Centre de Biochimie du Centre National de la Recherche Scientifique, Faculté des Sciences, Parc Valrose, 06034 Nice Cedex, France*

ROBERT LEVENSON (35), *Department of Cell Biology, Yale University School of Medicine, New Haven, Connecticut 06510*

HIDEO MATSUI (21), *Department of Biochemistry, Kyorin University School of Medicine, Mitaka, Tokyo 181, Japan*

ARVID B. MAUNSBACH (8, 38), *Department of Cell Biology, Institute of Anatomy, University of Aarhus, DK-8000 Aarhus C, Denmark*

HIROSHI MAYAHARA (37), *Drug Safety Research Laboratories, Takeda Chemical Industries, Takatsuki, Osaka 569, Japan*

ROBERT W. MERCER (35), *Department of Physiology, Yale University School of Medicine, New Haven, Connecticut 06510*

JENS G. NØRBY (11, 18), *Institute of Biophysics, University of Aarhus, DK-8000 Aarhus C, Denmark*

KAZUO OGAWA (37), *Department of Anatomy, Faculty of Medicine, Kyoto University, Sakyo-ku, Kyoto 606, Japan*

SANKARIDRUG M. PERIYASAMY (32), *Department of Pharmacology, Medical College of Ohio, Toledo, Ohio 43699*

GARY L. PETERSON (5), *Department of Biochemistry and Biophysics, Oregon State University, Corvallis, Oregon 97331*

MARILYN D. RESH (12), *Department of Biology, Princeton University, Princeton, New Jersey 08544*

D. E. RICHARDS (26), *Physiological Laboratory, University of Cambridge, Cambridge CB2 3EG, England*

JOSEPH D. ROBINSON (22), *Department of Pharmacology, State University of New York, Health Science Center, Syracuse, New York 13210*

BERNARD ROSSI (30), *Groupe INSERM U-145, Faculté de Médecine, Chemin de Vallombrose, 06100 Nice, France*

GEORGIOS SCHEINER-BOBIS (29), *Institut für Biochemie und Endokrinologie, Justus-Liebig-Universität Giessen, D-6300 Giessen, Federal Republic of Germany*

JAY W. SCHNEIDER (35), *Department of Internal Medicine and Human Genetics, Yale University School of Medicine, New Haven, Connecticut 06510*

WILHELM SCHONER (28, 29), *Institut für Biochemie und Endokrinologie, Justus-Liebig-Universität Giessen, D-6300 Giessen, Federal Republic of Germany*

ARNOLD SCHWARTZ (19), *Department of Pharmacology and Cell Biophysics, University of Cincinnati College of Medicine, Cincinnati, Ohio 45267-0575*

JENS CHRISTIAN SKOU (1, 3, 25), *Institute of Biophysics, University of Aarhus, DK-8000 Aarhus C, Denmark*

ELISABETH SKRIVER (8, 38), *Department of Cell Biology, Institute of Anatomy, University of Aarhus, DK-8000 Aarhus C, Denmark*

THOMAS WOODWARD SMITH (4), *Cardiovascular Division and the Department of Medicine, Brigham and Women's Hospital, and Harvard Medical School, Boston, Massachusetts 02115*

KATHLEEN J. SWEADNER (6), *Neurosurgical Research, Massachusetts General Hospital, Boston, Massachusetts 02114*

EARL T. WALLICK (19), *Department of Pharmacology and Cell Biophysics, University of Cincinnati College of Medicine, Cincinnati, Ohio 45267-0575*

Preface

The transport volumes of Biomembranes were initiated with Volumes 125 and 126 (Transport in Bacteria, Mitochondria, and Chloroplasts) of *Methods in Enzymology*. Biological transport represents a continuation of methodology regarding the study of membrane function, Volumes 96–98 having dealt with membrane biogenesis, assembly, targeting, and recycling.

This is a particularly good time to cover the topic of biological membrane transport because a strong conceptual basis for its understanding now exists. Membrane transport has been divided into five topics. Topic 2 is covered in Volumes 156 and 157. The remaining three topics will be covered in subsequent volumes of the Biomembranes series.

1. Transport in Bacteria, Mitochondria, and Chloroplasts
2. ATP-Driven Pumps and Related Transport
3. General Methodology of Cellular and Subcellular Transport
4. Cellular and Subcellular Transport: Eukaryotic (Nonepithelial) Cells
5. Cellular and Subcellular Transport: Epithelial Cells

We are fortunate to have the good counsel of our Advisory Board. Their input ensures the quality of these volumes. The same Advisory Board has served for the complete transport series. Valuable input on the outlines of the five volumes was also provided by Qais Al-Awqati, Ernesto Carafoli, Halvor Christiansen, Isadore Edelman, Joseph Hoffman, Phil Knauf, and Hermann Passow.

The names of our board members and advisors were inadvertantly omitted in Volumes 125 and 126. When we noted the omission, it was too late to rectify the problem. For Volumes 125 and 126, we are also pleased to acknowledge the advice of Angelo Azzi, Youssef Hatefi, Dieter Oesterhelt, and Peter Pedersen.

Additional valuable input to Volumes 156 and 157 was obtained from Jens Skou, Peter Jørgensen, Steve Karlish, Gerhard Meissner, and Giuseppi Inesi. The enthusiasm and cooperation of the participants have enriched and made these volumes possible. The friendly cooperation of the staff of Academic Press is gratefully acknowledged.

These volumes are dedicated to Professor Sidney Colowick, a dear friend and colleague, who died in 1985. We shall miss his wise counsel, encouragement, and friendship.

SIDNEY FLEISCHER
BECCA FLEISCHER

METHODS IN ENZYMOLOGY

EDITED BY

Sidney P. Colowick and Nathan O. Kaplan

VANDERBILT UNIVERSITY
SCHOOL OF MEDICINE
NASHVILLE, TENNESSEE

DEPARTMENT OF CHEMISTRY
UNIVERSITY OF CALIFORNIA
AT SAN DIEGO
LA JOLLA, CALIFORNIA

I. Preparation and Assay of Enzymes
II. Preparation and Assay of Enzymes
III. Preparation and Assay of Substrates
IV. Special Techniques for the Enzymologist
V. Preparation and Assay of Enzymes
VI. Preparation and Assay of Enzymes (*Continued*)
   Preparation and Assay of Substrates
   Special Techniques
VII. Cumulative Subject Index

METHODS IN ENZYMOLOGY

EDITORS-IN-CHIEF

Sidney P. Colowick and Nathan O. Kaplan

VOLUME VIII. Complex Carbohydrates
Edited by ELIZABETH F. NEUFELD AND VICTOR GINSBURG

VOLUME IX. Carbohydrate Metabolism
Edited by WILLIS A. WOOD

VOLUME X. Oxidation and Phosphorylation
Edited by RONALD W. ESTABROOK AND MAYNARD E. PULLMAN

VOLUME XI. Enzyme Structure
Edited by C. H. W. HIRS

VOLUME XII. Nucleic Acids (Parts A and B)
Edited by LAWRENCE GROSSMAN AND KIVIE MOLDAVE

VOLUME XIII. Citric Acid Cycle
Edited by J. M. LOWENSTEIN

VOLUME XIV. Lipids
Edited by J. M. LOWENSTEIN

VOLUME XV. Steroids and Terpenoids
Edited by RAYMOND B. CLAYTON

VOLUME XVI. Fast Reactions
Edited by KENNETH KUSTIN

VOLUME XVII. Metabolism of Amino Acids and Amines (Parts A and B)
Edited by HERBERT TABOR AND CELIA WHITE TABOR

VOLUME XVIII. Vitamins and Coenzymes (Parts A, B, and C)
Edited by DONALD B. MCCORMICK AND LEMUEL D. WRIGHT

VOLUME XIX. Proteolytic Enzymes
Edited by GERTRUDE E. PERLMANN AND LASZLO LORAND

VOLUME XX. Nucleic Acids and Protein Synthesis (Part C)
Edited by KIVIE MOLDAVE AND LAWRENCE GROSSMAN

VOLUME XXI. Nucleic Acids (Part D)
Edited by LAWRENCE GROSSMAN AND KIVIE MOLDAVE

VOLUME XXII. Enzyme Purification and Related Techniques
Edited by WILLIAM B. JAKOBY

VOLUME XXIII. Photosynthesis (Part A)
Edited by ANTHONY SAN PIETRO

VOLUME XXIV. Photosynthesis and Nitrogen Fixation (Part B)
Edited by ANTHONY SAN PIETRO

VOLUME XXV. Enzyme Structure (Part B)
Edited by C. H. W. HIRS AND SERGE N. TIMASHEFF

VOLUME XXVI. Enzyme Structure (Part C)
Edited by C. H. W. HIRS AND SERGE N. TIMASHEFF

VOLUME XXVII. Enzyme Structure (Part D)
Edited by C. H. W. HIRS AND SERGE N. TIMASHEFF

VOLUME XXVIII. Complex Carbohydrates (Part B)
Edited by VICTOR GINSBURG

VOLUME XXIX. Nucleic Acids and Protein Synthesis (Part E)
Edited by LAWRENCE GROSSMAN AND KIVIE MOLDAVE

VOLUME XXX. Nucleic Acids and Protein Synthesis (Part F)
Edited by KIVIE MOLDAVE AND LAWRENCE GROSSMAN

VOLUME XXXI. Biomembranes (Part A)
Edited by SIDNEY FLEISCHER AND LESTER PACKER

VOLUME XXXII. Biomembranes (Part B)
Edited by SIDNEY FLEISCHER AND LESTER PACKER

VOLUME XXXIII. Cumulative Subject Index Volumes I–XXX
Edited by MARTHA G. DENNIS AND EDWARD A. DENNIS

VOLUME XXXIV. Affinity Techniques (Enzyme Purification: Part B)
Edited by WILLIAM B. JAKOBY AND MEIR WILCHEK

VOLUME XXXV. Lipids (Part B)
Edited by JOHN M. LOWENSTEIN

VOLUME XXXVI. Hormone Action (Part A: Steroid Hormones)
Edited by BERT W. O'MALLEY AND JOEL G. HARDMAN

VOLUME XXXVII. Hormone Action (Part B: Peptide Hormones)
Edited by BERT W. O'MALLEY AND JOEL G. HARDMAN

VOLUME XXXVIII. Hormone Action (Part C: Cyclic Nucleotides)
Edited by JOEL G. HARDMAN AND BERT W. O'MALLEY

VOLUME XXXIX. Hormone Action (Part D: Isolated Cells, Tissues, and Organ Systems)
Edited by JOEL G. HARDMAN AND BERT W. O'MALLEY

VOLUME XL. Hormone Action (Part E: Nuclear Structure and Function)
Edited by BERT W. O'MALLEY AND JOEL G. HARDMAN

VOLUME XLI. Carbohydrate Metabolism (Part B)
Edited by W. A. WOOD

VOLUME XLII. Carbohydrate Metabolism (Part C)
Edited by W. A. WOOD

VOLUME XLIII. Antibiotics
Edited by JOHN H. HASH

VOLUME XLIV. Immobilized Enzymes
Edited by KLAUS MOSBACH

VOLUME XLV. Proteolytic Enzymes (Part B)
Edited by LASZLO LORAND

VOLUME XLVI. Affinity Labeling
Edited by WILLIAM B. JAKOBY AND MEIR WILCHEK

VOLUME XLVII. Enzyme Structure (Part E)
Edited by C. H. W. HIRS AND SERGE N. TIMASHEFF

VOLUME XLVIII. Enzyme Structure (Part F)
Edited by C. H. W. HIRS AND SERGE N. TIMASHEFF

VOLUME XLIX. Enzyme Structure (Part G)
Edited by C. H. W. HIRS AND SERGE N. TIMASHEFF

VOLUME L. Complex Carbohydrates (Part C)
Edited by VICTOR GINSBURG

VOLUME LI. Purine and Pyrimidine Nucleotide Metabolism
Edited by PATRICIA A. HOFFEE AND MARY ELLEN JONES

VOLUME LII. Biomembranes (Part C: Biological Oxidations)
Edited by SIDNEY FLEISCHER AND LESTER PACKER

VOLUME LIII. Biomembranes (Part D: Biological Oxidations)
Edited by SIDNEY FLEISCHER AND LESTER PACKER

VOLUME LIV. Biomembranes (Part E: Biological Oxidations)
Edited by SIDNEY FLEISCHER AND LESTER PACKER

VOLUME LV. Biomembranes (Part F: Bioenergetics)
Edited by SIDNEY FLEISCHER AND LESTER PACKER

VOLUME LVI. Biomembranes (Part G: Bioenergetics)
Edited by SIDNEY FLEISCHER AND LESTER PACKER

VOLUME LVII. Bioluminescence and Chemiluminescence
Edited by MARLENE A. DELUCA

VOLUME LVIII. Cell Culture
Edited by WILLIAM B. JAKOBY AND IRA PASTAN

VOLUME LIX. Nucleic Acids and Protein Synthesis (Part G)
Edited by KIVIE MOLDAVE AND LAWRENCE GROSSMAN

VOLUME LX. Nucleic Acids and Protein Synthesis (Part H)
Edited by KIVIE MOLDAVE AND LAWRENCE GROSSMAN

VOLUME 61. Enzyme Structure (Part H)
Edited by C. H. W. HIRS AND SERGE N. TIMASHEFF

VOLUME 62. Vitamins and Coenzymes (Part D)
Edited by DONALD B. MCCORMICK AND LEMUEL D. WRIGHT

VOLUME 63. Enzyme Kinetics and Mechanism (Part A: Initial Rate and Inhibitor Methods)
Edited by DANIEL L. PURICH

VOLUME 64. Enzyme Kinetics and Mechanism (Part B: Isotopic Probes and Complex Enzyme Systems)
Edited by DANIEL L. PURICH

VOLUME 65. Nucleic Acids (Part I)
Edited by LAWRENCE GROSSMAN AND KIVIE MOLDAVE

VOLUME 66. Vitamins and Coenzymes (Part E)
Edited by DONALD B. MCCORMICK AND LEMUEL D. WRIGHT

VOLUME 67. Vitamins and Coenzymes (Part F)
Edited by DONALD B. MCCORMICK AND LEMUEL D. WRIGHT

VOLUME 68. Recombinant DNA
Edited by RAY WU

VOLUME 69. Photosynthesis and Nitrogen Fixation (Part C)
Edited by ANTHONY SAN PIETRO

VOLUME 70. Immunochemical Techniques (Part A)
Edited by HELEN VAN VUNAKIS AND JOHN J. LANGONE

VOLUME 71. Lipids (Part C)
Edited by JOHN M. LOWENSTEIN

VOLUME 72. Lipids (Part D)
Edited by JOHN M. LOWENSTEIN

VOLUME 73. Immunochemical Techniques (Part B)
Edited by JOHN J. LANGONE AND HELEN VAN VUNAKIS

VOLUME 74. Immunochemical Techniques (Part C)
Edited by JOHN J. LANGONE AND HELEN VAN VUNAKIS

VOLUME 75. Cumulative Subject Index Volumes XXXI, XXXII, XXXIV–LX
Edited by EDWARD A. DENNIS AND MARTHA G. DENNIS

VOLUME 76. Hemoglobins
Edited by ERALDO ANTONINI, LUIGI ROSSI-BERNARDI, AND EMILIA CHIANCONE

VOLUME 77. Detoxication and Drug Metabolism
Edited by WILLIAM B. JAKOBY

VOLUME 78. Interferons (Part A)
Edited by SIDNEY PESTKA

VOLUME 79. Interferons (Part B)
Edited by SIDNEY PESTKA

VOLUME 80. Proteolytic Enzymes (Part C)
Edited by LASZLO LORAND

VOLUME 81. Biomembranes (Part H: Visual Pigments and Purple Membranes, I)
Edited by LESTER PACKER

VOLUME 82. Structural and Contractile Proteins (Part A: Extracellular Matrix)
Edited by LEON W. CUNNINGHAM AND DIXIE W. FREDERIKSEN

VOLUME 83. Complex Carbohydrates (Part D)
Edited by VICTOR GINSBURG

VOLUME 84. Immunochemical Techniques (Part D: Selected Immunoassays)
Edited by JOHN J. LANGONE AND HELEN VAN VUNAKIS

VOLUME 85. Structural and Contractile Proteins (Part B: The Contractile Apparatus and the Cytoskeleton)
Edited by DIXIE W. FREDERIKSEN AND LEON W. CUNNINGHAM

VOLUME 86. Prostaglandins and Arachidonate Metabolites
Edited by WILLIAM E. M. LANDS AND WILLIAM L. SMITH

VOLUME 87. Enzyme Kinetics and Mechanism (Part C: Intermediates, Stereochemistry, and Rate Studies)
Edited by DANIEL L. PURICH

VOLUME 88. Biomembranes (Part I: Visual Pigments and Purple Membranes, II)
Edited by LESTER PACKER

VOLUME 89. Carbohydrate Metabolism (Part D)
Edited by WILLIS A. WOOD

VOLUME 90. Carbohydrate Metabolism (Part E)
Edited by WILLIS A. WOOD

VOLUME 91. Enzyme Structure (Part I)
Edited by C. H. W. HIRS AND SERGE N. TIMASHEFF

VOLUME 92. Immunochemical Techniques (Part E: Monoclonal Antibodies and General Immunoassay Methods)
Edited by JOHN J. LANGONE AND HELEN VAN VUNAKIS

VOLUME 93. Immunochemical Techniques (Part F: Conventional Antibodies, Fc Receptors, and Cytotoxicity)
Edited by JOHN J. LANGONE AND HELEN VAN VUNAKIS

VOLUME 94. Polyamines
Edited by HERBERT TABOR AND CELIA WHITE TABOR

VOLUME 95. Cumulative Subject Index Volumes 61–74, 76–80
Edited by EDWARD A. DENNIS AND MARTHA G. DENNIS

VOLUME 96. Biomembranes [Part J: Membrane Biogenesis: Assembly and Targeting (General Methods; Eukaryotes)]
Edited by SIDNEY FLEISCHER AND BECCA FLEISCHER

VOLUME 97. Biomembranes [Part K: Membrane Biogenesis: Assembly and Targeting (Prokaryotes, Mitochondria, and Chloroplasts)]
Edited by SIDNEY FLEISCHER AND BECCA FLEISCHER

VOLUME 98. Biomembranes (Part L: Membrane Biogenesis: Processing and Recycling)
Edited by SIDNEY FLEISCHER AND BECCA FLEISCHER

VOLUME 99. Hormone Action (Part F: Protein Kinases)
Edited by JACKIE D. CORBIN AND JOEL G. HARDMAN

VOLUME 100. Recombinant DNA (Part B)
Edited by RAY WU, LAWRENCE GROSSMAN, AND KIVIE MOLDAVE

VOLUME 101. Recombinant DNA (Part C)
Edited by RAY WU, LAWRENCE GROSSMAN, AND KIVIE MOLDAVE

VOLUME 102. Hormone Action (Part G: Calmodulin and Calcium-Binding Proteins)
Edited by ANTHONY R. MEANS AND BERT W. O'MALLEY

VOLUME 103. Hormone Action (Part H: Neuroendocrine Peptides)
Edited by P. MICHAEL CONN

VOLUME 104. Enzyme Purification and Related Techniques (Part C)
Edited by WILLIAM B. JAKOBY

VOLUME 105. Oxygen Radicals in Biological Systems
Edited by LESTER PACKER

VOLUME 106. Posttranslational Modifications (Part A)
Edited by FINN WOLD AND KIVIE MOLDAVE

VOLUME 107. Posttranslational Modifications (Part B)
Edited by FINN WOLD AND KIVIE MOLDAVE

VOLUME 108. Immunochemical Techniques (Part G: Separation and Characterization of Lymphoid Cells)
Edited by GIOVANNI DI SABATO, JOHN J. LANGONE, AND HELEN VAN VUNAKIS

VOLUME 109. Hormone Action (Part I: Peptide Hormones)
Edited by LUTZ BIRNBAUMER AND BERT W. O'MALLEY

VOLUME 110. Steroids and Isoprenoids (Part A)
Edited by JOHN H. LAW AND HANS C. RILLING

VOLUME 111. Steroids and Isoprenoids (Part B)
Edited by JOHN H. LAW AND HANS C. RILLING

VOLUME 112. Drug and Enzyme Targeting (Part A)
Edited by KENNETH J. WIDDER AND RALPH GREEN

VOLUME 113. Glutamate, Glutamine, Glutathione, and Related Compounds
Edited by ALTON MEISTER

VOLUME 114. Diffraction Methods for Biological Macromolecules (Part A)
Edited by HAROLD W. WYCKOFF, C. H. W. HIRS, AND SERGE N. TIMASHEFF

VOLUME 115. Diffraction Methods for Biological Macromolecules (Part B)
Edited by HAROLD W. WYCKOFF, C. H. W. HIRS, AND SERGE N. TIMASHEFF

VOLUME 116. Immunochemical Techniques (Part H: Effectors and Mediators of Lymphoid Cell Functions)
Edited by GIOVANNI DI SABATO, JOHN J. LANGONE, AND HELEN VAN VUNAKIS

VOLUME 117. Enzyme Structure (Part J)
Edited by C. H. W. HIRS AND SERGE N. TIMASHEFF

VOLUME 118. Plant Molecular Biology
Edited by ARTHUR WEISSBACH AND HERBERT WEISSBACH

VOLUME 119. Interferons (Part C)
Edited by SIDNEY PESTKA

VOLUME 120. Cumulative Subject Index Volumes 81–94, 96–101

VOLUME 121. Immunochemical Techniques (Part I: Hybridoma Technology and Monoclonal Antibodies)
Edited by JOHN J. LANGONE AND HELEN VAN VUNAKIS

VOLUME 122. Vitamins and Coenzymes (Part G)
Edited by FRANK CHYTIL AND DONALD B. MCCORMICK

VOLUME 123. Vitamins and Coenzymes (Part H)
Edited by FRANK CHYTIL AND DONALD B. MCCORMICK

VOLUME 124. Hormone Action (Part J: Neuroendocrine Peptides)
Edited by P. MICHAEL CONN

VOLUME 125. Biomembranes (Part M: Transport in Bacteria, Mitochondria, and Chloroplasts: General Approaches and Transport Systems)
Edited by SIDNEY FLEISCHER AND BECCA FLEISCHER

VOLUME 126. Biomembranes (Part N: Transport in Bacteria, Mitochondria, and Chloroplasts: Protonmotive Force)
Edited by SIDNEY FLEISCHER AND BECCA FLEISCHER

VOLUME 127. Biomembranes (Part O: Protons and Water: Structure and Translocation)
Edited by LESTER PACKER

VOLUME 128. Plasma Lipoproteins (Part A: Preparation, Structure, and Molecular Biology)
Edited by JERE P. SEGREST AND JOHN J. ALBERS

VOLUME 129. Plasma Lipoproteins (Part B: Characterization, Cell Biology, and Metabolism)
Edited by JOHN J. ALBERS AND JERE P. SEGREST

VOLUME 130. Enzyme Structure (Part K)
Edited by C. H. W. HIRS AND SERGE N. TIMASHEFF

VOLUME 131. Enzyme Structure (Part L)
Edited by C. H. W. HIRS AND SERGE N. TIMASHEFF

VOLUME 132. Immunochemical Techniques (Part J: Phagocytosis and Cell-Mediated Cytotoxicity)
Edited by GIOVANNI DI SABATO AND JOHANNES EVERSE

VOLUME 133. Bioluminescence and Chemiluminescence (Part B)
Edited by MARLENE DELUCA AND WILLIAM D. MCELROY

VOLUME 134. Structural and Contractile Proteins (Part C: The Contractile Apparatus and the Cytoskeleton)
Edited by RICHARD B. VALLEE

VOLUME 135. Immobilized Enzymes and Cells (Part B)
Edited by KLAUS MOSBACH

VOLUME 136. Immobilized Enzymes and Cells (Part C)
Edited by KLAUS MOSBACH

VOLUME 137. Immobilized Enzymes and Cells (Part D)
Edited by KLAUS MOSBACH

VOLUME 138. Complex Carbohydrates (Part E)
Edited by VICTOR GINSBURG

VOLUME 139. Cellular Regulators (Part A: Calcium- and Calmodulin-Binding Proteins)
Edited by ANTHONY R. MEANS AND P. MICHAEL CONN

VOLUME 140. Cumulative Subject Index Volumes 102–119, 121–134

VOLUME 141. Cellular Regulators (Part B: Calcium and Lipids)
Edited by P. MICHAEL CONN AND ANTHONY R. MEANS

VOLUME 142. Metabolism of Aromatic Amino Acids and Amines
Edited by SEYMOUR KAUFMAN

VOLUME 143. Sulfur and Sulfur Amino Acids
Edited by WILLIAM B. JAKOBY AND OWEN GRIFFITH

VOLUME 144. Structural and Contractile Proteins (Part D: Extracellular Matrix)
Edited by LEON W. CUNNINGHAM

VOLUME 145. Structural and Contractile Proteins (Part E: Extracellular Matrix)
Edited by LEON W. CUNNINGHAM

VOLUME 146. Peptide Growth Factors (Part A)
Edited by DAVID BARNES AND DAVID A. SIRBASKU

VOLUME 147. Peptide Growth Factors (Part B)
Edited by DAVID BARNES AND DAVID A. SIRBASKU

VOLUME 148. Plant Cell Membranes
Edited by LESTER PACKER AND ROLAND DOUCE

VOLUME 149. Drug and Enzyme Targeting (Part B)
Edited by RALPH GREEN AND KENNETH J. WIDDER

VOLUME 150. Immunochemical Techniques (Part K: *In Vitro* Models of B and T Cell Functions and Lymphoid Cell Receptors)
Edited by GIOVANNI DI SABATO

VOLUME 151. Molecular Genetics of Mammalian Cells
Edited by MICHAEL M. GOTTESMAN

VOLUME 152. Guide to Molecular Cloning Techniques
Edited by SHELBY L. BERGER AND ALAN R. KIMMEL

VOLUME 153. Recombinant DNA (Part D)
Edited by RAY WU AND LAWRENCE GROSSMAN

VOLUME 154. Recombinant DNA (Part E)
Edited by RAY WU AND LAWRENCE GROSSMAN

VOLUME 155. Recombinant DNA (Part F)
Edited by RAY WU

VOLUME 156. Biomembranes (Part P: ATP-Driven Pumps and Related Transport: The Na,K-Pump)
Edited by SIDNEY FLEISCHER AND BECCA FLEISCHER

VOLUME 157. Biomembranes (Part Q: ATP-Driven Pumps and Related Transport: Calcium, Proton, and Potassium Pumps) (in preparation)
Edited by SIDNEY FLEISCHER AND BECCA FLEISCHER

VOLUME 158. Metalloproteins (Part A) (in preparation)
Edited by JAMES F. RIORDAN AND BERT L. VALLEE

VOLUME 159. Initiation and Termination of Cyclic Nucleotide Action (in preparation)
Edited by JACKIE D. CORBIN AND ROGER A. JOHNSON

VOLUME 160. Biomass (Part A: Cellulose and Hemicellulose) (in preparation)
Edited by WILLIS A. WOOD AND SCOTT T. KELLOGG

VOLUME 161. Biomass (Part B: Lignin, Pectin, and Chitin) (in preparation)
Edited by WILLIS A. WOOD AND SCOTT T. KELLOGG

VOLUME 162. Immunochemical Techniques (Part L: Chemotaxis and Inflammation) (in preparation)
Edited by GIOVANNI DI SABATO

VOLUME 163. Immunochemical Techniques (Part M: Chemotaxis and Inflammation) (in preparation)
Edited by GIOVANNI DI SABATO

[1] Overview: The Na,K-Pump

By JENS CHRISTIAN SKOU

Introduction

Most mammalian cells have Na,K-pumps in their membranes, which pump Na^+ ions out of and K^+ ions into the cell, thereby creating an electrochemical gradient of Na^+ ions into and K^+ ions out of the cell. The pumping requires energy, and the Na,K-pump is an energy transducer, which converts chemical energy from the hydrolysis of ATP to an electrochemical gradient of the cations. Transport coupled to a chemical reaction is by definition active transport.

The transport is electrogenic: for each ATP molecule hydrolyzed three Na^+ ions are transported out of and two K^+ ions into the cell (see Ref. 1). The net movement of positive charges to the outside of the membrane adds a few millivolts to the membrane potential.[2]

In the steady state the back diffusion of the cations, driven by the electrochemical gradients set up by the pumping, keeps pace with the pumping.

Physiological Function

Membrane Potential

Due to a higher permeability of the cell membrane for K^+ than for Na^+ ions the passive back diffusion of the cations, driven by the gradients, leads to a diffusion potential across the cell membrane, the membrane potential, which is negative on the inside. The gradients for the cations, which are a free energy source, and the membrane potential are the basis for the action potential in excitable tissue.

Osmotic Regulation

With a negative membrane potential the intracellular concentration of diffusible anions is lower than the extracellular. This compensates for the osmotic effect of intracellular anions which cannot pass the membrane.

---

[1] I. M. Glynn, *Br. Med. Bull.* **24**, 156 (1968).
[2] R. C. Thomas, *Physiol. Rev.* **52**, 563 (1972).

Uphill Transport

The gradient for Na^+ is used as a free energy source for cotransport, or symport, of other substances such as sugars, amino acids, Cl^-, or K^+ plus Cl^- and for countertransport, or antiport, of Ca^{2+} or H^+ against gradients across cell membranes. These transports via membrane-located carrier systems are called uphill or secondary active transport.

High Intracellular K^+ Ion Concentration

High intracellular K^+ ion concentration is of importance for a number of intracellular enzymatic reactions.

Reabsorption Processes

The coupling between a Na^+ gradient-driven, uphill transport across the luminal membrane and an active Na^+ transport across the basolateral membrane of epithelial cells is of importance for reabsorption processes in intestine and kidney.

Na^+,K^+-ATPase

The Na,K-pump is identified as a membrane-bound ATP-hydrolyzing enzyme system, which is activated by the combined effect of Na^+ on the cytoplasmic side and of K^+ on the extracellular side of the membrane.[3] Because of its enzymatic properties it is identified as Na^+,K^+-ATPase. Using broken membrane pieces, ATP hydrolysis, but not transport, can be measured in a test tube. Provided the isolation procedure has not interfered with coupling between the ATP hydrolysis reaction and the transport, ATP hydrolysis can be taken as an indication of the transport reaction.

Effects of Na^+ and K^+ Ions

Competition exists between Na^+ and K^+ for Na^+ activation on the cytoplasmic side of the membrane and for K^+ activation on the extracellular side. The apparent $Na^+:K^+$ affinity ratio for Na^+ activation on the cytoplasmic sites is about 3:1 with saturating concentrations of ATP, and the apparent $K^+:Na^+$ affinity ratio for K^+ activation on the extracellular sites is about 100:1 (Fig. 1). Because of this large difference in apparent affinity ratios on the two sides of the system it is possible, even with broken membranes in the test tube, to measure the activating effect of

---

[3] J. C. Skou, *Biochim. Biophys. Acta* **23**, 394 (1957).

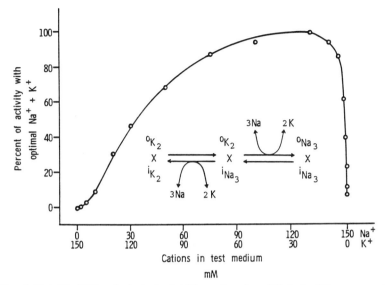

FIG. 1. Na^+,K^+-ATPase hydrolysis of ATP as a function of Na^+ plus K^+ concentrations. Enzyme isolated from ox brain. Tested in 30 mM histidine HCl buffer, pH 7.4, 37°, with 3 mM ATP and 3 mM Mg^{2+}.

Na^+ on the cytoplasmic sites with the extracellular sites saturated with K^+ (left-hand side of the curve in Fig. 1) and to measure the activating effect of K^+ on the extracellular sites with the cytoplasmic sites saturated with Na^+ (right-hand side of the curve in Fig. 1). With 130 mM Na^+ and 20 mM K^+ the cytoplasmic sites are almost saturated with Na^+ and the extracellular sites with K^+ (Fig. 1).

For activation on the cytoplasmic sites Na^+ is necessary and cannot be replaced by any other monovalent cation, whereas for activation on the extracellular sites K^+ can be replaced by any of the other monovalent cations. The apparent affinity for activation is $K^+ > Rb^+ > NH_4^+ > Cs^+ > Li^+ > Na^+$.[4] With Na^+, which has both a cytoplasmic and an extracellular activating effect, the activity in the test tube is 3–5% of the activity with optimal Na^+ plus K^+ (Fig. 1). The steepness of the curve in Fig. 1 varies for enzymes from different tissues. For a given enzyme it is dependent on the ATP concentration; at a given concentration of ATP it is dependent on pH.[5]

For a given K^+ concentration, the activation by Na^+ follows an S-shaped curve, as does the activation by K^+ at a given Na^+ concentration,

[4] J. C. Skou, *Biochim. Biophys. Acta* **42,** 6 (1960).
[5] J. C. Skou, *Biochim. Biophys. Acta* **567,** 421 (1979).

suggesting that more than one Na^+ and one K^+ ion is necessary for activation.

It has not been determined whether the cation-binding sites on the system alternate between an exposure to the two sides of the membrane, i.e., a consecutive or Ping-Pong reaction, or whether the sites exist simultaneously on the two sides of the membrane, i.e., the reaction with Na^+ and K^+ is due to a simultaneous binding of Na^+ to the cytoplasmic sites, and of K^+ to the extracellular sites. To understand the transport mechanism it is important to resolve this problem (for a discussion, see Refs. 6–8).

From Fig. 1, it can be seen that with normal intracellular 20–30 mM Na^+ and ~120 mM K^+, the Na^+ activation is about 30–40% of maximum, and with normal extracellular 4 mM K^+ and 140 mM Na^+, the K^+ activation on the extracellular sites is about 85% of maximum. This means that the transport system (isolated from ox brain) in the intact cell membrane operates with an activity which is about 25–35% of maximum. The turnover rate of ATP hydrolysis of Na^+,K^+-ATPase is about 10,000/min at 37°, pH 7.4 (see Ref. 9).

Conformational Transition

Na^+,K^+-ATPase exists in two distinct conformations denoted as E_1 and E_2.[10–16] With no cations in the medium and under conditions where the Na^+ ion effect of buffer cations can be minimized, the enzyme is in the E_2 conformation.[16,17] Addition of Na^+ leads to a transition of E_2 to E_1Na_n (n and m are numbers) with a $K_{0.5}$ for Na^+ which is in the millimolar range.[17] Addition of K^+ leads to transition from E_2 into a conformation from which K^+ has a very low rate of release, $E_2(K_m)$,[18] and which is denoted a K^+-occluded conformation; it is K^+ on the cytoplasmic sites which are occluded. The affinity of E_2 for K^+ is low, but as the equilibrium

[6] P. J. Garrahan and R. P. Garay, *Curr. Top. Membr. Transp.* **8**, 37 (1976).
[7] P. J. Garrahan, A. Horenstein, and A. F. Rega, in "Na,K-ATPase: Structure and Kinetics" (J. C. Skou and J. G. Nørby, eds.), pp. 261–274. Academic Press, London, 1979.
[8] J. R. Sachs, *J. Physiol. (London)* **302**, 219 (1980).
[9] P. L. Jørgensen, *Q. Rev. Biophys.* **7**, 239 (1975).
[10] J. G. Nørby and J. Jensen, *Biochim. Biophys. Acta* **233**, 104 (1971).
[11] C. Hegyvary and R. L. Post, *J. Biol. Chem.* **246**, 5234 (1971).
[12] P. L. Jørgensen, *Biochim. Biophys. Acta* **401**, 394 (1975).
[13] S. J. D. Karlish, D. W. Yates, and I. M. Glynn, *Biochim. Biophys. Acta* **525**, 252 (1978).
[14] S. J. D. Karlish and D. W. Yates, *Biochim. Biophys. Acta* **527**, 115 (1978).
[15] S. J. D. Karlish, *J. Bioenerg. Biomembr.* **12**, 111 (1980).
[16] J. C. Skou and M. Esmann, *Biochim. Biophys. Acta* **601**, 386 (1980).
[17] J. C. Skou and M. Esmann, *Biochim. Biophys. Acta* **746**, 101 (1983).
[18] I. M. Glynn and D. E. Richards, *J. Physiol. (London)* **330**, 17 (1982).

between E_2K_m and $E_2(K_m)$ is poised toward $E_2(K_m)$,[14] the apparent affinity for K^+ for the transition from E_2 to $E_2(K_m)$ is high, with $K_{0.5}$ in the micromolar range.[19]

$$E_2(K_m) \rightleftharpoons E_2K_m \rightleftharpoons E_2 \rightleftharpoons E_2Na_n \rightleftharpoons E_1Na_n \qquad (1)$$

E_1Na_n has a high affinity for ATP, K_D is 0.1–0.2 μM, whereas $E_2(K_m)$ has a low affinity.[10,11] They have a different reactivity toward tryptic digestion[12] and toward a number of inhibitors (see Ref. 20). The E_1 conformation can also be distinguished from the E_2 conformation using fluorescence measurements; this has been used for measuring the rate of transition in between the conformations. Intrinsic tryptophan fluorescence is 1–2% higher of E_2 than of E_1.[14] Extrinsic probes used include fluorescein isothiocyanate,[15] formycin triphosphate,[13] and eosin.[21] Extrinsic probes give a larger difference in fluorescence between the two conformations than intrinsic fluorescence. Fluorescein isothiocyanate binds covalently to a lysine group near the ATP site and has a higher fluorescence in the E_1 than in the E_2 conformation. Formycin triphosphate, an ATP analog, binds to E_1 with a high affinity, resulting in an increase in fluorescence, as does eosin, which binds noncovalently to what seems to be the ATP site.

The rate of transition from E_1Na_n to $E_2(K_m)$ is high; at 22° $t_{1/2}$ is a few milliseconds. In contrast the rate of the transition from $E_2(K_m)$ to E_1Na_n is low and $t_{1/2}$ is a fraction of a second.[13–15,17,22] The rate of transition from $E_2(K_m)$ to E_2K_m is low [see Eq. (1)], whereas the rate of transition from E_2K_m to E_1Na_n is relatively high.[17]

ATP on a low-affinity site (K_D 0.45 mM)[14] increases the rate of transition from $E_2(K_m)$ to E_1Na_n. At a given $Na^+ : K^+$ ratio, an increase in ATP concentration shifts the distribution between the two forms toward the Na^+ form.[5] An increase in pH, like an increase in ATP concentration, increases the rate of the transition from $E_2(K_m)$ to E_1Na_n and decreases the rate of the reverse reaction.[5,16,22]

Reaction with ATP

In the presence of Mg^{2+}, Na^+, and ATP the enzyme forms acid-stable phospho intermediates,[23] due to the effect of Na^+ on the cytoplasmic sites of the system. The rate of phosphorylation is high, while the rate of dephosphorylation is low, which means that in the steady state with Na^+,

[19] M. Esmann and J. C. Skou, *Biochim. Biophys. Acta* **748**, 413 (1983).
[20] I. M. Glynn and S. J. D. Karlish, *Annu. Rev. Physiol.* **35**, 13 (1975).
[21] J. C. Skou and M. Esmann, *Biochim. Biophys. Acta* **647**, 232 (1981).
[22] J. C. Skou, *Biochim. Biophys. Acta* **688**, 369 (1982).
[23] R. L. Post, A. K. Sen, and A. S. Rosenthal, *J. Biol. Chem.* **240**, 1437 (1965).

but no K^+, the main part of the enzyme is in the phospho form. The ratio between the E_1 and E_2 phospho forms (see below) at steady state varies for enzymes from different tissues, but the E_2 phospho forms is predominant for most enzymes. For a given enzyme the fraction of E_1 phospho forms increases with Na^+ concentration.[24]

Besides the phosphoenzyme with bound ADP, at least three consecutive acid-stable phosphoenzymes are formed[25] (Mg^{2+}, which is necessary for phosphorylation, is omitted):

$$E_1ATPNa_n \rightleftharpoons E_1 \sim PADPNa_n \rightleftharpoons E_1' \sim P(Na_n)$$
$$\rightleftharpoons E_1'' \sim PNa_n \rightleftharpoons E_2\text{-}PNa_n \quad (2)$$

The E_1 and E_2 phospho forms can be distinguished by their reactivity toward ADP and K^+. When phosphorylation from [^{32}P]ATP is stopped by addition of cold ATP, the addition of ADP leads to a high rate of dephosphorylation of the E_1 phospho forms but a low rate of dephosphorylation of the E_2 phospho form.[25] ADP reverses the phosphorylation of the ADP-sensitive phosphoenzyme $E_1' \sim P(Na_n)$ with reformation of ATP. $E_1'' \sim PNa_n$ has a high rate of transition to $E_1' \sim P(Na_n)$ and this rate is about 10 times higher (at 0°) than the rate of transition of $E_2\text{-}PNa_n$ to $E_1'' \sim PNa_n$, which means that the ADP-induced disappearance of $E_1' \sim P(Na)_n$ leads to a fast disappearance of $E_1'' \sim PNa_n$ but not of $E_2\text{-}PNa_n$.[25] Therefore, even if $E_1'' \sim PNa_n$ does not directly react with ADP, it is considered an ADP-sensitive phosphoenzyme. The existence of ADP-sensitive phosphoenzymes can be observed as an ADP–ATP exchange reaction.[4,26]

Addition of K^+ in extracellular concentrations instead of ADP gives a high rate of dephosphorylation of E_2–P, and a low rate of dephosphorylation of the E_1 phosphoenzymes. The rate of dephosphorylation of E_2–P by K^+ is much higher than the rate of transition of the ADP-sensitive phosphoenzymes to E_2–P.[25] E_2–P is denoted the K^+-sensitive phosphoenzyme.

Phosphorylation leads to a conformation of the enzyme from which Na^+ is slowly released, to an occlusion of Na^+ [shown in parentheses in Eq. (2)].[27] As $E_1 \sim PADP$ exists in insignificant amounts,[25] the measured occlusion probably is on the phosphoenzyme which follows formation of $E_1 \sim PADP$. It is not possible to determine whether the occlusion is due to the phosphorylation, or as shown in Eq. (2), it is the phosphorylation and the following release of ADP which leads to occlusion. It seems likely that it is the Na^+ on the cytoplasmic sites which becomes occluded.

[24] Y. Kuriki and E. Racker, *Biochemistry* **15**, 4951 (1976).
[25] J. G. Nørby, I. Klodos, and N. O. Christiansen, *J. Gen. Physiol.* **82**, 725 (1983).
[26] S. Fahn, G. J. Koval, and R. W. Albers, *J. Biol. Chem.* **241**, 1882 (1966).
[27] I. M. Glynn, Y. Hara, and D. E. Richards, *J. Physiol. (London)* **351**, 531 (1984).

The dephosphorylation of E_2–P by K^+ in extracellular concentrations leads to an occlusion of K^+,[28] which probably means an occlusion of K^+ from the extracellular sites. As discussed above K^+ is occluded from the cytoplasmic sites of the nonphosphorylated enzyme.

$$E_2-PNa_n \rightleftharpoons E_2-PK_m \rightleftharpoons E_2K_m \rightleftharpoons E_2(K_m) \tag{3}$$

With Na^+, the K_m for ATP hydrolysis is a micromolar fraction (see Ref. 29) and the Lineweaver–Burk plot is rectilinear. With Na^+ plus K^+, the K_m for ATP hydrolysis increases to a millimolar fraction. The Lineweaver–Burk plot is not linear and can be resolved with a high affinity for ATP, which is equal to the affinity in the presence of Na^+ but no K^+, and a low affinity for ATP (see Ref. 29). As mentioned above, ATP by a low-affinity effect increases the rate of deocclusion of K^+ from $E_2(K_m)$; the higher K_m for ATP for the hydrolysis of ATP in the presence of Na^+ plus K^+ than of Na^+ is explained from the requirement for ATP to speed up the rate of transition from $E_2(K_m)$ to E_1Na_n in the ATP hydrolysis reaction.[14] It has not been determined whether the high and low affinity for ATP in the presence of Na^+ plus K^+ means two ATP-binding sites on a molecule or the consecutive effect of two ATP molecules.

The Na^+,K^+-ATPase hydrolyzes not only ATP but also other triphosphates (see Ref. 30), but with a lower affinity.[30]

Phosphatase Activity

Na^+,K^+-ATPase has two phosphatase activities (see Ref. 31). One is activated by K^+ and inhibited by Na^+ along a curve, which suggests that Na^+ inhibits an activating effect of K^+ on the cytoplasmic sites.[31] This finding is supported by experiments on inverted red blood cells.[32] The activity is also inhibited by ATP, probably due to a competition with the phosphatase substrate. With p-nitrophenyl phosphate as substrate, maximum activity is obtained with 100 mM K^+, 10 mM substrate, and 20 mM Mg^{2+}. At 37° the turnover is 14–18% of the turnover of the Na^+,K^+-ATPase activity. As discussed above, in the presence of K^+, but no Na^+, the enzyme is in the K^+-occluded conformation, $E_2(K_m)$, suggesting that it is this conformation which has the phosphatase activity.

[28] R. L. Post, C. Hegyvary, and S. Kume, *J. Biol. Chem.* **247**, 6530 (1972).
[29] J. D. Robinson, *Curr. Top. Membr. Transp.* **19**, 485 (1983).
[30] J. Jensen and J. G. Nørby, *Biochim. Biophys. Acta* **233**, 395 (1971).
[31] J. C. Skou, *Biochim. Biophys. Acta* **339**, 258 (1974).
[32] R. Blostein, H. A. Pershadsingh, P. Drapeau, and L. Chu, in "Na,K-ATPase: Structure and Kinetics" (J. C. Skou and J. G. Nørby, eds.), pp. 233–245. Academic Press, London, (1979).

The other phosphatase activity is activated by low, extracellular concentrations of K^+ in the presence of a high concentration of Na^+. As in Na^+,K^+-ATPase activity, this suggests that Na^+ on the cytoplasmic sites and K^+ on the extracellular sites are necessary for this phosphatase activity.[31] The activity is low and varies for enzymes from different sources. With ox brain enzyme and with 10 mM p-nitrophenyl phosphate, 20 mM Mg^{2+}, 100 mM Na^+, 2 mM K^+ it is about 15% of the maximal K^+-phosphatase activity.[31] ATP in low concentrations enhances the activity (see Ref. 31) and with the optimum ATP concentration, 0.1–0.2 mM, the activity is about 80% of the K^+-phosphatase activity.[31] ATP in higher concentrations acts as an inhibitor.

It has also been proposed that this phosphatase activity is due to $E_2(K_m)$, which is formed as an intermediate in the turnover with Mg^{2+}, Na^+, K^+, and ATP.[28] However, this does not explain the activity without ATP and with a combination of cations (a high Na^+ and low K^+ concentration) which gives the E_1 conformation of the enzyme. Neither does it explain that oligomycin, like ATP, enhances the K^+ plus Na^+-dependent phosphatase activity.[33] Oligomycin occludes Na^+ in an E_1 conformation, $E_1(Na_n)$[34] (see below). This suggests that this phosphatase activity is due to an effect of K^+ on extracellular sites of the conformation of the enzyme with Na^+ occluded, $E_1(Na_n)$. The effect of Li^+ instead of K^+ on the phosphatase activity supports this view.[35] If this is correct, it leads to the conclusion that with a high Na^+, low K^+ (and a high Mg^{2+} [31]) concentration and without oligomycin, a small fraction of the enzyme is in the Na^+-occluded conformation, and that extracellular sites and sites with Na^+ occluded exist simultaneously.[35]

The Na^+-occluded conformation, $E_1' \sim P(Na_n)$, as well as the K^+-occluded conformation, $E_2(K_m)$, are intermediates in the turnover of the enzyme with Na^+ and K^+ and with ATP as substrate. This means that the ATP enhancement of the phosphatase activity with a high Na^+ and a low K^+ concentration may be due to both conformations.

A tentative conclusion is that the enzyme has phosphatase activity in an E_2 conformation with occluded K^+ or congeners as well as in an E_1 conformation with Na^+ occluded but with K^+ or congeners on simultaneously existing extracellular sites. K^+ or congeners on the extracellular sites appear not to be necessary for phosphatase activity of $E_2(K_m)$.[32]

[33] A. Askari and D. Koyal, *Biochim. Biophys. Acta* **225**, 20 (1971).
[34] M. Esmann and J. C. Skou, *Biochem. Biophys. Res Commun.* **127**, 857 (1985).
[35] J. C. Skou, in "The Sodium Pump—Proceedings of the Fourth Na,K-ATPase Conference" (I. M. Glynn and J. C. Ellory, eds.), pp. 575–588. Co. of Biologists, Cambridge, England (1985).

Inhibitors

Cardiac Glycosides

In 1953 it was observed that active Na^+-K^+ transport is inhibited by cardiac glycosides.[36] Cardiac glycosides have become an important tool in experiments on transport of cations, in the identification of transport systems, and in experiments on isolated systems. The most widely used cardiac glycoside has been ouabain (g-strophanthin) which is the most water soluble. The cardiac glycoside binds to the extracellular side of the system (see Refs. 37 and 38) and binding and inhibition are dependent on the ligands present. Mg^{2+} is necessary, but with Mg^{2+} alone the rate of binding is low. Phosphorylation of the enzyme with formation of the K^+-sensitive phosphoenzyme E_2–P promotes binding. E_2–P is formed either with Mg^{2+}, Na^+, and ATP [see Eq. (2)] or with Mg^{2+} and P_i. With Mg^{2+}, Na^+, and ATP, extracellular K^+ protects against inhibition by low concentrations of cardiac glycosides, probably due to the dephosphorylating effect. With Mg^{2+} and P_i, K^+, as well as Na^+, protects against binding.

Cardiac glycosides differ in their binding affinity. With a given cardiac glycoside and a given combination of ligands, the inhibitory effect varies for different tissues. For enzymes from most tissues the $K_{0.5}$ for ouabain for inhibition of the Na^+,K^+-ATPase activity is of the order of 10^{-7}–10^{-5} M. Na^+,K^+-ATPase from rat heart or crab nerves is much less sensitive.

Vanadate

Vanadate, which was found as a contaminant in certain commercial preparations of ATP, inhibits Na^+,K^+-ATPase in micromolar concentrations.[39] Mg^{2+} is necessary for binding, and in contrast to the cardiac glycosides, vanadate binds to the cytoplasmic side of the system.[40,41] P_i, ATP, and p-nitrophenyl phosphate compete for binding, suggesting that vanadate binds to the site which releases P_i.[41] Since P_i is released from E_2–P, vanadate seems to stabilize the enzyme in the E_2 conformation. K^+ on the cytoplasmic sites of the system promotes vanadate binding, whereas K^+ on the extracellular site has no effect. Na^+ on the extracellu-

---

[36] H. J. Schatzmann, *Helv. Physiol. Acta* **11**, 346 (1953).
[37] A. Schwartz, D. E. Lindenmayer, and J. C. Allen, *Pharmacol. Rev.* **27**, 3 (1975).
[38] T. Akera, *Handb. Exp. Pharmakol* **56/I**, 287 (1981).
[39] L. C. Cantley, L. Josephson, R. Warner, M. Yanagisawa, C. Lechene, and G. Guidotti, *J. Biol. Chem.* **252**, 7421 (1977).
[40] L. C. Cantley, M. Resh, and G. Guidotti, *Nature (London)* **272**, 552 (1978).
[41] L. C. Cantley, L. G. Cantley, and L. Josephson, *J. Biol. Chem.* **253**, 7361 (1978).

lar sites protects against vanadate binding, and the effect can be overcome by displacement of Na^+ from the sites by K^+ or congeners (see Ref. 42). This means that, in the test tube with no sidedness, K^+ in the presence of Mg^{2+} promotes binding; Na^+ prevents binding, but the effect can be overcome by K^+.

Vanadate is not as specific as the cardiac glycosides which inhibit only Na^+,K^+-ATPase since vanadate also inhibits the Ca^{2+}-ATPase and other ATPases.

Oligomycin

In contrast to the cardiac glycosides and vanadate, oligomycin does not completely inhibit the enzyme, but decreases the rate of turnover of transport and of the hydrolysis reaction (see Ref. 42). With Na^+ plus K^+, the maximum effect on the hydrolysis of ATP is obtained with 10–20 μg oligomycin/ml. Fractional inhibition increases with the ATP concentration, and with saturating ATP concentrations the activity with oligomycin is about 20% of the activity without oligomycin.[22]

Oligomycin affects the extracellular side of the system,[43] and Na^+ on the cytoplasmic side of the system[8] is necessary for the effect.[34]

Oligomycin decreases the rate of transition from the Na^+ form to the K^+ form of the enzyme, but has no effect on the rate of the reverse reaction.[22] This is due to an occlusion of Na^+, which means that it is due to a transition of E_1Na_n to a conformation, $E_1(Na_n)$, which has a low rate of release of Na^+.[34] Oligomycin has either no effect on or increases the ATP–ADP exchange reaction seen in the presence of Na^+,[26,44] but it decreases the ADP-dependent exchange of intracellular Na^+ for extracellular Na^+ (see Ref. 45 and below).

Isolation and Purification

Na^+,K^+-ATPase is an integral part of the plasma membrane. The number of Na^+,K^+-ATPase molecules per cell varies from tissue to tissue. In a red blood cell there are about 250 molecules[46] while in a white blood cell there are about 25,000–35,000.[47] Plasma membranes of the outer medulla of kidney, of rectal glands from dogfish, of electric eel tissue, and salt

[42] I. M. Glynn, *Enzymes Biol. Membr.* **3**, 28 (1985).
[43] F. Cornelius and J. C. Skou, *Biochim. Biophys. Acta* **818**, 211 (1985).
[44] R. Blostein, *J. Biol. Chem.* **245**, 270 (1970).
[45] P. De Weer, *Curr. Top. Membr. Transp.* **19**, 599 (1983).
[46] E. Erdmann and W. Hasse, *J. Physiol. (London)* **251**, 671 (1975).
[47] N. A. Boon, V. M. S. Cu, E. A. Taylor, T. Johansen, I. K. Aronson, and D. G. Grahame-Smith, *Br. J. Clin. Pharmacol.* **18**, 153 (1984).

glands from marine birds (see Ref. 48) are especially rich in transport molecules.

Isolated plasma membrane fragments in the test tube show Na^+,K^+-ATPase activity. However, the plasma membrane fragments tend to form vesicles, and in order to measure the Na^+,K^+-ATPase activity in the test tube it is necessary that the ligands have access to both sides of the membrane. The vesicles can be opened by mild treatment with detergents.

The enzyme can be purified from plasma membrane fragments isolated from tissues having a high content of Na^+,K^+-ATPase (see Chapter 2 and Ref. 48). One approach is to remove all the other proteins from the membrane but not the Na^+,K^+-ATPase protein. This can be done by a careful treatment (pH, temperature, concentration of detergent, and time) with ionic detergents like deoxycholate (DOC) and sodium dodecyl sulfate (SDS) followed by differential and density gradient centrifugation. Besides removing non-Na^+,K^+-ATPase protein, the detergents also open the vesicles formed from the plasma membranes and thereby give the ligands access to both sides of the membrane.

Another approach is to extract Na^+,K^+-ATPase from partially purified membrane preparations using a nonionic detergent like Lubrol[49] or $C_{12}E_8$ (octaethylene glycol mono-n-dodecyl ether).[50] The nonionic detergent $C_{12}E_8$ selectively extracts Na^+,K^+-ATPase as a soluble complex with the detergent and leaves other proteins in the membrane. The detergent-solubilized Na^+,K^+-ATPase can be separated from the membrane pieces by centrifugation.

A problem in both procedures is that detergents in higher concentrations inactivate Na^+,K^+-ATPase. The detergent treatment is a balance between a detergent concentration high enough to purify but not so high as to inactivate. A concentration of detergent which apparently does not inactivate may have an effect, however, on some of the intermediate steps, on the kinetic parameters, or on the coupling between the hydrolysis of ATP and transport. This is a side of the detergent effect which deserves attention.

Na^+,K^+-ATPase can be purified to about 95% in the membrane. Gel electrophoresis of the purified membranes dissolved in SDS shows two polypeptide bands: an α subunit, with a molecular weight of about 100,000, and a glycoprotein β subunit, with a molecular weight of about

---

[48] P. L. Jørgensen, *Biochim. Biophys. Acta* **694**, 27 (1982).
[49] L. E. Hokin, J. L. Dahl, J. D. Deupree, J. F. Dixon, J. F. Hackney, and J. F. Perdue, *J. Biol. Chem.* **248**, 2593 (1973).
[50] M. Esmann, J. C. Skou, and C. Christiansen, *Biochim. Biophys. Acta* **567**, 410 (1979).

38,000. The ratio is about 1 : 1 on a mole basis. The 100K protein is labeled with ^{32}P when the enzyme is phosphorylated from [^{32}P]ATP in the presence of Na$^+$ and the labeling disappears after dephosphorylation with K$^+$, indicating that this protein takes part in the catalytic reaction. There is no specific labeling of the β protein, but all active preparations contain the β protein and cross-linking experiments suggest that the β protein is in close contact with the α protein, suggesting that the β protein is part of the enzyme molecule (see Ref. 48).

Besides the α and β proteins a low-molecular-weight polypeptide (M_r about 10,000) is isolated by gel electrophoresis and this protein, called the γ protein, may also be part of the enzyme molecule.[51]

Lipids

When the enzyme is purified in the membrane, the lipid core of the membrane is preserved. The lipids are necessary for enzyme activity. Removal of the lipids leads to inactivation and readdition of the lipids results in reactivation of enzyme (see Ref. 52). When the enzyme is dissolved in $C_{12}E_8$, it is purified with respect to the α and β proteins. The specific activity is comparable to the enzyme purified in the membrane. About 50 molecules of phospholipid and about 40 molecules of cholesterol per $(\alpha,\beta)_2$ protein[53] are bound to this enzyme. These lipids may represent the amount necessary for activity. However, besides the lipids $C_{12}E_8$ is bound to the enzyme and even if $C_{12}E_8$ cannot reactivate a delipidated enzyme it may replace lipids necessary for activity and therefore more than the measured number of lipid molecules may be necessary. Acid phospholipid seems to be necessary for activity.[54] One acid phospholipid molecule out of the 50 phospholipids is bound to the $C_{12}E_8$-dissolved enzyme, suggesting that one is enough. When $C_{12}E_8$-dissolved enzyme is reincorporated into the lipid vesicles, the presence of the acid phospholipid, phosphatidylinositol, is necessary for full recovery of activity after reconstitution.[55]

Besides being important for anchoring of proteins in the membrane, it is not known why the lipids are necessary. How and where are the lipids bound? Do they affect the structure and/or are they of importance for the hydrophobicity of the molecule?

[51] B. Forbush and J. F. Hoffman, *Biochemistry* **17**, 3667 (1978).
[52] P. Ottolenghi, *Eur. J. Biochem.* **99**, 113 (1979).
[53] M. Esmann, C. Christiansen, K.-A. Karlsson, G. C. Hansson, and J. C. Skou, *Biochim. Biophys. Acta* **603**, 1 (1980).
[54] B. Roelofson and L. L. M. van Deenen, *Eur. J. Biochem.* **40**, 245 (1973).
[55] F. Cornelius and J. C. Skou, *Biochim. Biophys. Acta* **772**, 357 (1984).

Molecular Weight

As discussed above, the enzyme can be purified so that it consists of only α and β polypeptide chains (and maybe γ) which seem to be part of the enzyme molecule. It is not known how many lipid molecules are part of the system. For this reason the molecular weight of the Na^+,K^+-ATPase is usually expressed as the molecular weight of the protein part of the molecule.

The number of moles of ^{32}P bound from $[^{32}P]ATP$ in the presence of Mg^{2+} and Na^+ is equal to the number of high-affinity ATP-, ouabain-, and vanadate-binding sites (see Ref. 56). The amount of protein per binding site of one of these ligands in a pure preparation gives a minimum molecular weight. The values vary from about 3.7^{57} to about $5.7^{58,59}$ nmol ligand bound/mg protein, suggesting molecular weights from about 270,000 to 175,000.

Sedimentation equilibrium analysis of purified enzyme from shark rectal glands dissolved in Lubrol gives a molecular weight of 379,000[60] and dissolved in $C_{12}E_8$ it gives molecular weights of 265,000[50] and 271,000.[53] Sedimentation velocity analysis of mammalian kidney enzyme dissolved in $C_{12}E_8$ gives a molecular weight of 170,000[61] and low-angle neutron scattering analysis of mammalian kidney enzyme dissolved in Brij gives a molecular weight of 310,000.[62]

Radiation inactivation measurements on partly purified membrane preparations give a molecular weight for Na^+,K^+-ATPase from red blood cells of 300,000[63] and 330,000,[64] from mammalian kidney of 190,000[63] and 261,000,[65] and from mammalian brain of 264,000.[65]

Based on a molecular weight of about 100,000 for the α chain and about 38,000 for the β chain and on a 1:1 mole ratio of α and β (see Ref. 48), the values for molecular weight from 250,000 to 330,000 from the radiation inactivation experiments, the sedimentation equilibrium analysis, and low-angle neutron scattering analysis of dissolved enzyme sug-

---

[56] J. G. Nørby, *Curr. Top. Membr. Transp.* **19**, 281 (1983).
[57] P. L. Jørgensen, *Biochim. Biophys. Acta* **466**, 97 (1977).
[58] E. G. Modzydlowski and P. A. G. Fortes, *J. Biol. Chem.* **256**, 2346 (1981).
[59] W. H. M. Peters, J. J. H. H. M. De Pont, A. Koppers, and S. L. Bonting, *Biochim. Biophys. Acta* **641**, 55 (1981).
[60] D. F. Hastings and J. A. Reynolds, *Biochemistry* **18**, 817 (1979).
[61] J. R. Brotherus, J. D. Møller, and P. L. Jørgensen, *Biochem. Biophys. Res. Commun.* **100**, 146 (1981).
[62] J. M. Pachence and B. P. Schoenenborn, *J. Biol. Chem.* **262**, 702 (1986).
[63] G. R. Kepner and R. I. Macey, *Biochim. Biophys. Acta* **163**, 188 (1968).
[64] J. C. Ellory, J. R. Green, S. M. Jarvis, and J. D. Young, *J. Physiol. (London)* **295**, 10 (1979).
[65] P. Ottolenghi and J. C. Ellory, *J. Biol. Chem.* **258**, 14895 (1983).

gest an $(\alpha,\beta)_2$ structure [the 379,000 value suggests an $(\alpha,\beta_2)_2$]. The lower value of about 170,000 from the sedimentation velocity experiments suggests that the main part of the kidney enzyme dissolved in $C_{12}E_8$ has dissociated into an α,β structure. This enzyme preparation has retained about three-fourths of its hydrolytic activity, suggesting hydrolytic activity of α,β. An activity of α,β has also been observed in experiments on rectal gland enzyme where $(\alpha,\beta)_2$ has been dissociated into α,β by $C_{12}E_8$ and separated by chromatography.[66] A problem is whether the measured activity is due to a reaggregation of α,β into $(\alpha,\beta)_2$ in the test tube.

The values of minimum molecular weight obtained from measurements of ligand binding capacity which vary from 270,000[57] to 175,000[58,59] need to be explained. The 270,000 value based on protein determination by the Lowry method is probably too high. The 175,000 values are based on amino acid analysis. As pointed out by Craig and Kyte,[67] the Lowry method gives protein values for the Na^+,K^+-ATPase that are too high. It seems unlikely, however, that a correction factor can bring the 270,000 value down to 175,000.

A minimum molecular weight of 175,000 is higher than that for α,β. The ligand-binding measurements give values which are higher than one binding site per $(\alpha,\beta)_2$, but lower than one binding site per α,β. Dissociation of $C_{12}E_8$-solubilized $(\alpha,\beta)_2$ into α,β as a result of increasing the concentration of the detergent does not increase the number of vanadate-binding sites.[68] This shows that the ligand does not bind to a mixture of $(\alpha,\beta)_2$ with one binding site and α,β, which in the dissociated form has one binding site. The most likely explanation is that there is one binding site per α,β and that the membranes which appear to be practically pure α and β proteins also contain inactive protein, either detergent-inactivated Na^+,K^+-ATPase or non-Na^+,K^+-ATPase protein, which comigrates with α and β on SDS gels. This means that the so-called pure preparations either are not pure or if they are pure they are partly inactivated, which raises several questions. How extensively can preparations be purified with detergents before enzyme activity is lost? Is it possible to obtain a pure preparation with all enzyme molecules fully active or is it only possible to get partial purification without losing activity? So far there are no answers.

The Active Unit

ATP binding shows negative cooperativity in the presence of K^+ but not in the presence of Na^+, and ouabain binding shows negative

---

[66] M. Esmann, *Biochim. Biophys. Acta* **787**, 81 (1984).
[67] W. S. Craig and J. Kyte, *J. Biol. Chem.* **255**, 6262 (1978).
[68] M. Esmann and J. C. Skou, *Biochim. Biophys. Acta* **787**, 71 (1984).

cooperativity in the presence of Na^+ but not of K^+.[69] This means that two or more binding sites for the ligands must interact, which is in agreement with the $(\alpha,\beta)_2$ structure suggested by radiation inactivation experiments, sedimentation equilibrium centrifugation analysis, and low-angle neutron-scattering analysis, and with one binding site per α,β suggested from the ligand binding experiments. However, V_{max} for the $Na^+ + K^+$-dependent hydrolysis of ATP decreases in proportion to the number of ouabain molecules prebound. This shows that ouabain bound to one α,β has no effect on V_{max} for the hydrolytic activity of another α,β, and suggests no cooperativity between two α,β units for hydrolysis. This agrees with the observation that α,β obtained by detergent dissociation of $(\alpha,\beta)_2$ has $Na^+ + K^+$-dependent ATPase activity. Thus, under certain conditions there is interaction between binding sites on different α,β units which have Na^+,K^+-ATPase activity, but apparently there is no interaction between the units for the Na^+,K^+-ATPase activity. If this is correct it means that the low and high affinity effect of ATP for hydrolysis with Na^+ and K^+ is not due to negative cooperativity for ATP binding on two α,β units but to a consecutive effect of two ATP molecules on the same α,β.

A tentative conclusion is that the transport system in the membrane is an $(\alpha,\beta)_2$ structure, and that α,β has Na^+,K^+-ATPase activity. It is not known if the activity of α,β is due to an uncoupling between ATP hydrolysis and transport, i.e., only $(\alpha,\beta)_2$, but not α,β, has transport activity. The apparent lack of interaction between the two units in $(\alpha,\beta)_2$ for hydrolysis and the cooperativity for ligand binding have not been explained.

Sequence of Amino Acids

The complementary DNA of the alpha[69a–c] as well as of the β subunit[69d–g] has been isolated and characterized and from this has been de-

---

[69] P. Ottolenghi and J. Jensen, *Biochim. Biophys. Acta* **727**, 89 (1983).
[69a] G. E. Shull, A. Schwartz, and J. B. Lingrell, *Nature (London)* **316**, 91 (1985).
[69b] K. Kawakami, S. Noguchi, M. Noda, H. Takahashi, T. Ohta, M. Kawamura, H. Nojima, K. Nagano, T. Hirose, S. Inayama, H. Hayashida, T. Miyata, and S. Numa, *Nature (London)* **316**, 733 (1985).
[69c] Y. A. Ovchinnikov, N. N. Modyanov, N. E. Broude, K. E. Petrukhin, A. V. Grishin, N. M. Arzamazova, N. A. Aldanova, G. S. Monastyrskaya, and E. D. Sverlov, *FEBS Lett.* **201**, 237 (1986).
[69d] S. Noguchi, M. Noda, H. Takahashi, K. Kawakami, T. Ohta, K. Nagano, T. Hirose, S. Inayama, M. Kawamura, and S. Numa, *FEBS Lett.* **196**, 315 (1986).
[69e] K. Kawakami, H. Nojima, T. Ohta, and K. Nagano, *Nucleic Acids Res.* **14**, 2833 (1986).
[69f] G. E. Shull, L. K. Lane, and J. B. Lingrel, *Nature (London)* **321**, 429 (1986).
[69g] T. A. Brown, B. Horowitz, R. P. Miller, A. A. McDonough, and R. A. Farley, *Biochim. Biophys. Acta* **912**, 244 (1987).

duced the amino acid sequence. The alpha chain consists of 1.016 amino acids and has eight hydrophobic regions suggesting eight transmembrane segments. The N-terminal hydrophilic region is on the cytoplasmic side. Three isoforms of the α chain have been identified.[69h] The protein part of the β subunit consists of 302 amino acids and seems to have one transmembrane segment located near the cytoplasmic N-terminal with the hydrophilic part of the molecule on the extracellular side.

Crystallization

The enzyme has been crystallized in the membrane[70] but not in three dimensions. Dependent on the conditions for crystallization, the enzyme exists in a form which suggests an (α,β) and an $(\alpha,\beta)_2$ structure, respectively,[71] supporting the view that the enzyme in the purified membrane can dissociate from $(\alpha,\beta)_2$ to (α,β).

From a Fourier analysis of tilted membrane preparations of negatively stained $(\alpha,\beta)_2$ crystals, a three-dimensional model of the Na^+,K^+-ATPase has been constructed[71a,b] The molecule consists of two symmetric rodlike structures which each seem to consist of an α–β protomer. The height perpendicular to the membrane is about 100 Å. There is an ~20 Å wide cleft between the rods, and they are connected by an ~20 Å high area in that half of the molecule which faces the cytoplasmic side. The molecule seems to protrude about 40 Å on the cytoplasmic side of the lipid bilayer and about 20 Å on the extracellular side. The intramembrane part of each protomer is about 25% of the mass of the α as well as of the β subunit. Practically all the rest, 75%, of the β unit is on the extracellular side, whereas the cytoplasmic part of the alpha subunit is about three times larger than the extracellular part.

Results of Purification

The Na^+,K^+-ATPase is the Na^+,K^+-transport system. However, with broken membranes in the test tube, only the enzymatic activity, but not

---

[69h] G. E. Shull, J. Greeb, and J. B. Lingrel, *Biochemistry* **25**, 8125 (1986).
[70] E. Skriver, A. B. Maunsbach, and P. L. Jørgensen, *FEBS Lett.* **131**, 219 (1981).
[71] H. Hebert, P. L. Jørgensen, E. Skriver, and A. B. Maunsbach, *Biochim. Biophys. Acta* **689**, 571 (1982).
[71a] H. Hebert, E. Skriver, and A. B. Maunsbach, *FEBS Lett.* **187**, 182 (1985).
[71b] Y. A. Ovchinnikov, V. V. Demin, A. N. Barnakov, A. P. Kuzin, A. V. Lunev, N. N. Modyanov, and K. N. Dzhandzhugazyan, *FEBS Lett.* **190**, 73 (1985).

transport, can be measured. Therefore, can purification of the catalytic activity be taken as an indication of purification of the transport molecule, or does purification lead to dissociation of the coupling between the two activities?

This can be tested by reincorporation of the purified enzyme into the membranes of lipid vesicles and measuring the transport activity as well as the catalytic activity (see Ref. 55 and Chapter 15). It is necessary that reconstitution does not lead to a decrease in the activity and that the reconstituted enzyme has the same specific activity and also the same turnover number as the purified enzyme before reconstitution. This can be achieved by using $C_{12}E_8$ to dissolve the enzyme and by choosing the proper composition of the lipids for formation of the vesicles and the proper protein:lipid ratio for the reconstitution procedure (see Ref. 55 and Chapter 15). With a reconstituted enzyme which has the same specific activity as the dissolved purified enzyme about three Na^+ are exchanged for two K^+ per ATP molecule hydrolyzed, just as in the intact membrane, indicating that the purified enzyme not only contains the catalytic activity but is the intact transport system. This does not necessarily mean that every purified Na^+,K^+-ATPase preparation has retained its transport properties.

Coupling between Chemical Reactions and Transport

In order to exchange Na^+ from the cytoplasmic medium, which has a high K^+ and a low Na^+ concentration, with K^+ from the extracellular medium, which has a high Na^+ and a low K^+ concentration, the transport system must undergo a change in affinities from a high K^+–low Na^+ affinity to a high Na^+–low K^+ affinity on the cytoplasmic side and vice versa on the extracellular side of the system. Furthermore, the translocation must involve a change in exposure of the cation binding sites from one side of the membrane to the other, i.e., a gating reaction. These reactions must be governed by the reaction with the chemical substrate, ATP.

The E_2-E_1 transition shows that the system can undergo changes in affinity for the cations from a high K^+–low Na^+ affinity, E_2, to a high Na^+–low K^+ affinity, E_1. ATP facilitates the transition from E_2 to E_1,[14] and phosphorylation the transition from E_1 to E_2 (see Ref. 25).

Occlusion of the cations can be looked on as a transfer of the cations to the membrane phase, a closing of a gate. The occlusion shows that in this conformation of the enzyme, the exposure of the cation sites is neither to the cytoplasmic nor to the extracellular side of the system, but the cations are inside gates closed toward the cytoplasmic as well as toward the extracellular medium. As discussed above, binding of K^+ to the cytoplasmic sites of the nonphosphorylated enzyme leads to an occlusion of K^+,[18] while Na^+ on the cytoplasmic sites becomes occluded by the phosphorylation.[27] Dephosphorylation of E_2–P by extracellular K^+ gives occlusion of K^+,[28] which probably means that it is the K^+ on the extracellular sites which becomes occluded.

On the assumption that the reactions with Na^+ and K^+ are consecutive, the coupling between the chemical reactions, the conformational transitions, and the transport of the cations may be illustrated by the scheme shown in Fig. 2. Figure 2 is based on the most widely used scheme to describe the transport reaction, the Albers–Post scheme,[72,73] on the modified scheme by Karlish et al.,[13] and on the scheme for phosphorylation in the presence of Na^+ by Nørby et al.[25] The exchange of three Na^+ for two K^+ (see Ref. 1) suggests that n and m used in Eq. (1–3) is 3 and 2, respectively.

It must be emphasized that although this scheme is one way to combine the results, it has shortcomings (see comments below). For alternative schemes, see Refs. 35 and 74.

Na^+–K^+ Exchange and ADP-Dependent Na^+–Na^+ Exchange

ATP increases the rate of deocclusion of K^+ from $E_2(K_2)$,[14] the opening of the gate, and in the presence of Na^+ it shifts the equilibrium toward E_1Na_3. This is followed by phosphorylation from ATP[28] with an occlusion of Na^+,[27] the closing of the gate. This reaction may describe the transfer of K^+ from the membrane phase to the cytoplasmic medium, followed by the exchange of the translocated K^+ for outgoing Na^+ and the transfer of Na^+ to the membrane phase (Fig. 2).

The transfer of Na^+ from the membrane phase to the extracellular medium must involve a deocclusion, which means a transition from the Na^+ occluded form, $E_1' \sim P(Na_3)$ to a conformation which can exchange Na^+ with the extracellular medium, an opening of the gate to the extracellular side.

In the absence of extracellular K^+, the transport system can engage in

---

[72] R. W. Albers, *Annu. Rev. Biochem.* **36**, 727 (1967).
[73] R. Post, S. Kume, T. Tobin, and B. Orcutt, *J. Gen. Physiol.* **54**, 306 (1969).
[74] I. W. Plesner, L. Plesner, J. G. Nørby, and I. Klodos, *Biochim. Biophys. Acta* **643**, 483 (1981).

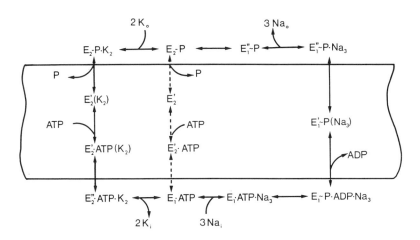

FIG. 2. A scheme for the Na^+–K^+ exchange by the Na^+,K^+-ATPase. The reaction with the cations is consecutive and the scheme is based on the Albers–Post scheme,[72,73] the modifications of this scheme by Karlish et al.,[13] and on the scheme for formation of the phosphoenzymes by Nørby et al.[25] The symbols E_1 and E_2 refer to different conformations of the enzyme; E_1 representing the form with three cation sites (the Na^+ form) and E_2 the conformation with two cation sites (the K^+ form). E, E′, and E″ represent different enzyme conformations, E ~ P represents the high-energy bonded and E–P the low-energy bonded phosphoenzyme. Parentheses indicate that the cations are occluded, i.e., in the membrane phase inside gates closed toward the cytoplasmic and the extracellular medium. For further explanation see text.

an Na^+–Na^+ exchange which requires ADP in addition to ATP and which involves phosphorylation, but with no net hydrolysis of ATP (see Ref. 42). The reaction seems to be a phosphorylation with an outward transport of Na^+, followed by an inward transport of Na^+ due to an ADP-induced reversal of the phosphorylation with reformation of ATP. In the reaction shown in Fig. 3A, the conformation, which can exchange the outward transported Na^+ with Na^+ from the extracellular medium, must be ADP-sensitive. Therefore, it cannot be E_2–P, but must be the phosphoenzyme which precedes formation of E_2–P and which follows the phosphoenzyme with Na^+ occluded. This phosphoenzyme, which in Fig. 3A is denoted $E_1'' \sim PNa_3$, does not react directly with ADP. However,

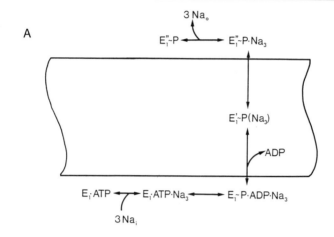

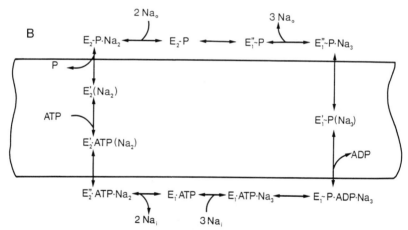

Fig. 3. Schemes for the ADP-dependent (A) and for the ATP hydrolysis-dependent (B) Na^+–Na^+ exchange based on the scheme in Fig. 2.

as discussed in relation to the scheme shown in Eq. (2), it has a high rate of transition to the phosphoenzyme which reacts with ADP, $E_1' \sim P(Na_n)$. Therefore it is considered ADP-sensitive, as it disappears quickly when ADP is added.[25] E_2–P has a low rate of transition to $E_1'' \sim P$.[25] It is only $E_1'' \sim P$ with three Na^+ bound which has the high rate of transition, not $E_1'' \sim P$ with two or one Na^+ bound.[25] $E_1'' \sim PNa_3$ is denoted an E_1 form because it is K^+ insensitive and has three cation binding sites.

In a consecutive reaction with the cations, it must be the same conformation of the enzyme which delivers Na^+ to the extracellular medium in the Na^+–Na^+ and in the Na^+–K^+ exchange reaction, i.e., $E_1'' \sim PNa_3$ (Fig. 2). According to kinetic experiments[25] $E_1'' \sim PNa_3$ is followed by a transition to the conformation which is ADP insensitive, but K^+ sensitive, E_2–P. The Na^+–K^+ exchange is electrogenic with an exchange of two K^+ for three Na^+ (see Ref. 1), suggesting that E_2–P has two cation binding sites instead of the three on $E_1'' \sim P$. E_2–P has a high affinity for K^+. Binding of K^+ to the extracellular sites triggers a fast rate of dephosphorylation, which leads to K^+ occlusion,[28] a transfer of K^+ from the extracellular medium to the membrane phase. The resulting deocclusion of K^+ to the cytoplasmic side due to a low-affinity effect of ATP[14] terminates the cycle (Fig. 2).

In the ADP-sensitive Na^+–Na^+ exchange the enzyme shuttles between E_1ATPNa_3 and $E_1'' \sim PNa_3$ (Fig. 3A). Oligomycin inhibits the ADP-dependent Na^+–Na^+ exchange (see Ref. 45) but has either no effect on or increases the ADP–ATP exchange.[26,44] Oligomycin occludes Na^+ and the effect of oligomycin can be explained by a decreased rate of deocclusion of Na^+ from $E_1' \sim P(Na_3)$.[34]

The Na^+–K^+ exchange is reversible. With no extracellular K^+, with a gradient for Na^+ into and for K^+ out of the cell, and with ADP plus P_i intracellular the transport runs backward with synthesis of ATP (see Ref. 42).

ATP Hydrolysis-Dependent Na^+–Na^+ Exchange

The transport system can also engage in an Na^+–Na^+ exchange which is ATP hydrolysis dependent (see Ref. 42), which is inhibited by ADP and which is electrogenic with an outward transport of three Na^+ and an inward transport of one to two Na^+ for each ATP hydrolyzed.[43] The rate of Na^+ transport in this reaction is about 6% of the rate of Na^+ transport in the Na^+–K^+ exchange reaction.[43] The conformation which delivers Na^+ to the extracellular medium in a consecutive reaction must be the same in the ATP hydrolysis-dependent and in the ADP-stimulated Na^+–Na^+ exchange. However, the stoichiometry with three Na^+ transported out for one to two Na^+ transported in suggests that the conformation of the enzyme, which accepts Na^+ for inward transport in the ATP hydrolysis-dependent Na^+–Na^+ exchange is not the same as the conformation which delivers Na^+ to the extracellular medium. The K^+ behavior of extracellular Na^+ in the ATP hydrolysis-dependent Na^+–Na^+ exchange (the stoichiometry and the electrogenic effect) suggests that the phosphoenzyme which accepts Na^+ for the inward transport is the same as the phosphoenzyme which accepts K^+ for inward transport in the Na^+–K^+ exchange reaction, E_2–P (Fig. 3B; compare with Fig. 2).

Red blood cells show ADP-stimulated Na^+–Na^+ exchange (see Ref. 42), while with reconstituted Na^+,K^+-ATPase from shark rectal glands there is ATP hydrolysis-dependent Na^+–Na^+ exchange but no ADP-stimulated Na^+–Na^+ exchange.[43] Whether one or the other predominates depends on the steady state distribution between the ADP-sensitive and the K^+-sensitive phosphoenzymes. The ratio between these two forms varies for different enzyme preparations. With shark rectal gland enzyme the equilibrium is shifted toward E_2–P. The equilibrium has not been determined for enzymes from red blood cells.

In the presence of Na^+, but no K^+, K_m for ATP is a micromolar fraction. The higher K_m, a millimolar fraction in the presence of Na^+ plus K^+, is explained from the low-affinity effect of ATP necessary to increase the rate of deocclusion of K^+ from $E_2(K_m)$ (Fig. 2). At a low ATP concentration this step is rate limiting in the Na^+–K^+ exchange reaction.[14] In a consecutive reaction as shown in Fig. 2, the rate constants for the translocation of Na^+ from the cytoplasmic side of the membrane to the extracellular side are the same in the Na^+–K^+ and the Na^+–Na^+ exchange reactions, and the rate constants must be high enough to account for the rate of hydrolysis with Na^+ plus K^+. The rate of the ATP hydrolysis-dependent Na^+–Na^+ exchange, which is much lower than that of the Na^+–K^+ exchange, means that the rate-limiting step in Na^+–Na^+ exchange is the translocation of Na^+ from the extracellular to the cytoplasmic side. As the rate of the Na^+-dependent dephosphorylation of E_2–P is very low,[25] it is likely that this is the rate-limiting step.

Uncoupled Na^+ Efflux and $(ATP + P_i)$-Dependent K^+–K^+ Exchange

With no Na^+ and no K^+ in the extracellular medium, the transport system can engage in an uncoupled Na^+ efflux (see Ref. 42) (Fig. 4A). According to this scheme, transport should be electrogenic, but apparently it is not but is accompanied by cotransport of an anion.[75]

With K^+, but with no Na^+ extracellular and with K^+ intracellular the transport system can engage in a K^+–K^+ exchange (Fig. 4B), which requires ATP and P_i on the cytoplasmic side (see Ref. 42). There is no

---

[75] S. Dissing and J. F. Hoffman, *Curr. Top. Membr. Transp.* **19**, 693 (1983).

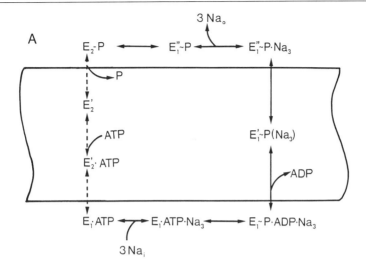

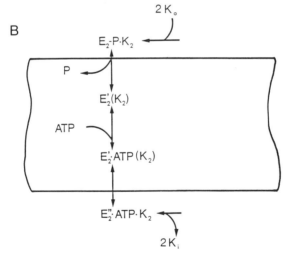

FIG. 4. Schemes for the uncoupled Na^+ efflux (A) and for the K^+–K^+ exchange (B) based on the scheme in Fig. 2.

hydrolysis of ATP and ATP can be replaced by a nonhydrolyzable ATP analog.[76]

Comments

In the scheme shown in Fig. 2, the enzyme has one binding site for ATP, which according to the ligand-binding experiments means an α,β structure. If the transport unit, as suggested from molecular weight determinations, is an $(\alpha,\beta)_2$ structure, what will then be the scheme? Does the transport system show half of the site reactivity? For a discussion see Refs. 20 and 42.

According to the scheme in Fig. 2 the phosphoenzymes formed in the presence of Na^+ are also part of the reaction with Na^+ plus K^+. This means that the rate of dephosphorylation by K^+ of phosphoenzymes formed in the presence of Na^+, but no K^+, should be high enough to explain the overall rate of hydrolysis in the presence of Na^+ plus K^+. This seems, however, not to be the case. When the enzyme is phosphorylated in the presence of Na^+ and [^{32}P]ATP and the phosphorylation with ^{32}P is stopped by addition of cold ATP, the addition of K^+ gives a rate of release of ^{32}P from the phosphoenzymes which is too low to account for the overall hydrolysis in the presence of Na^+ plus K^+.[25] The intermediary steps in the hydrolysis of ATP in the presence of Na^+ plus K^+ can therefore not be the same as in the presence of Na^+, suggesting that the enzyme hydrolyzes ATP along two different but interconnected cycles in the presence of Na^+ and of Na^+ plus K^+, respectively.[25,74] K^+ must have an effect not only on the rate of dephosphorylation but also on the formation of the phosphoenzyme.

In this connection it is of interest that K^+ in extracellular concentrations in the presence of a high concentration of Na^+ inhibits the transition from the ADP-sensitive phosphoenzyme to the K^+-sensitive phosphoenzyme; K^+ inhibits the transition $E_1' \sim P(Na_3)$ to $E_1'' \sim PNa_3$ (see Fig. 2). This is similar to the effect of oligomycin. As K^+, like oligomycin, in the presence of Na^+ under certain conditions stimulates ADP–ATP exchange,[77,78] it appears as if extracellular K^+ has an oligomycin-like effect on formation of the phosphoenzymes in the presence of Na^+; K^+ decreases the rate of deocclusion of Na^+ to the extracellular medium.

Such an effect of K^+ must mean that K^+ is bound to the enzyme, together with Na^+. Based on kinetic experiments (see Refs. 6 and 7) and

---

[76] T. J. B. Simmons, *J. Physiol. (London)* **244**, 731 (1975).
[77] S. P. Banerjee and S. M. E. Wong, *J. Biol. Chem.* **247**, 5409 (1972).
[78] J. D. Robinson, *Biochim. Biophys. Acta* **484**, 161 (1977).

from the effect of Li$^+$ on the phosphatase activity,[35] it has been postulated that there are simultaneously existing extracellular and cytoplasmic sites and that the reaction, which leads to the Na$^+$–K$^+$ exchange, occurs with simultaneous binding of K$^+$ to the extracellular sites and of Na$^+$ to the cytoplasmic sites. These postulates lead to a scheme for the reaction which is different from the one presented in Fig. 2. For a discussion of the problem see Refs. 25, 35, and 74.

Section I

Preparation of Na$^+$,K$^+$-ATPase and Subunits

[2] Purification of Na$^+$,K$^+$-ATPase: Enzyme Sources, Preparative Problems, and Preparation from Mammalian Kidney

By PETER LETH JØRGENSEN

Introduction

Purification of Na$^+$,K$^+$-ATPase to homogeneity requires starting material from tissues with exceptionally large capacities for active Na$^+$ transport like the outer renal medulla in mammalian kidney,[1–4] rectal glands of dogfish,[5,6] eel electroplax,[7,8] and salt glands of ducks in which Na$^+$,K$^+$-ATPase can be induced by salt treatment.[9]

To the extent that Na$^+$,K$^+$-ATPase can be purified in the membrane without breaking native lipoprotein associations, structural and functional pertubation can be kept at a minimum. Purification in membrane-bound form allows study of the pure Na$^+$,K$^+$-ATPase in its natural environment, the lipid bilayer. The pump protein retains its asymmetric orientation in the membrane during the purification procedure and formation of two-dimensional crystals suitable for image analysis can be induced by incubation in vanadate or phosphate solutions.[10,11] This preparation has proved ideal for studying structure and organization of the Na$^+$,K$^+$-ATPase protein in the membrane and for establishing structure–function relationships.[12,13]

Preparations of soluble Na$^+$,K$^+$-ATPase in homogeneous and stable form are required for examination of subunit structure and determination

---

[1] P. L. Jørgensen and J. C. Skou, *Biochem. Biophys. Res. Commun.* **37**, 39 (1969).
[2] P. L. Jørgensen and J. C. Skou, *Biochim. Biophys. Acta* **233**, 366 (1971).
[3] J. Kyte, *J. Biol. Chem.* **246**, 4157 (1971).
[4] P. L. Jørgensen, this series, Vol. 32, p. 277.
[5] J. R. Perrone, J. F. Hackney, J. F. Dixon, and L. E. Hokin, *J. Biol. Chem.* **250**, 4178 (1975).
[6] J. C. Skou and M. Esman, *Biochim. Biophys. Acta* **567**, 436 (1979).
[7] R. W. Albers, C. J. Koval, and G. J. Siegel, *Mol. Pharmacol.* **4**, 324 (1968).
[8] J. F. Dixon and L. G. Hokin, *Arch. Biochem. Biophys.* **163**, 749 (1974).
[9] S. Ernst, C. C. Goerte Miller, and R. A. Ellis, *Biochim. Biophys. Acta* **135**, 682 (1967).
[10] E. Skriver, A. B. Maunsbach, and P. L. Jørgensen, *FEBS Lett.* **131**, 219 (1981).
[11] H. Hebert, P. L. Jørgensen, E. Skriver, and A. B. Maunsbach, *Biochim. Biophys. Acta* **689**, 571 (1982).
[12] P. L. Jørgensen, *Biochim. Biophys. Acta* **694**, 27 (1982).
[13] P. L. Jørgensen, E. Skriver, H. Hebert, and A. B. Maunsbach, *Ann. N.Y. Acad. Sci.* **402**, 207 (1982).

of molecular weight by analytical ultracentrifugation and gel chromatography. In principle there are two approaches to these preparations. One is purification of Na^+,K^+-ATPase in membrane-bound form, followed by solubilization of the pure enzyme in nonionic detergent.[14,15] The other is extraction of Na^+,K^+-ATPase from crude membrane preparations with nonionic detergent and separation of Na^+,K^+-ATPase from contaminant proteins by centrifugation,[16] fractional precipitation with salt,[16] or column chromatography.[17-19]

Preparative Problems

Latency of Na^+,K^+-ATPase in Plasma Membrane Preparations

In tissues homogenates from kidney,[2] brain,[20] or salt glands,[5,6] a large fraction of Na^+,K^+-ATPase is latent. Ultrastructural studies[21] and analysis of ligand binding,[22] tryptic digestion,[22] and cation transport[23] show that latency in preparations from outer renal medulla is due to formation of right-side-out vesicles of plasma membranes. Membrane preparations from skeletal muscle contain a mixture of inside-out and right-side-out vesicles.[24] Centrifugation in sucrose gradient[2] or iodinated density gradient media[22,25] allows separation of closed vesicles from open membrane fragments. Demasking of latent activity is seen after incubation with ionic or nonionic detergents at concentrations close to the critical micelle concentration (cmc) where lysis or rupture of membranes is observed.[2,26,27] Higher concentrations of detergents usually inactivate the Na^+,K^+-ATPase activity. Careful control of the conditions for detergent treatment is therefore essential for determination of total activity of Na^+,K^+-ATPase in membrane preparations. For this purpose sodium deoxycho-

[14] J. B. Brotherus, J. V. Møller, and P. L. Jørgensen, *Biochem. Biophys. Res. Commun.* **100**, 146 (1981).
[15] J. R. Brotherus, L. Jacobsen, and P. L. Jørgensen, *Biochim. Biophys. Acta* **731**, 290 (1983).
[16] J. F. Dixon and L. E. Hokin, *Anal. Biochem.* **86**, 378 (1978).
[17] M. Esman, J. C. Skou, and C. Christiansen, *Biochim. Biophys. Acta* **567**, 410 (1979).
[18] D. F. Hastings and J. A. Reynolds, *Biochemistry* **18**, 817 (1979).
[19] M. Morohashi and M. Kawamura, *J. Biol. Chem.* **259**, 14928 (1984).
[20] K. Sweadner, *Biochim. Biophys. Acta* **508**, 486 (1978).
[21] E. Skriver, A. B. Maunsbach, and P. L. Jørgensen, *Curr. Top. Membr. Transp.* **19**, 119 (1983).
[22] B. Forbush, *J. Biol. Chem.* **257**, 12678 (1982).
[23] B. Forbush, *Anal. Biochem.* **140**, 495 (1984).
[24] S. Seiler and S. Fleischer, *J. Biol. Chem.* **259**, 8550 (1984).
[25] K. D. Dzhandzhugazyan and P. L. Jørgensen, *Biochim. Biophys. Acta* **817**, 165 (1985).
[26] P. L. Jørgensen, this series, Vol. 36, p. 434.
[27] J. B. Brotherus, P. C. Jost, O. H. Griffith, and L. E. Hokin, *Biochemistry* **18**, 5043 (1979).

late is an ideal detergent. Its cmc is relatively high and the optimum concentration for demasking is to a large extent independent of protein concentration.[2,26] Sodium dodecyl sulfate (SDS) has a 10-fold lower cmc than sodium deoxycholate.[2] The use of SDS for demasking of latent activity in quantitative assay of Na^+,K^+-ATPase therefore requires the presence of 1% bovine serum as buffer for the detergent.[22]

Detergent Inactivation

Purification of Na^+,K^+-ATPase in membrane-bound or soluble form requires treatment with detergent in concentrations close to or within the range where inactivation occurs. In spite of the extensive use of detergents in purification, solubilization, and reconstitution of Na^+,K^+-ATPase only a few systematic studies of the mechanism of detergent inactivation of Na^+,K^+-ATPase are available.[2,27] General rules for the use of detergents[28] are applicable, but optimum conditions for detergent treatment of different membrane preparations of Na^+,K^+-ATPase must be found empirically. Inactivation due to the disruptive effects of detergents on the membrane depends on their partition between the membrane lipid phase, the aqueous phase, and the detergent micelles.[27] Partition is often slow and the time course of inactivation is complex with several phases. The sigmoidal shape relating Na^+,K^+-ATPase activity to detergent concentration is seen with deoxycholate, SDS (Fig. 4), and $C_{12}E_8$. This indicates a cooperative process with several detergent molecules involved in inactivation of one enzyme unit.[12,27] As illustrated in Fig. 4, the slope of the linear curve relating detergent concentration at onset of inactivation to membrane concentration gives the value of the critical mole fraction of detergent in the membrane. The intercept at the ordinate gives the aqueous concentration of detergent in equilibrium with the membrane phase and the partition coefficient. Interaction of detergent with hydrophobic areas on the proteins of Na^+,K^+-ATPase are not well examined. A likely mechanism of inactivation is replacement of phospholipid with detergent at the binding positions on hydrophobic areas on the protein. The detergent alkyl chains may be inadequate for stabilization of protein domains that are normally associated with alkyl chains of lipids with 16 or 18 carbon atoms.

Criteria of Purity and Integrity of Na^+,K^+-ATPase

In an ideal pure preparation of Na^+,K^+-ATPase, the α subunit should form 65–70% of total protein and the molar ratio of α to β subunit should

---

[28] E. Helenius, D. R. McCaslin, E. Fries, and C. Tanford, this series, Vol. 56, p. 734.

be 1 : 1, corresponding to a mass ratio about 3 : 1.[12] Functionally the preparation should be fully active in the sense that each α subunit should bind ATP, P_i, cations, and the inhibitors vanadate and ouabain. The molecular activity should be close to a maximum value of 7000–8000 P_i/min. Problems related to determination of ligand binding and molecular activity have been thoroughly reviewed.[12,29,30] The highest reported binding capacities for ATP and phosphate are in the range of 5–6 nmol/mg protein. As shown in Table I, capacities of 6.2–6.7 nmol/mg protein can be obtained in optimum conditions for binding of vanadate or ouabain when fractions with exceptionally high specific activities of Na^+,K^+-ATPase are selected for assay.

For enzymatic or kinetic experiments it is essential that the enzyme preparation is homogeneous in the sense that all Na^+,K^+-ATPase molecules are uniform with respect to function. The presence of partially damaged enzyme units may give rise to nonlinear behavior. Another factor of importance for kinetic experiments is the density of enzyme molecules in the membrane-bound preparations. Ultrastructural studies show a dense packing of 12,000 $\alpha\beta$ units/μm^2 in the membrane fragments.[31] This corresponds to a concentration of α subunit in the lipid bilayer of about 7 mM or 1 g protein/ml of lipid phase. In these conditions of supersaturation, crystallization in two dimensions can be induced by incubation in phosphate or vanadate solutions. However, the extraordinary high concentration of sites in the membrane disks may result in apparent negative cooperativity that is relieved upon dispersion of the sites by solubilization in detergents as in the case of ATP binding.[32]

Among membrane-bound preparations, the highest specific activity and purity of Na^+,K^+-ATPase are obtained by incubation of membranes from mammalian kidney with SDS–ATP followed by zonal centrifugation.[4,26,33] The best preparations obtained by this procedure have purities in the range 95–100%.[12,33,34] The purity of preparations obtained from kidney by incubation with SDS–ATP followed by centrifugation in angle rotors is 50–70%.[33] In membranes from electroplax or salt gland of *Squalus acanthias,* Na^+,K^+-ATPase is very sensitive to detergents, in particular SDS. The highest purity of membrane-bound preparations from these tissues, 50–60%, is obtained after treatment of the membranes with sodium deoxycholate.[6]

[29] J. Kyte, *Nature (London)* **292,** 201 (1981).
[30] J. G. Nørby, *Curr. Top. Membr. Transp.* **19,** 281 (1983).
[31] N. Deguchi, P. L. Jørgensen, and A. B. Maunsbach, *J. Cell Biol.* **75,** 619 (1977).
[32] J. Jensen and P. Ottolenghi, *Biochim. Biophys. Acta* **731,** 282 (1983).
[33] P. L. Jørgensen, *Biochim Biophys. Acta* **356,** 36 (1974).
[34] G. Zamphigi, J. Kyte, and W. Freytag, *J. Cell Biol.* **98,** 1851 (1984).

Solubilization of membrane-bound Na^+,K^+-ATPase from kidney in $C_{12}E_8$ results in preparations with the same activity and purity as in the membrane-bound state. Analytical ultracentrifugation within a few hours after solubilization shows that the preparation consists predominantly (80–85%) of soluble $\alpha\beta$ units with M_r 140,000–170,000.[14,15] After storage or prolonged chromatography the particles aggregate to $(\alpha\beta)_2$ units or higher oligomers, presumably by detergent–detergent association.[12,15] Soluble preparations from rectal gland of dogfish or eel electroplax must be fractionated by chromatography after solubilization. This may explain the high molecular weight, 280,000–380,000, and oligomeric structure, $\alpha_2\beta_2$[17] or $\alpha_2\beta_4$.[18]

Preparative Procedures

Update of Procedures for Purification of Na^+, K^+-ATPase from Outer Renal Medulla

A survey of the procedures is given in Fig. 1. The principles of isolation of crude membranes, incubation with SDS, and isopycnic zonal centrifugation are the same as previously described.[4,26,33] Membranes with high concentration of Na^+,K^+-ATPase, 10–15% of total protein, are prepared from outer renal medulla. Extraneous protein is extracted from the membranes by incubation with SDS in the presence of ATP. Na^+,K^+-ATPase remains embedded in the membrane structure and is separated from soluble protein and other particles in a single isopycnic zonal centrifugation. The success of the procedure depends critically on dissection of tissue from the red outer medulla (inner stripe of the outer medullary zone). It is essential to obtain crude membrane fractions with specific activities of Na^+,K^+-ATPase in the range 3–5 μmol P_i/min · mg protein. Another critical factor is the technique for incubation with SDS.

The purification can be explained by two mechanisms: membrane fractionation and selective extraction of protein by SDS. As the membrane fractions are inhomogeneous, some membrane pieces in the mixture may be completely solubilized by SDS. The explanation for the selective extraction can be that SDS interacts with surface proteins more readily than with Na^+,K^+-ATPase in the presence of ATP.[12] The asymmetric arrangement of the protein particles in the membranes of the pure Na^+,K^+-ATPase show that the protein remains embedded in the membrane during this incubation.

The first step consists of isolation of plasma membranes from tubule cells of the inner stripe of the outer medulla. The predominant structure in this tissue is the thick ascending limb of Henle (MTAL, Fig. 2) which is

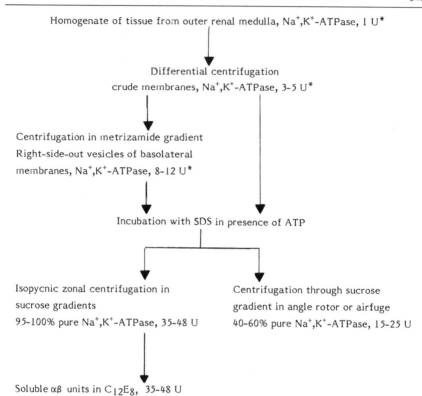

FIG. 1. Flow diagram of procedures for purification of Na^+,K^+-ATPase from outer renal medulla. The figures to the right indicate the specific activity of each preparation. U, μmol P_i/min · mg protein at 37°. *, In preparations consisting of native plasma membranes activities are determined after demasking of latent activity by preincubation with sodium deoxycholate.

specialized for active transcellular NaCl transport with large capacity. The basolateral cell membranes of the TAL are tightly packed with Na,K-pump sites in a concentration exceeding 40 million sites/cell as estimated by [^{3}H]ouabain binding.[35] A crude membrane fraction is prepared by differential centrifugation and this fraction forms the starting material.

Centrifugation of the crude membranes on metrizamide gradients separates membrane vesicles containing Na^+,K^+-ATPase in high concentration from open fragments and mitochondria. The membrane vesicles are useful in structural and transport studies.[22] After chemical labeling in vesicles with well-defined orientation, Na^+,K^+-ATPase can be purified

[35] G. E. Mernissi and A. Doucet, Am. J. Physiol. 247, F158 (1984).

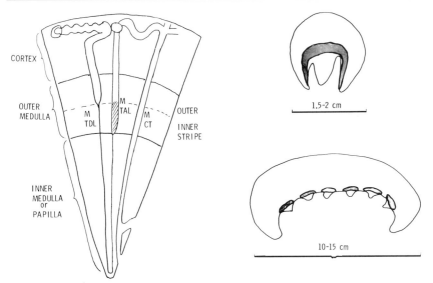

FIG. 2. Organization of the nephron with medullary thick ascending limb of Henle (MTAL), medullary thin descending limb of Henle (MTDL), and medullary collecting tubule (MCT) in inner stripe of outer medulla. To the right, the dark areas correspond to the inner stripe of outer medulla in transverse section of rabbit kidney (above) and longitudinal section of pig kidney (below). The gross structure of dog and lamb kidney is similar to that of rabbit kidney. Human kidney is similar to pig.

from the vesicles using a microscale version of the purification procedure in the Beckman airfuge.[24]

Dissection of Kidney Tissue, Homogenization, and Fractionation of Membranes

Procedure. Immediately after sacrifice and bleeding the kidneys are removed from the animals and stored in ice-cold imidazole–sucrose. For large-scale preparations rabbit or pig kidneys are obtained at a slaughterhouse. Kidneys from dog and sheep are good alternatives. Capsule and vessels are removed and rabbit kidneys are cut in transverse sections, 3–5 mm thick. Pig kidneys are cut in longitudinal sections 0.5–1 cm thick. Dissection is performed on an ice-cold stainless steel plate which is covered with filter paper and moistened with imidazole (25 mM), EDTA–Tris (1 mM), sucrose (250 mM), pH 7.2 (20°) (imidazole–sucrose). The light gray inner medulla or papilla is dissected away with a scalpel and the tissue from the dark red outer medulla is separated from the light brown outer stripe of outer medulla as illustrated in Fig. 2. The tissue is placed in tared vials with ice-cold imidazole–sucrose.

After weighing on an analytical balance the tissue is homogenized in 10 ml imidazole–sucrose/g of tissue with five strokes in a tightly fitting Teflon–glass homogenizer (Braun, Melsungen), operated at 1000 rpm in an ice bath. One stroke means that the Teflon pestle has moved from the meniscus of the fluid to the bottom of the tube.

Preparation of Crude Membranes by Differential Centrifugation

Differential centrifugation in isotonic sucrose separates particles by differences in sedimentation velocity and equilibrium density. The purpose is to prepare a crude membrane fraction which contains the major part, 50–60%, of Na^+,K^+-ATPase in the homogenate.

Procedure. The homogenate is centrifuged at 7000 rpm for 15 min in the SS-34 rotor of the Sorvall high-speed centrifuge to sediment nuclei, large fragments, and most mitochondria. The sediment after this centrifugation is rehomogenized in 5 ml imidazole–HCl sucrose/g tissue and centrifuged again for 15 min at 7000 rpm. The two supernatants are mixed and centrifuged for 30 min at 20,000 rpm to sediment most cell membranes and 50–60% of Na^+,K^+-ATPase in the homogenate. The supernatant contains small light fragments of intracellular membranes and the soluble cytoplasmic proteins.

Separation of Membrane Vesicles from Open Membrane Fragments in Metrizamide Gradients

In metrizamide (Nyegård, Oslo, Norway) open membrane fragments sediment to their equilibrium density, ~1.14 g/ml, while closed membrane vesicles stop at lower densities (1.05–1.07 g/ml), presumably because the cell membrane is impermeable to metrizamide. Purity of Na^+,K^+-ATPase in these vesicles is 15–25% (Fig. 3).

Procedure. Using a gradient mixer, a linear gradient from 5 g/100 ml to 15 g/100 ml metrizamide in 25 mM imidazole–HCl, 250 mM sucrose, pH 7.2, is layered on top of a cushion of 1.5 ml 30 g/100 ml metrizamide. A sample containing 15–25 mg membrane protein in imidazole–sucrose is layered on top of the gradient. Centrifugation is for 16 hr at 20,000 rpm ($\omega^2 t = 2.5 \times 10^{11}$) in a SW 27 swinging bucket rotor or for 2 hr at 60,000–70,000 rpm in an angle rotor.

Determination of Specific Activity of Na^+,K^+-ATPase in Crude Membrane Fractions[2,4,26]

Na^+,K^+-*ATPase Assay.* For demasking of latent Na^+,K^+-ATPase activity, the membranes are incubated at 0.25–1 mg protein/ml with 0.065%

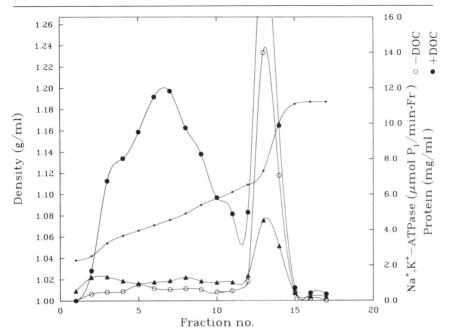

FIG. 3. Distribution of Na$^+$,K$^+$-ATPase activity after centrifugation in a metrizamide gradient. (▲) protein; (○) Na$^+$,K$^+$-ATPase activity without incubation with sodium deoxycholate; (●) activity after incubation with sodium deoxycholate. The activity in the broad peak in fractions 4–10 is located in right-side-out vesicles of basolateral membranes where Na$^+$,K$^+$-ATPase can be demasked by treatment with sodium deoxycholate. The peak at fractions 12–15 consists mainly of open membrane fragments and mitochondria.

deoxycholate in 50 mM imidazole–HCl, 2 mM EDTA, pH 7.0. After incubation for 30 min at 20°, 25-μl aliquots are transferred to test tubes containing 1 ml of 130 mM NaCl, 20 mM KCl, 3 mM MgCl$_2$, 3 mM ATP, 25 mM imidazole–HCl, pH 7.5. After incubation for 1, 3, 5, or 10 min at 37° the reaction is stopped with 1 ml ice-cold 0.5 M HCl containing 30 mg ascorbic acid, 5 mg ammonium heptamolybdate, 10 mg SDS. The tubes are transferred to an ice bath. For color development, 1.5 ml containing 30 mg sodium metaarsenite, 30 mg sodium citrate, 30 μl acetic acid is added. The tubes are heated for 10 min at 37° and absorbance is read at 850 nm.[36]

Protein Determination. Protein is determined by the method of Lowry *et al.*[37] after precipitation with trichloroacetic acid as before.[4] As stan-

---

[36] P. Ottolenghi, *Eur. J. Biochem.* **99**, 113 (1979).
[37] O. H. Lowry, N. J. Rosebrough, A. L. Farr, and R. J. Randall, *J. Biol. Chem.* **193**, 265 (1951).

dards for the analysis membrane-bound purified Na^+,K^+-ATPase preparations are used, in which the protein concentration has been determined by quantitative amino acid analysis after removal of imidazole by repeated washing of the membranes in 50 mM ammonium carbaminate.[16]

Determination of Optimum Concentration of SDS for the Incubation

Since SDS has a low critical micelle concentration the bulk of detergent added to the medium will partition into the membrane and the weight ratio of SDS/membrane mass will determine the inactivation. The sigmoid-shaped inactivation curves in Fig. 4 indicate that several SDS molecules are involved in the inactivation of one unit of Na^+,K^+-ATPase. At the optimum concentration for the purification Na^+,K^+-ATPase is inactivated to 85–90% of maximum activity. The inset in Fig. 4 shows the linear relationship between the SDS concentration at 85% of maximum activity and the membrane concentration. The equation for this line is used for calculation of the optimum concentration. Inactivation is reversible to some extent, suggesting that SDS perturbs the lipid environment in addition to removing peripheral proteins. After centrifugation the concentration of SDS in the pure Na^+,K^+-ATPase is very low, 9 μg/mg protein.[33]

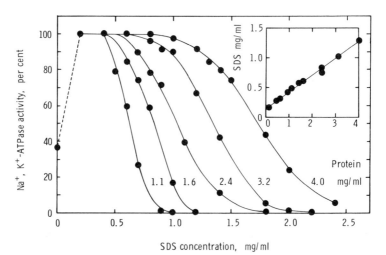

FIG. 4. Effect of membrane concentration on concentration of SDS for incubation prior to purification. Aliquots of membrane fractions from outer renal medulla containing 0.52 mg phospholipid and 0.14 mg cholesterol/mg protein were incubated at 20° in 25 mM imidazole–HCl, 1 mM EDTA, 3 mM Na$_2$ATP, pH 7.5, with the concentrations of SDS shown on the abscissa. After 30 min, 10 μl was transferred to test tubes for assay of Na^+,K^+-ATPase. Equation for the straight line of the curve in the inset is as follows:

$$\text{mg SDS/ml} = 0.146 + 0.282 \text{ (mg protein/ml)}$$

Preparation of Sample for Sucrose Gradient Centrifugation

In a series of trial centrifugations we found that the maximum capacity of the Ti-14 rotor for this separation is about 150 mg protein and that the optimum concentration of protein in the sample is 2 mg/ml. A sample of 75 ml is prepared by mixing 150 mg crude membrane protein with solutions containing Na_2ATP, EDTA, and SDS to obtain the final concentration of 2 mg/ml protein, 2 mM Na_2ATP, 2 mM EDTA, 0.71 mg/ml SDS in 25 mM imidazole–HCl, pH 7.5, at 20°. The sequence of the mixing is important. Crude membranes are first mixed with Na_2ATP and EDTA and finally SDS is added dropwise from a solution of 2 mg/ml under continuous stirring with a magnetic bar. After incubation for 30 min at 20° on a water bath the sample is injected into the Ti-14 zonal rotor.

Centrifugation in the Zonal Rotor

In our experience the Beckman zonal centrifuge system (Ti-14 or Ti-15 zonal rotors with seal assembly) or the Kontron system (TZT 48 rotor) is well suited for this preparation. With both systems we used the model 141 Beckman gradient pump with the gradient cam shown in Fig. 5. The pump is adjusted so that the scale 0–10 on the abscissa corresponds to 500 ml.

Procedure. Using a light solution of 15 mg/100 ml sucrose and a heavy solution of 40 g/100 ml sucrose in 25 mM imidazole–HCl, 1 mM EDTA, pH 7.5, 450 ml of gradient is pumped in at a rate of 28 ml/min. After reaching No. 9 at the abscissa of the cam (Fig. 5) the heavy solution is exchanged for 45 g/100 ml sucrose and pumping is continued until the rotor has been filled. The sample, 75 ml, is then injected from the center of the rotor, followed by 30 ml of overlay consisting of 25 mM imidazole–HCl, 1 mM EDTA, pH 7.5, without sucrose. Centrifugation is for 2 hr at 48,000 rpm ($\omega^2 t = 2.2 \times 10^{11}$). The gradient is displaced through the center in 14- to 16-ml fractions. Aliquots of 10–25 μl are taken out for Na^+,K^+-

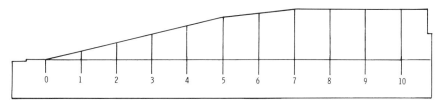

FIG. 5. Shape of gradient cam for model 141 Beckman gradient pump for producing the sucrose gradient in the Ti-14 (Beckman) or TZT 48 (Kontron) zonal rotor. One unit on the abscissa corresponds to 50 ml. Using 15% and 40% (w/v) sucrose solutions for the gradient pump and a 105-ml sample plus overlay, the peak of Na^+,K^+-ATPase will appear in the gradient corresponding to No. 6 on the abscissa.

ATPase assay. Pure Na^+,K^+-ATPase (30–48 μmol P_i/min · mg protein) is collected in one or two fractions at the peak of Na^+,K^+-ATPase activity and in three fractions to the denser side of the center of the peak. The peak is found at densities 1.13–1.14 g/ml corresponding to a volume of 3–400 ml when the gradient is displaced through the center of the rotor. Fractions of 16 ml containing the purified Na^+,K^+-ATPase are diluted to 35 ml with 25 mM imidazole–HCl/1 mM EDTA, pH 7.5, and centrifuged in a Ti-60 rotor at 40,000 rpm overnight (16–17 hr) to collect the enzyme in the pellet.

The yield per centrifugation is 5–8 mg protein with specific activity above 30 μmol P_i/min · mg protein and 3–5 mg protein with specific activity above 20 μmol P_i/min · mg protein. The purified Na^+,K^+-ATPase consists of equal amounts of protein and lipid, 0.7 mg phospholipid, and 0.28 mg cholesterol/mg protein. The protein is organized in membrane disks where $\alpha\beta$ units are observed as distinct surface particles.[31] The preparation is 90–100% pure with respect to the content of the specific proteins,

TABLE I
Binding Capacity of Vanadate and Ouabain[a]

Na^+,K^+-ATPase activity (μmol P_i/mg protein · min)	Ligand binding (nmol/mg protein)	Molecular activity (min^{-1})	Molecular mass (kDa)
[^{48}V]Vanadate binding			
49.8	6.4	7830	157
48	6.3	7660	159
46.8	6.7	6944	148
43.3	5.9	7377	170
42.7	5.3	8060	189
[^{3}H]Ouabain binding			
48.3	6.1	7890	163
46.6	6.2	7580	163
45.3	5.8	7810	172
42.0	5.2	8150	194
40.7	5.3	7680	189

[a] [^{3}H]Ouabain binding supported by vanadate or [^{48}V]vanadate binding supported by ouabain in selected fractions of Na^+,K^+-ATPase obtained by zonal centrifugation. Fractions with particularly high specific activities were selected among the fractions of a series of zonal centrifugations of membrane preparations from pig kidney. For determination of [^{48}V]vanadate binding aliquots containing 75 μg protein were incubated with 3 mM MgCl$_2$, 10^{-5} M ouabain, 10 mM TES, 1 mM EDTA, 1 mM DTT, pH 7.0, and 0.2–1.4 μM [^{48}V]vanadate. For determination of [^{3}H]ouabain binding incubation was with 100 μM vanadate, 3 mM MgCl$_2$, 10 mM TES, 1 mM EDTA, 1 mM DTT, pH 7.0, and 0.2–1.2 μM [^{3}H]ouabain. Bound and free ligand was separated by centrifugation as before.[33]

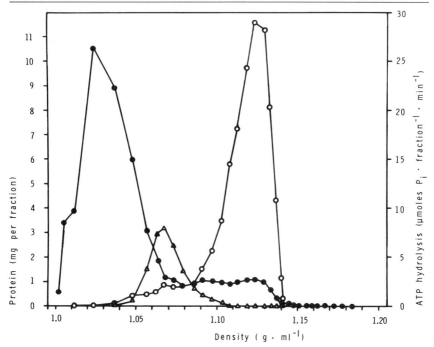

FIG. 6. Distribution of protein (●), Mg^{2+}-ATPase (▲), and Na^+,K^+-ATPase (○) after isopycnic zonal centrifugation in the Ti-14 Beckman zonal rotor as described in the text.

the α subunit with M_r 93,000–112,000, and the β subunit with M_r 36,000 plus carbohydrate. Binding capacities for ouabain and vanadate correspond to one site per αβ unit (Table I).

Centrifugation in Angle Rotor

Figure 6 shows that only small amounts of protein were found in fractions peripheral to the peak of Na^+,K^+-ATPase after the isopycnic zonal centrifugation. It is therefore an obvious modification to collect the enzyme as a pellet in the tubes of an angle rotor after sedimentation through a gradient that prevents passage of Mg^{2+}-ATPase and soluble proteins. Another alternative to the zonal rotor is centrifugation in sucrose gradients in angle rotors with large capacities (Ti-45, with six tubes of 110 ml).[38]

Procedure. Eighty milliliters of the medium for incubation with SDS is prepared. Discontinuous density gradients of 25 ml are made in the 2.5 ×

[38] K. B. Munson, *J. Biol. Chem.* **256**, 3223 (1981).

9 cm tubes of the Ti-60 Beckman fixed angle rotor: 12.5 ml–29.4 g/100 ml, 7.5 ml–15 g/100 ml, and 5 ml–10 g/100 ml sucrose in 25 mM imidazole–HCl, 1 mM EDTA, pH 7.5 (20°). Portions of 10 ml of the sample are layered on each tube. Centrifuge at 60,000 rpm for 90 min ($\omega^2 t = 2.4 \times 10^{11}$) at 4°. The pellets are resuspended in imidazole–EDTA and stored frozen.

To collect the enzyme in a band in the tube the gradient is supplemented with a layer of 38.5 g/100 ml sucrose and the membranes are collected from the interface between the two heavier layers.

Centrifugation in Airfuge

A microscale version of the procedure has been developed in structural studies on Na$^+$,K$^+$-ATPase in membrane vesicles obtained from metrizamide gradients[25] as described above.

Procedure. For purification of Na$^+$,K$^+$-ATPase from labeled samples of sealed and open vesicles, the membranes are incubated with SDS and ATP at 2 mg protein/ml as described above. Aliquots of 50 μl (100 μg protein) are layered on discontinuous gradients consisting of 100 μl 25 g/100 ml sucrose and 25 μl 15 g/100 ml sucrose in 25 mM imidazole–HCl, 1 mM EDTA, pH 7.5, in the 200 μl tubes of the Beckman airfuge. After centrifugation for 45 min at 100,000 rpm, the pure Na$^+$,K$^+$-ATPase (20–30 μg protein) is collected in the pellet.

Preparation of Pure Na$^+$, K$^+$-ATPase in Large Membrane Fragments for Crystallization in Two Dimensions[11,34]

Preparation of large and uniform sheets of membrane crystals requires modifications of the purification procedure. Systematic exploration[34] showed that the largest fragments of membrane with the highest density of particles are obtained when the concentration of SDS in the incubation medium is lowered to 0.6 mg/ml with 2.0 mg/ml of membrane protein and 3 mM Na$_2$ATP, 1 mM EDTA, 25 mM imidazole–HCl, pH 7.5. After isopycnic zonal centrifugation the largest membranes with the highest density of particles are found in the two or three fractions to the denser side of the center of the peak of Na$^+$,K$^+$-ATPase activity.

Preparation of Soluble αβ Units with Full Na$^+$, K$^+$-ATPase Activity

Soluble and fully active αβ units are prepared by mixing pure membrane-bound Na$^+$,K$^+$-ATPase with C$_{12}$E$_8$ in proper conditions. The preparation of soluble αβ units is stable for 3–5 hr at 0°. Aggregation is observed after prolonged incubation or column chromatography.[15]

Procedure. Sediment pure membrane-bound Na$^+$,K$^+$-ATPase in a

Beckman airfuge (10 min, 100,000 rpm). Resuspend in 40 mM KCl, 260 mM NaCl, 20 mM TES, 2 mM Tris–EDTA, pH 7.5, at 4.0 mg/ml protein. Mix with an equal volume of $C_{12}E_8$ in water to a final concentration of 5.0 mg $C_{12}E_8$/ml, 20 mM KCl, 130 mM NaCl, 10 mM TES, 1 mM Tris–EDTA, 2.0 mg/ml protein. Incubate for 30 min at 20°. Centrifuge for 5 min at 100,000 rpm in the airfuge. For Na^+,K^+-ATPase assay mix 40-μl aliquots of the supernatant with 4 μl Mg-ATPase (100 mM) and stop after 20, 40, or 60 sec with 1 ml ice-cold 0.5 M HCl containing 30 mg ascorbic acid, 5 mg ammonium heptamolybdate, and 10 mg SDS and determine P_i as above. For the potassium phosphatase assay, KCl, 20 mM/NaCl, 130 mM is replaced with KCl, 150 mM, and Mg-ATPase with 4 μl 200 mM magnesium-p-nitrophenylphosphate.[14,15]

Stability of Purified Preparations

In membrane-bound form the Na^+,K^+-ATPase activity is stable for hours at 37°, for days at 20°, for weeks at 0°, and indefinitely at −80°. Na^+,K^+-ATPase is inactivated by freezing at −20–25° in imidazole–EDTA without cryoprotectant (10% sucrose or glycerol). Soluble preparations of Na^+,K^+-ATPase in $C_{12}E_8$ are more stable in KCl (E_2 form) than in NaCl (E_1 form).[15,19]

[3] Preparation of Membrane Na^+,K^+-ATPase from Rectal Glands of *Squalus acanthias*

By JENS CHRISTIAN SKOU and MIKAEL ESMANN

Introduction

The rectal gland of *Squalus acanthias* is a rich source of Na^+,K^+-ATPase.[1,2] The purification procedure used here employs the bile salt deoxycholate (DOC) to extract protein impurities selectively from the Na^+,K^+-ATPase-containing membranes.[3] Preparation of Na^+,K^+-ATPase from salt glands has the advantage of not involving tedious dissection of tissue.

---

[1] L. E. Hokin, J. L. Dahl, J. D. Deupree, J. F. Dixon, J. F. Hackney, and F. Perdue, *J. Biol. Chem.* **248**, 2593 (1973).
[2] J. C. Skou and M. Esmann, *Biochim. Biophys. Acta* **567**, 436 (1979).
[3] P. L. Jørgensen and J. C. Skou, *Biochem. Biophys. Res. Commun.* **37**, 39 (1969).

Purification Procedure

Preparation of Microsomes

Rectal salt glands are excised from *Squalus acanthias* and washed in seawater. The glands are stored on ice on the fishing boats at 0° in 0.9% NaCl with 50 mg polymyxin B, 0.5 g streptomycin, and 600 mg penicillin/liter, for up to 1 week after the sharks are caught.

The glands are homogenized in a Waring blender in small portions for 60 sec in 0.25 M sucrose, 5 mM EDTA, 30 mM histidine (pH 6.8, adjusted with propanediol).

The homogenate, with a protein content of about 20 g/100 ml, is centrifuged for 15 min at 5900 g. The supernatant is collected and the pellet is rehomogenized as above with one-half volume of buffer. The second homogenate is also centrifuged for 15 min at 5900 g. The two supernatants are combined and centrifuged for 30 min at 48,000 g. The resulting supernatant is discarded and the pellet resuspended in 30 mM histidine, 0.25 M sucrose, and 1 mM EDTA (pH 6.8), at a protein concentration of about 6 mg/ml. The pellet (termed the microsomal fraction) is stored at $-70°$.

DOC Treatment

The microsomal fraction is heated slowly to room temperature and diluted 1:1 (v/v) with 25 mM imidazole, 1 mM EDTA (pH 7.7) containing DOC at a concentration giving maximum activity after 30 min incubation at room temperature. The optimum concentration of DOC is about 0.15% with 2.5 mg protein/ml (see Fig. 1). The optimum DOC concentration is determined for each portion of microsomes.

Saponin (0.05% final concentration) is routinely included together with DOC; the reason for this is the broader peak of activation (see Fig. 1). The purity of the final enzyme preparation is not affected by the presence of saponin (see below).

Differential Centrifugation

After incubation for 30 min at room temperature the DOC-treated microsomal fraction is cooled to 4° and subjected to a differential centrifugation. All the following steps are performed at 4° unless otherwise stated. The first centrifugation is at 5900 g for 15 min, the pellet is discarded, and the supernatant is centrifuged for 30 min at 31,000 g. The supernatant after this centrifugation is centrifuged for 3 hr at 48,000 g and the resulting supernatant is discarded. The pellets after the 31,000 and 48,000 g centrifugations are combined and homogenized in 30 mM histidine, 25% glycerol

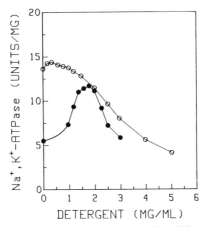

FIG. 1. The effect of DOC and saponin on the Na⁺,K⁺-ATPase activity in microsomes from the rectal gland of *Squalus acanthias*. Microsomes are incubated for 30 min at 23° at 2.5 mg protein/ml and varying concentrations of DOC plus 0.05% saponin (open symbols) or DOC alone (filled symbols).

(pH 6.8, buffer A) at about 4 mg protein/ml. This suspension is centrifuged at 220,000 g for 1 hr at 12°. The pellet after the 220,000 g centrifugation has a firm yellowish center and the rest is white and more loose. By gentle stirring with buffer A it is possible to separate the two parts with the white part having a higher specific activity than the yellow part (see Table I).

TABLE I
PURIFICATION OF Na⁺,K⁺-ATPase FROM RECTAL GLAND OF *Squalus acanthias*

Step	Specific activity (units/mg)		Protein recovery (%)	Phosphorylation level (nmol/mg)
	Na⁺,K⁺-ATPase	pNPPase		
Microsomes	4.75	1.00	100	0.50
DOC-treated microsomes	12.5	1.87	100	—
6,100 g pellet	11.3	1.55	6	—
31,000 g pellet	22.6	3.05	18	—
48,000 g pellet	22.1	2.95	33	—
220,000 g pellet	25.3	3.35	36	2.02
220,000 g yellow pellet	20.3	2.91	—	—
220,000 g white pellet	27.6	3.87	—	—
Sucrose gradient peak	33.7	4.98	12	—

Density Gradient Centrifugation

Enzyme (3 ml) is put on a 28-ml linear gradient of 15–45% sucrose, with 2 ml 65% sucrose in the bottom of the centrifuge tube and centrifuged at 98,000 g at 4° overnight. The protein around the activity peak is collected and centrifuged at 220,000 g for 1 hr after dilution with buffer A. The pellet is resuspended in buffer A and stored at $-20°$.

Assay of Enzyme Activity and Purity

The methods of measurement of Na^+,K^+-ATPase and K^+-pNPPase activities are treated elsewhere in this volume.[4] The protein purity of the enzyme preparation at different stages of purification can be estimated from the content of α and β peptide on Coomassie blue-stained SDS gels.[2]

The concentration of enzyme at the different stages of preparation is estimated from the phosphorylation capacity.[5]

[4] M. Esmann, this volume [10].
[5] J. C. Skou and C. Hilbert, *Biochim. Biophys. Acta* **185**, 198 (1969).

[4] Purification of Na^+,K^+-ATPase from the Supraorbital Salt Gland of the Duck

By THOMAS WOODWARD SMITH

Na^+,K^+-ATPase of specific activity 2000 to 2300 μmol P_i/mg protein/ hr can be prepared from the supraorbital salt-secreting gland of the domestic duck (*Anas platyrhynchos*) in 1 day[1] using an approach analogous to that devised by Jørgensen for kidney enzyme.[2]

White Peking ducks 6 to 8 weeks old are fed standard mash and allowed only 0.2 M NaCl solution to drink for 10 to 12 days. They are then decapitated and the supraorbital salt-secreting glands dissected free and frozen at once in liquid nitrogen. Within an hour they are thawed, minced finely with scissors, added to 10 vol of 0.25 M sucrose/20 mM Tris/1 mM EDTA, pH 7.5 at 0°, and homogenized with a Polytron PT10 (Brinkmann) set at 4.5 for two periods of 15 sec separated by 2 min on ice. The

[1] B. E. Hopkins, H. Wagner, Jr., and T. W. Smith, *J. Biol. Chem.* **251**, 4365 (1976).
[2] P. L. Jørgensen, *Biochim. Biophys. Acta* **356**, 36 (1974).

homogenate is strained through three layers of gauze and centrifuged for 15 min at 5000 g in the SS34 rotor of a Sorvall RC-5 centrifuge. The supernatant is saved and the pellet resuspended in 10 ml of the Tris/sucrose buffer described above by two strokes for a Potter homogenizer at 3500 rpm, cooled, and again centrifuged at 5000 g for 15 min. The supernatants are combined and centrifuged at 48,000 g in the SS34 rotor for 1 hr. The pellet is resuspended in Tris/sucrose buffer to a protein concentration of 10 to 15 mg/ml for subsequent treatment with sodium dodecyl sulfate (SDS) (Pierce Chemical Co., Sequenal grade).

SDS (2.0 mg/ml) is added dropwise at 20° through a 27-gauge needle with constant stirring over 2 min to a suspension of 1.4 mg enzyme protein/ml of solution containing 3 mM Na$_2$ATP/2 mM EDTA/20 mM Tris, pH 7.5. The final SDS concentration that yields the highest specific activity enzyme is 0.55 mg/ml. After standing for 30 min at 20°, 1.0-ml aliquots are layered over sucrose density gradients in 9-ml polycarbonate tubes for the Beckman 40 fixed angle rotor. Gradients of sucrose in 20 mM Tris/1 mM EDTA, pH 7.5, consist of 4.0 ml of 29.4% (w/v) sucrose, 2.4 ml of 15% sucrose, and 1.6 ml of 10% sucrose. After centrifugation at 105,000 g for 210 min at 4° in a Beckman L5-65 centrifuge, the supernatants are aspirated and discarded and the pellets resuspended in 0.25 M sucrose/20 mM Tris/1 mM EDTA, pH 7.5, and rehomogenized with two strokes of a Duall glass homogenizer.

Na$^+$,K$^+$-ATPase activity is measured by a linked enzyme assay or by release of P$_i$ as described.[1] Protein is measured according to the method of Lowry et al.[3] using bovine serum albumin as standard. Activity is stable for periods of months when stored at $-70°$. There is no measurable loss of activity when stored for 3 months at $-20°$, but the preparation typically loses 10% of its activity after 1 week at 4°.

Na$^+$,K$^+$-ATPase prepared as described is essentially pure by criteria of specific enzymatic activity, specific [^{3}H]ouabain binding, phosphorylation yield, and homogeneity of yield in Edman degradation used to determine the N-terminal amino acid sequence of the α subunit.[1]

The yield can be increased substantially, with some loss of purity, by scaling up the density gradient centrifugation steps.[4] Ten milliliters of SDS-treated enzyme (1.4 mg protein/ml) is layered on a sucrose step gradient consisting of 20 ml of 29.4%, 12 ml of 15%, and 8 ml of 10% sucrose in 20 mM Tris/1 mM EDTA, pH 7.5, in 50-ml tubes for the

---

[3] O. H. Lowry, N. J. Rosebrough, A. L. Farr, and R. J. Randall, *J. Biol. Chem.* **193**, 265 (1951).
[4] E. T. Fossel, R. L. Post, D. S. O'Hara, and T. W. Smith, *Biochemistry* **20**, 7215 (1981).

Beckman 35 fixed angle rotor. After centrifugation at 35,000 rpm (100,000 g) for 5 hr, the pellets are combined and processed further as described above. Specific activity of enzyme prepared by this modification averages 1600 to 1700 μmol P_i/mg protein/hr. Yields of 100 mg protein from salt glands from 20 to 24 ducks are sufficient to permit observation of the ^{31}P NMR resonance of the β-aspartyl phosphate at +17.4 ppm in the presence of ATP and Na^+ (zero K^+) or P_i, Mg^{2+}, and ouabain (zero Na^+).[4]

[5] Preparation of Na^+,K^+-ATPase from Brine Shrimp

By GARY L. PETERSON and LOWELL E. HOKIN

Introduction

Na^+,K^+-ATPase from the brine shrimp, *Artemia salina,* is of general interest from at least three viewpoints. First, it is a developmental system in which relatively high concentrations of the enzyme (3 μmol P_i/hr mg homogenate protein) are synthesized *de novo* over a relatively short period of time.[1,2] Brine shrimp embryos at 6–8 hr after hydration and initiation of development contain no detectable Na^+,K^+-ATPase activity, whereas at 30–36 hr activity is maximal. These conditions are well suited for biogenesis studies of the Na^+,K^+-ATPase. Second, the brine shrimp Na^+,K^+-ATPase contains two molecular forms of the α subunit ($α_1$ and $α_2$) which have recently been shown by two independent methods to be due at least in part to differences in primary sequence[3,4] and hence must represent different gene products. Such forms, which are distinguished by differences in electrophoretic mobility of the α subunit, were first described in the brine shrimp[5] and have subsequently been reported in brain[6] and other tissues (see Ref. 3). It is likely that a genetic explanation will

[1] G. L. Peterson, R. D. Ewing, and F. P. Conte, *Dev. Biol.* **67**, 90 (1978).
[2] G. L. Peterson and L. E. Hokin, in "Biochemistry of Artemia Development" (J. G. Bagshaw and A. H. Warner, eds.), p. 120. Univ. Microfilms Int., Ann Arbor, Michigan, 1979.
[3] J. A. Fisher, L. A. Baxter-Lowe, and L. E. Hokin, *J. Biol. Chem.* **259**, 14217 (1984).
[4] M. Morohashi and M. Kawamura, *J. Biol. Chem.* **259**, 14928 (1984).
[5] G. L. Peterson, R. D. Ewing, S. R. Hootman, and F. P. Conte, *J. Biol. Chem.* **253**, 4762 (1978).
[6] K. J. Sweadner, *J. Biol. Chem.* **254**, 6060 (1979).

also be found for the multiple α subunit forms in these other tissues,[3] and furthermore, it is expected that such forms will have important physiological relevance. Third, the brine shrimp represents the most phylogenetically distant source of Na^+,K^+-ATPase available for study in a purified form, and is thus of interest from an evolutionary point of view.

Purification of the brine shrimp Na^+,K^+-ATPase has been more difficult than that from the kidney, electric eel, and dogfish rectal gland[7-11] because it is not as rich a source as these latter cases (about 5-fold lower specific activity), and has resisted solubilization in an active form by all of the conventional detergents. This disadvantage is partially offset by the fact that virtually unlimited amounts of brine shrimp can be reared in the laboratory in 1–2 days with little expense and with simple equipment. A partially purified preparation (60% pure) of brine shrimp Na^+,K^+-ATPase[12] has been sufficient for many biosynthesis and structural studies since the contaminating proteins do not comigrate or overlap with either the α or β subunits on SDS-polyacrylamide gels. This partially purified preparation of brine shrimp Na^+,K^+-ATPase has been recently further purified to homogeneity by Morohashi and Kawamura.[4] These techniques as well as procedures for storage and preparation of brine shrimp nauplii and for isolation of the pure subunits are described below.

Brine Shrimp Storage and Rearing

The Na^+,K^+-ATPase is purified from the larval form of the brine shrimp, *Artemia salina*. The brine shrimp are obtained in a dehydrated cyst form in which embryonic development is arrested in the gastrula stage. They are purchased in dry, sealed containers such as from San Francisco Bay Brand (Newark, California) and are stored frozen at $-20°$. Viability can be retained for years in this state. Development is initiated by hydration of the encysted embryo in 3–4 vol of ordinary tap water. To maximize synchronization of development, this hydration process is carried out at 4° for 4 hr or overnight. Cysts which remain floating after the hydration period are largely nonviable and are discarded. The remainder are washed 5–10 min under continuous-running cold tap water in a bag formed by lining a wire mesh container with a filtering cloth (Miracloth

[7] J. Kyte, *J. Biol. Chem.* **246**, 4157 (1971).
[8] L. E. Hokin, J. L. Dahl, J. D. Deupree, J. F. Dixon, J. F. Hackney, and J. F. Perdue, *J. Biol. Chem.* **248**, 2593 (1973).
[9] L. K. Lane, J. H. Copenhaver, Jr., G. E. Lindenmayer, and A. Schwartz, *J. Biol. Chem.* **248**, 7197 (1973).
[10] P. L. Jørgensen, *Biochim. Biophys. Acta* **356**, 36 (1974).
[11] J. F. Dixon and L. E. Hokin, *Arch. Biochem. Biophys.* **163**, 749 (1974).
[12] G. L. Peterson and L. E. Hokin, *Biochem. J.* **192**, 107 (1980).

from Calbiochem-Boehring). It is usually unnecessary to render the cysts aseptic although this can be accomplished by treatment with 7% antiformin [3.2% Na_2CO_3, 7.8% NaOH in NaOCl (Chlorox) diluted to 7% in H_2O][13] for 15–20 min and rearing the nauplii in autoclaved artificial seawater fortified with 100 µg/ml each of penicillin and streptomycin. Although there is some loss of viability with antiformin treatment, the cysts survive this procedure since they are essentially impermeable to everything but water and gases such as CO_2 and O_2 at this stage. After the cysts have been hydrated and washed, development is allowed to proceed by incubation in half-strength seawater (Instant Ocean Sea Salts, Aquarium Systems, Inc., Eastlake, OH) at 28–30° with vigorous aeration for about 36 hr. Two-liter separatory funnels work well for brine shrimp originating from up to 25 g dry cysts. For quantities up to 200 g dry cysts, 20-liter plastic carboys with spigots at the bottom work well. At full capacity, at least six air stones producing fine air bubbles at high flow rates must be used to supply adequate aeration. Inadequate aeration results in delayed and poorly synchronized development. At the end of the incubation period, the air stones are removed and the hatched nauplii are allowed to settle to the bottom of the container (empty cyst shells float to the top) and are collected through the spigots, washed through a Miracloth filter, and rinsed three times with water.

The appearance of hatched brine shrimp nauplii with time after hydration and the relationship of the development of Na^+,K^+-ATPase activity is shown in Fig. 1. Hatching begins after 12 hr of development and proceeds rapidly over the next 6 hr, followed by a gradual recruitment of newly hatched nauplii over the subsequent 24-hr period. Na^+,K^+-ATPase activity first appears after 8 hr of development and is maximal by 30 hr. Experience has shown that purification of brine shrimp Na^+,K^+-ATPase from animals younger than 18 hr is difficult, owing to incomplete membrane differentiation, resulting in poor yields of Na^+,K^+-ATPase-enriched membranes. Purification procedures are most satisfactory at 36 hr and acceptable between 24 and 48 hr.

Preparation of Membrane-Bound and Partially Purified Brine Shrimp Na^+,K^+-ATPase

It is useful to describe the preparation of membrane-bound and partially purified brine shrimp Na^+,K^+-ATPase on the basis of small-scale and large-scale procedures. Small-scale procedures involve use of 10–100 g of dry cysts and will be described in the greatest detail as this is the more

---

[13] Y. H. Nakanishi, T. Iwasaki, T. Okigaki, and H. Kato, *Annot. Zool. Jpn.* **35,** 223 (1962).

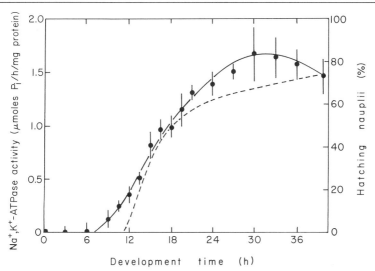

FIG. 1. Development of Na$^+$,K$^+$-ATPase activity (solid line) during embryogenesis of the San Franciso variety of brine shrimp. The symbols and bars represent the mean ± standard deviation for four samples. The broken line is the hatching curve for free-swimming nauplii.

common experimental condition. Large-scale procedures involving 300–900 g of brine shrimp are similar except for the modifications, such as use of zonal rotors, necessary for handling the larger volumes of material. The methods in this section are taken largely from Peterson and Hokin.[12]

Common Solutions

Homogenizing buffer: 250 mM sucrose, 50 mM imidazole/HCl, 2 mM Na$_2$EDTA, 2 mM 2-mercaptoethanol, pH 7.2

Imidazole buffer: 25 mM imidazole/HCl, 1 mM Na$_2$EDTA, pH 7.2

ATP buffer: 15 mM Na$_2$ATP, 250 mM imidazole/HCl, 10 mM Na$_2$EDTA, pH 7.2

16% Lubrol WX: Prepared from Lubrol WX obtained from Sigma Chemical Company; The detergent (16%, w/v) is first dissolved in water at room temperature or slightly warmer, then allowed to sit at 4° overnight. Any cold-precipitated material is then removed by centrifugation at 100,000 g for 30 min at 4°

Na$^+$,K$^+$-ATPase Assay

Brine shrimp Na$^+$,K$^+$-ATPase is assayed in the presence of 5 mM Na$_2$ATP, 10 mM MgCl$_2$, 40 mM imidazole/HCl, pH 7.2, 160 mM NaCl, and either 40 mM KCl or 1 mM ouabain. Incubations are carried out for 5, 10, or 20 min at 37° and enzyme activity determined from the difference in phosphate release between the reactions containing KCl and those containing ouabain. Phosphate is determined by the method of Peterson.[14] Protein is determined by the Peterson[15,16] modification of Lowry et al.[17] using bovine serum albumin as the standard. SDS-polyacrylamide gel electrophoresis is performed according to Laemmli[18] and the proteins stained with Coomassie blue. Phospholipid content is measured according to Bartlett.[19]

Preparation of Membrane-Bound Na^+,K^+-ATPase (Small Scale). The harvested and washed nauplii are homogenized in a Waring blender containing 60 ml homogenizing buffer/10 g of original dry cysts for 10 sec at low speed. The homogenate is then centrifuged at 25,000 g for 2 hr at 4° (12,000 rpm with Sorvall GSA rotor). The desired membranes form a top layer on the resulting pellet and are tan in color, becoming more orange in color with older brine shrimp. The supernatant is decanted except for the last few milliliters to avoid any loss of the pellet and the membrane layer is resuspended with the aid of a glass rod. There is a soft fibrous textured underlying pellet layer which is not as easily removable and should be left with the pellet. The pellet is then homogenized and centrifuged a second time using the same procedure as above and the membranes combined and resuspended in homogenizing buffer to a final volume of 25 ml/10 g original dry cysts. Alternatively, this crude membrane fraction can be obtained by centrifuging the homogenate at 1800 g for 5 min at 4°, decanting and saving the supernatant, rehomogenizing the pellet, and repeating the 1800 g centrifugation step. The desired membranes are then centrifuged from the combined supernatants at 25,000 g for 2 hr at 4°.[4] These membranes may be stored frozen at -20 or $-80°$ with no loss of activity over a 2- to 3-month period.

The crude membranes are further processed by centrifugation through a sucrose step gradient to remove lighter endoplasmic reticulum membranes and heavier mitochondrial membranes. The gradient consists of 9 ml of 40% (w/v) sucrose in imidazole buffer, 9 ml of 20% (w/v) sucrose in imidazole buffer, and 20 ml of crude membrane suspension in 25 × 89 mm swinging bucket rotor tubes. The gradients are centrifuged in a Beckman

[14] G. L. Peterson, *Anal. Biochem.* **84**, 164 (1978).
[15] G. L. Peterson, *Anal. Biochem.* **83**, 346 (1977).
[16] G. L. Peterson, this series, Vol. 91, p. 95.
[17] O. H. Lowry, N. J. Rosebrough, A. L. Farr, and R. J. Randall, *J. Biol. Chem.* **193**, 265 (1951).
[18] U. K. Laemmli, *Nature (London)* **227**, 680 (1970).
[19] G. R. Bartlett, *J. Biol. Chem.* **234**, 466 (1959).

SW28 (or SW27) rotor at 27,000 rpm for 1 hr at 4°, and the desired membranes banding at the 20–40% sucrose interface are collected by aspiration. The membranes are then isolated from the high sucrose medium by diluting with 1–1.5 vol of imidazole buffer (to assure that sucrose is less than 20%) and centrifuging at 100,000 g for 1 hr at 4° (Beckman 30 rotor). The membranes are resuspended to a final protein concentration of 6 mg/ml.

Representative results for the preparation of Na^+,K^+-ATPase-enriched membranes from homogenates of nauplii after 24- and 42-hr development is shown in Table I. The crude membrane fraction represents about a 5-fold enrichment of the Na^+,K^+-ATPase at a yield of 50–70% of the total enzyme. Further purification of the membranes by sucrose gradient centrifugation results in an additional 4- to 5-fold enrichment at a loss of one-third to one-half of the enzyme present in the crude membranes. The specific activity of the gradient-purified membrane fraction is typically about 50 μmol P_i/hr/mg protein or higher.

Partial Purification of Na^+,K^+-ATPase (Small Scale). The membrane-bound brine shrimp Na^+,K^+-ATPase can be purified to about 60% purity (by criteria of SDS-polyacrylamide gel analysis) through treatment with Lubrol WX and SDS. The enzyme remains in the particulate form in both treatments. Lubrol treatment effects a 1.6-fold activation of the brine shrimp Na^+,K^+-ATPase, but usually less than a 2-fold enrichment. The subsequent SDS-purification step, however, is ineffective without prior

TABLE I
SMALL-SCALE PURIFICATION OF Na^+,K^+-ATPase FROM BRINE SHRIMP[a]

Age (hr)	Purification step	Total activity (μmol P_i/hr)	Total protein (mg)	Specific activity (μmol P_i/hr per mg of protein)	Enzyme yield (%)	Relative purification (-fold)
24	Homogenate	8010	2960	2.70	100	1.0
	Crude membranes	5680	411	13.8	71	5.1
	Gradient membranes	3070	45.8	66.9	38	24.8
	Lubrol pellet	4230	44.3	95.3	53	35.3
	SDS-treated enzyme	1900	3.90	486	24	180
42	Homogenate	9720	3490	2.79	100	1.0
	Crude membranes	4950	380	13.1	51	4.7
	Gradient membranes	2480	51.6	48.1	26	17.2
	Lubrol pellet	3580	27.5	130	37	46.6
	SDS-treated enzyme	2370	4.44	534	24	191

[a] Reared from 25 g of dry cysts. Reproduced from Peterson and Hokin,[12] with permission of the publisher.

treatment of the membranes with Lubrol. For Lubrol treatment, the gradient membranes (6 mg protein/ml) are mixed with 0.01 vol of ATP buffer followed by 0.12 vol of 16% Lubrol WX while stirring. The Lubrol-treated membranes are then recovered by centrifugation at 100,000 g for 2 hr at 4° (Beckman 40 rotor). The supernatant is decanted completely and the pellet is resuspended in homogenizing buffer at a final protein concentration of 3.0 mg/ml (4.0 mg/ml if prepared from <24-hr brine shrimp). If not used directly, this preparation should be stored frozen at −80°.

For SDS treatment, the Lubrol-treated membranes are first mixed with 0.4 vol of ATP buffer and allowed to sit on ice for 15 min. Then 0.6 vol of 1.67 mg/ml SDS is added at about 1 ml/min while stirring at 0° and the mixture incubated for 30 min at 30°. The material is then cooled on ice and centrifuged through a sucrose step gradient according to one of the protocols given in Table II.

The gradients are centrifuged at 27,000 rpm (all rotor types) for 4 hr at 4°. The desired material bands at the 15–35% sucrose interface and is most easily collected by aspiration through a small-bore tubing connected to a peristaltic pump. The gradient fraction is diluted at least 2-fold to assure the sucrose concentration is reduced to less than 15%, and the enzyme is then pelleted by centrifuging at 100,000 g for 4 hr at 4°. The final pellet is resuspended in homogenizing buffer to a final protein concentration of 1–10 mg/ml and stored frozen at −80°.

Representative samples of the Na^+,K^+-ATPase partially purified from 24- and 42-hr brine shrimp are shown in Table I. Lubrol treatment results in a particulate fraction with increased Na^+,K^+-ATPase-specific activity which is due in part to enzyme activation and in part to extraction of a few contaminating proteins. Nearly all of the enzyme activity is recovered in the particulate fraction, indicating near complete recovery of Na^+,K^+-ATPase at the Lubrol step. The SDS treatment, which is done under low, nondenaturing concentrations of SDS, is a modification of the procedure

TABLE II
SUCROSE STEP GRADIENT PROTOCOL OF SDS-TREATED MEMBRANES FOR VARIOUS BECKMAN ROTORS

Preparation	Rotor type		
	SW28(27)	SW28.1(27.1)	SW40
Sample (plus any imidazole buffer overlay):	22 ml	11 ml	7 ml
15% (w/v) sucrose in imidazole buffer:	10 ml	4 ml	3 ml
35% (w/v) sucrose in imidazole buffer:	6 ml	3 ml	3 ml

developed by Jørgensen[10] for the mammalian kidney Na^+,K^+-ATPase purification. The modifications include changes in the concentrations of both SDS and protein, and the use of an elevated incubation temperature (30 vs 20°) which proved necessary for successful purification with the brine shrimp system. Purification at the SDS step is typically 4- to 5-fold with a yield of 40%. This then gives an overall yield of about 4 mg partially purified enzyme from 25 g starting material (dry cysts), representing 25% of the homogenate activity. The specific activity of the partially purified enzyme is about 500 μmol P_i/hr/mg protein, about a 200-fold enrichment over the homogenate.

Preparation of Membrane-Bound Na^+,K^+-ATPase (Large Scale). Homogenization procedures are similar to those for the small-scale preparations except a larger blender flask (capable of handling 300 ml buffer and nauplii from 50 g original dry cysts) is used. Crude membranes are prepared as described above for the small-scale preparation and further processing is performed after accumulation of crude membranes from 300 g of original dry cysts. Because of the large volume of crude membranes (720 ml), a zonal rotor is used for the 20–40% sucrose step gradient. The steps below are designed for the Beckman Ti-15 (1650-ml capacity zonal rotor). The rotor containing the thawed crude membranes (720 ml) is brought to loading speed and the sucrose solutions (in imidazole buffer) are loaded from the wall at about 25 ml/min using a peristaltic pump. The sucrose gradient consists of 350 ml 20% sucrose, 500 ml 40% sucrose, and 100–150 ml 60% sucrose as needed to fill the rotor. The rotor is centrifuged at 29,000–33,000 rpm for 1.5 hr at 4° and allowed to return to loading speed without the brake. Fractions (about 25 ml) are collected from the wall by pumping 2% sucrose into the center at 25 ml/min (initially slower speeds, 15 ml/min, are required until the viscous 60% sucrose is removed). The Na^+,K^+-ATPase membranes banding at the 20–40% sucrose interface can, with experience, be detected by the appearance of the fractions. This experience should be mastered by analyzing both protein and Na^+,K^+-ATPase activity from several runs. The pooled high specific activity fractions total 300–350 ml and must be concentrated and the high sucrose must be removed. This is most easily and rapidly achieved by diluting with buffer to 500 ml and concentrating overnight on a thin-channel (Amicon TCF10) ultrafiltration system using a 90-mm-diameter Amicon XM300 Diaflo membrane. Standard Amicon ultrafiltration cells are considerably slower due to inadequate stirring, resulting in clogged filters. Alternatively, the pooled membranes may be diluted 2- to 2.5-fold and centrifuged in a large-volume high-speed preparative rotor such as the Beckman type 35 or 45 Ti rotors. The concentrated membranes are adjusted to a protein concentration of 6 mg/ml.

Partial Purification of Na^+,K^+-ATPase (Large Scale). The first detergent treatment using Lubrol WX is performed the same as described in the procedures for small-scale purification except for the use of a larger capacity high-speed rotor (such as the Beckman type 30). The second detergent treatment using SDS is also similar except that the SDS solution is added to the enzyme at a faster rate (3 ml/min). From 300 g original dry cysts the volume of the SDS-treated enzyme will be about 120–150 ml, which will largely or entirely fit into one centrifugation with the SW28 (or SW27) rotor (see Table II), and is processed as described for the small-scale procedures. With material accumulated from up to 900 g original dry cysts, it is more convenient to use the large-volume Beckman Ti-15 zonal rotor. In this case the gradient consists of 650–750 ml SDS-treated enzyme, 500 ml of 15% sucrose (in imidazole buffer), and 450–550 ml of 35% sucrose (in imidazole buffer) as needed to fill the rotor. The procedures for loading and unloading the zonal gradient are otherwise the same as described above for the large-scale membrane isolation. The gradient is centrifuged at 29,000–33,000 rpm for at least 4 hr at 4° (overnight is often more convenient) and the fractions are analyzed for enzyme activity and protein to determine which contain the high specific activity Na^+,K^+-ATPase. The pooled fractions are then concentrated overnight by ultrafiltration from an Amicon XM300 Diaflo membrane using a 400-ml (76 cm diameter) standard Amicon cell. The concentrated material is then diluted at least 2-fold with imidazole buffer and centrifuged at 100,000 g (Beckman 30 rotor) for 4 hr at 4°. The pellet is resuspended to a protein concentration of 5–10 mg/ml and stored frozen at $-80°$.

TABLE III
LARGE-SCALE PURIFICATION OF Na^+,K^+-ATPase FROM BRINE SHRIMP[a]

Age (hr)	Purification step	Total activity (μmol P_i/hr)	Total protein (mg)	Specific activity (μmol P_i/hr per mg of protein)	Enzyme yield (%)	Relative purification[b] (-fold)
20	Crude membranes	245,000	19,500	12.5	100	4.2
	Gradient membranes	155,000	3,740	42.5	63	14.2
	Lubrol pellet	135,000	1,450	93.2	55	31.1
	SDS-treated enzyme	49,800	132	377	20	126
48	Crude membranes	168,000	11,100	15.2	100	5.1
	Gradient membranes	106,000	1,800	58.6	63	19.5
	Lubrol pellet	153,000	968	158	91	30.3
	SDS-treated enzyme	57,200	96.2	594	34	198

[a] Reared from 900 g of dry cysts. Reproduced from Peterson and Hokin,[12] with permission of the publisher.
[b] Calculated assuming a homogenate specific activity of 3 μmol P_i/hr/mg protein.

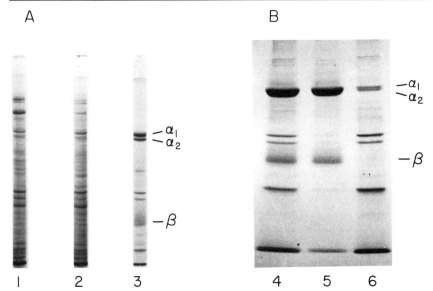

FIG. 2. SDS-polyacrylamide gel protein patterns of the various fractions obtained during brine shrimp Na^+,K^+-ATPase purification. (A) Patterns of the gradient membrane fraction (lane 1), Lubrol pellet fraction (lane 2), and SDS-treated fraction (lane 3) prepared from 24-hr brine shrimp. The positions of the Na^+,K^+-ATPase subunits are as indicated. Reproduced from Peterson and Hokin,[12] with permission of the publisher. (B) Patterns of the partially purified and $C_{12}E_8$-solubilized brine shrimp Na^+,K^+-ATPase. Lane 4 represents the particulate enzyme partially purified through the SDS-treatment step. Lane 5 represents the 100,000 g supernatant fraction after solubilization of the SDS-treated enzyme with $C_{12}E_8$. Lane 6 represents the particulate fraction after extraction with $C_{12}E_8$. The positions of the Na^+,K^+-ATPase subunits are indicated. Reproduced from Morahashi and Kawamura,[4] with permission of the authors and publisher.

Representative examples of the large-scale (900 g dry cysts) preparation of Na^+,K^+-ATPase-enriched membranes and the partial purification of the enzyme from 20- and 48-hr brine shrimp are shown in Table III. Comparable yields and specific activities are obtained at each step for the large-scale preparation as compared to the small-scale preparation (Table I). Final specific activities and yields of partially purified Na^+,K^+-ATPase are less in brine shrimp harvested prior to 24 hr and this is evident in the example of the 20-hr animals shown in Table III.

Figure 2A shows the SDS-polyacrylamide gel protein patterns of the fractions obtained during partial purification of the brine shrimp Na^+,K^+-ATPase. The α subunit doublet bands are detectable in the membrane and Lubrol pellet fractions, but there is little enrichment with Lubrol treatment in agreement with the specific activities reported in Tables I and III.

Lubrol treatment appears to principally remove some higher molecular weight proteins. The α and β subunit proteins of the SDS-treated fraction, however, are enriched to the point that they are the major protein constituents, together representing usually 60% or more of the total.

Purification of the Brine Shrimp Na^+,K^+-ATPase

Final purification of the brine shrimp Na^+,K^+-ATPase is achieved by solubilization with the nonionic detergent $C_{12}E_8$ (octaethyleneglycol mono-n-dodecyl ether from Nikko Chemicals, Tokyo).[4] The SDS-treated enzyme is adjusted to 1.5 mg protein/ml and 3.0 mg/ml $C_{12}E_8$ in a medium containing 40 mM KCl, 125 mM sucrose, 0.05 mM phenylmethylsulfonyl fluoride (PMSF) in imidazole buffer. The mixture is allowed to stand for 20 min at 25° and then centrifuged at 100,000 g for 1 hr at 4°. The supernatant is then concentrated 2- to 3-fold (to 2-3 ml) by ultrafiltration through an Amicon YM5 Diaflo filter. The concentrated solubilized enzyme is then further purified by gel filtration of a 2.1 × 100 cm BioGel A-1.5m (200-400 mesh, Bio-Rad Laboratories) column equilibrated with 150 μg/ml $C_{12}E_8$, 40 mM KCl, 1 mM EDTA, 1 mM 2-mercaptoethanol, 0.1 mM PMSF, and 25 mM Tris/HCl, pH 7.2. The column is run at 4° at a flow rate of 6.2-6.5 ml/hr. Contaminating proteins elute at the void volume and in fractions preceding the Na^+,K^+-ATPase peak (see Fig. 3). After treatment with $C_{12}E_8$, the Na^+,K^+-ATPase is more thermally unstable and enzyme reactions should be run at 25° for 2-3 min.[4]

Figure 2B shows the SDS-polyacrylamide gel electrophoresis protein patterns of the partially purified and $C_{12}E_8$-solubilized brine shrimp Na^+,K^+-ATPase. In the partially purified preparation (SDS enzyme), the α and β subunits represent about 60% of the total protein and are relatively free of comigrating contaminating bands. Most of the contaminating proteins remain in the particulate fraction after solubilization with $C_{12}E_8$. Remaining minor contaminants are removed by gel filtration over BioGel A-1.5m (Fig. 3), which also removes the bulk of detergent-extracted phospholipids.

Isolation of Brine Shrimp Na^+,K^+-ATPase Subunits

The Na^+,K^+-ATPase subunits are isolated from SDS-denatured enzyme by gel filtration over columns of BioGel A-1.5m (200-400 mesh, Bio-Rad Laboratories). Either purified or partially purified (SDS-treated) Na^+,K^+-ATPase is denatured with 2.5-5.0% SDS and 2.5% 2-mercaptoethanol. For sample loads of 1-5 mg protein, a 1.4 × 90 cm BioGel A-1.5m column is used, and for loads up to 30-40 mg a 2.5 × 90 cm

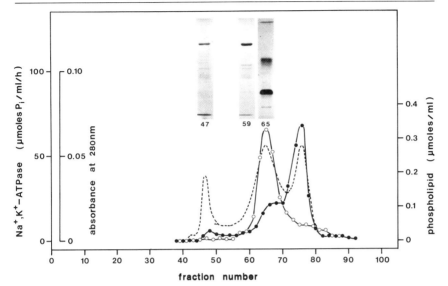

FIG. 3. Gel filtration of $C_{12}E_8$-solubilized brine shrimp Na^+,K^+-ATPase. $C_{12}E_8$-solubilized protein (5.5 mg) was loaded on a 2.1 × 100 cm BioGel A-1.5m column and eluted with $C_{12}E_8$–KCl buffer as described in the text. Na^+,K^+-ATPase activity (solid circles), phospholipid content (open circles), and absorbance at 280 nm (dashed line) were measured on the eluted fractions. SDS-polyacrylamide gel protein patterns are also shown for the fractions indicated. Reproduced from Morohashi and Kawamura,[4] with permission of the authors and publisher.

column is used. The column is equilibrated with 0.1% SDS, 0.2 M Tris/HCl, pH 7.4, 1 mM Na_2EDTA, and after sample loading is eluted with the equilibration buffer at 4 ml/hr (1.4 × 90 cm column) or 20 ml/hr (2.5 × 90 cm column) at room temperature. Five 280-nm absorption peaks are observed (Fig. 4). The position of the Na^+,K^+-ATPase subunits in the eluate fractions are verified by SDS-polyacrylamide gel electrophoresis.[18] Peaks 1 and 5 contain no significant proteins. When partially purified samples are chromatographed, peak 4 contains some low-molecular-weight proteins ($M_r \leq 10,000$). Peak 2 contains pure α subunit and peak 3 contains pure β subunit when starting with purified Na^+,K^+-ATPase. With partially purified Na^+,K^+-ATPase, peak 3 contains contaminating proteins in addition to the β subunit. In the latter case, the β subunit can be purified by elution from SDS-polyacrylamide gels (6–7% acrylamide).[18] The position of the β subunit in the SDS gels can be determined by staining a guide strip (slab gels)[12] or by UV scanning.[20] The gel section containing the β

[20] G. L. Peterson and L. E. Hokin, *J. Biol. Chem.* **256**, 3751 (1981).

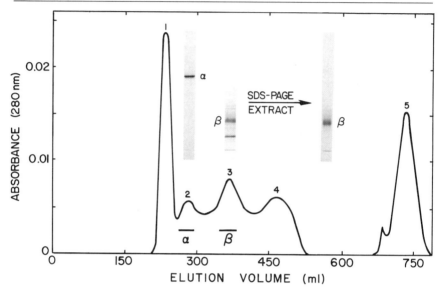

FIG. 4. Brine shrimp Na$^+$,K$^+$-ATPase subunit isolation. Partially purified (SDS-treated) Na$^+$,K$^+$-ATPase (36.5 mg) prepared from 48-hr animals was denatured and chromatographed on a 2.5 × 90 cm BioGel A-1.5m column in the presence of 0.1% SDS as described in the text. The indicated fractions underlying the 280-nm absorption peaks 2 and 3 were pooled and examined by SDS-polyacrylamide gel electrophoresis (gel patterns above peaks 2 and 3). Contaminating proteins cochromatographing with the β subunit were removed by SDS-polyacrylamide gel electrophoresis (SDS-PAGE) and extraction (far right gel pattern). Reproduced from Peterson and Hokin,[12] with permission of the publisher.

subunit is macerated by ejection through the bore of a syringe and extracted three times with 2–3 vol of 0.1% SDS, 0.1% NaHCO$_3$, 0.1% 2-mercaptoethanol at 30° over a 36-hr period. The gel extract of the β subunit or the column fractions containing the α and β subunits can be concentrated by lyophilization or ultrafiltration and dialyzed against water or water containing low concentrations of SDS (0.05%).

Properties of the Purified Brine Shrimp Na$^+$,K$^+$-ATPase

The brine shrimp Na$^+$,K$^+$-ATPase subunits examined by SDS-polyacrylamide gel electrophoresis in gradients of acrylamide and over a range (5–15%) of single acrylamide concentrations showed similar molecular weight estimates (or more precisely chain length, calibrated using the number of amino acids, N, per chain) regardless of the analysis method.[20] Ferguson plots showed no marked deviations in gel migration behavior for the brine shrimp Na$^+$,K$^+$-ATPase subunits in SDS and the resulting molecular weight estimates agree closely with the sedimentation equilibrium values obtained for the Na^+,K^+-ATPase subunits of other species (see discussions in Refs. 20 and 21). These findings indicate that the molecular size of the brine shrimp Na^+,K^+-ATPase subunits has been reliably estimated by the SDS-polyacrylamide gel methods.

Table IV summarizes some of the molecular properties of the brine shrimp Na^+,K^+-ATPase. The mass ratio of the two subunits determined by analysis of the total amino acid released by alkaline hydrolysis of the SDS-gel separated subunits was 2.44 $\alpha/1\ \beta$. This agrees very closely with the value obtained from the molecular size measurements of the two subunits assuming an equimolar ratio (2.50 $\alpha/1\ \beta$). Compositional analysis shows that both subunits contain carbohydrate in the form of amino sugars and neutral sugars. Sialic acid is absent from both subunits.[12]

The brine shrimp Na^+,K^+-ATPase α subunit consists of two closely migrating bands when examined by SDS-polyacrylamide gel electrophoresis[5] (see Fig. 2). In younger nauplii (12–15 hr), the two bands are in approximately equal proportion, whereas in older animals (48 hr) the proportion of the faster migrating band α_2 is reduced to about 10% of the total, and is often obscured in overloaded gels. The proportions of α_1 and α_2 in the brine shrimp during development are shown in Fig. 5. Biosynthetic studies including pulse–chase experiments and measurements of the rates of synthesis and degradation of the brine shrimp Na^+,K^+-ATPase subunits do not support the existence of a precursor–product relationship between α_1 and α_2.[22] Isolation of polysomal RNA from 16-hr nauplii and *in vitro* translation in a rabbit reticulocyte lysate system results in anti-subunit immunoprecipitable products.[3] The products identified as α subunit are derived entirely from membrane-bound polysomal mRNA and consist of two bands which migrate slightly slower (estimated 5000 Da larger molecular weight) than mature α_1 and α_2 (see Table V). Since posttranslational processing systems were not present in these *in vitro* studies, the two α precursor proteins must represent products of two different mRNAs. Amino acid sequence analysis of mature brine shrimp Na^+,K^+-ATPase α subunit shows two amino acids, glycine and lysine, at position 3 for preparations containing near-equal proportions of α_1 and α_2, and only one, glycine, at the same position for the α subunit in a preparation where the proportion of α_2 is low.[4] Thus α_1 and α_2 are separate gene products. There is also some evidence that the two α subunits represent isoenzymes that may differ in function. α_2 appears to have a lower affinity

---

[21] G. L. Peterson, L. Churchill, J. A. Fisher, and L. E. Hokin, *Ann. N.Y. Acad. Sci.* **402**, 185 (1982).

[22] G. L. Peterson, L. Churchill, J. A. Fisher, and L. E. Hokin, *J. Exp. Zool.* **221**, 295 (1982).

TABLE IV
MOLECULAR PROPERTIES OF THE BRINE SHRIMP Na^+,K^+-ATPase

Property	Subunit	
	α	β
Measured relative mass (analysis of total amino acids)[a]	2.44	1
Number of amino acid residues by SDS gels[a]	888	347
Molecular weight by SDS gels[a]	97,800	39,100
Calculated equimolar relative mass	2.50	1
Composition (nearest integer in mol/mol subunit)[b]		
Alanine	74	20
Arginine	27	9
Aspartic acid	93	39
Cysteine	24	12
Glutamic acid	87	36
Glycine	68	29
Histidine	9	3
Isoleucine	50	17
Leucine	80	27
Lysine	44	21
Methionine	29	9
Phenylalanine	46	19
Proline	39	19
Serine	66	26
Threonine	62	21
Tryptophan	19	13
Tyrosine	26	15
Valine	47	20
Amino sugars	6	4
Neutral sugars	26	20
Total amino acid residues	890	355
Partial specific volume (cm³/g)	0.727	0.721
Molecular weight from composition		
Protein portion	98,100	40,000
Carbohydrate portion	5,400	4,100
Total	103,500	44,100
Calculated equimolar relative mass	2.45	1
Percentage carbohydrate by weight	5.2	9.2

[a] From Peterson and Hokin.[20]
[b] From Peterson and Hokin.[12]

for both ATP and cardiac glycosides.[5,21] It would be useful for further enzymatic studies related to the difference in the two isoenzymes if it were possible to obtain pure α_1 Na^+,K^+-ATPase and pure α_2 Na^+,K^+-ATPase. Since the two α subunit forms have some sequence differences,

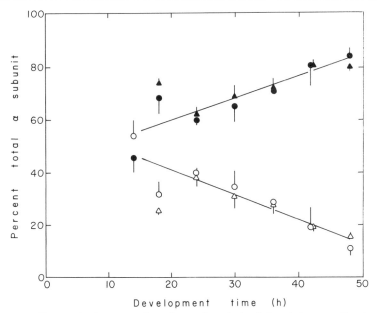

FIG. 5. Changes in the proportion of α_1 and α_2 during brine shrimp naupliar development. Brine shrimp were harvested at the times shown and the Na^+,K^+-ATPase partially purified from lighter (triangles) and heavier (circles) sedimenting membranes[22] to the SDS-treatment step. The relative proportion of α_1 (closed symbols) and α_2 (open symbols) was determined by densitometry of Coomassie blue-stained SDS gels in which electrophoresis was continued until the α subunit neared the bottom of the gel to effect greater separation of the two bands. Reproduced from Peterson et al.[22] with permission of the publisher.

it may be possible to isolate monoclonal antibodies that would recognize sequence domains specific to one form only. Such antibodies would aid in mapping out the cellular and tissue distribution of each form, their developmental programming, and the distribution of each specific mRNA, in addition to providing a means of isolating each isoenzyme.

Table V summarizes the known biosynthetic properties of the brine shrimp Na^+,K^+-ATPase subunits. All subunits are degraded at similar rates (within experimental error) with a half-life between 10 and 20 hr. The α subunit is synthesized on membrane-bound polysomes and pre-α subunit is approximately 5000 Da larger than the mature α subunit, which is sufficient to allow for the presence of a leader sequence (usually about 3000 Da). The β subunit, however, is synthesized on free polysomes and pre-β subunit is synthesized as a single discrete band (on SDS gels) of slightly lower molecular weight than the diffuse-appearing mature β subunit. The mature β subunit contains a higher percentage by weight of carbohydrate than the α subunit (see Table IV), which is apparently re-

TABLE V
BIOSYNTHETIC PROPERTIES OF THE BRINE SHRIMP Na^+,K^+-ATPase

Property	Subunit		
	α_1	α_2	β
Average degradation rate ($t_{1/2}$, in hours)[a] (minimum half-life)	13.2 (8.4)	17.4 (11.0)	23.0 (10.0)
Site of synthesis[b]	Membrane-bound polysomes	Membrane-bound polysomes	Free polysomes
Precursor SDS-gel molecular weight[b]	104,000	97,000	36,000
Band appearance	Discrete	Discrete	Discrete
Mature SDS-gel molecular weight[b]	99,000	92,000	36,000–43,000
Band appearance	Discrete	Discrete	Diffuse

[a] From Peterson et al.[22] These values are apparent degradation rates which are not corrected for precursor reutilization.
[b] From Fisher et al.[3]

sponsible for the diffuse-banding appearance of this subunit. Thus the molecular size of the mature β subunit (protein portion) is probably only slightly overestimated by SDS-polyacrylamide gels. This is in contrast to the behavior of the β subunit from higher animals,[20] where the molecular

TABLE VI
NH_2-TERMINAL SEQUENCES OF THE Na^+,K^+-ATPase α SUBUNIT

Source	Sequence
	1 10
Lamb kidney[a]	Gly-Arg-Asp-Lys-Tyr-Glu-Pro-Ala-Ala-
Hog kidney[b]	Gly-Arg-Asp-Lys-Tyr-Glu-Pro-Ala-
Dog kidney[c]	Gly-Arg-Asn-Lys-Tyr-Glu-Pro-Ala-Ala-(?)-Ser-Glu-
Duck salt gland[d]	Gly-Arg-Asn-Lys-Tyr-Glu-Thr-Thr-Ala-(?)-Ser-Glu-
Brine shrimp[e]	1 10
α_1	Ala-Lys-Gly-Lys-Gln-Lys-Lys-Gly-Lys-Asp-Leu-Asn-Glu-
α_2	Ala-Lys-Lys-Lys-Gln-Lys-Lys-Gly-Lys-Asp-Leu-Asn-Glu-
	14 20
α_1	Leu-Lys-Lys-Glu-Leu-Asp-Ile-Asp-Phe-His-Lys-Ile-Pro-
α_2	Leu-Lys-Lys-Glu-Leu-Asp-Ile-Asp-Phe-His-Lys-Ile-Pro-

[a] From J. H. Collins and A. S. Zot, IRCS Med. Sci.: Libr. Compend. 11, 799 (1983).
[b] From R. H. McParland and R. R. Becker, personal communication (1984).
[c] From L. C. Cantley, Curr. Top. Bioenerg. 11, 201 (1981).
[d] From B. E. Hopkins, H. Wagner, Jr., and T. W. Smith, J. Biol. Chem. 251, 4365 (1976).
[e] From M. Morohashi and M. Kawamura, J. Biol. Chem. 259, 14928 (1984).

weight is seriously overestimated by SDS-polyacrylamide gel electrophoresis. Synthesis of the Na^+,K^+-ATPase subunits at two different sites suggests that posttranslational processing of the two subunits occurs by different pathways and that studies of these processes and the mechanism of holoenzyme assembly will continue to be an interesting and fruitful area for further research.

The NH_3-terminal sequences for the brine shrimp Na^+,K^+-ATPase α subunit are compared in Table VI with the available sequences from other sources of the enzyme. The brine shrimp sequence over the first 10–12 residues differs from that of the sequences for higher animals, which, among themselves, show a great deal of homology. Perhaps a notable exception is lysine at position 4, which is conserved in all the sequences, and may therefore play some fundamental role important to all Na^+,K^+-ATPase. The sequences for the mature α subunits do not contain the compositions typical of leader sequences, indicating that such sequences if existent are cleaved after translation.

[6] Preparation of the $\alpha(+)$ Isozyme of the Na^+,K^+-ATPase from Mammalian Axolemma

By KATHLEEN J. SWEADNER

There are at least two different isozymes of Na^+,K^+-ATPase, whose catalytic subunits have different apparent molecular weights,[1] different sensitivities to ouabain,[1] and distinct antigenic determinants.[2] One isozyme of the catalytic subunit, α (M_r 92,000), is found in the kidney, while the other, $\alpha(+)$ (M_r 95,000), is found in brain stem axolemma.[1] The brain has both isozymes, and their proportions vary from region to region.[3] While brain nonneuronal cells in culture express only the α form, they are not a very enriched or abundant source, and the isozyme containing α can be most easily purified from the kidney.[4] A method is presented here for the purification of the isozyme containing $\alpha(+)$ from rat brain stem axolemma, substantially free of α. An α_{III} isozyme, identified by molecular genetics,[5] has not yet been described as a protein.

[1] K. J. Sweadner, *J. Biol. Chem.* **254**, 6060 (1979).
[2] K. J. Sweadner and R. C. Gilkeson, *J. Biol. Chem.* **260**, 9016 (1985).
[3] S. C. Specht and K. J. Sweadner, *Proc. Natl. Acad. Sci. U.S.A.* **81**, 1234 (1984).
[4] P. L. Jørgensen, this volume [38].
[5] G. E. Shull, J. Greeb, and J. B. Lingrel, *Biochemistry* **25**, 8125 (1986).

Isolation of Brain Stem Axolemma

Principle

The brain stem is rich in myelinated axons. Long fragments of axons can survive gentle tissue disruption in Dounce homogenizers, and they can be purified by virtue of their unique sedimentation properties.[6] The myelin can then be removed from the axons, and the low-density myelin can be separated from the denser membranes of the axon on sucrose gradients.[6] The resulting axolemma preparation has a Na^+,K^+-ATPase-specific activity as high as that of a membrane preparation from the renal medulla, and thus is a good starting material for purification of the Na^+,K^+-ATPase. One might expect that pure white matter would be an even better source, but in practice it is more difficult to homogenize because of its gluey consistency.

Procedure

All manipulations are carried out at the benchtop, keeping samples chilled on ice wherever possible. Fresh brains are taken from rats, and the cerebral cortex and cerebellum are discarded. The resulting crudely dissected brain stems are minced with a razor blade in solution 1, which contains 1.0 M sucrose, 150 mM NaCl, and 10 mM Tris–HCl, pH 7.4. Twelve to 18 brain stems are processed at a time in 220 ml, enough to fill 1 rotor. The minced material is homogenized in 45-ml batches with 15 strokes with the loose-fitting pestle of a large Dounce homogenizer. The homogenate is then centrifuged at 25,000 rpm for 15 min in a large-capacity swinging bucket rotor (Beckman SW27 or Sorvall 627). Myelinated axons and myelin float to the top and appear as a light pink, cohesive wafer at the top of the tube, which can be easily lifted out on a spatula. A pellet containing the bright red meninges and other tan brain material is discarded, along with whatever remains suspended in solution. The wafer of myelinated axons is chopped lightly with a razor blade and rehomogenized in solution 1, and the homogenate is centrifuged again. The resulting second wafer of floated myelinated axons is whiter than the first, and a pinkish tan pellet is discarded. The second wafer is chopped lightly and homogenized in solution 2, which is identical to solution 1 except that it contains only 0.85 M sucrose, and the homogenate is centrifuged again. The resulting third wafer is nearly white, and a pinkish tan pellet is again discarded.

The second stage of the procedure is to cause the myelin to dissociate from the axolemma by reducing the ionic strength.[6] This is accomplished

---

[6] G. H. DeVries, *Res. Methods Neurochem.* **5**, 3 (1981).

by chopping the wafers and suspending them in 10 mM Tris–HCl, pH 7.4, at 25 ml/wafer. Similar results have been obtained by using 0.315 M sucrose, 1 mM EDTA. The suspension is homogenized with three strokes of the tight-fitting pestle of the Dounce homogenizer, and is allowed to stir gently at 4° for 1 hr or overnight. It is then homogenized with 10 more strokes with the tight-fitting pestle. The membranes and myelin are pelleted by centrifugation in an angle rotor for 30 min at 30,000 rpm. All traces of the supernatant are carefully removed.

The third stage of the procedure is to separate low-density myelin from higher density axolemma on a sucrose step gradient.[6] Each of the sucrose solutions for the gradient contains 10 mM Tris Cl and 1 mM EDTA, at pH 7.4. The pellets of the previous step (each resulting from two or three brain stems) are resuspended directly in 11 ml of the 1.0 M sucrose solution, and layered over 11 ml of the 1.2 M sucrose solution. Then 11 ml of the 0.8 M sucrose solution is layered over the sample, and the 0.32 M sucrose solution is used to fill and balance the tubes. The gradients are centrifuged for 1 hr at 25,000 rpm in a 627 or SW27 swinging bucket rotor. The majority of the turbidity, which is due to myelin, rises to the 0.32/0.8 M sucrose interface. Two turbid axolemma fractions are found at the 0.8/1.0 M sucrose interface and the 1.0/1.2 M sucrose interface. Any material pelleting through the 1.2 M sucrose is discarded. The 0.8/1.0 and 1.0/1.2 interface axolemma fractions are collected separately, diluted with 2 vol of water, and pelleted by centrifugation at 30,000 rpm for 30 min. They are resuspended in 0.315 M sucrose, 10 mM Tris Cl, 1 mM EDTA, quick frozen, and stored at $-70°$.

The yield of the two axolemma fractions is typically 0.25–0.4 mg of protein each per rat. The specific activity of the Na^+,K^+-ATPase is 1.6–3.0 μmol of ATP hydrolyzed/min/mg protein in the 1.0/1.2 fraction and 1.0–1.4 μmol/min/mg in the 0.8/1.0 fraction; apparent specific activities can be increased approximately 40% if the preparations are incubated with a low concentration of SDS (0.1–0.15 mg SDS/mg protein), which is believed to open sealed membrane fragments. Analysis of the protein composition by gel electrophoresis in SDS reveals many protein bands, of which the $\alpha(+)$ subunit of the Na^+,K^+-ATPase is the most prominent (Fig. 1). Only a trace of the α subunit can generally be observed.

Practical Considerations

The use of a Dounce homogenizer is indispensible, because other homogenizers will disrupt the integrity of the myelinated axons. If an angle rotor is used for the flotation steps, the myelinated axon fraction adheres to the side of the tube and is difficult to recover. The discarded pellets from the first three flotation steps contain a high proportion of the

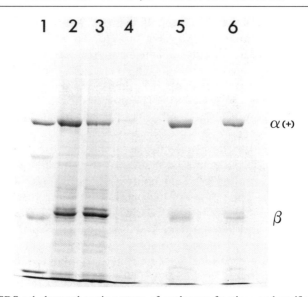

FIG. 1. SDS-gel electrophoretic pattern of axolemma fractions and purified axolemma Na^+,K^+-ATPase. Electrophoresis was performed in a 7.5% polyacrylamide gel using the Laemmli buffer system, as previously described.[1] Lane 1: molecular weight markers of 97,400, 66,000, 43,000, and 23,000. Lane 2: axolemma from the 1.0/1.2 M sucrose interface of the step gradient. The band comigrating with the 97,400 M_r marker is the $\alpha(+)$ subunit of the Na^+,K^+-ATPase. Lane 3: axolemma from the 0.8/1.0 M sucrose interface of the step gradient. Lane 4: particulate material from the pellets of the flotation steps. Lanes 5 and 6 both show purified Na^+,K^+-ATPase recovered from a 7–30% sucrose gradient after extraction of the 1.0/1.2 M sucrose axolemma fraction with SDS. Lane 5: lower (denser) half of the peak of Na^+,K^+-ATPase activity. The band at approximately 46,000 M_r contains the β subunit of the Na^+,K^+-ATPase and some contaminating protein of the same apparent molecular weight. Lane 6: upper (less dense) half of the peak. The material in lane 6 is less pure.

total particulate fraction of the brain stems, and they contain some Na^+,K^+-ATPase, albeit at a low specific activity. The flotation steps serve to greatly reduce the proportion of the α isozyme in the floated fraction, and thus are essential to purify the $\alpha(+)$ isozyme separately.

We have attempted to apply the procedure to canine brain tissue, but the result was always an axolemma fraction containing both α and $\alpha(+)$. Species differences, either in the way the membranes fractionate or in the axonal localization of the two isozymes of the Na^+,K^+-ATPase, may limit the applicability of the technique as a way to separate the isozymes.

DeVries, who originated the procedure, has subsequently developed an abbreviated procedure for rat brain stem axolemma involving only one flotation step, followed by large-scale fractionation in a zonal rotor.[7] The

[7] G. H. DeVries, M. G. Anderson, and D. Johnson, *J. Neurochem.* **40**, 1709 (1983).

Na^+,K^+-ATPase-specific activities obtained in the axolemma fraction were much lower, however, only 0.33–0.67 μmol/min/mg protein, and the α(+) subunit of the Na^+,K^+-ATPase was not the most prominent band seen after gel electrophoresis.[7] This makes the simplified procedure disadvantageous for the purification of the Na^+,K^+-ATPase.

Purification of the Na^+,K^+-ATPase from Axolemma

Procedure

The α(+) isozyme of the Na^+,K^+-ATPase can be substantially purified from axolemma by extraction of the membranes with sodium dodecyl sulfate (SDS),[2] essentially by the procedure developed by Jørgensen.[8] Axolemma from the 1.0/1.2 M sucrose interface is suspended at 1.4 mg protein/ml at room temperature in a solution whose final concentrations are 0.16 M sucrose, 1.5 mM EDTA, 3 mM ATP, 25 mM histidine, pH 7.2. SDS (recrystallized from 95% ethanol) is added from a 10% stock solution to a final concentration of 0.56 mg/ml, and the solution is rapidly mixed to distribute the detergent evenly before local high concentrations can denature some of the Na^+,K^+-ATPase. After 30 min at room temperature, the mixture is chilled and layered onto 7–30% (w/v) linear sucrose gradients containing 1 mM EDTA and 10 mM histidine, pH 7.2, and centrifuged at 27,000 rpm for 6 hr at 4° in the 627 or SW27 swinging bucket rotor.

After centrifugation, there are normally two bands of turbidity in the gradients. The low-density band near the top of the gradient has little measurable Na^+,K^+-ATPase, whereas the denser band contains the Na^+,K^+-ATPase. Its density is less than that of axolemma not treated with SDS, indicating that more protein than lipid is removed from the membranes during the extraction step.[9] Figure 2 illustrates how the density and yield of membranes containing Na^+,K^+-ATPase activity is affected as a function of the amount of SDS used.

The purified Na^+,K^+-ATPase is collected by puncturing the bottom of the tube just above any pelleted material, and collecting fractions by drops. This procedure minimizes contamination of the product by the SDS and solubilized proteins which remain at the top of the gradient. The visible denser turbid fraction is not homogeneous, however. The lower half is collected separately from the upper half, and it routinely has a higher Na^+,K^+-ATPase-specific activity, 10–20 μmol/min/mg protein as opposed to 7–9 μmol/min/mg protein. When analyzed by gel electropho-

---

[8] P. L. Jørgensen, this series, Vol. 32, p. 277.
[9] K. J. Sweadner, *Biochim. Biophys. Acta* **508,** 486 (1978).

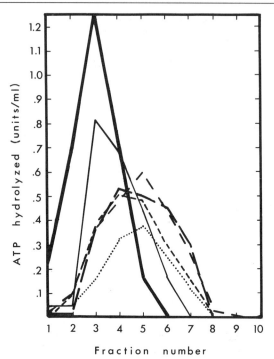

FIG. 2. Sedimentation of the axolemma Na$^+$,K$^+$-ATPase on sucrose gradients before and after extraction with SDS. Samples of axolemma extracted with different amounts of SDS were sedimented on six identical 7–30% sucrose gradients. Fractions were collected and assayed for total Na$^+$,K$^+$-ATPase activity; fraction 1 was from the bottom of the gradient, and fraction 10 from the top. As the amount of SDS used per milligram of protein was increased, the density of the resulting membrane preparation was decreased, as well as the yield of Na$^+$,K$^+$-ATPase activity. (———), no SDS; (———), 0.25 mg SDS/mg protein; (— —), 0.3 mg SDS/mg protein; (– – –), 0.35 mg SDS/mg protein; (-----), 0.4 mg SDS/mg protein; (······), 0.45 mg SDS/mg protein.

resis, the chief difference between the lower and upper fractions from the sucrose gradients is a higher proportion in the upper fraction of a doublet of contaminating proteins that comigrates with the β subunit of the Na$^+$,K$^+$-ATPase (Fig. 1). Typical yields from 10 mg of axolemma are 350–400 μg in the lower half of the peak and 500 μg of the upper half of the peak.

Practical Considerations

The axolemma fraction originally collected from the 0.8/1.0 M sucrose interface of the step gradient in the first part of the procedure is not successfully purified by SDS extraction, with either the same or higher concentrations of detergent. This lower density fraction is saved for use in experiments where purification of the Na^+,K^+-ATPase is not required. If myelin is not removed at all from the axon fragment-enriched wafers obtained by flotation, there is negligible purification of the Na^+,K^+-ATPase by SDS extraction. The inference is that the ratio of detergent to lipid, as well as of detergent to protein, influences the success of the SDS extraction.

Purification of Na^+,K^+-ATPase from Whole Brain Microsomes

Jørgensen's SDS extraction technique can also be applied to membrane fractions from whole brain, which contain both isozymes of the Na^+,K^+-ATPase.[9,10] In this case, the specific activity of the final product after SDS extraction is limited by the specific activity of the starting membrane material. When microsomes with specific activities of 0.8–1.3 μmol/min/mg protein are used, final specific activities approach those obtained by extraction of purified axolemma. When crude synaptosomes are used as the starting material, however, with specific activities of only 0.3 μmol/min/mg protein, final specific activities are only 1.6–1.9 μmol/min/mg protein.[3] This underscores the general observation that purification of the Na^+,K^+-ATPase is only really successful with highly enriched starting materials.

Mayrand et al. have introduced a simpler modification of the Jørgensen SDS extraction procedure applicable to rat brain microsomes, in which the extracted membranes are floated on 1.2 M sucrose instead of being centrifuged through a sucrose gradient.[10] The specific activities obtained are 7.0–13.0 μmol/min/mg protein, with yields of protein that average 3%, equivalent to yields of Na^+,K^+-ATPase activity of 30%.[10] When rat brain microsomes extracted with SDS are applied to sucrose gradients as described here, specific activities obtained are 9.0–12.0 μmol/min/mg protein, with similar yields.[9,11] The ratio of the two Na^+,K^+-ATPase isozymes present in the final preparation depends on whether whole rat brain or just cortex is used as the starting material[3]; cortex has roughly equal proportions of the two, while brain stem contributes a higher proportion of $\alpha(+)$.

---

[10] R. R. Mayrand, D. S. Fullerton, and K. Ahmed, *J. Pharmacol. Methods* **7**, 279 (1982).
[11] The specific activities reported in Ref. 9 were erroneously low because of the use of vanadate-contaminated ATP.

[7] Solubilization of Na^+,K^+-ATPase

By MIKAEL ESMANN

Introduction

Use of Detergents

Detergents have been used extensively to solubilize Na^+,K^+-ATPase from various tissues. Three main goals of solubilization can be considered. The first is selective solubilization of Na^+,K^+-ATPase from membranes containing impurities. The solubilized fraction can be purified by standard chemical methods and later precipitated to give a membranous pure Na^+,K^+-ATPase preparation.[1,2] The second aim is to achieve a well-defined solubilized preparation for incorporation into artificial lipid vesicles for transport studies, as exemplified by the cholate solubilization procedure[3] and $C_{12}E_8$-solubilized enzyme.[4] The third aim of solubilization is to measure the molecular weight of the solubilized enzyme using standard classical physicochemical methods such as sedimentation equilibrium centrifugation.[5,6]

This chapter will deal solely with solubilization of Na^+,K^+-ATPase in the nonionic polyoxyethylene detergent $C_{12}E_8$. Attention will be focused on the inactivating effect of detergent on the ATPase activity, on the measurement of the ATPase activity in the presence of detergent, as well as on defining conditions in which solubilized and active enzymes are obtained. The second part will deal with purification of the solubilized enzyme using size-exclusion chromatography and affinity chromatography.

Materials

$C_{12}E_8$ (octaethylene glycol dodecyl monoether) was obtained from Nikko Chemicals, and was used without further purification.

---

[1] L. E. Hokin, J. L. Dahl, J. D. Deupree, J. F. Dixon, J. F. Hackney, and J. F. Perdue, *J. Biol. Chem.* **248,** 2593 (1973).
[2] L. K. Lane, J. H. Copenhaver, Jr., G. E. Lindenmayer, and A. Schwartz, *J. Biol. Chem.* **248,** 7197 (1973).
[3] S. M. Goldin and S. W. Tong, *J. Biol. Chem.* **249,** 5907 (1974).
[4] F. Cornelius and J. C. Skou, *Biochim. Biophys. Acta* **772,** 357 (1984).
[5] D. F. Hastings and J. A. Reynolds, *Biochemistry* **18,** 817 (1979).
[6] M. Esmann, J. C. Skou, and C. Christiansen, *Biochim. Biophys. Acta* **567,** 410 (1979).

$C_{12}E_{10}$ was obtained from Sigma and used without further purification. Na^+,K^+-ATPase was purified in the membranous form according to Skou and Esmann (see Chapter 3).
All other chemicals were analytical grade.

Methods

Precautions

As will be evident from the experiments presented below, three factors are very important for successful solubilization of active Na^+,K^+-ATPase. First, working at a low temperature will prevent inhibition to a large extent whereas high temperatures (e.g., 37°) lead to a rapid inactivation of the solubilized enzyme. Second, a low $C_{12}E_8$/protein weight ratio will also give a more active solubilized enzyme. Third, a time-dependent aggregation of the solubilized particles has been observed which means that it is important to control the time between solubilization and the actual physical measurement (e.g., specific activity, molecular weight) to be performed.

Solubilizing Efficiency of $C_{12}E_8$

A solubilized enzyme is defined as not sedimenting during centrifugation for 60 min at 100,000 g. Figure 1 shows the amount of protein in the supernatant (in percentage of total) when Na^+,K^+-ATPase is solubilized in the presence of 150 mM KCl or with low ionic strength. At high ionic strength about 4 g of detergent is required per gram of protein, whereas only 1 g of detergent per gram of protein is required at low ionic strength for full solubilization of the enzyme.[7] The nonsolubilized (30–40 wt %) protein consists of impurities in the membrane preparation.[6]

Solubilization Procedure

Solubilization is carried out as follows: To 9 vol of membranous enzyme in buffer (20 mM histidine, pH 7.0, and 25% glycerol) is added 1 vol $C_{12}E_8$ in water, while the suspension is stirred rapidly on ice. An instantaneous clarification of the solution is observed and the enzyme is solubilized immediately.

Supernatant enzyme is prepared by centrifugation of the solubilized enzyme for 1 hr at 8° either in a Beckman airfuge at full speed or in a Beckman L80 centrifuge at 200,000 g. The protein concentration in the

[7] M. Esmann and J. C. Skou, *Biochim. Biophys. Acta* **787**, 71 (1984).

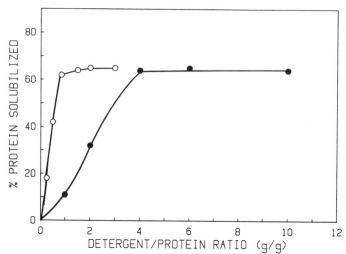

FIG. 1. Solubilization of Na$^+$,K$^+$-ATPase by $C_{12}E_8$. The amount of protein solubilized by $C_{12}E_8$ at low ionic strength [20 mM histidine (pH 7.0), open symbols] and at high ionic strength [150 mM KCl and 20 mM histidine (pH 7.0), filled symbols]. The solubilized protein is defined as the protein remaining in the supernatant after a centrifugation for 60 min at 100,000 g. The supernatant was more than 95% pure with respect to the α and β subunits. The initial protein concentration was kept at 0.8 mg/ml and the detergent concentration varied to give the detergent/protein weight ratios indicated. Protein concentrations were determined using the Peterson modification of the Lowry method.[8]

supernatant is measured using the Peterson modification of the Lowry method.[8] This modification of the Lowry method is necessary in order to achieve a complete precipitation of the solubilized protein for the subsequent protein analysis. A more rapid method which gives estimates of the protein concentration with an accuracy of about 10% is to make the supernatant 1% in SDS and measure the optical density of 280 nm[9] against a blank containing buffer, $C_{12}E_8$, and SDS. A 1 mg protein/ml solution will absorb about 1.1 absorbancy units.

At sufficiently high detergent/protein ratios all of the Na$^+$,K$^+$-ATPase activity will remain in the supernatant. The centrifugation step does not change the properties or stability of the solubilized enzyme, so this step can be avoided (if, of course, the degree of solubilization has been determined once).

[8] G. L. Peterson, *Anal. Biochem.* **83**, 346 (1977).
[9] M. Esmann, this volume [10].

Test of Na^+,K^+-ATPase Activity

The temperature at which the Na^+,K^+-ATPase activity is measured is a compromise between the inactivating effect of the detergent at high temperatures and the low specific activity at lower temperatures. We have routinely used 23° as a compromise with protein concentrations ranging from 5 to 20 µg/ml. Figure 2A shows the phosphate liberation as a function of test time at 23° at a given protein concentration (20 µg/ml) and different concentrations of $C_{12}E_8$. The slope of the curve remains constant with time for only a small range of the detergent concentration (0.1–0.2 mg $C_{12}E_8$/ml). The initial specific activity, i.e., the slope at time zero, is about the same for all $C_{12}E_8$ concentrations above the critical micelle concentration (60 µg/ml).

The K^+-dependent phosphatase activity of the Na^+,K^+-ATPase is less susceptible to detergent inactivation (Fig. 2B). A larger range of $C_{12}E_8$

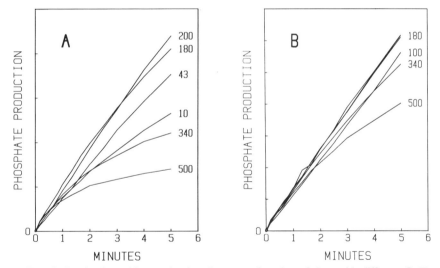

FIG. 2. Production of inorganic phosphate as a function of time with different $C_{12}E_8$ concentrations. $C_{12}E_8$-solubilized supernatant enzyme was prepared as previously described.[6] Incubation took place at 23°, in a 30 mM histidine buffer (pH 7.1) containing 130 mM NaCl, 20 mM KCl, 4 mM MgCl$_2$, 3 mM ATP (Tris salt), and $C_{12}E_8$ concentrations as indicated in the figure (in µg/ml). The protein concentration was 14 µg/ml. Panel A shows the amount of phosphate produced from the hydrolysis of ATP as a function of time. Panel B shows the corresponding time course for phosphate production when the K^+-dependent p-nitrophenylphosphatase reaction was tested. The incubation medium contained 30 mM histidine (pH 7.1), 150 mM KCl$_2$, 20 mM MgCl$_2$, 10 mM p-nitrophenyl phosphate, $C_{12}E_8$ as indicated, and 14 µg protein/ml.

concentrations can therefore be used to give linear phosphate liberation versus time curves.

Since the phosphatase activity is less susceptible to inactivation than the Na^+,K^+-ATPase activity, it is important to measure *both* activities in order to detect a partial denaturation.

Detergent/Protein Weight Ratio

The weight ratio of $C_{12}E_8$ to protein has proved to be a critical parameter in irreversible detergent inactivation of the Na^+,K^+-ATPase.[7] At low ionic strength (Fig. 1) all of the Na^+,K^+-ATPase activity is solubilized at 1 g detergent/g protein. Figure 3 shows that the specific activity of the solubilized enzyme (measured as shown in Fig. 2A) remains constant and high for detergent/protein ratios of 2–5. At higher detergent/protein ratios the specific activity is lower and it has not been possible to restore the enzymatic activity, i.e., the enzyme is irreversibly inactivated.

Protective Agents against Detergent Inactivation

The deleterious effect of detergents can be reduced by addition of albumin to the test solution, presumably due to albumin binding of the

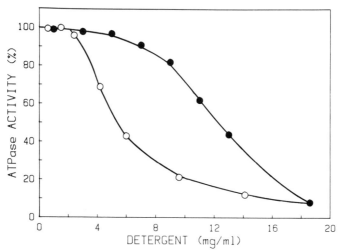

FIG. 3. Irreversible inactivation of Na^+,K^+-ATPase by $C_{12}E_8$. Membranous enzyme (3.5 mg/ml) is diluted 10-fold with $C_{12}E_8$ to give the indicated concentrations of $C_{12}E_8$ (and a protein concentration of 0.35 mg/ml). The Na^+,K^+-ATPase activity is measured at 23° after 2 hr at 4° by a further 10-fold dilution of the enzyme $C_{12}E_8$ suspension into the assay medium [130 mM NaCl, 20 mM KCl, 4 mM MgCl$_2$, 3 mM ATP, 1 mg/ml $C_{12}E_{10}$, and 30 mM histidine (pH 7.1)]. In one experiment the ionic strength was low (open symbols) and in the other 150 mM KCl was added to the enzyme/$C_{12}E_8$ suspension (filled symbols).

detergent.[6] For example, for 33 μg albumin/ml test solution, up to 0.4 mg $C_{12}E_8$ can be tolerated and linear P_i vs time curves are still obtained for Na^+,K^+-ATPase. It is, however, difficult to ascertain that the protein is in true solution under these conditions.

The nonionic detergent $C_{12}E_{10}$ can also be used as a protective agent in the test solution.[7] $C_{12}E_{10}$, which is a mixture of polyoxyethylenes with the average oxyethylene chain length of about 10, does not inactivate the Na^+,K^+-ATPase in the same way as $C_{12}E_8$. Linear P_i vs time curves can, for example, be obtained at a $C_{12}E_{10}$ concentration of 1 mg/ml at 23°. $C_{12}E_{10}$ can "buffer" excess $C_{12}E_8$, therefore more $C_{12}E_8$ can be tolerated in the test solution.[7]

Since the inactivating effect of detergent probably is through delipidation of the enzyme rather than by denaturation of the protein,[10] protection against inactivation can also be offered by addition of phospholipids. However, this is equivalent to a lowering of the "free" detergent concentration, so the effect of phospholipids is difficult to analyze.

Purification and Analysis of Solubilized Na^+,K^+-ATPase

Affinity Chromatography on Concanavalin A-Sepharose

Since the β chain of the Na^+,K^+-ATPase from shark rectal glands is glycoprotein, the sugar-binding lectin concanavalin A can be used to bind the solubilized enzyme particles.[11] Figure 4 shows such an experiment with the column equilibrated with 0.1 mg $C_{12}E_8$/ml. It should be noted that this method does not separate solubilized particles after size and therefore a mixture of protomers (consisting of 1 α subunit and 1 β subunit) and diprotomers will not be separated with this procedure. However, mixed phospholipid/detergent micelles will be separated from the Na^+,K^+-ATPase-containing detergent micelles and a delipidation is therefore obtained. It should be noted that the binding of the Na^+,K^+-ATPase to the lectin is "loose" and the enzyme should therefore be eluted with an appropriate sugar (for example, 0.2 M glucose) as soon as possible after the nonbinding species has emerged from the column. The solubilized enzyme is fully active after chromatography on Con A.[11]

Size-Exclusion Chromatography

The solubilized particles can be separated by size in conventional gel filtration media. We have used Sephacryl S-300 (Pharmacia) because of

[10] M. Esmann, *Biochim. Biophys. Acta* **787,** 81 (1984).
[11] M. Esmann, *Anal. Biochem.* **108,** 83 (1980).

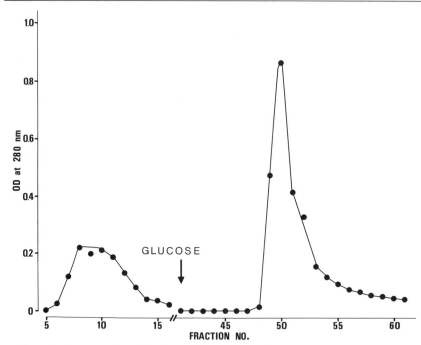

FIG. 4. Concanavalin A-Sepharose chromatography of $C_{12}E_8$-solubilized Na^+,K^+-ATPase. Membraneous Na^+,K^+-ATPase was solubilized with $C_{12}E_8{}^6$ and about 9 mg supernatant protein was applied to a column of Con A-Sepharose (1.2 × 8 cm, Pharmacia). The column was equilibrated in 30 mM histidine (pH 7.0), 100 mM KCl, and 0.3 mg/ml $C_{12}E_8$. The column was run at 4° with 12 ml/hr and 2.5-ml fractions were collected and assayed for protein (optical density at 280 nm). Glucose (0.2 M) was added to the elution buffer at the point indicated by the arrow.

the high stability of the gel matrix and of the high elution rates that can be employed.

Figure 5 shows such an experiment of $C_{12}E_8$-solubilized shark Na^+,K^+-ATPase. The protein peak of solubilized particles is broad and clearly inhomogeneous, but the specific Na^+,K^+-ATPase activity (about 500 μmol/mg per hour at 23°) is high across the peak, indicating that the differently sized particles have about the same specific Na^+,K^+-ATPase activity. The calculated elution position of the protomer and the diprotomer is indicated in Fig. 5.

The optimal conditions for purification using gel filtration—i.e., with minimal loss of activity—is essentially as when the Na^+,K^+-ATPase activity is to be measured: a low temperature will protect against the inactivation and a low detergent concentration (e.g., 0.1–0.2 mg/ml) in the column will give a higher specific activity of the purified particles.[10]

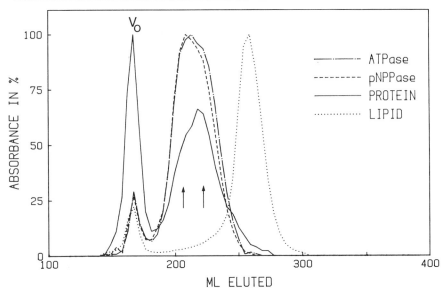

FIG. 5. Sephacryl S-300 gel filtration of solubilized Na^+,K^+-ATPase. A Sephacryl S-300 column (2.5 × 90 cm, Pharmacia) was equilibrated in 150 mM KCl, 0.2 mM CDTA, and 5 mM Tris (pH 7.0) at 6°. The dilution buffer also contained 1 mg $C_{12}E_8$/ml. Flow rate was 1 ml/min and the void volume and total volume were about 170 and 450 ml. Sample size was 5 ml and 4-ml fractions were collected. The elution position corresponding to the α-β protomer and diprotomer are indicated by arrows.[10] The protein concentration, the Na^+,K^+-ATPase activity, the pNPPase activity, and the phospholipid content of the fractions are given in percentage of maximum to ease comparison of the elution position of the peaks. The void volume (V_0) is indicated. Protein analysis shows that the α and β peptides elute in the broad peak around 220 ml, whereas the protein composition of the peak eluting at the void volume showed no α or β peptides.[10]

More sophisticated techniques such as HPLC can be used to obtain a separation of the α-β protomer and the diprotomer.[12]

Comments

The advantage of using $C_{12}E_8$ over other detergents such as Lubrol WX and Brij 56 is the chemical purity of $C_{12}E_8$. It should be noted that the inactivating effect of impure detergents such as Lubrol[5] or $C_{12}E_{10}$[7] is much less pronounced than with $C_{12}E_8$, and they are therefore easier to use. We have not yet been able to solubilize the shark enzyme in a fully active form with other types of detergents such as Triton, digitonin, or octylglucosides.

[12] Y. Hayashi, T. Tagaki, S. Maezewa, and H. Matsui, *Biochim. Biophys. Acta* **748**, 153 (1983).

[8] Crystallization of Membrane-Bound Na^+,K^+-ATPase in Two Dimensions

By ELISABETH SKRIVER, ARVID B. MAUNSBACH, HANS HEBERT, and PETER L. JØRGENSEN

Introduction

Electron microscopy analysis of regular two-dimensional crystalline arrays of membrane proteins can provide structural information that is not obtainable from studies of the nonordered proteins. Image reconstruction techniques based on Fourier filtering of electron micrographs of molecules which form two-dimensional crystalline arrays disclose the projection structure of the protein molecules.[1-3]

Membrane-bound Na^+,K^+-ATPase can be purified from outer renal medulla without perturbing lipoprotein associations.[4] When examined by electron microscopy after negative staining or freeze–fracture the purified enzyme in the membrane has the appearance of randomly distributed surface or intramembrane particles.[5,6] The orientation of the protein remains asymmetric in the purified, membrane-bound Na^+,K^+-ATPase and the protein units are densely packed in the membrane fragments in concentrations corresponding to almost 1 g protein/ml lipid phase. Under these conditions of supersaturation we found that incubation of the membranes with different ligands, in particular, vanadate in combination with magnesium, can rearrange the particles within the membrane to form two-dimensional crystalline arrays.[7] The crystallization can be monitored by electron microscopy after negative staining of the preparation. Further information about projection structure of the enzyme protein is then obtained by optical diffraction and image analysis of the electron micrographs of the crystals and shows that the crystals are either protomeric (monomeric) or dimeric.[8] From tilt series of two-dimensional crystals the

---

[1] A. Klug, *Chem. Scr.* **14**, 245 (1979).
[2] D. L. Misell, Pract. Methods Electron Microsc. **7** (1978).
[3] H. P. Erickson, W. A. Voter, and K. Leonard, this series, Vol. 49, p. 39.
[4] P. L. Jørgensen, *Biochim. Biophys. Acta* **356**, 53 (1974).
[5] N. Deguchi, P. L. Jørgensen, and A. B. Maunsbach, *J. Cell Biol.* **75**, 619 (1977).
[6] A. B. Maunsbach, E. Skriver, and P. L. Jørgensen, in "International Cell Biology 1980–1981" (H. G. Schweiger, ed.), p. 711. Springer-Verlag, Berlin and Heidelberg, 1981.
[7] E. Skriver, A. B. Maunsbach, and P. L. Jørgensen, *FEBS Lett.* **131**, 219 (1981).
[8] H. Hebert, P. L. Jørgensen, E. Skriver, and A. B. Maunsbach, *Biochim. Biophys. Acta* **689**, 571 (1982).

three-dimensional structure of the protein units can be obtained.[9] Induction of two-dimensional crystals with vanadate has also been shown for Ca^{2+}-ATPase.[10] In this chapter we describe the induction of two-dimensional crystals of Na^+,K^+-ATPase and the main steps in their analysis.

Enzyme Preparation

Membrane-bound Na^+,K^+-ATPase is purified from outer medulla of kidney by selective extraction of a crude membrane fraction with SDS (sodium dodecyl sulfate) in the presence of ATP followed by isopycnic centrifugation.[4,11] The specific activity of the enzyme used for crystallization is 28–40 μmol $P_i \cdot min^{-1} \cdot mg$ protein^{-1} and ouabain-insensitive ATPase activity is not detectable. Polyacrylamide gel electrophoresis in SDS shows that the preparation is 90–100% pure with respect to the content of the specific proteins, the α subunit and the β subunit. The preparations are stored before use at 0–4° in 25 mM imidazole, 1 mM EDTA–Tris (pH 7.5).

Crystallization with Vanadate/Magnesium

Incubation of purified membrane-bound Na^+,K^+-ATPase with vanadate in the presence of magnesium is the most efficient method to form two-dimensional crystalline arrays both with respect to extent and rate of crystallization.[7]

Procedure. Aliquots of enzyme (10–100 μg) are incubated at a final concentration of 0.1 mg protein/ml in 0.25–1 mM NaVO$_3$ and 3 mM MgCl$_2$ or 1 mM NH$_4$VO$_3$ and 3 mM MgCl$_2$ in 10 mM Tris–HCl or 10 mM imidazole buffer, pH 7.0–7.5. The suspensions are kept at 4–8°. Incubation solutions may be sterilized before use by filtration through 0.1 μm Millipore filter.

Formation of crystalline arrays, as followed by electron microscopy after negative staining, is usually observed within minutes after the start of incubation in some membrane fragments. It increases gradually over the following hours and days and two-dimensional crystals are present in almost all membrane fragments after 4 weeks of incubation (Fig. 1), although differences in rate and extent of crystallization can be seen from preparation to preparation.

Both protomeric (monomeric) and dimeric membrane crystals can be observed at the same time in the preparations. However, during the first

[9] H. Hebert, E. Skriver, and A. B. Maunsbach, *FEBS Lett.* **187**, 182 (1985).
[10] L. Dux and A. Martonosi, *J. Biol. Chem.* **258**, 2599 (1983).
[11] P. L. Jørgensen, this volume [2].

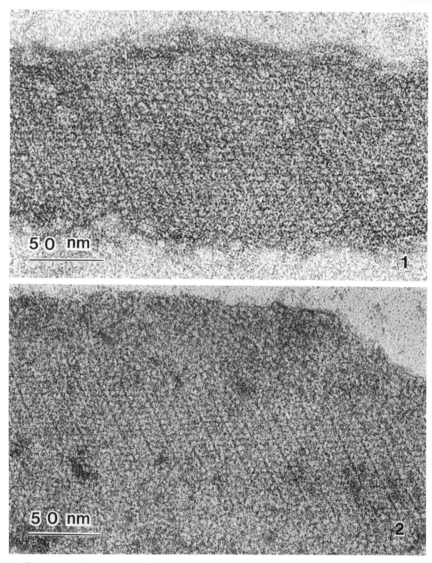

FIG. 1. *Top:* Electron micrograph of two-dimensional protomeric (monomeric) crystal of purified membrane-bound Na^+,K^+-ATPase formed during incubation of 0.25 mM sodium monovanadate, 1 mM magnesium chloride, and 10 mM Tris–HCl, pH 7.5. Negative staining with uranyl acetate. ×400,000. From Hebert *et al.*[8]

FIG. 2. *Bottom:* Electron micrograph of two-dimensional dimeric crystals of Na^+,K^+-ATPase formed during incubation in 12.5 mM phosphate, 3 mM magnesium chloride, and 10 mM Tris–HCl, pH 7.5. ×400,000. From Hebert *et al.*[8]

hours and days of incubation the largest proportion of crystals are of the dimeric type. Subsequently the frequency of protomeric type crystals increases, while the frequency of dimeric crystals gradually diminishes. Thus, by appropriate choice of incubation time it is possible to prepare samples where there is an overweight of either dimeric or protomeric crystals of Na^+,K^+-ATPase.

The rate and extent of crystallization is the same in Tris and imidazole buffers and no differences are observed when crystals are induced with $NaVO_3$ or NH_4VO_3.[12] Optimal pH for formation of two-dimensional crystalline arrays with vanadate/magnesium is 7.0–7.5. Addition of 150 mM KCl induces more crystallization centers within each membrane and often results in several small lattice systems in one membrane fragment.[13]

Crystallization with Other Ligands

In addition to vanadate other ligands of Na^+,K^+-ATPase also induce crystalline arrays of the protein in the membrane (Table I), although the extent of crystallization varies among the ligands.

Phosphate in the presence of magnesium (12.5 mM Tris–PO_4 or K_2HPO_4 in presence of 3 mM $MgCl_2$, pH 7.0–7.5) induces crystalline arrays mainly of the dimeric type (Fig. 2) but, after a longer incubation time, crystals of the protomeric type are also present. Addition of ouabain to phosphate/magnesium medium does not change the extent or type of crystal formation. Magnesium in the presence of 0.5 mM EGTA induces membrane crystals at a rate and extent almost similar to that observed in vanadate/magnesium medium,[13] but crystalline arrays are rare in membranes exposed to only EGTA or to magnesium. Crystallization in the presence of a high (50 mM) concentration of $MgCl_2$ and 0.5 mM EGTA is more extensive than with a low (3 mM) concentration of $MgCl_2$ and 0.5 mM EGTA.

Electron Microscopy of Membrane Crystals

Formation of two-dimensional crystals is monitored by negative staining with 1 or 0.5% uranyl acetate in water on hydrophilic carbon films as described separately.[14] Some stains, such as phosphotungstic acid, disturb the regularity of the crystalline arrays. Negative staining is carried out at a final protein concentration of about 0.1 mg protein/ml, which

[12] E. Skriver and A. B. Maunsbach, in "Structure and Function of Membrane Proteins" (E. Quagliariello and F. Palmieri, eds.), p. 211. Elsevier, Amsterdam, 1983.
[13] E. Skriver, H. Hebert, and A. B. Maunsbach, in "The Sodium Pump" (I. Glynn and J. C. Ellory, eds.), p. 37. Co. of Biol. Ltd., Cambridge, England, 1985.
[14] A. B. Maunsbach, E. Skriver, and P. L. Jørgensen, this volume [38].

TABLE I
CONDITIONS FOR FORMATION OF TWO-DIMENSIONAL CRYSTALS OF Na^+,K^+-ATPase

Vanadate (mM)	Phosphate (mM)	$MgCl_2$ (mM)	Other (mM)	Buffer (mM)	pH	Species	Reference
0.25 ($NaVO_3$)		1		10 (Tris–HCl)	7.5	Pig	7
1.0 (NH_4VO_3)		3		10 (Imidazole)	6.5–7.5	Rabbit	12
	12.5 (Tris–PO_4)	3		10 (Tris–HCl)	7.5	Pig	8
			150 (KCl)	10 (Imidazole)	6.1–7.5	Rabbit	12
0.25–0.5 (Na_3VO_4)		1	10 (KCl)	10 (Tris–HCl)	7.5	Pig, dog	20
		1	10 (KCl), 0.25–0.5 (K_3CrO_4)	10 (Tris–HCl)	7.5	Pig, dog	20
	5–10 (K_2HPO_4)	1	10 (KCl)	10 (Tris–HCl)	7.5	Pig, dog	20
	5 (H_3PO_4)	10	1 ($MnCl_2$)	25 (TES)	6.8	Dog	15
		3–50	0.5 (EGTA)	10 (Imidazole)	7.0	Rabbit	13

gives a suitable frequency of membranes on the grid. The enzyme is also incubated with the crystal-inducing medium at this concentration since dilution before staining may cause distortions of the crystalline arrays.

Crystalline arrays can also be outlined in thin sections when membranes are oriented parallel to the section.[13,15] In order to visualize the crystalline arrays, we find it advantageous to fix the membranes in suspension and to add tannic acid to the fixative media. Crystalline patterns are also present in unfixed or fixed membranes following freeze–fracture.

Electron microscopy is carried out using direct magnification of 20,000–66,000×.

Image Analysis

Image analysis of two-dimensional crystals of Na^+,K^+-ATPase can be carried out using Fourier filtering or correlation averaging[8,16] using methods previously applied to other two-dimensional protein crystals.[17,18] Selection of well-ordered crystalline arrays is first made by optical diffraction of the electron micrograph negatives. Images suitable for further analysis are densitometered at 20-μm intervals. Projection maps are then calculated either using the Fourier transform amplitudes and phases collected at the reciprocal lattice points or by superposition of motives localized by correlation functions. The processing of the electron micrographs in our investigations was carried out on a VAX 11/780 computer using the image analysis system EM.[19] In the projection maps the protein regions (positive regions) are drawn with unbroken contour lines while negatively stained regions have dashed lines (Figs. 3 and 4).

Comments and Interpretations

The two types of crystalline arrays which are induced in preparations of membrane-bound Na^+,K^+-ATPase can be observed with all the ligands that induce crystals, but the proportions between them depend upon the length of incubation and vary with the ligand combinations. The two types of crystals have also been confirmed in other studies.[15,20,21] The very efficient crystallization of membrane-bound Na^+,K^+-ATPase in vanadate/

---

[15] G. Zampighi, J. Kyte, and W. Freytag, *J. Cell Biol.* **98,** 1851 (1984).
[16] H. Hebert, E. Skriver, R. Hegerl, and A. B. Maunsbach, *Proc. Eur. Congr. Electron Microsc., 8th* **2,** 1497 (1984).
[17] L. A. Amos, R. Henderson, and P. N. T. Unwin, *Prog. Biophys. Biol.* **39,** 183 (1982).
[18] W. O. Saxton and W. Baumeister, *J. Microsc.* **127,** 127 (1982).
[19] R. Hegerl and A. Altbauer, *Ultramicroscopy* **9,** 109 (1982).
[20] M. Mohraz and P. R. Smith, *J. Cell Biol.* **98,** 1836 (1984).
[21] V. V. Demin, A. N. Barnakov, A. V. Lunev, K. N. Dzhandzhugazyan, A. P. Kuzin, N. N. Modyanov, S. Hovmöller, and G. Farrants, *Biol. Membr.* **1,** 831 (1984). (In russian).

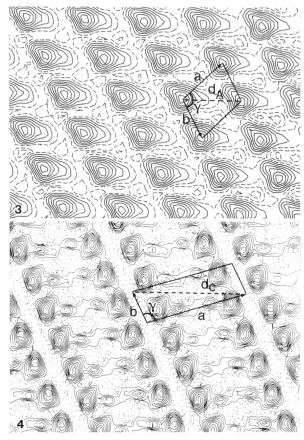

FIGS. 3 and 4. Computer-reconstructed images from the two-dimensional crystals shown in Figs. 1 and 2, respectively. The protein-rich regions (positive regions) are drawn with unbroken contour lines while negative stain regions have dashed lines. The unit cell dimensions in Fig. 3 (*top*) are $a = 69$ Å, $b = 53$ Å, and $\gamma = 105°$ and in Fig. 4 (*bottom*) $a = 135$ Å, $b = 44$ Å, and $\gamma = 101°$. Each unit cell in Fig. 3 contains only one $\alpha\beta$ unit while the unit cell in Fig. 4 contains 2 $\alpha\beta$ units. The two-sided plane group symmetry is $P1$ for Fig. 3 and $P21$ in Fig. 4. From Hebert *et al.*[8]

magnesium medium suggests that formation of two-dimensional crystals is favored when the protein is in the E_2-conformation which is stabilized by vanadate. At the concentrations used the equilibrium between the E_1 and E_2 form is heavily shifted toward the E_2 form due to the very high affinity of Na$^+$,K$^+$-ATPase for vanadate.[22,23]

[22] L. C. Cantley, L. G. Cantley, and L. Josephson, *J. Biol. Chem.* **253**, 7361 (1978).
[23] P. L. Jørgensen and S. J. D. Karlish, *Biochim. Biophys. Acta* **597**, 305 (1980).

The very first step in the formation of two-dimensional crystals seems to be the formation of linear arrays of paired protein units, which then associate laterally.[7] The dimeric crystals show in one direction a spacing of about 50–60 Å and in the other direction lattice lines with a spacing of 120–140 Å. The protomeric crystals show lattice lines which in both directions have spacings of 50–70 Å (Fig. 2). The spacings of the protomeric crystals vary slightly, apparently depending upon the maturity of the crystals, since crystals incubated for long periods usually have smaller spacings than newly formed crystals.

Image analyses of the two types of crystals[8,16] reveal that the unit cell in the protomeric crystal contains one positive unit. On the basis of size this unit is interpreted as an $\alpha\beta$ unit of the enzyme. In the dimeric crystals, on the other hand, the projection map demonstrates that the unit cell contains two strong positive regions, suggesting that two $\alpha\beta$ units occupy one unit. The positive regions show a 2-fold rotational relationship and each is subdivided in one large and one small peak. The dimensions and shape of the large peak are similar to the positive peak in the monomeric crystals (Figs. 3 and 4).

Acknowledgments

Supported by the Danish Medical Research Counsil and the Research Foundation at the University of Aarhus.

[9] Preparation of Antibodies to Na$^+$,K$^+$-ATPase and Its Subunits

By WILLIAM J. BALL, JR., TERENCE L. KIRLEY, and LOIS K. LANE

Introduction

Antibodies have served as a primary tool in identifying and characterizing cell-surface antigens. They can also serve as useful tools in studying the function and structure of membrane-bound enzymes. Indeed, antisera have been raised to many different preparations of the membrane-bound Na$^+$,K$^+$-ATPase. While the results of many of the initial studies utilizing rabbit polyclonal antibodies were confusing and sometimes contradictory,[1] the recent development of many new immunochemical techniques

[1] P. K. Lauf, *in* "Membrane Transport in Biology" (D. C. Tosteson, H. H. Ussing, and G. Giebisch, eds.), p. 291. Springer-Verlag, Berlin, 1978.

has greatly enhanced the suitability of using antibodies to study Na^+,K^+-ATPase. In our laboratory, we have utilized rabbit polyclonal antibodies raised to purified lamb kidney Na^+,K^+-ATPase and to its α and β subunits. Monoclonal antibodies directed against the α subunit of the lamb enzyme have also been obtained. Although the antibody populations generated in response to an immunizing antigen obviously vary with the exact nature of the antigen and the particular animal immunized, the immunochemical techniques used can be standardized and applied to different problems in a reproducible manner.

In this chapter, we describe the techniques routinely used in our laboratories for immunological studies of Na^+,K^+-ATPase, as well as the techniques employed in the preparation of native enzyme, subunits, and α peptides.

Preparation of Lamb Kidney Na^+,K^+-ATPase

In this procedure, Na^+,K^+-ATPase is purified from the outer medulla of frozen lamb kidneys, but we have used essentially the same procedure with fresh tissue and with kidneys from other species.[2,3] The only obvious differences between the preparations from fresh and frozen kidneys are the Na^+,K^+-ATPase activities of the initial membrane fraction (microsomes) and the degrees of activation with NaI and detergent treatments. Fresh kidneys tend to yield microsomes which exhibit lower Na^+,K^+-ATPase activities and higher degrees of activation than do kidneys that have been stored frozen. It is generally accepted that the degree of detergent activation of Na^+,K^+-ATPase is a reflection of the amount of sealed, right-side-out vesicles in the membrane preparation. The frozen whole kidneys are thawed in 0.25 M sucrose, 1 mM EDTA, and 25 mM imidazole, pH 7.3 (medium H). The cortex is removed using scissors, and the exposed outer medulla is then collected in and washed with ice-cold medium H before being minced at low speed in a Waring blender. For smaller amounts of tissue, the mincing is done by hand with scissors. The minced outer medulla is then diluted with medium H (approximately 7 ml/g tissue) and homogenized with two 30- to 45-sec bursts of a Polytron (Brinkmann Instruments; PT-35) at two-thirds maximum speed. The homogenate is centrifuged at 11,300 g_{max} for 45 min and the supernatant is decanted and centrifuged at 30,000 g_{max} for 60 min. This supernatant is removed by aspiration and the soft pellet is resuspended and homogenized in 1 mM EDTA, adjusted to pH 7.3 with imidazole, with a glass–

[2] L. K. Lane, J. H. Copenhaver, G. E. Lindenmayer, and A. Schwartz, *J. Biol. Chem.* **248**, 7197 (1973).

[3] L. K. Lane, J. D. Potter, and J. H. Collins, *Prep. Biochem.* **9**, 157 (1979).

Teflon homogenizer to a protein concentration of 10–15 mg/ml. This microsomal fraction can be stored at $-20°$ after freezing using a dry ice–ethanol mixture, or it can be treated immediately with NaI as follows. The microsomal fraction is diluted to 6 mg protein/ml with 1 mM EDTA, pH 7.3, and 0.2 vol of a solution containing 6 M NaI, 7.5 mM MgCl$_2$, 150 mM Tris, and 15 mM EDTA, pH 8.3, is added slowly with thorough stirring. The mixture is stirred on ice for 30 min, diluted with 1.5–2 vol of 1 mM EDTA, pH 7.3, and centrifuged at 30,000 g_{max} for 60 min. The supernatant is removed by aspiration, with care being taken not to disturb the very loose top layer of the pellet. The pellet is resuspended in a small volume (15–25 ml) of 1 mM EDTA, pH 7.3, homogenized by hand with two passes in a glass–Teflon homogenizer, diluted to the same volume, centrifuged as in the previous step with 1 mM EDTA, pH 7.3, and centrifuged for 60 min at 30,000 g_{max}. This wash is repeated twice, and the NaI-treated microsomal fraction is then resuspended at 11–15 mg protein/ml in 25 mM imidazole, pH 7.0, 1 mM EDTA, 0.4 M NaCl, and 0.04 M KCl (medium S), frozen in a dry ice–ethanol bath, and stored overnight at $-20°$.

After thawing, the NaI-treated microsomes are adjusted to 0.1% sodium deoxycholate or to a total detergent concentration of 0.1% [sodium deoxycholate : sodium cholate (3 : 1)], with 0.1 mg detergent/mg protein, stirred gently for 15 min, and centrifuged for 1 hr at 142,800 g_{max}. The supernatant is removed and the pellet is gently homogenized to the same volume in medium S and centrifuged as before. The washed pellet is homogenized to 8–15 mg protein/ml in medium S and either frozen as described above or extracted immediately with detergent. For extraction, the fraction is adjusted to 0.35% with a mixture of sodium deoxycholate and sodium cholate (3 : 1) with approximately 0.5 mg detergent/mg protein. After stirring for 10 min, the suspension is centrifuged for 45 min at 142,800 g_{max}, and the resultant supernatant is carefully decanted and diluted with 0.25 vol of glycerin. This suspension is stirred slowly for 15 min, diluted with an equal vol of 1 mM EDTA, pH 7.3, and centrifuged for 2 hr at 142,800 g_{max}. The pellet is resuspended in 1 mM EDTA, pH 7.3, and centrifuged at the same speed for 45 min to reduce the amount of detergent in the preparation prior to its being dialyzed for 1 to 2 days against 1 mM EDTA, pH 7.3. The washed glycerin-precipitated Na$^+$,K$^+$-ATPase is stable for months at 3°.

The critical step in the above procedure is the extraction with detergent, and it is the ratio of sodium deoxycholate and sodium cholate to protein that is important. The optimal detergent-to-protein ratio is readily determined by carrying out a series of miniextractions at varying ratios. The supernatants of the extractions are diluted about 2-fold with 1 mM

TABLE I
PURIFICATION OF Na^+,K^+-ATPase FROM 600 g OF LAMB KIDNEY OUTER MEDULLA

Fraction	Protein (mg)	Ouabain-insensitive ATPase (μmol P_i/min)	Na^+,K^+-ATPase (ouabain sensitive)	
			μmol P_i/min/mg	μmol P_i/min/fraction
Microsomes	3080	1232	1.40	4312
NaI treated	1242	310	4.15	5154
DOC washed	1050	175	3.90	4095
DOC extracted and glycerin precipitated	198	62	21.9	4336

EDTA, pH 7.3, containing 25% glycerin, and the amount of Na^+,K^+-ATPase activity in the supernatants is measured. Once determined, the optimal ratio is normally constant unless the method of protein determination, brand of detergent, or source of tissue is altered. As illustrated in Table I, approximately 200 mg of Na^+,K^+-ATPase can be purified from 600 g of lamb kidney outer medulla. The specific activity of the enzyme ranges from 16 to 23 μmol P_i/mg/min and it binds 2 to 2.8 nmol of ouabain/mg. As visualized by SDS–polyacrylamide gel electrophoresis (SDS–PAGE), the preparation consists almost solely of the M_r 100,000 α subunit and the M_r 44,000 β subunit.

Isolation of Na^+,K^+-ATPase Protein Subunits

To separate the catalytic (α) and the glycoprotein (β) subunits of the Na^+,K^+-ATPase, the enzyme–lipid complex is first solubilized with sodium dodecyl sulfate (SDS) at 5 to 10 mg/mg protein and incubated for 30 min at approximately 50°. To minimize aggregation of the α subunit, the enzyme–SDS mixture should not be boiled. Although there do not appear to be any disulfide linkages between the α and β subunits, 2-mercaptoethanol (10–100 mM) is generally included in the solubilization buffer. The protein subunits are routinely separated by gel filtration on columns of 4 or 6% agarose (BioGel A-15m or A-5m, 100–200 mesh) or Sepharose Cl-6B (Pharmacia) in 25 mM Tris–HCl, pH 7.5, with 0.1% SDS, 1 mM 2-mercaptoethanol, and 0.01% sodium azide.[2,4] Depending upon the size of the column, approximately 10 mg (1.5 × 190 cm) or 40 mg (2.5 × 190 cm)

[4] J. H. Collins, B. Forbush, L. K. Lane, E. Ling, A. Schwartz, and A. Zot, *Biochim. Biophys. Acta* **686**, 7 (1982).

of SDS-solubilized Na^+,K^+-ATPase can be resolved into its α and β subunits. Each run requires 18–24 hr. The recovered subunits are essentially lipid free, since most of the phospholipids elute behind the β subunit.

A similar separation of α and β subunits can be achieved much more rapidly using gel filtration high-performance liquid chromatography (HPLC). Figure 1 illustrates the resolution of the α and β subunits obtained by size-exclusion chromatography of 3 mg of lamb kidney Na^+,K^+-ATPase on two Bio-Sil TSK-250 columns (300 × 7.5 mm; Bio-Rad) connected in series. The columns are equilibrated and run at a flow rate of 1 ml/min with 100 mM sodium phosphate, pH 6.9, containing 0.1% SDS, 0.01% NaN_3, and 1 mM 2-mercaptoethanol. Each run requires 25 min.

Preparation of α Subunit Peptides

In addition to the preparation of the enzyme and its subunits, we have also developed a simple method for the rapid isolation of specific soluble peptides of the α subunit of Na^+,K^+-ATPase. The Na^+,K^+-ATPase is first labeled covalently with a binding site or amino acid-specific probe and then the soluble, labeled polypeptide is isolated from a supernatant fraction after limited trypsin digestion of the enzyme, using an approach

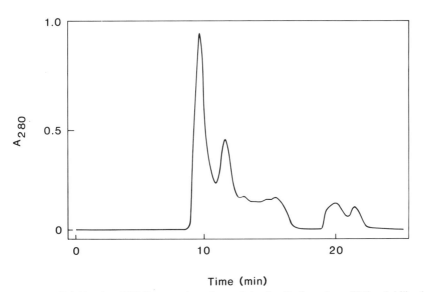

FIG. 1. Gel filtration HPLC separation of α and β subunits from 3 mg SDS-solubilized lamb kidney Na^+,K^+-ATPase. The peaks eluting from the column at approximately 9.5 and 11.5 min contain the α and β subunits, respectively. The later eluting peaks contain trace amounts of low-molecular-weight proteins, phospholipids, and 2-mercaptoethanol.

similar to that described by Mitchinson et al.[5] for the Ca^{2+},Mg^{2+}-ATPase. We have labeled the ATP-binding site with fluorescein isothiocyanate (FITC),[6,7] the phosphorylation site with [γ-^{32}P]ATP, and the ouabain-binding site with a 2-nitro-5-azidobenzyl(NAB)-ouabain photoaffinity label.[4,8] Then an aliquot is first resolved by 7.5% SDS–PAGE to demonstrate α specific labeling. The enzyme is then partially digested with proteolytic enzymes. We have found TPCK-trypsin to be suitable due to its relatively high specificity. Typically, the Na^+,K^+-ATPase (2 mg/ml) is digested for 90 min at 37° in 25 mM imidazole–HCl, 1 mM EDTA, pH 7.2, with 0.02 mg/ml TPCK-trypsin. The reaction is stopped by the addition of 0.5 mM phenylmethylsulfonyl fluoride (PMSF). The optimum digestion time and trypsin/Na^+,K^+-ATPase ratio should be determined empirically. We have varied the trypsin/Na^+,K^+-ATPase ratio from 1:100 to 1:10, depending upon the particular label used and the source of the enzyme. For example, rat kidney Na^+,K^+-ATPase requires more trypsin (1:10) than the lamb kidney enzyme (1:100) to reach an equivalent state of digestion. Following the protease digestion, the mixture is centrifuged at 150,000 g for 30 min. The supernatant is collected and lyophilized and the amounts of protein and label in both the supernatant and the pellet are determined. Typically, 45–50% of the total protein and 80% of the bound FITC are released into the supernatant fraction. It is important that a substantial amount of the label be released into the soluble fraction if the labeled peptide is to be isolated and identified. The soluble peptides are then resolved using reversed-phase HPLC (LDC–Milton Roy HPLC). We use a 15.-cm, 300-Å pore size C_{18} Vydac column (Separations Group Cat. #218TP5415) and a gradient of 0.1% trifluoracetic acid (TFA) in HPLC grade H_2O to 0.1% of TFA in HPLC grade acetonitrile, pH 2.0, with the peptides dissolved in 0.5 ml 0.1% TFA/H_2O. The solvent flow rate is 0.7 ml/min and the gradient is run in 30 min. If the low pH causes a problem of stability in retaining the label, a 10 mM ammonium acetate/ H_2O, pH 7.0, with 10 mM ammonium acetate/methanol gradient can be used. The Na^+,K^+-ATPase tryptic peptides generally elute between 20 and 35% acetonitrile, with 0.1% TFA, although the optimal gradient needed for a particular peptide needs to be determined experimentally. Since the peptides behave as if they stack at the top of the column and are

---

[5] C. Mitchinson, A. F. Wilderspin, B. J. Trinnaman, and N. M. Green, *FEBS Lett.* **146**, 87 (1982).

[6] R. A. Farley, C. M. Tran, C. T. Carilli, D. Hawke, and J. E. Shively, *J. Biol. Chem.* **259**, 9532 (1984).

[7] T. L. Kirley, E. T. Wallick, and L. K. Lane, *Biochem. Biophys. Res. Commun.* **125**, 767 (1984).

[8] B. Forbush, J. H. Kaplan, and J. F. Hoffman, *Biochemistry* **17**, 3667 (1978).

then displaced by a specific organic solvent concentration, there are essentially no restrictions on the volume of sample that can be injected. Detection of the peptides can be done by monitoring the absorbance in the 210–230 nm range. If the label absorbs in a different spectral region from that of the peptide bonds, this wavelength should also be used to specifically monitor the labeled peptide(s).

After an initial injection and fractionation, the partially purified labeled peptide(s) can then be directly reinjected (via a 1.0-ml sample loop) and rechromatographed using a modified gradient, or a series of HPLC columns having different bonded phases (i.e., C_4, C_8, phenyl, etc.). The approximate solvent composition at which the peptide elutes can be determined by noting the solvent composition at the pumps when the peptide elutes and calculating the delay time (in solvent composition) between the pumps and the detector. This can be determined by running a blank gradient containing an abrupt solvent composition change which is monitored by following the increase in solvent absorption with the detector set at a sensitive setting. We generally use a 30-min total run time and then repurify the eluted peptides by rechromatography using progressively shallower gradients on a series of different bonded phase columns. For the last repurification of the FITC peptide, a gradient of less than 1.0% acetonitrile/16 min is used, and a single 220 nm peak is obtained (Fig. 2). The recovery of various labeled peptides with this system has been near 100%. The purified peptides are then dried down (lyophilized or evaporated using a Speed-Vac concentrator) and used directly for amino acid analysis, molecular weight determination, and amino acid sequencing.

We have found that molecular weight determinations of peptides less than 4000 cannot be accurately determined by SDS–PAGE systems such as Laemmli[9] or Swank and Munkries[10] gels. The only method that gave a molecular weight of the FITC peptide which agreed with the actual sequence was a Bio-Sil TSK-125 column (Bio-Rad Laboratories) using 6.0 M guanidine as the eluant. The disadvantages of this system are a high solvent absorbance at 220 nm and the possible corrosion of the stainless steel pumps of the HPLC apparatus. Sequence analysis[11] of the isolated peptides is done using a gas-phase sequenator (Applied Biosystems Protein Sequencer 470A). The peptides are coupled in a single step and analyzed for 15 cycles (65 min each). The labeled amino acid–phenylthio-

---

[9] U. K. Laemmli, *Nature (London)* **227**, 680 (1970).
[10] R. T. Swank and K. D. Munkries, *Anal. Biochem.* **39**, 462 (1971).
[11] The FITC peptides were generously sequenced by T. Vanaman at the Biochemistry Department of the University of Kentucky, Lexington.

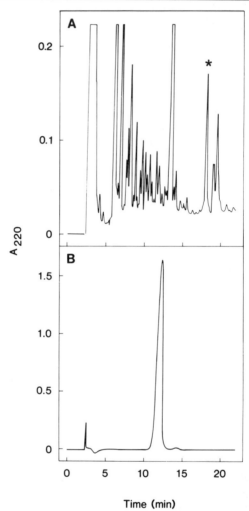

FIG. 2. Reversed-phase HPLC separation of soluble tryptic peptides of FITC-labeled lamb kidney Na^+,K^+-ATPase. (A) The total tryptic supernatant from 0.4 mg of ATPase. The gradient is a 20 to 35% acetonitrile gradient with a running time of 2–18 min. The asterisk (*) indicates the position of the FITC-labeled peptide. (B) The final repurification of the FITC-labeled peptide isolated from tryptic digestion of 40 mg of Na^+,K^+-ATPase. The gradient used is a 28.2 to 28.8% acetonitrile gradient with a 2- to 18-min running time.

hydantoin (PTH) derivatives obtained from the sequenator do not, of course, elute at times which coincide with the unlabeled PTH amino acids but these residues can sometimes be identified by the residual color, radioactivity, or known amino acid specificity of the label.

Preparation of α and β Subunits for Immunization

For the immunological studies, the subunits eluted from the gel filtration columns are concentrated by lyophilization and dissolved in a buffer containing 6 M urea, 10 mM Tris–HCl, and 0.01% NaN$_3$, pH 7.8. The SDS is removed from the proteins by chromatography on an AG-1-X2 resin column (Bio-Rad) equilibrated in the same buffer containing 6 M urea. We use 1.5 ml of settled bed volume of resin for each 100 mg of SDS in the sample. After removal of SDS, the protein sample is dialyzed extensively against a 10 mM Tris, 0.5 mM EGTA, 0.01% NaN$_3$ buffer, pH 7.3, to remove the urea. The samples are then assayed for residual levels of SDS using a methylene blue colorimetric assay as described by Hayashi.[12] The subunit samples are generally stored at 5° since freezing can cause an aggregation of the protein and it cannot be easily resolubilized.

Preparation of Rabbit Polyclonal Antibodies

Immunization. Rabbits (New Zealand, pathogen free) are injected with an emulsified suspension of Freund's complete adjuvant (Gibco laboratories) and the antigen (0.75–1.0 mg/ml). The holoenzyme or the prepared subunit at 1.5–2.0 mg/ml in 50 mM Tris, 0.15 M NaCl, pH 7.3, is mixed in aliquots with the Freund's complete adjuvant to give a thick (1 : 1) emulsified suspension. This suspension (1 ml) is injected subcutaneously (10 sites) on the back of rabbits, weekly for 4 weeks, followed by 0.2 ml injected intramuscularly in each haunch for 2 weeks. Animals are then bled bimonthly from the ear marginal vein, with the intramuscular booster injections being administered on the weeks between bleedings. The blood samples are allowed to clot overnight at 5° and the serum is separated from the clot by centrifugation for 30 min at 1500 g. The serum samples, containing 1–2 drops of 1% Merthiolate (thimerosal) (~0.005%) as preservative, are then divided into aliquots and stored at −80°. Before using the antiserum the complement is heat inactivated for 30 min at 57° and the samples are centrifuged and filtered through both 0.45- and 0.22-μm Millipore filters.

Preparation of Antiserum

We routinely use the antiserum for determinations of antibody titer, immunoblot, and immunofluorescence work, but we prepare the IgG fraction when testing antibody effects on enzyme function or for immunoprecipitation procedures. The immunoglobulin fraction is precipitated from

---

[12] K. Hayashi, *Anal. Biochem.* **67**, 503 (1975).

the serum by adding a saturated $(NH_4)_2SO_4$ solution to bring the serum to 40% $(NH_4)_2SO_4$ saturation and then pelleting the immunoglobulins by centrifugation for 30 min at 1000 g and resuspending the pellet in Tris–saline buffer and reprecipitating and resuspending the protein three more times. The immunoglobulin fraction is then dialyzed first against 50 mM Tris, 0.15 M NaCl, 0.5 mM EGTA, and 0.01% sodium azide (pH 7.4), then extensively dialyzed against 50 mM Tris, pH 7.4.

When testing antibody effects on enzyme function, it is also important to test the effects of similarly prepared preimmune serum, nonspecific globulins, and the final dialysis solution. Endogenous serum ATPase activity, as well as nonspecific protein, and cation effects on the Na^+,K^+-ATPase activity and ouabain binding, have been observed.

Preparation of Monoclonal Antibodies to Na^+, K^+-ATPase

Immunization and Cell Fusion. In our laboratory, we use BALB/cJ × C57BL/6J, F_1 hybrid mice (CB6F_1/J from Jackson Laboratories). These animals receive 4–6 weekly subcutaneous injections of the purified Na^+,K^+-ATPase (100 μg) emulsified in Freund's complete adjuvant. They are then given biweekly or monthly injections until shortly before they are used as the source of the spleen cells for a cell fusion. One week after an injection, the animals are bled from the retrobulbar vein and their antisera titers are determined using an enzyme-linked immunoadsorbent (ELISA) assay. Mice with high titers to the Na^+,K^+-ATPase are then given a final booster injection of 100 μg of enzyme suspended in 10 mM Tris, pH 7.2, administered intravenously (retrobulbar) 3–4 days before the animals are killed.

The spleen cells from the immunized mice are fused with the mouse myeloma cell line SP2/O-Ag14 (obtained from Schulman et al.[13]) according to the method of Galfré et al.[14] This cell line is especially suited for cell hybrids because it is an IgG nonproducer variant derived from the MOPC myeloma (P3). A detailed description of the methods and strategies for obtaining hybridomas by Galfré and Milstein can be found in an earlier volume of this series.[15] The average spleen contains approximately 1 × 10^8 isolatable cells and these are mixed with 2 × 10^7 myeloma cells (5 : 1 ratio) which are in log phase growth. A pelleted mixture of these cells is then fused through the addition of 1 ml of a 40% polyethylene glycol (1500, Baker), 9% DMSO, phosphate-buffered saline (PBS) solution, pH

---

[13] M. Schulman, C. Wilde, and G. Köhler, *Nature (London)* **276**, 269 (1978).
[14] G. Galfré, S. Howe, C. Milstein, G. Butcher, and J. Howard, *Nature (London)* **266**, 550 (1977).
[15] G. Galfré and C. Milstein, this series, Vol. 73, p. 1.

7.2, for 2 min at room temperature. This mixture is then diluted with 15 ml of Dulbecco's modified Eagle's medium, centrifuged, and the cells rediluted to $1-2 \times 10^6$ cells/ml with complete selective medium containing hypoxanthine/aminopterin/thymidine (10% fetal calf serum, pretested for its ability to support hybridoma growth). The diluted cells are then distributed into approximately 300–600 culture wells which contain selective medium and mouse peritoneal exudate cells (2×10^4 cells/ml/well) that were plated 1–7 days earlier. The fused cells are plated out ($\sim 2-5 \times 10^5$ cells/ml) such that we expect viable hybridoma colonies in approximately 60–80% of the culture wells and, according to Poisson statistics, an average of 1–2 hybridoma colonies per well. This ensures a reasonable probability of cloning the hybridomas from antibody-positive culture wells.

After 10–16 days, when sufficient cell growth has occurred and acidification of the medium has begun, duplicate aliquots of the medium are removed from the wells and tested for Na^+,K^+-ATPase-directed antibodies using an ELISA assay (Fig. 3). Hybridoma cells from positive testing wells are then transferred to multiple 1-ml wells containing selection media and peritoneal exudate cells. These cells are grown, tested for antibody production, and then cloned and recloned by limiting dilution into 96-well plates. We do not attempt to clone hybridomas directly from the

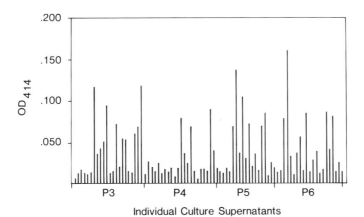

FIG. 3. Screening of cell culture supernatants for anti-Na^+,K^+-ATPase antibodies. The β-galactosidase sheep anti-mouse $F(ab)_2'$ antibody conjugate was used to identify Na^+,K^+-ATPase-specific antibody secretion by hybridomas into approximately 50 μl of culture medium. The extent to which the antibody–β-galactosidase conjugate hydrolyzes the substrate o-nitrophenyl-1-β-galactoside is given in absorbance units at 414 nm for a 1 : 40 dilution of the assay solution. In this particular experiment, 21 of the 84 samples were considered positive (greater than one standard deviation from mean), but only 7 samples were designated as suitable for cloning. Reprinted from Ball et al.[24]

initial cell fusion platings because we often find that originally positive testing cultures do not test positive upon replating. Therefore, it is inefficient to try to clone the cells without at least one replating to test the stability of the cell line. This lack of antibody production by first transfer cell colonies could be caused by the overgrowth of producer cells by nonproducers or by the presence of unstable hybridomas. Also, it appears that some nonhybrid spleen cells survive in the selection media for a few days with residual antibody production. These cell culture samples test positive initially, but test negative upon being transferred.

After recloning, the hybridomas can be grown in T-flasks and antibody is obtained from the culture medium. In addition, the hybridoma cells are injected (0.3 ml of 3–5 × 10^6 cells/ml) into F_1 hybrid mice primed 2 weeks earlier with 0.3 ml of pristane (tetramethylpentadecane, Aldrich). The ascites fluid is collected or "tapped" 2–3 weeks later, after the tumor cells have grown and considerable ascites fluid production is evident.

Identification of Antibody Isotype

The isotype of the secreted monoclonal antibody is determined using the double diffusion immunoprecipitation method of Ouchterlony.[16] Samples of the hybridoma culture supernatant are concentrated using a minicon concentrator (Amicon) and mouse Ig subclass-directed immunoglobulins (anti-IgG_1, IgG_2, IgG_{2b}, IgG_3, IgM, and IgA) purchased from Litton Bionetics are used. The diffusion agar consist of 0.8% agarose A-45 (L'Industrie Biologique Francaise) dissolved in 1 mM EGTA–Tris, pH 7.4.

Preparation of Monoclonal Antibodies

The IgG monoclonal antibodies can be prepared by Na_2SO_4 precipitation of the immunoglobulin in the hybridoma culture medium. Alternatively, larger amounts can be obtained from the ascites fluid of mice inoculated with the hybridoma cells. The collected ascites fluid is allowed to clot overnight at 5°, centrifuged using a clinical tabletop centrifuge, and the supernatant collected and filtered through 0.45- and 0.22-μm Millipore filters. The immunoglobulin is precipitated at room temperature by slowly adding solid Na_2SO_4 to 18% (w/v) with continuous stirring and allowing the precipitation to occur for 30 min after the addition of the Na_2SO_4.[17] The precipitate is collected by centrifugation at 30,000 g for 30 min. After removing the supernatant by aspiration, the pellet is resuspended in 25

---

[16] O. Ouchterlony, *Lancet* **1**, 680 (1970).
[17] H. Deutsch, *Methods Immunol. Immunochem.* **1**, 315 (1967).

mM Tris, pH 7.4, and reprecipitated. The immunoglobulin is resuspended in the Tris buffer, dialyzed against 25 mM Tris, 0.15 M NaCl, and 0.02% NaN_3, pH 7.4, and then stored at $-80°$. We then purify the immunoglobulin fraction by affinity column chromatography using *Staphylococcus aureus* protein A–Sepharose CL-4B (Pharmacia). The immunoglobulins (adjusted to pH 8.0) are loaded on the column, incubated for 60 min, and then the serum proteins and the IgM and IgA immunoglobulins are eluted with 0.14 M sodium phosphate, pH 8.0. The IgG_1 globulins are eluted with 0.14 M sodium phosphate at pH 6.5; IgG_{2a} with 0.2 M sodium citrate, pH 5.0, and IgG_{2b} with 0.2 M glycine–HCl, pH 3.0, as described by Ey *et al.*[18] The antibody fractions eluted from the column are dialyzed against several changes of 25 mM Tris–HCl, pH 7.4, and then antibody titer values and their effects on enzyme function determined.

Determination of Antibody Binding

The antibody binding to antigen is determined using a solid-surface adsorption method similar to that developed by Engvall.[19] The test antigen is adsorbed onto the plastic surface of flat-bottom microtiter plate wells (Cooke, flexible 96-well plates). The adsorbed protein is then exposed to antibody and the level of antibody–antigen complex formed is detected with a second antibody or an antibody-specific probe. We routinely treat the plates for 30 min with 100 μl/well of test protein (0.01 to 0.5 mg/ml) in 1 mM EGTA–Tris, pH 7.4. The protein solution is removed (for reuse) and the plate wells are washed three times with buffer A (5 mg/ml bovine serum albumin, 10 mM Tris, 0.15 M NaCl, and 0.05% NaN_3, pH 7.4). The plates are then incubated for 10 min with an excess of buffer A. The buffer is removed and 100-μl aliquots of antiserum, or immunoglobulin diluted in buffer A, are added to each well and incubated at room temperature for 90 min. When rabbit immunoglobulins are used, the wells are emptied, rinsed three times with buffer A, and then 100 μl/well of ^{125}I-labeled protein A (Pharmacia) in buffer A is added for 30 min. The ^{125}I-labeled protein A solution is removed and the plates are rinsed and dried. The wells are separated and the bound radioactivity is determined using a gamma counter. The protein A is iodinated using the chloramine-T procedure of Dorvall.[20] Alternatively, a β-galactosidase protein A conjugate (Amersham) diluted in buffer B is used (as described for the mouse immunoglobulins) to detect bound antibody.

When mouse immunoglobulins are used, the wells are rinsed three

---

[18] P. Ey, S. Prowse, and C. Jenkin, *Immunochemistry* **5**, 429 (1978).
[19] E. Engvall, *Scand. J. Immunol.* **8** (Suppl. 7), 25 (1978).
[20] G. Dorvall, K. Welsch, and H. Wigzell, *J. Immunol. Methods* **7**, 237 (1975).

times with buffer B (5 mg/ml bovine serum albumin, 10 mM sodium phosphate, 1.5 mM MgCl$_2$, 2 mM 2-mercaptoethanol, 0.05% Triton X-100, and 0.05% NaN$_3$, pH 7.2) and the immunoglobulins are detected using a β-galactosidase sheep anti-mouse IgG F(ab)$_2'$ conjugate (Bethesda Research Laboratories). The conjugate is diluted 1 : 200 in buffer B (50 μl/well) and allowed to bind for 90 min. The plates are then washed three times with buffer B and the bound second antibody (conjugate) is detected using 50 μl/well of a substrate solution consisting of 1 mg/ml o-nitrophenyl-β-galactoside, 1.5 mM MgCl$_2$, 100 mM 2-mercaptoethanol, and 50 mM sodium phosphate, pH 7.2. After a 10-min incubation period, the reaction is terminated by adding 50 μl of freshly prepared 0.5 M Na$_2$CO$_3$. Each sample is then transferred to a fresh microtiter plate well and the absorbancy of each sample is determined using an ELISA plate reader (Dynatech) at 404 nm.

When using these procedures, it is important to do a number of control experiments to check the specificity of antibody–protein binding. We routinely use human, bovine, turkey, and chicken egg albumins and Dulbecco's modified Eagle's (DME) medium containing 10% fetal calf serum as control proteins. In addition, antiovalbumin serum (Cappel Lab), mouse, or rabbit preimmune serum, and Sp2/O-Ag14 myeloma cell culture medium are used as immunoglobulin controls. The ELISA assay conditions must also be standardized. Different concentrations of the second antibody–conjugate should be tried to ensure an excess in all experiments. The concentration(s) of antigen used to coat the microtiter plates should be such that there is a direct relationship between the protein concentration and the level of antibody bound when an excess of antibody is used. In addition, the assay step should be terminated after a standard time period during which the rate of substrate hydrolysis is linearly proportional to the time of reaction. Finally, nonrelated proteins and nonspecific immunoglobulins should be used to establish the appropriate background controls and to test the specificity of antibody binding.

Comments

As discussed above, the nature of the antibodies raised to the Na$^+$,K$^+$-ATPase will, of course, vary with the source of the enzyme and its method of preparation. Our studies[21,22] and those of Dzhandzhugazyan *et al.*[23] with polyclonal antibodies as well as more recent work with mono-

---

[21] W. Ball and A. Schwartz, *Arch. Biochem. Biophys.* **217**, 110 (1982).
[22] W. Ball, J. Collins, L. Lane, and A. Schwartz, *Arch. Biochem. Biophys.* **221**, 371 (1983).
[23] K. Dzhandzhugazyan, N. Modyanov, and Y. Ouchinnikov, *Bioorg. Khim.* **7**, 1790 (1981).

clonal antibodies[24,25] suggest that the preponderance of antigenic determinants of the holoenzyme are retained by the isolated and denatured subunits (SDS free) of the enzyme. In working with antibodies raised to the isolated subunits, generally we find that α-directed antibodies cross-react with the holoenzyme, while the β-directed antibodies show little binding to the holoenzyme. In addition, although our studies with hybridomas suggest that the Na^+,K^+-ATPase is strongly antigenic, with 2–10% of the hybridomas initially screened testing positive for enzyme-directed antibodies, we have found that virtually all of the hybridomas produce only anti-α antibodies. From these results, it appears that the α subunit of the Na^+,K^+-ATPase is more amenable to immunochemical studies than is the β subunit.

Acknowledgment

This work was supported by Grants R01HL-32214 (WJB), R01HL-25545 (LKL), P01HL-22619, and T32HL-07382 from the National Institutes of Health. W.J.B. is an Established Investigator of the American Heart Association.

[24] W. Ball, A. Schwartz, and J. Lessard, *Biochim. Biophys. Acta* **719,** 413 (1982).
[25] D. Schenk and H. Leffert, *Proc. Natl. Acad. Sci. U.S.A.* **80,** 5231 (1983).

Section II

Assay of Na^+,K^+-ATPase Activity

[10] ATPase and Phosphatase Activity of Na^+,K^+-ATPase: Molar and Specific Activity, Protein Determination

By MIKAEL ESMANN

Introduction

The aim of this chapter is to describe the procedures involved in the characterization of an Na^+,K^+-ATPase preparation. The first part is concerned with measurement of the overall Na^+,K^+-ATPase activity and the partial reactions catalyzed by the enzyme. The second part deals with the problem of determining the protein concentration, which must be known accurately for calculation of specific activity. A third section is devoted to the molar activity and the purity of the preparation in terms of the protein components, the α and β peptides. The final section contains some comments on the problem of inactive or partially active enzyme molecules in the preparations.

The presentation in this chapter is simplified for the benefit of investigators new to the study of Na^+,K^+-ATPase.

Measurement of Enzyme Activity

Na^+,K^+-ATPase

Na^+,K^+-ATPase activity is defined as the ouabain-sensitive hydrolysis of ATP in the presence of Na^+, K^+, and Mg^{2+}. The following requirements are to be met in order to achieve an optimal Na^+,K^+-ATPase activity.

Cations. Both the ratio of Na^+ to K^+ and the total concentration of Na^+ plus K^+ are important parameters in achieving optimum Na^+,K^+-ATPase activity. The optimal Na^+ plus K^+ concentration is about 150 mM, and the Na^+/K^+ ratio giving optimal activity is about 6.5.[1] This ratio can vary between 5 and 7 without appreciably smaller activity.

About 1 mM free Mg^{2+} is required for optimum activity (see below). The pH optimum for Na^+,K^+-ATPase activity at 37° is about 7.2, but the pH may range between 6.9 and 7.4 in the test medium without appreciable loss of activity. Cations such as Tris and choline are antagonistic to K^+ and should be omitted where possible.

[1] J. C. Skou, *Biochim. Biophys. Acta* **567**, 421 (1979).

TABLE I
COMPOSITION OF Na$^+$,K$^+$-ATPase TEST MEDIUM^a

Reagent solution	Amount (μl)		Concentration in assay
	Assay	Blank	
NaCl (1300 mM)	100	100	130 mM
KCl (200 mM)	100	100	20 mM
MgCl$_2$ (40 mM)	100	100	4 mM
ATP^b (30 mM)	100	100	3 mM
Histidine (150 mM, pH 7.4)	200	200	30 mM
Ouabain (10 mM)	0	100	1 mM
H$_2$O	300	200	
Na$^+$,K$^+$-ATPase	100	100	5 μg/ml

^{a} One milliliter of test medium.
^{b} Tris salt. If Na-ATP is used the amount of NaCl added should be adjusted to give a final Na$^+$ concentration of 130 mM.

Nucleotides. The K_m for the Na$^+$,K$^+$-ATPase activity is about 400 μM[2] and millimolar concentrations of ATP are thus required to achieve maximum Na$^+$,K$^+$-ATPase activity. An ATP concentration of about 3 mM is used routinely, and an Mg^{2+} concentration of 4 mM is employed, giving about 1.2 mM free Mg^{2+}. One of the products of the Na$^+$,K$^+$-ATPase reaction, ADP, is inhibitory and should not exceed 10–15% of the ATP concentration. The ADP concentration can be kept low using the phosphoenolpyruvate/pyruvate kinase system (see [11], this volume).

Ouabain. Inclusion of 1 mM ouabain in the Na$^+$,K$^+$-ATPase reaction medium will inhibit the Na$^+$,K$^+$-ATPase completely in most tissues. Ouabain therefore serves as a convenient reagent blank as well as a blank for nonspecific ATPase activity.

Temperature. Most membrane-bound Na$^+$,K$^+$-ATPase preparations are stable for at least 10 min at 37° which—apart from being a physiologically relevant temperature to work at—also is optimum for enzyme function in most tissues. The activity is therefore conveniently expressed as the amount of phosphate produced at 37° in micromoles per minute.

Incubation. The incubation time at 37° should be adjusted so that no more than 10–15% of the ATP is hydrolyzed in order to avoid ADP inhibition. Details for measuring the Na$^+$,K$^+$-ATPase activity of a 50% pure preparation of shark rectal gland enzyme[3] are as follows: 900 μl of a test medium (except the enzyme, see Table I) is brought to 37°, and 100 μl

[2] A. H. Neufeld and H. M. Levy, *J. Biol. Chem.* **245**, 4962 (1970).
[3] J. C. Skou and M. Esmann, *Biochim. Biophys. Acta* **567**, 436 (1979).

of enzyme (50 μg/ml in 20 mM histidine and 25% glycerol) is added. After 5 min the reaction is terminated by addition of 100 μl 50% trichloroacetic acid (TCA) and the test tube is cooled to 4°. Nonspecific phosphate production is measured using the same test medium containing 1 mM ouabain (Table I). Samples can be kept at 4° for some time, but due to the spontaneous acid-catalyzed hydrolysis of ATP the inorganic phosphate should be measured within 2–5 hr to avoid an unnecessary high background.

Measurement of Liberated Phosphate

Two simple methods[4,5] for measuring the amount of inorganic phosphate will be described below.

Fiske–Subbarow Method

Reagents

Reagent A: 5% sodium dodecyl sulfate (SDS)
Reagent B: 0.6 g 2,4-diaminophenol dihydrochloride in 600 ml 1% sodium sulfite
Reagent C: 100 g ammonium heptamolybdate in 1000 ml 1 M sulfuric acid

Reagent A is stable indefinitely, reagent B must be prepared daily (lasts 2 days at 4°), and reagent C keeps for 1 month at 4°.

A sample of 1.1 ml containing 60–300 nmol inorganic phosphate (see above) is brought to room temperature. To this is added 1 ml reagent A (SDS in order to achieve solubilization of precipitated protein that might interfere with the colorimetric measurement). Thereafter 1 ml reagent B is added. The color reaction is initiated by addition of 1 ml reagent C. After 20 min the absorbance is read at 660 nm, and the phosphate content is calculated using an extinction coefficient of 860 per micromole (1-cm light path in the cuvette). An inorganic phosphate standard can be included, but the method is so reproducible that it can be omitted in routine use.

Baginski Method. The colorimetric method for detection of inorganic phosphate developed by Baginski *et al.*[5] is about eight times more sensitive than the Fiske–Subbarow method described above. In the following paragraph the details of the method will be described, from the point where the incubation at 37° with all the ligands present (see above) is to be terminated.

---

[4] C. H. Fiske and Y. Subbarow, *J. Biol. Chem.* **66,** 375 (1925).
[5] E. S. Baginski, P. P. Foa, and B. Zak, *Clin. Chim. Acta* **15,** 155 (1967).

Reagents

Reagent A: 0.75 g ascorbic acid is dissolved in 25 ml 0.5 N HCl. To this is added 1.25 ml of a 10% ammonium heptamolybdate solution with stirring

Reagent B: 20 g sodium-m-arsenite and 20 g trisodium citrate-2-hydrate are dissolved in 980 ml H_2O. To this is added 20 ml concentrated acetic acid

Reagent A must be prepared daily, whereas reagent B is stable for at least 1 year.

The enzymatic hydrolysis of ATP is stopped by addition of 0.5 ml of reagent A to 1 ml of test solution (see above) and the sample is cooled. After at least 6 min at 4° the sample is brought to 37° and 1.5 ml of reagent B is added. After 10 min at 37° the reaction is complete, and the sample can be brought to room temperature. The absorbance at 850 mm is read against an H_2O blank, and the amount of inorganic phosphate produced is estimated from a calibration curve of an inorganic phosphate standard in the range 10 to 100 nmol phosphate per sample.

Other methods for determining ATP hydrolysis such as the coupled assay and ^{32}P-labeling are dealt with in Chapters 11 and 12, this volume.

Inhibitory and Protective Agents. Almost anything can inhibit Na^+,K^+-ATPase. A common contaminant of solutions is Ca^{2+}, which can be chelated by the addition of 0.2 mM EGTA to the test solution (see above). The addition of defatted albumin (obtained from Behringwerke AG) to the test solution increases the Na^+,K^+-ATPase activity of the shark enzyme by about 30%, presumably due to binding of inhibitory free fatty acids to the albumin. The protective effect of albumin of the kidney enzyme is less pronounced, which could be due to the different lipid composition of the two membranes.

Glycerol inhibits the Na^+,K^+-ATPase activity, and the concentration of glycerol should therefore be kept smaller than 5% in the test solution. This holds also for sucrose and DMSO.

Na^+-ATPase. The Na^+-ATPase activity of Na^+,K^+-ATPase is defined as the ouabain-sensitive hydrolysis of ATP in the presence of Na^+ and Mg^{2+}, and in the absence of K^+.

Test solution is as above for the Na^+,K^+-ATPase, only it is 130 mM NaCl substituted with 150 mM NaCl and KCl omitted (see Table I). Since even low concentrations of KCl will activate the Na^+,K^+-ATPase activity it is essential that the KCl contamination of the solutions is kept below 20–25 μM.

K_m for ATP is less than 1 μM for the Na^+-ATPase reaction[6] so ATP

---

[6] L. Plesner and I. W. Plesner, *Biochm. Biophys. Acta* **643**, 449 (1981).

concentrations of 100–200 μM will give maximal activity. The Mg^{2+} concentration should be adjusted to give about 1 mM free Mg^{2+}.

The maximal specific Na^+-ATPase activity is about 2–5% of the maximal Na^+,K^+-ATPase activity. The above-mentioned Baginski method is conveniently used for phosphate measurement because it has a higher sensitivity than the Fiske–Subbarow method. Alternatively ATP hydrolysis can be followed using [γ-^{32}P] ATP.[7]

K^+-Dependent Phosphatase Activity (K^+-pNPPase)

Na^+,K^+-ATPase catalyzes ouabain-sensitive K^+-activated hydrolysis of phosphoric anhydrides such as acetyl phosphate and p-nitrophenyl phosphate (pNPP) in the absence of Na^+. Details of how to measure a maximal K^+-pNPPase activity are described below.

Cations. $K_{0.5}$ for activation of K^+-pNPPase by K^+ is in the millimolar range,[8] and a concentration of 150 mM K^+ is therefore saturating. The K^+-pNPPase requires about 20 mM Mg^{2+} for maximal activity. Na^+ inhibits K^+-pNPPase, and the Na^+ concentration should therefore be kept low in the test solution. Cations with an Na^+-like effect (e.g., Tris and choline) counteract the effect of K^+, and should therefore be omitted from the test solution (see Ref. 8 for a discussion of this).

Substrate. We routinely use p-nitrophenyl phosphate as a substrate, and a concentration of 10 mM is saturating. ATP or ADP will inhibit the K^+-pNPPase (but see below, Na^+,K^+-pNPPase), and should be omitted from the test solution.

Reaction Blind. Ouabain is not as effective an inhibitor of K^+-pNPPase as of Na^+,K^+-ATPase; therefore inclusion of ouabain in the test solution does not give a good measurement of nonspecific hydrolysis of pNPP. We routinely either use an Na^+ blank (i.e., substitution of 150 mM KCl for 150 mM NaCl) or simply do not add enzyme to the test solution.

Incubation. The incubation of enzyme at 37° is performed as above for the Na^+,K^+-ATPase activity, with a test solution containing (final concentrations) the following: 30 mM histidine (pH 7.4), 150 mM KCl, 20 mM MgCl$_2$, 10 mM pNPP (prepared daily in H$_2$O), 0.2 mM EGTA, 0.33 mg albumin/ml, and 5 μg enzyme/ml.

Measurement of p-Nitrophenol. At high pH one of the products of the K^+-pNPPase reaction, p-nitrophenol, is intensely yellow, and this can conveniently be used to determine colorimetrically the hydrolysis of pNPP:

To 1.1 ml of TCA-quenched reaction medium is added 2 ml Tris base (0.5 M), and the absorbance is read at 410 nm. The specific K^+-pNPPase

---

[7] J. C. Skou and C. Hilberg, *Biochim. Biophys. Acta* **185**, 198 (1969).
[8] J. C. Skou, *Biochim. Biophys. Acta* **339**, 258 (1974).

activity is calculated after subtraction of the reagent blank using an extinction coefficient of 5835 per micromole (1-cm light path in the cuvette).

Comments on K^+-pNPPase Activity. K^+-pNPPase activity is less susceptible to inactivation than Na^+,K^+-ATPase activity. Albumin has only a very small activating effect on K^+-pNPPase, and glycerol increases the specific K^+-pNPPase activity, in contrast to its effect on Na^+,K^+-ATPase activity.

The specific K^+-pNPPase activity is usually 5- to 7-fold lower than the Na^+,K^+-ATPase activity (see below).

$Na^+ + K^+$-Activated Phosphatase (Na^+,K^+-pNPPase)

In the presence of Na^+ and K^+ the hydrolysis of pNPP is increased by low concentrations of ATP (e.g., 1–20 μM) and inhibited by high concentrations of ATP. The reader is referred to the original observations[8,9] on this partial reaction for details.

Criteria Defining the Purity of an Na^+,K^+-ATPase Preparation

There are (at least) three experimental criteria defining the purity of a given Na^+,K^+-ATPase preparation.

These are (1) the specific activity, expressed in micromoles phosphate/milligrams protein per minute, (2) the molar activity, which is the activity per catalytic site (and is given in min^{-1}), and (3) the protein composition evaluated from scans of Coomassie blue-stained SDS–polyacrylamide gels.

The problems involved with each of these methods will be discussed below.

Specific Activity

Specific enzyme activity is a direct measure of the purity of an enzyme preparation with respect to the number of active enzyme molecules per milligram protein (Table II[3,10–16]). A "low" specific activity is therefore an

[9] H. Yoshida, K. Nagai, T. Ohashi, and Y. Nakagawa, *Biochim. Biophys. Acta* **171,** 178 (1969).
[10] P. L. Jørgensen, *Biochim. Biophys. Acta* **356,** 53 (1974).
[11] L. K. Lane, J. H. Copenhaver, G. E. Lindenmayer, and A. Schwartz, *J. Biol. Chem.* **248,** 7197 (1973).
[12] W. H. M. Peters, H. G. P. Swartz, J. J. H. H. M. De Pont, F. M. A. H. Schuurmans Stekhoven, and S. L. Bonting, *Nature (London)* **290,** 338 (1981).
[13] I. Klodos, P. Ottolenghi, and A. Boldyrev, *Anal. Biochem.* **67,** 397 (1975).
[14] J. R. Perrone, J. F. Hackney, J. F. Dixon, and L. E. Hokin, *J. Biol. Chem.* **250,** 4178 (1975).
[15] E. G. Moczydlowski and P. A. G. Fortes, *J. Biol. Chem.* **256,** 2346 (1981).
[16] A. Yoda, A. W. Clark, and S. Yoda, *Biochim. Biophys. Acta* **778,** 332 (1984).

TABLE II
SPECIFIC AND MOLAR Na^+,K^+-ATPase ACTIVITY AND LIGAND-BINDING CAPACITY OF Na^+,K^+-ATPase FROM VARIOUS PREPARATIONS

Tissue	Na^+,K^+-ATPase (units/mg)	Site concentration (nmol/mg)	Molar activity (min^{-1})	Protein purity (% $\alpha + \beta$)	Ref.
Renal medulla	37	4.0	9300	95–100	10
Renal medulla	25	3–4	6300–8300	95–100	11
Renal medulla	50	5.6	8900	95–100	12
Ox brain	3.33	0.36	9300	ND[a]	13
Rectal glands	25	4	6300	95–100	3
Rectal glands	25	2.5	10,000	60–70	14
Eel electroplax	34	5.7	6000	95–100	15
Eel electroplax	20	1.5	13,300	ND	16

[a] ND, Not determined.

indication that the preparation either is not fully active (contains inactive enzyme) or contains proteins other than the enzyme.

However, measurement of a specific activity (micromoles phosphate/milligrams protein per minute) relies heavily on the ability to measure a protein concentration accurately. The following paragraphs are devoted to this problem

Lowry Method and Quantitative Amino Acid Analysis. The most popular method is that of Lowry et al.,[17] which has been modified for membranous and detergent-solubilized proteins by Peterson.[18] Here only the basic problem of quantifying the results will be considered.

Albumin is usually the standard employed in the assay, and the amount of Na^+,K^+-ATPase protein is expressed in albumin units. The albumin concentration (mg protein/ml) is usually measured either by drying the albumin prior to weighing or by using an extinction coefficient of 0.67 at 280 nm for albumin in water (i.e., 1 mg albumin/ml will give a reading of 0.67 absorbance units).

There are, however, two problems. The first is whether the extinction coefficient (amount of blue color formed per milligram protein) is the same for albumin and the Na^+,K^+-ATPase preparation in hand. The second problem is how to measure the true protein concentration in the albumin standard to be used.

The only rigorous method that will solve the above-mentioned problems is quantitative amino acid analysis (QAAA), i.e., simply to count the number of individual amino acids per volume unit of an Na^+,K^+-ATPase

[17] O. H. Lowry, N. J. Rosebrough, A. L. Farr, and R. J. Randall, *J. Biol. Chem.* **193**, 265 (1951).
[18] G. L. Peterson, *Anal. Biochem.* **84**, 164 (1978).

solution using an amino acid analyzer and, for example, the procedures described by Moczydlowsky and Fortes.[15]

Once this is done for a given preparation of Na^+,K^+-ATPase, the sample on which the QAAA was performed can be used as a standard. A less expensive method in terms of Na^+,K^+-ATPase is to calibrate an albumin solution against the QAAA sample, and use the calibrated albumin solution as a standard.

Obviously the procedure for obtaining a "true" protein concentration is vital for comparison of data on specific activities and binding site concentrations with results from other laboratories.

Optical Density at 280 nm. A very quick and reproducible method for estimating the protein concentration is to measure the optical density at 280 nm of a protein sample, solubilized in 2% SDS. In our laboratory a 1 mg Na^+,K^+-ATPase/ml solution (calibrated with QAAA as above) in 2% SDS will have an absorbancy of about 1 U at 280 nm (after subtraction of a buffer and SDS blank). This method is of course very sensitive toward light scattering and ultraviolet light-absorbing contaminants, and requires more protein per sample than the Lowry method (a 0.1 mg protein/ml solution is sufficient to give a reliable reading at 280 nm whereas only 10–20 μg is required for the Lowry method). These drawbacks are compensated for by the speed and reproducibility of the measurement.

The absorption spectra in the ultraviolet region, arising from the tryptophan, tyrosine, and phenylalanine residues, for SDS-solubilized Na^+,K^+-ATPase as well as for the purified α and β subunits, are shown in Fig. 1 for reference.

Note that the 280-nm extinction coefficient must be obtained for each particular preparation, i.e., determined by QAAA. It is also useful to take the spectrum between 330 and 240 nm to check that the ratio between the absorbance at 280 and 255 nm is constant (about 2.1 for Na^+,K^+-ATPase in Fig. 1). A decrease in the ratio frequently arises from the presence of light-scattering substances and will give rise to an overestimation of the protein concentration.

Molar Activity

The molar activity of the Na^+,K^+-ATPase is independent of the method of protein determination and is therefore a convenient parameter for comparison of different preparations of enzyme. In contrast to the specific activity, the molar activity is (or should be) independent of the state of purity of the preparation. Molar activity is expressed in min^{-1}:

$$\frac{\text{micromoles phosphate liberated}/(\text{mg protein} \cdot \text{min})}{\text{micromoles binding site/mg protein}}$$

$$= \text{molar activity (min}^{-1})$$

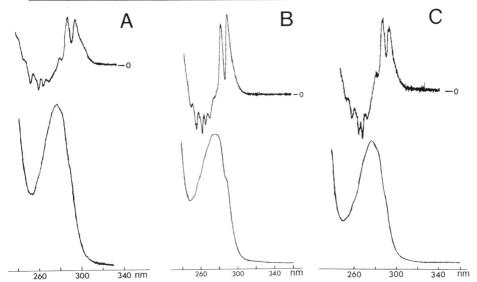

FIG. 1. Ultraviolet absorption spectra of an SDS-solubilized Na^+,K^+-ATPase preparation (A) and of isolated α (B) and β (C) subunits. The spectra were taken with a Varian-Cary model 219 spectrophotometer and the following extinction coefficients at 280 nm were obtained for a 1 mg protein/ml solution: 1.01 for Na^+,K^+-ATPase, 0.83 for the α subunit, and 1.15 for the β subunit. The first derivatives of the curves are also shown, which clearly reveal the difference between the α and β subunits in the 290-nm region: the shoulder at 290 nm, attributable to the tryptophan content, is more pronounced in the α subunit than in the β subunit (the dip at 290 nm in the first derivative spectrum is deeper).

where the site concentration (μmol/mg) is estimated from equilibrium binding of the ligands ATP, ADP, ouabain, or vanadate, or from the maximal phosphorylation level. Since the site concentration for each of these ligands usually is the same in a given preparation,[19] the molar activity can be given in terms of the site concentration of any of these ligands.

The molar activity gives an estimate of how much the enzyme has been "tampered" with during purification. A molar activity cannot be artificially high, whereas a low molar activity can reflect partial denaturation of the enzyme (loss of Na^+,K^+-ATPase activity but not binding sites) or the presence of closed vesicles, etc.

Table II lists the molar activity of a number of different enzyme preparations from various species and tissue. Although there is a considerable variation in specific activity and site concentrations, even for "pure" preparations, the molar activity is consistently about 9000–10,000 min^{-1} (37°).

[19] O. Hansen, J. Jensen, J. G. Nørby, and P. Ottolenghi, *Nature (London)* **280**, 410 (1979).

Protein Purity

The Na$^+$,K$^+$-ATPase consists of two proteins, the α subunit (M_r about 100,000) and the β subunit (M_r about 40,000) in equimolar amounts.

The protein purity of a given preparation can be estimated from the protein composition as revealed by SDS–polyacrylamide gel electrophoresis. Very pure and active preparations[10,11,14] contain only the α and β subunits on SDS gels (and maybe a small γ subunit with M_r about 10,000). The methods for electrophoresis and quantitation of bands on SDS gels have been dealt with earlier in this series.[20]

SDS gel scanning of an enzyme preparation can give an estimate of how impure the preparation is, i.e., set an "upper limit" for the specific activity. Obviously the preparation can be more pure judged from SDS gels than from the specific activity or binding site concentrations, which indicates that an inactivation of the enzyme during purification has occurred. It is therefore vital to correlate the specific activity and the protein purity during purification (see Fig. 2). The upper tracing in Fig. 2 represents a 60% pure (in terms of α and β content) membrane-bound enzyme with a specific Na$^+$,K$^+$-ATPase activity of about 25 U and the lower tracing a 95–100% pure enzyme with an activity of about 40 U.[21] Correct determination of the purity requires a well-established baseline in the gel scans, (indicated by the dotted line in Fig. 2[22,23]), otherwise serious overestimation (or, less frequent, underestimation) of the purity may occur.

General Comments

A "good" enzyme preparation will have a molar Na$^+$,K$^+$-ATPase activity of 9000–10,000 min^{-1}, a specific Na$^+$,K$^+$-ATPase activity of about 40 U/mg, a K$^+$-pNPPase activity of about 7 U/mg, and about 4–5 nmol binding sites/mg for ligands (ATP, etc.). SDS gels should reveal only two bands, the α and β subunits.

Closed vesicles in the preparation will lead to lower specific activities, and (maybe) to a situation where the binding site concentration for the ligands is unequal (i.e., an inside-out vesicle will bind ATP but not ouabain, etc.). Whether vesicles are present can be established from the effect of nonsolubilizing concentrations of mild detergents such as Lubrol or saponin.

Protein impurities are revealed by SDS gel electrophoresis, low specific activities, and binding site concentrations.

[20] W. L. Zahler, this series, Vol. 32. [7].
[21] M. Esmann, J. C. Skou, and C. Christiansen, *Biochim. Biophys. Acta* **567**, 410 (1979).
[22] G. Fairbanks, T. L. Steck, and D. H. Wallach, *Biochemistry* **10**, 2606 (1971).
[23] K. Weber and M. Osborn, *J. Biol. Chem.* **244**, 4406 (1969).

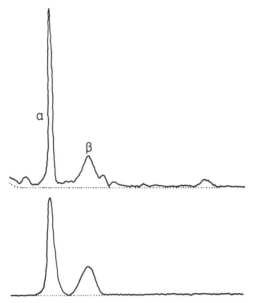

FIG. 2. A scan at 555 nm of two Coomassie blue-stained[22] SDS–polyacrylamide gels.[23] The upper tracing is for an about 60% pure enzyme preparation and the lower tracing is for a detergent-solubilized purified enzyme.[21]

Partially active or inactive enzyme present major problems with preparations that look "pure" on SDS gels. With partially active enzyme molecules the molar activity may be lower than 9000–10,000 min^{-1} (if, for example, the partially inactivated enzyme molecules can bind ATP but do not hydrolyze it at maximum speed). A simple test for partial inactivation is to determine the Na$^+$,K$^+$-ATPase/K$^+$-pNPPase ratio since in many cases the K$^+$-pNPPase is less susceptible to inactivation than the Na$^+$,K$^+$-ATPase. A ratio of less than 5–6 indicates partial inactivation. With fully inactive enzyme molecules present in the preparation the molar activity will be normal and SDS gels look "pure." The presence of inactive enzyme is then only determined by the specific activity, and relies therefore, as mentioned above, on the ability to produce a correct protein determination. Another problem is that the number of ligand-binding sites must be stoichiometric with the molecular weight of the enzyme and the number of α subunits, a point which indicates that many enzyme preparations contain substantial amounts of inactive enzyme (see, for example, Refs. 24 and 25 for a discussion of this).

[24] J. Jensen and P. Ottolenghi, *Biochim. Biophys. Acta* **731**, 282 (1983).
[25] M. Esmann and J. C. Skou, *Biochim. Biophys. Acta* **787**, 71 (1984).

[11] Coupled Assay of Na^+,K^+-ATPase Activity

By JENS G. NØRBY

The mathematical models and analysis of coupled enzyme systems have been treated in a comprehensive paper by Rudolph et al.[1] Their study includes the more complicated assays using two auxiliary enzymes like the one to be described here, and also contains valuable references to earlier work. A detailed study of the coupled ATPase assay has been described previously.[2] The coupled ATPase assay has been used for determination of mitochondrial,[3,4] Ca^{2+},[5–7] and Na^+,K^+-ATPase.[8–10]

Principle

Three enzymatic reactions are operative in the assay:

$$ATP + H_2O \xrightleftharpoons{Na^+,K^+\text{-ATPase}} P_i + ATP \qquad (1)$$

$$ADP + PEP \xrightleftharpoons{(PK)} ATP + Py \qquad (2)$$

$$Py + NADH + H^+ \xrightleftharpoons{(LDH)} La + NAD^+ \qquad (3)$$

P_i, inorganic phosphate; PEP, phosphoenolpyruvate; PK, pyruvate kinase; Py, pyruvate; LDH, lactate dehydrogenase; La, lactate. Reaction (1) requires Mg^{2+} (and $Na^+ + K^+$); reaction (2) requires Mg^{2+} and K^+ (or NH_4^+). During the assay, which is started by the addition of Na^+,K^+-ATPase, all the reactions proceed to the right. The rate of reaction (3) is determined as dA_{340}/dt under conditions where this rate becomes constant and equal to the rates of reactions (2) and (1) within 30 sec after the initiation of the assay.[2] The steady-state value of dA_{340}/dt measures the uninhibited rate of reaction (1) since (a) the steady state concentration of

---

[1] F. B. Rudolph, B. W. Bangher, and R. S. Beissner, this series, Vol. 63, p. 22.
[2] J. G. Nørby, *Acta Chem. Scand.* **25**, 2717 (1971).
[3] M. E. Pullman, H. S. Penefsky, A. Datta, and E. Racker, *J. Biol. Chem.* **235**, 3322 (1960).
[4] B. C. Monk and G. M. Kellerman, *Anal. Biochem.* **73**, 187 (1976).
[5] G. B. Warren, P.A. Toon, N. J. M. Birdsall, A. G. Lee, and J. C. Metcalfe, *Proc. Natl. Acad. Sci. U.S.A.* **71**, 622 (1974).
[6] D. W. Martin, *Biochemistry* **22**, 2276 (1983).
[7] D. W. Martin, C. Tanford, and J. A. Reynolds, *Proc. Natl. Acad. Sci. U.S.A.* **81**, 6623 (1984).
[8] R. W. Albers, G. J. Koval, and G. J. Siegel, *Mol. Pharmacol.* **4**, 324 (1968).
[9] W. Schoner, C. von Ilberg, R. Kramer, and W. Seubert, *Eur. J. Biochem.* **1**, 334 (1967).
[10] J. G. Nørby and J. Jensen, *Biochim. Biophys. Acta* **233**, 104 (1971).

ADP is always much lower than the inhibitory concentrations and (b) the auxiliary enzymes and their reactants are without influence on the Na^+,K^+-ATPase reaction.[2] The steady state rate is proportional to the Na^+,K^+-ATPase concentration within wide limits (see Comments), and the rate remains constant during the assay almost to the point of exhaustion of NADH ($+ H^+$), provided, of course, that the enzyme hydrolyzing ATP is stable under the assay conditions.

Reagents

Pyruvate kinase (PK), Boehringer No. 109045, 10 mg/ml solution in 50% glycerol, pH ~6, specific activity ~300 U/mg (37°)

Lactate dehydrogenase (LDH), Boehringer No. 127221, 25 mg/2.5 ml solution in 50% glycerol, pH ~7, specific activity ~850 U/mg (37°)

Phosphoenolpyruvate (PEP), Boehringer 128112, tricyclohexylammonium salt (M_r 465.3). Prepare fresh solution, 15 mg/ml H_2O = ~30 mM, each day

NADH, Boehringer 128023, disodium salt (M_r 709.4). Prepare fresh solution, 10.5 mg/ml H_2O = ~12 mM, each day

ATP, Boehringer 127531, disodium salt $3H_2O$ (M_r 605.2) converted to Tris salt on a Dowex 1-X2, Cl^- column (BioRad). Eluted with Tris–HCl and HCl and buffered with Tris base. A 30 mM ATP solution, pH 7.4 (37°), contains about 150 mM Cl^-, 270 mM $Tris^+$, and 70 mM Tris base

Ouabain (*g*-strophanthin), 10 mM in H_2O

Histidine, 180 mM in H_2O, pH ~7.5 (37°)

$MgCl_2$, 30 mM in H_2O

KCl, 400 mM in H_2O

NaCl, 2600 mM in H_2O

Procedure

Aliquots of the stock solutions (except Na^+,K^+-ATPase) are pipetted into ordinary glass cuvettes, 1-cm light path, as shown in Table I. The mixture, and the thermostatted cuvette compartment of a recording spectrophotometer, are heated to ensure 37° in the cuvette during the assay. The reference cuvette contains distilled H_2O. Automatic recording at 340 nm is then initiated and the Na^+,K^+-ATPase suspension is added and mixed into the assay solution by a few seconds' stirring with a plastic rod. With the assay conditions as in Table I, a recorder with paper width of 20 cm corresponding to 1 absorbance unit and running with a speed of 3 cm/min is convenient.

TABLE I
COMPOSITION OF THE COUPLED Na$^+$,K$^+$-ATPase ASSAY[a]

Reagent solution[b]	Amount (μl) Assay	Amount (μl) Blank	Approximate concentration in assay	
NaCl	150	150	130	mM
KCl	150	150	20	mM
MgCl$_2$	400	400	4	mM
ATP	300	300	3	mM
PEP	100	100	1	mM
NADH	50	50	0.2	mM
PK	10	10	10	U/ml
LDH	10	10	30	U/ml
Histidine	500	500	30	mM
Ouabain	0	250	1	mM (blank)
H$_2$O	1250	1000		
Na$^+$,K$^+$-ATPase	100	100	0.03	U/ml
Total volume, pH 7.30 at 37°	3020	3020		

[a] One unit of enzyme activity, U, is the amount of enzyme producing 1 μmol of product/min.
[b] See text for composition of stock solutions.

Calculation of Na$^+$, K$^+$-ATPase Activity. The difference in extinction coefficient at 340 nm between NADH(+ H$^+$) and NAD$^+$, ε(NADH) − ε(NAD), is 6.22 × 10^3 cm^{-1} · M^{-1}. The decrease in absorbance at 340 nm/min, ΔA, is converted to Na$^+$,K$^+$-ATPase activity by the expression:

$$\frac{\Delta A \times (\text{total volume of assay, in ml})}{6.22 \times 10^3 \times (\text{volume of Na}^+,\text{K}^+\text{-ATPase solution added, in ml})} \times 10^6$$

= Na$^+$,K$^+$-ATPase activity in U/liter Na$^+$,K$^+$-ATPase suspension

The ouabain-sensitive activity is obtained as the difference between U/liter in the absence and presence of ouabain (see Table I). If the specific activity is desired the above number must be divided by the concentration of protein (mg/liter) to give U/mg protein.

Comments

It should be noted that the ATP in the assay usually will be contaminated with ADP (e.g., about 1–2%), and that some of the NADH (+ H$^+$) therefore will be oxidized before the Na$^+$,K$^+$-ATPase is added. This will, under the conditions described, bring A_{340} down to about 0.8–0.9 from the theoretical value from the 0.2 mM NADH (+ H$^+$), which is 1.25. Other ATP preparations than the mentioned Tris–ATP may be used. If Na$_2$H$_2$ATP is used, Tris base or NaOH must be added to neutralize about 2H$^+$/mol and it may be necessary to correct the Na$^+$ concentration of the NaCl stock.

With about 0.03 U Na$^+$,K$^+$-ATPase/ml (Table I) the NADH (+ H$^+$) is used up in 4–5 min. Together with the fact that there is an initial, transient phase of about 30 sec, this shows that the practical upper limit of Na$^+$,K$^+$-ATPase concentration is about 0.1 U/ml, or 0.3 U total. The concentration can, of course, be lowered considerably provided that the enzyme preparation is stable under the assay conditions.

Other methods for determination of Na$^+$,K$^+$-ATPase activity are described by Esmann in this volume.[11]

[11] M. Esmann, this volume [10].

[12] Identification and Quantitation of Na$^+$,K$^+$-ATPase by Back-Door Phosphorylation

By MARILYN D. RESH

Active transport of sodium and potassium ions across the plasma membrane of most eukaryotic cells is mediated by the Na$^+$ and K$^+$-activated adenosinetriphosphatase (Na$^+$,K$^+$-ATPase or Na-pump). Under physiological conditions, this membrane-bound enzyme couples the influx of two K$^+$ ions and efflux of three Na$^+$ ions to the hydrolysis of one molecule of ATP.[1] The Na-pump consists of a catalytic subunit of 100,000 Da, denoted α, and a 50,000- to 60,000-Da glycoprotein subunit, denoted β. During the reaction sequence, the 100,000-Da catalytic subunit becomes phosphorylated. When the reaction conditions favor the forward direction of the enzymatic mechanism, i.e., Na$^+$ + ATP + Mg^{2+}, the phosphorylated intermediate is formed by incorporation of radiolabel from [γ-^{32}P]ATP into the α subunit.[2] However, in crude membrane preparations, several other ATPases become phosphorylated as well, and it is often difficult to identify and quantitate the Na-pump α polypeptide. Alternatively, in the presence of ouabain and Mg^{2+}, the reaction sequence

[1] L. C. Cantley, *Curr. Top. Bioenerg.* **11**, 201 (1981).
[2] R. L. Post, G. Toda, and F. N. Rogers, *J. Biol. Chem.* **250**, 691 (1975).

can be forced to operate in reverse. Phosphorylation of the α subunit then occurs from inorganic phosphate[3] through the "back door."

Recently, the presence of another molecular form of the catalytic subunit, designated $\alpha(+)$, has been described in brain, muscle, and fat tissues.[4-6] The $\alpha(+)$ form exhibits a slightly larger apparent molecular weight on polyacrylamide gels and a higher affinity for cardiac glycoside inhibitors than the α subunit. Moreover, it appears that $\alpha(+)$ is the hormonally sensitive form of the enzyme.[6,7] Based on this information, the investigator who seeks to study Na^+,K^+-ATPase in a given cell or tissue type should also determine whether one or both forms of the α subunit are present. In this chapter, I first briefly describe three methodological approaches which can be utilized to identify α and $\alpha(+)$ forms of Na^+,K^+-ATPase. The use of any of these approaches (ion transport, antibodies, front door phosphorylation) can complement the identification of the enzyme by the back door method, described in detail herein. However, for accurate quantitation of the number of active Na-pumps, the use of backdoor phosphorylation is recommended.

Transport Assays

One of the more straightforward methods for determining whether multiple species of Na-pumps are present in a particular cell type is to measure ion transport activity as a function of ouabain concentration. These methods have been described in detail elsewhere.[6-8] Those cells which contain α and $\alpha(+)$ forms of the enzyme exhibit biphasic ouabain inhibition curves. The relative affinities of the two forms for ouabain can be estimated from nonlinear least-squares analyses. In addition, transport assays enable one to determine the ion-pumping rate in intact cells, which is important for calculating pumping efficiency *in vivo* (see below).

Immunological Methods

Several laboratories have described antibodies which recognize either α and/or $\alpha(+)$ forms of Na^+,K^+-ATPase.[6,9] Cellular proteins are sepa-

[3] F. M. A. H. Schuurmans Stekhoven, H. G. P. Swarts, J. J. H. H. M. DePont, and S. L. Bonting, *Biochim. Biophys. Acta* **597**, 100 (1980).
[4] K. J. Sweadner, *J. Biol. Chem.* **254**, 6060 (1979).
[5] T. Matsuda, H. Iwata, and J. R. Cooper, *J. Biol. Chem.* **259**, 3858 (1984).
[6] J. Lytton, J. C. Lin, and G. Guidotti, *J. Biol. Chem.* **260**, 1177 (1985).
[7] M. D. Resh, R. A. Nemenoff, and G. Guidotti, *J. Biol. Chem.* **255**, 10938 (1980).
[8] M. D. Resh, *J. Biol. Chem.* **257**, 6978 (1982).
[9] D. M. Fambrough and E. K. Bayne, *J. Biol. Chem.* **258**, 3926 (1983).

rated by SDS–polyacrylamide gel electrophoresis, electrophoretically transferred to nitrocellulose, and probed with anti-Na$^+$,K$^+$-ATPase antibody followed by peroxidase-conjugated second antibody. In brain, muscle, and fat tissue, two separate polypeptide bands corresponding to α and $\alpha(+)$ are visible.[6]

Front-Door Phosphorylation from ATP

In the presence of Na$^+$ and Mg^{2+}, a covalent phosphorylated intermediate of the Na$^+$,K$^+$-ATPase can be formed from [γ-^{32}P]ATP.[2] The addition of ouabain inhibits this reaction. When the phosphorylation reactions are performed in the presence of varying concentrations of ouabain, the two molecular forms, α and $\alpha(+)$, of the Na-pump are evident on polyacrylamide gels.[4–6] Quantitation of the relative ^{32}P incorporation into each of the two radiolabeled bands as a function of ouabain concentration will yield the ouabain affinities of the α and $\alpha(+)$ species.

Back-Door Phosphorylation

In the presence of Mg^{2+} and ouabain, an alkali-labile covalent phosphorylated intermediate of the Na$^+$,K$^+$-ATPase is formed from inorganic [^{32}P]phosphate, which is chemically identical to that formed by phosphorylation with ATP.[2,3] Formation of a phosphorylated intermediate from phosphate in the presence of ouabain, rather than from ATP, essentially eliminates the contribution from other ATPases.[10] Since the enzyme must turn over at least partially in order to incorporate phosphate, only active Na-pumps are detected. Moreover, based on the stoichiometry of 1 mol P$_i$ incorporated/mol enzyme, and the Michaelis–Menten dependence of phosphate incorporation on phosphate incorporation, measurement of ^{32}P incorporation at several phosphate concentrations allows extrapolation of the total number of Na-pumps in the membrane. This method can be applied to quantitate the number of Na-pumps in cell types in which Na$^+$,K$^+$-ATPase comprises only a minor fraction of the total membrane ATPase species.[8,10,11]

Assay Method[10]

Membrane suspensions are adjusted to a concentration of 50–100 μg protein in 100 μl buffer (100 mM HEPES, pH 7.4, with Tris base/5 mM

[10] M. D. Resh, *J. Biol. Chem.* **257**, 11946 (1982).
[11] J. F. Ash, R. M. Fineman, T. Kalka, M. Morgan, and B. Wire, *J. Cell Biol.* **99**, 971 (1984).

MgCl$_2$). Aliquots of 100 μl are incubated with concentrations of phosphate ranging from 10 to 100 μM H$_3$PO$_4$ in the presence or absence of ouabain, for 30 min to 1 hr at room temperature. The radioactive phosphate is prepared by diluting ortho[^{32}P]phosphate with buffer to 1 mCi/ml followed by filtration through a Millipore Millex-GS 0.22-μm filter unit attached to a 1-ml hypodermic syringe. This step is necessary to remove phosphate polymers which contaminate the ^{32}P and often give increased background levels. The phosphorylation reaction is then initiated by the addition of 10–20 μl [^{32}P]phosphate (10–20 μCi) to the membrane suspensions with frequent vortexing. The incubation is continued until steady state incorporation of radiolabel is achieved; for the rat adipocyte Na-pump, this occurs after 10 min at room temperature.[10] The reaction is then quenched by the successive addition of 50 μg bovine serum albumin as carrier, and 1.0 ml ice-cold 5% trichloroacetic acid/0.1 M H$_3$PO$_4$ (TCA/P$_i$), and the tubes are placed on ice for 5 min. The tubes are spun for 2 min in a microfuge (10,000 g) and the radioactive supernatant is removed. The pellets are washed and spun three additional times with 1 ml ice-cold TCA/P$_i$. The phosphorylated reaction products in the final pellet can be analyzed for ^{32}P incorporation by liquid scintillation counting or by gel electrophoresis. The entire microfuge tube can be counted by placing it inside an empty 20-ml scintillation vial and counting low-energy Cerenkov emission. If determination of the absolute counts per minute incorporated is desired, the final pellet can be dispersed by adding 300 μl 5% sodium dodecyl sulfate and sonicating for a few seconds in a bath sonicator. Approximately 4 ml of liquid scintillation fluid (Aquasol, New England Nuclear) is added, and ^{32}P cpm are measured.

Data Analysis

The total number of Na-pumps in the membrane sample is calculated from the back-door phosphorylation data. At each phosphate concentration (10–100 μM), the amount of ouabain-dependent ^{32}P incorporation is determined, i.e., cpm with ouabain minus cpm without ouabain. A Lineweaver–Burk double-reciprocal plot (1/mol phosphate incorporated vs 1/phosphate concentration) is then constructed. One should obtain a straight line by least-squares linear regression analysis. The reciprocal of the y-axis intercept (infinite phosphate concentration) represents the maximal phosphorylation capacity of the membranes, and thus the number of Na-pumps per milligram membrane protein can be calculated. The reciprocal of the x-axis intercept represents the K_m for phosphate, which is usually 20–30 μM.[6,10]

Several control experiments should be performed to ensure that the activity of the Na^+,K^+-ATPase itself is being measured. First, phosphorylation in the absence of ouabain should be less than 10% of the level attained in the presence of ouabain. Second, addition of 100 mM NaCl to the reaction should inhibit greater than 80% of the ouabain-dependent phosphorylation. Third, addition of 2 mM ATP for 5 min after the initial phosphate incubation should release at least 80% of the radioactivity from the reaction products. Finally, incorporation of radiolabel into the catalytic subunit should be visualized by acidic pH polyacrylamide gel electrophoresis followed by autoradiography (see below).

The back-door phosphorylation method can be used to distinguish between and quantitate the α and $\alpha(+)$ forms of the Na-pump as described by Lytton et al.[6] Two different concentrations of ouabain should be chosen for the preincubation. The first is a low ouabain concentration, which will saturate almost all of the $\alpha(+)$ sites and very few of the α sites. This concentration is determined from biphasic ouabain inhibition curves of either ion pumping in intact cells or ATP-dependent "front-door" phosphorylation (described above). For example, in the rat adipocyte plasma membrane, Lytton et al.[6] determined that at $3-5 \times 10^{-6}$ M ouabain, the $\alpha(+)$ form is 90% saturated with ouabain, while α is less than 10% occupied. The second ouabain concentration is a high concentration, usually 1×10^{-3} M, where both $\alpha(+)$ and α are saturated with ligand. Extrapolation of the data obtained with the lower ouabain concentration will yield the maximal phosphorylation capacity of the $\alpha(+)$ form, while the same calculation at high ouabain will yield the total number of Na-pumps [α and $\alpha(+)$], and the differences between these two numbers will represent the number of pumps in the α form.

Once one has determined the total number of Na-pump sites several additional calculations can be made. The turnover number of the enzyme for ATP is calculated by dividing Na^+,K^+-ATPase activity by Na-pump sites. This number is usually on the order of 10,000/site/min at 37°.[1] If the amount of membrane protein per cell can be estimated, then the number of Na-pumps per cell is expressed by multiplying Na-pump sites/mg membrane protein by milligrams membrane protein per cell. Based on the rate of K^+ uptake in cells, and assuming $2K^+$ transported per ATP hydrolyzed, the turnover number for ATP can be calculated for the Na-pump in the intact cell. Comparison of this number in cells with the value obtained for membranes will yield the efficiency of ion pumping under physiological conditions. It is interesting to note that the Na-pump in several different cell types operates at only a fraction of its maximal activity.[6,10,12]

---

[12] T. Clausen and O. Hansen, *Biochim. Biophys. Acta* **345**, 387 (1974).

Acidic pH Polyacrylamide Gel Electrophoresis

Since the phosphorylated intermediate of the Na^+,K^+-ATPase is alkali labile, acidic pH gel electrophoresis is employed to resolve the phosphorylated protein species. The gel system is based on the method of Amory et al.,[13] modified as I have previously described[8] for the use of the cationic detergent 1-hexadecylpyridinium chloride (HPCl). The recipe for this gel system has been published.[14]

The final pellet from the back-door phosphorylation reaction is rapidly rinsed with 0.3 ml of 0.15 M KH_2PO_4, and is resuspended in sample buffer: 0.25 M sucrose, 35 mM HPCl, 100 mM KH_2PO_4, pH 4.0, 128 mM 2-mercaptoethanol, 10 µg/ml pyronin Y. The gel is connected with the negative electrode at the bottom (since the detergent is positively charged), and is run at 25–30 mA for 3–4 hr. The pyronin Y dye front often separates into two colors and runs behind the solvent front. Following electrophoresis, the gel is soaked in destain solution (10% acetic acid, 10% 2-propanol) to rinse away free ^{32}P, dried, and exposed to X-ray film. Incorporation of radiolabel into a polypeptide of 90,000–100,000 Da should be evident, and this reaction should be ouabain dependent and Na^+ and ATP sensitive. A significant amount of phosphate hydrolysis still occurs during electrophoresis, even though the gel is run at acid pH. Thus, the amount of ^{32}P recovered in the catalytic subunit may be less than 10% of the maximal phosphorylation capacity. Moreover, the resolution of polypeptide bands of similar molecular weight in this gel system is far below that obtained by the method of Laemmli,[15] often obscuring the difference between α and $\alpha(+)$ forms under these conditions.

[13] A. Amory, F. Foury, and A. Goffeau, J. Biol. Chem. **255**, 9353 (1980).
[14] P. J. Blackshear, this series, Vol. 104, p. 237.
[15] U. K. Laemmli, Nature (London) **227**, 680 (1970).

Section III

Reconstitution of Na,K-Pump Activity

[13] Reconstitution of Na,K-Pump Activity by Cholate Dialysis: Sidedness and Stoichiometry

By STANLEY M. GOLDIN, MICHAEL FORGAC, and GILBERT CHIN

Introduction

Racker and co-workers originated the cholate dialysis technique for functional reconstitution of membrane proteins, and applied it to the study of mitochondrial oxidative phosphorylation[1] and the Ca^{2+}-translocating ATPase of sarcoplasmic reticulum.[2] Cholate dialysis was also used for the first successful reconstitution of the Na,K-pump,[3] and a number of investigators have used this method to study the properties of the Na,K-pump.[4-8] Electron microscopy[9] has directly confirmed biochemical and biophysical evidence[6] that cholate dialysis results in the incorporation of individual molecules of the Na^+,K^+-ATPase into unilamellar phospholipid vesicles, 400–600 Å in diameter. The reconstitution process appears to result—to a first approximation—in the random distribution of the reconstituted Na^+,K^+-ATPase among these vesicles, and under certain conditions equal amounts of the transport enzyme are oriented inside-out versus right-side-out with respect to its *in vivo* orientation in the cell membrane.[6,10]

Transport activity of the inside-out fraction of the reconstituted Na,K-pump is most directly monitored by radioisotopic methods. *In vivo*, the transport enzyme hydrolyzes ATP at the inner surface of the cell membrane, and employs this energy to pump Na^+ out of the cell and K^+ into the cell[11,12]; the pump is inhibited by cardiac glycosides that bind to the enzyme at a site located on the extracellular surface. What is observed in the reconstituted system[3-10] is the activation, by externally added ATP, of

[1] Y. Kagawa and E. Racker, *J. Biol. Chem.* **246**, 5477 (1971).
[2] E. Racker, *J. Biol. Chem.* **249**, 8198 (1972).
[3] S. M. Goldin and S. W. Tong, *J. Biol. Chem.* **249**, 5907 (1974).
[4] S. Hilden and L. E. Hokin, *J. Biol. Chem.* **250**, 6296 (1975).
[5] K. J. Sweadner and S. M. Goldin, *J. Biol. Chem.* **250**, 4022 (1975).
[6] S. M. Goldin, *J. Biol. Chem.* **252**, 5630 (1977).
[7] B. M. Anner, L. K. Lane, A. Schwartz, and B. J. R. Pitts, *Biochim. Biophys. Acta* **467**, 340 (1977).
[8] M. Forgac and G. Chin, *J. Biol. Chem.* **256**, 3645 (1981).
[9] E. Skriver, A. B. Maunsbach, and P. L. Jørgensen, *J. Cell Biol.* **86**, 746 (1980).
[10] G. Chin and M. Forgac, *J. Biol. Chem.* **259**, 5255 (1984).
[11] P. J. Garrahan and I. M. Glynn, *J. Physiol. (London)* **192**, 217 (1967).
[12] R. C. Thomas, *J. Physiol. (London)* **201**, 495 (1969).

the pumping of ^{22}Na$^+$ into the vesicles, and of ^{42}K$^+$ (or that of ^{86}Rb$^+$ [5] or ^{137}Cs$^+$ [10], which substitute for K$^+$) out of the vesicles. Consistent with its inside-out orientation, the transport activity is inhibited by cardiac glycosides such as ouabain only when they gain access to the vesicle interior. For the efficient reconstitution procedure[6,8] presented here, the stoichiometry of Na$^+$/K$^+$ pumped/ATP hydrolyzed for the reconstituted enzyme is $\sim 3:2:1$, as observed *in vivo* in red cell[11] and nerve.[12]

Cholate Dialysis Reconstitution of the Na,K-Pump

Preparation of Egg Phosphatidylcholine

The lipid preparation employed is critical for quantitative and reproducible reconstitution. Highly oxidized lipid raises the cation permeability of lipid vesicles, and adversely affects the reproducibility of the reconstitution procedure. For that reason, the antioxidant 2-mercaptoethanol is used in the solvents employed for preparation and storage of lipid, and care is taken to prevent lipid oxidation by minimizing exposure of phospholipid to air, especially at temperatures above 0°. Many commercially available preparations of phosphatidylcholine are unsatisfactory for reconstitution of the Na,K-pump. However, Avanti Biochemical Co. (Birmingham, AL) produces purified phosphatidylcholine of satisfactory quality; we suggest that you specify that it be shipped on dry ice, as an acetone precipitate under nitrogen or argon.

Phosphatidylcholine from egg yolk can be purified by silicic acid chromatography, essentially by the method of Litman.[13] We have also observed that a cruder preparation of egg phosphatidylcholine—namely, the product of five successive acetone precipitations by the technique described below—can be used,[8] and results in reconstitution of the Na,K-pump exhibiting properties very similar to those obtained[6] with the chromatographically purified lipid. The major contaminants (10–15% of total lipid, based on TLC analysis), are phosphatidylethanolamine and, to a lesser extent, cholesterol.

Extraction and Acetone Precipitation. Egg yolk (150 ml) is obtained from extremely fresh jumbo eggs purchased from a local supermarket on the date of delivery. The yolk is placed in a Waring blender and to this is added 500 ml of chloroform/methanol (2/1, v/v), containing 1 mM 2-mercaptoethanol. The top of the blender is lined with aluminum foil, and the mixture, at room temperature, is blended for 10 sec at high speed. Add 38 ml of distilled water, and blend for another 10 sec. Pour this into a 1-liter beaker, and cover with aluminum foil. Let the mixture settle for ~ 5

[13] B. J. Litman, *Biochemistry* **12**, 2545 (1973).

min. Decant about two-thirds of the upper (aqueous) phase. Filter the remainder, under an aspirator-generated vacuum, through #1 Whatman filter paper on a Büchner funnel, ~20-cm diameter; the filtration should take no more than several minutes—keep the Büchner funnel covered with aluminum foil to minimize evaporation.

Pour the filtrate into a 1000-ml separatory funnel. Let the two phases separate for 30 min. Remove the lower phase, and dry it in a 1000-ml round-bottom flask, in a rotary evaporator under aspirator-driven vacuum. Start the evaporation with the bath water at 0°, and slowly increase the thermostatted bath temperature to 30° (do not exceed). By controlling the temperature during the drying process, one can avoid excessive bubbling and foaming. When the volume of this extract has been reduced to ~40 ml, cool the water bath with ice before breaking the vacuum.

Accurately measure the volume of the concentrate, and transfer it to a 500-ml glass-stoppered bottle. Add 10 vol of acetone, containing 5 mM 2-mercaptoethanol. This addition is made over a 30-sec time period, with periodic swirling of the bottle to mix it. Cap the bottle under nitrogen, and immerse it in ice water, tilted ~35° from vertical. After 3 hr in ice water, remove the supernatant by aspiration. To the precipitate, add 30 ml of diethyl ether, 5 mM 2-mercaptoethanol. After loosely capping under nitrogen, gently warm this suspension to 30°. In 5–10 min, with occasional swirling, the precipitate should dissolve; add another 5–10 ml of ether if the solution is still turbid. Add an additional 10 vol of acetone/5 mM 2-mercaptoethanol and chill to 0° as above. After 45 min, remove the supernatant.

Repeat the above ether resuspension/acetone precipitation step three more times. During successive acetone precipitation steps, the color of the precipitate will change from yellow to milky white. The final precipitate is dissolved in 80 ml of benzene and vacuum evaporated. In preparation for silicic acid chromatography, this material is suspended and flash evaporated three to five times in 100 ml of 4:1 benzene/ethanol, to remove residual water. The final dry solid is dissolved in 75 ml of chloroform containing 2 mM 2-mercaptoethanol.

Silicic Acid Chromatography. This step employs Unisil (200–325 mesh, Clarkson Chemical Co., Williamsport, PA). Unisil is treated by the Hirsch and Ahrens solvent activation procedure,[14] as follows: place 300–350 g of Unisil in a large Büchner funnel, and wash under aspirator vacuum in the following:

(1) Diethyl ether—1 liter
(2) Acetone : ether (1 : 1, v/v)—1 liter

---

[14] J. Hirsch and E. H. Ahrens, Jr., *J. Biol. Chem.* **233**, 311 (1958).

(3) Diethyl ether—0.5 liter
(4) Chloroform—1 liter

Break the vacuum before the Unisil dries out. Immediately transfer the Unisil to a beaker containing chloroform.

Add sand and a small amount of chloroform to the bottom of a 29 × 3.5 cm glass column (the support should be a glass fritted disk). Pour the Unisil in chloroform onto the column, and wash it with 1 liter of chloroform containing 2 mM 2-mercaptoethanol. Place the 75 ml of chloroform/lipid solution on the column, and wash it into the column with 10 ml of chloroform. Elute the column stepwise at a 2.4-m head with successively increasing concentrations of methanol in chloroform (see tabulation below). Collect ~115-ml fractions in Erlenmayer flasks; the flasks are capped with aluminum foil and kept on dry ice during elution. The lipid composition of each fraction is visualized by silica gel thin-layer chromatography in chloroform/methanol/water/acetic acid, 65 : 43 : 3 : 1 by volume.[15] Fractions containing only phosphatidylcholine elute at ~20–25% methanol; 7–10 of the most concentrated fractions are pooled, and dried down in a vacuum evaporator. The resulting solid is most conveniently and safely stored (at −70°) as a homogenized suspension in 30 ml distilled water containing 5 mM 2-mercaptoethanol.

Methanol (volume %)[a]	Volume on column (ml)
0	250
10	250
15	500
20	900
25	2000

[a] All the above solvents contain 2 mM 2-mercaptoethanol.

This procedure yields several grams of pure phosphatidylcholine. The concentration of phosphatidylcholine is determined by total phosphate analysis,[16] given an average molecular weight of 700 for the lipid.

Cholate Solubilization and Hollow Fiber Dialysis

Materials. Na$^+$,K$^+$-ATPase is prepared from dog kidney as purified membranes, using the method of Jørgensen as detailed in this volume,[17]

---

[15] K. Owens, *Biochem. J.* **100**, 354 (1966).
[16] B. N. Ames, this series, Vol. 8, p. 115.
[17] P. L. Jørgensen, this volume [2].

and stored frozen at $-70°$ in 30 mM imidazole, pH 6.8, 10 mM 2-mercaptoethanol, 2 mM EDTA, 0.25 M sucrose.

The product previously employed for hollow fiber dialysis, Bio-Fibers (Bio-Rad), is no longer commercially available. Spectra/Por hollow fiber bundles, molecular weight cutoff 6000 (Fisher Scientific, Springfield, NJ), are a satisfactory substitute. For the four different sizes that are available for this product, the total internal volumes of the hollow fibers range from 0.25 to 2.0 ml. Another potential replacement is the Amicon 3P10 cartridge (Amicon Corp., Danvers, MA, Cat. #650100), which contains polysulfone fibers instead of cellulose. This unit has the advantage of already being configured in a countercurrent dialysis arrangement (it is necessary to construct a chamber to provide this configuration for the Spectra/Por product, as described below).

In both of these substitutes for the Bio-Rad product, the recovery of phospholipid and protein and the extent of removal of cholate are similar to the values originally reported.[6] For the configuration described here, the total volume of reconstituted vesicles that are produced is about twice the internal fiber volume; the procedure detailed below is for the production of ~ 2 ml of vesicles, and can be directly scaled up or down.

The Spectra/Por hollow fiber bundle (Fisher Cat. #08-670-150C) is prepared for countercurrent hollow fiber dialysis as follows. The fibers are first soaked for 15 min in distilled water; one end is then connected to a source of 95% ethanol and the other end to a water aspirator. By this means, the interior of the fibers is rinsed for 3 min with 50 ml of 95% ethanol followed immediately by a 10-min rinse with distilled water. This operation is performed with the fibers continually submerged in distilled water and allows removal of the myristic acid present within the fibers. (*Note:* it appears that more prolonged exposure of the fibers to 95% ethanol causes them to become leaky.) The fibers may then be inserted into the chamber—readily fabricated from Lucite tubing—as shown in Fig. 1. If care is exercised in the handling of this hollow fiber unit and the fibers are not allowed to dry out, it should be reusable many times. Contaminated hollow fibers can usually be reused after dialyzing them against 1% SDS in the configuration described below.

Cholic acid (Sigma Chemical Co. or Nutritional Biochemicals) is decolorized with activated charcoal and recrystallized from 95% ethanol as follows. A saturated solution of cholic acid in 95% ethanol is obtained by heating the solution in a boiling water bath with stirring. Activated charcoal is then added, and the solution is heated and stirred for an additional 5 min and then filtered through a Whatman #1 filter using a Büchner funnel. The filtrate is allowed to cool slowly to room temperature, and then transferred to a 4° cold room and allowed to sit overnight. The

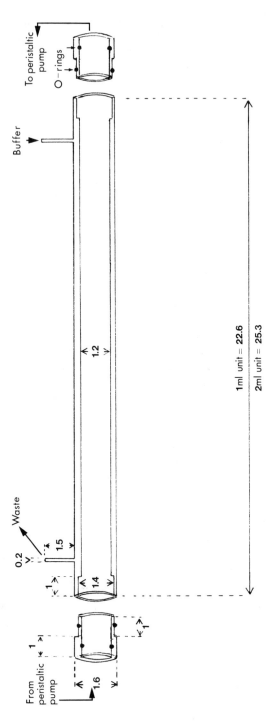

FIG. 1. Apparatus for containing Spectra/Por hollow fiber units for countercurrent hollow fiber dialysis. All parts except the O-rings were machined from cylinders of Lexan (General Electric, Schenectady, NY). Numbers refer to critical dimensions in centimeters. The two alternative dimensions for the length of the cylinder are for containment of Spectra/Por units of fiber interval volumes of 1 ml (Cat. #132285) and 2 ml (Cat. #132290), respectively. See the text for details of the assembly of the countercurrent dialysis system.

crystals are removed by filtration, washed with a small amount of cold ethanol, and then dried for 1 hr under vacuum. The crystals are titrated into a 100 mg/ml stock solution in distilled water with NaOH, to pH 7 (the acid form is not freely soluble in water).

Solubilization of Na^+,K^+-ATPase. The cholate solubilization buffer including phosphatidylcholine is prepared as a 1.11-fold concentrate, so that the eventual addition of 0.25 ml of purified enzyme to 2.25 ml of solubilization buffer will result in a final protein concentration of 0.4 mg/ml in 30 mM NaCl, 20 mM KCl, 10 mM 2-mercaptoethanol, 30 mM imidazole (pH 6.8), 0.25 M sucrose, 30 mM EDTA, 20 mg/ml cholate, 20 mg/ml phospholipid, and 0.4 mg/ml protein. As described above, the lipid stock is stored frozen as a concentrated suspension; the suspension should be rehomogenized after thawing to ensure uniformity. To completely dissolve the lipid, it may be necessary to incubate it in the solubilization buffer (without added Na^+,K^+-ATPase) for a few hours at room temperature. Aliquots of the solubilization buffer with lipid may be stored at $-70°$ for a month or more.

Solubilization and reconstitution were found to be sensitive to both ionic strength and the type of buffer employed. Thus, solubilization of the enzyme did not occur if phosphate was substituted for imidazole, and the enzyme could not be successfully reconstituted if both NaCl and KCl were removed from the solubilization and dialysis buffer (even after equilibration of the resulting vesicles with the normal concentrations of NaCl and KCl).

The 2.25 ml of solubilization buffer with lipid is chilled on ice. A 0.25-ml aliquot of 4 mg/ml enzyme is rapidly vortexed into this solution. The resulting suspension is incubated for 2 min at 23°, then incubated on ice for an additional 25 min. The mixture is centrifuged at 30,000 rpm for 5 min in a Beckman 50 Ti rotor at 2°. The supernatant is removed with care to avoid the small quantity of turbid material immediately above the solid pellet (this step is important because the pellet contains unsolubilized Na^+,K^+-ATPase: unsolubilized enzyme does not become reconstituted into sealed phospholipid vesicles but does contribute to the total Na^+,K^+-ATPase activity, thus decreasing the observed coupling ratio of Na^+ and K^+ ions pumped to ATP hydrolyzed). This supernatant is then immediately loaded into the internal space of the hollow fiber unit and dialyzed in a cold room at 4° against 2 liters of 0.25 M sucrose, 30 mM imidazole, 1 mM EDTA, 30 mM KCl, 10 mM 2-mercaptoethanol. The procedures for setup of the hollow fiber unit and the loading of the sample are described below.

As noted above, the hollow fibers have an internal volume of 1.0 ml; the ends of the unit (see Fig. 1) are connected to $\frac{3}{32}$ in. (outer diameter) × $\frac{1}{16}$ in. (inner diameter) Tygon tubing, to form a closed loop of ~2.0 ml total internal volume. The hollow fibers are prepared for the loading of the supernatant by equilibrating the space surrounding the fibers with dialysate, followed by purging the internal fiber volume of fluid by gently depressing the plunger of a 10-ml plastic syringe attached to the hollow fiber unit by a nested series of tightly fitting Tygon tubing of serially decreasing diameter. One end of the hollow fiber unit is attached to one end of the $\frac{1}{16}$-in. i.d. tygon tubing loop by the same means, and the hollow fiber unit is attached vertically to a ring stand with the free end pointing up. A fitting of Tygon tubing, identical to that used on the other end, is attached to the free end. The center of the length of $\frac{1}{16}$-in. i.d. tubing is attached to a Buchler Polystaltic peristaltic pump, or another pump capable of maintaining a flow rate of 0.5–2.0 ml/min with $\frac{1}{16}$-in. i.d. tubing.

The free end of the clean, dry $\frac{1}{16}$-in. tubing loop is inserted into the bottom of a test tube containing the supernatant. The peristaltic pump is turned on, adjusted to suck the supernatant into the tubing and hollow fiber unit at ~1 ml/min. After ~1 min, the fluid will begin to enter the hollow fibers themselves; this process is not uniform—some of the individual hollow fibers will fill more rapidly than others, so that when fluid begins to leave the free end of the hollow fiber unit, there will initially be a great deal of bubbling resulting from air escaping from those fibers that are still in the process of being filled. Accordingly, place the Pasteur pipet into the space in the fitting on the free end of the unit and draw up the bubbling fluid as it leaves the unit. Recycle this fluid into the 12 × 75 mm tube; continue this process until the bubbling largely subsides—at this point, the hollow fiber unit is filled and the free end of the $\frac{1}{16}$-in. Tygon tubing is inserted into the fitting on the free end of the hollow fiber unit. The speed of the peristaltic pump is raised, to cycle the supernatant through the hollow fibers at ~1.5 ml/min.

The dialysate flow rate is 80–90 ml/hr, and is scaled up or down in proportion to the volume of reconstituted vesicles to be produced. A convenient way to control the flow rate is by attaching a 10- or 20-μl Clay/Adams disposable micropipet or equivalent to the end of the length of tubing leaving the dialysate chamber of the hollow fiber unit (Fig. 1). This will restrict the flow to the above rate, if a fluid head of several feet is maintained. Prior to dialysis, the dialysis buffer is degassed while stirring, using a water aspirator. The degassing—together with the heat subsequently generated by the stirplate—will prevent the possible formation of air bubbles in the hollow fibers due to degassing *within* the fibers which results from the pressure differential across the fibers.

During dialysis, the initially clear supernatant will begin to turn turbid within the first hour or two. After 16–20 hr of dialysis, the supernatant is pumped out of the hollow fibers; additional supernatant trapped in the fibers is extruded by gentle pressure of a clean 10-ml syringe as described above for purging the hollow fiber unit.

The resulting material constitutes the unfractionated, reconstituted Na,K-pump vesicles. This preparation consists of an Na,K-pump reconstituted into sealed vesicles in both inside-out and right-side-out orientation, together with a variable (but smaller) amount of Na,K-pump that is unreconstituted or associated with unsealed lipid vesicles.

Comments on Cholate Removal

One difference between the Spectra/Por hollow fiber bundles and the discontinued Bio-Fiber apparatus is in the kinetics of cholate removal. The Spectra-Por hollow fibers produce a much more rapid initial removal, which greatly slows after 2 hr. In contrast, cholate removal in the Bio-Fibers follows a single exponential decline which at first lags behind but then overtakes the Spectra/Por unit, with the crossover point at ~8 hr. Thus the Bio-Fibers removes more cholate during the standard 16-hr (overnight) dialysis regime—99.5% vs 98–99% for the Spectra/Por.

Although not systematically compared for reconstitution of the Na^+,K^+-ATPase, removal of cholate has also been achieved using standard dialysis tubing[3,4] (M. Forgac, unpublished observations). The use of standard dialysis tubing has the advantage of allowing a larger number of reconstitutions to be performed simultaneously. This procedure is slower than the hollow fiber method, but removes sufficient detergent to give tightly sealed reconstituted vesicles: after dialysis in 6.4-mm Spectra/Por No. 2 tubing (M_r cutoff 12,000–14,000) for 4 days against four changes of 300 ml of dialysate/ml of sample volume, more than 99% of the cholate is removed. The tubing is not filled to capacity in order to increase the ratio of tubing surface area to internal fluid volume.

ATPase Assays on the Reconstituted Vesicles

The assay described here involves monitoring release of radioactive inorganic phosphate from [γ-^{32}P]ATP. The [γ-^{32}P]ATP can be obtained from New England Nuclear (Cat. #NEG-002), or alternatively can be prepared by the method of Glynn and Chappell.[18] It is added to 100 mM unlabeled Tris–ATP in reconstitution medium to a final specific activity of ~10^8 dpm/ml. The pH of this solution is adjusted to 6.8 by the addition of

[18] I. M. Glynn and J. B. Chappell, *Biochem. J.* **90,** 147 (1964).

solid Tris base. For ATPase activity determinations on reconstituted Na,K-pump vesicles at 23° under the same conditions used to determine active transport activity, 25- to 50-μl aliquots of vesicles are assayed undiluted. MgCl$_2$ is added to a concentration of 4.5–5 mM. The reaction is conducted in 6 × 50 mm tubes; it is initiated by the addition of 1 μl of [γ-^{32}P]ATP solution. The sample is quenched at the appropriate time, with 0.9 ml of 1.11% ammonium molybdate. This material is then immediately pipetted into a 12 × 75 mm tube containing 100 μl of 10 N H$_2$SO$_4$ and vortexed briefly; 1.0 ml of 1 : 1 (v/v) 2-methyl-1-propanol/benzene is immediately added. This two-phase mixture is capped with plastic caps for half-dram shell vials (Rochester Scientific), vortexed thoroughly for 15 sec to extract the [^{32}P]phosphomolybdate into the aqueous phase, and placed on ice for no more than 10 min, until further processed. Accumulated samples, processed as above (normally no more than four at a time) are spun for 2 min at the highest speed on a standard clinical centrifuge. A portion (0.7 ml) of the upper (organic) phase is pipetted into 6 ml of scintillation fluid and counted.

Standards constitute various amounts (0–50 nmol) of ^{32}P-labeled inorganic phosphate (New England Nuclear) prepared in unlabeled phosphate so as to be of the same specific activity as the [γ-^{32}P]ATP solution. They are added in aliquots of no more than 10 μl to the same volume of buffer as employed above, and processed by an identical procedure. The sensitivity of the reconstituted preparation to externally added ouabain is determined at 23° by determining the ratio of ATP hydrolyzed in 0.3 mM ouabain between 2 and 4 min of incubation, to that during the same time interval in the absence of ouabain. This is done because controls on the inhibition of unreconstituted, purified Na$^+$,K$^+$-ATPase indicate that the enzyme had to be incubated in the presence of ATP for 2 min to ensure complete inhibition. Addition of external ouabain typically causes inhibition of 30–40% of the ATPase activity, indicating the presence of an appreciable amount of unreconstituted ATPase. The sensitivity of ATPase activity to internal ouabain is determined by reconstituting the Na,K-pump in the presence of 0.3 mM ouabain in the dialysis during vesicle formation, by an otherwise identical procedure.

This assay was found to be linear in the 0–50 nmol range and capable of detecting as little as 1.5 nmol of ATP hydrolyzed with 5% reproducibility.

Transport Assays

Determination of intravesicular content of radiolabeled materials involves separation of vesicles from incubation medium on columns of

Sephadex G-50 medium, poured within Pasteur pipets to a height of ~8.4 cm. The columns are equilibrated and eluted at 4° with nonradioactive medium identical in all other respects to that in which the samples are incubated (see below). The fluid head is ~12 cm (±20%), and is maintained with a reservoir made of Tygon tubing, attached to the top of the column. Many fraction collectors (e.g., an ISCO model 270 with a reel for 12 × 75 mm tubes) can be adapted to accept ½-dram cappable shell vials; 20 fractions per column are collected at 1-min intervals. With such open real fraction collectors or with the Gilson "racetrack" fraction collector, it is possible to time the assay procedure so that four or five columns are simultaneously delivering fractions into one collector; this makes it practical to collect up to 50 transport assay "data points" per day. The scintillation fluid [2 ml Hydrofluor (National Diagnostics) or equivalent] is pipetted directly into the ½-dram vials, which are capped and counted in plastic outer vials.

The transport properties of the reconstituted vesicles do not significantly vary during the time course of the assay. The presence of unlabeled salts in the column elution buffer eliminates significant radioisotope eluting with the vesicles in the column void volume that are merely bound to the vesicle surface, rather than trapped intravesicularly. Extravesicular counts elute in the column volume. The raw data are obtained as the ratios of these two well-defined peaks (the intravesicular counts are no more than a few percent of the total counts), and the data are expressed as nanomoles of intravesicular radiolabeled ion per milligram of protein and/ or phospholipid.

Uptake of Externally Added $^{22}Na^+$

Aliquots of ~10^6 dpm of carrier-free ^{22}NaCl in HCl (New England Nuclear) are added to 6 × 50 mm tubes. The tubes are dried at 90° for several hours. Reconstituted vesicles are diluted from 1- to 10-fold in medium identical with the reconstitution medium employed for their formation. $MgCl_2$ is added to 50-μl aliquots of the diluted vesicles, to a concentration of 4.5–5.0 mM. Uptake is measured immediately following addition of the vesicle aliquot to the dried, ^{22}Na-containing tube on ice; the tube is immediately vortexed to dissolve the ^{22}Na. Uptake is initiated by adding either 2 μl of 100 mM Tris–ATP in reconstitution dialysate buffer, or (for controls) 2 μl of buffer alone. The tubes are tightly capped with Parafilm to retard evaporation, and incubated for various times at 23°. Incubation is terminated by immediately chilling the tubes in ice water for ~10 sec; the vesicles are then immediately loaded onto the Sephadex column. Vesicles are always diluted so that no more than 30% of the ATP is used during the time course of the transport assay.

Double-Label Determinations of ATP-Dependent Changes in Intravesicular Content of K^+ and Na^+

The procedure employed in early studies[6] involved the simultaneous trapping of ^{22}Na and ^{42}K inside the vesicles during hollow fiber dialysate; this employs undesirably large quantities of radioisotope, and has been superseded by the observation that it is possible to equilibrate the vesicles with externally added radioisotope by prolonged incubation at 37°. Thus, to achieve complete equilibration, vesicles are incubated for 2–3 days at 37°. It is advisable to do this incubation under nitrogen in the presence of a reducing agent (i.e., 2 mM 2-mercaptoethanol) to prevent oxidation of lipid and protein. Sodium azide (0.02%) can also be added if bacterial growth is a concern. The vesicles equilibrated in this manner are still capable of normal ATP-dependent ^{22}Na$^+$ and ^{86}Rb$^+$ transport, attesting to the stability of the reconstituted enzyme.

Effect of Readdition of Cholate on ATPase Activity: Evidence for Inside-Out and Right-Side-Out Populations of Reconstituted Enzyme

One method of determining the relative proportions of inside-out- and right-side-out-oriented vesicles in a reconstituted preparation is to measure stimulation of Na$^+$,K$^+$-ATPase activity resulting from the addition of low concentrations of cholate. Because cholate will make the vesicles permeable to ATP, the ATPase activity measured in the presence of cholate will be the sum of the ATPase activity due to inside-out vesicles (i.e., with the ATP side exposed on the outer face of the vesicle) and right-side-out vesicles (with the ATP side normally sequestered within the vesicle). It is important to measure ATP hydrolysis in the presence of a range of cholate concentrations (generally 0.01 to 1.0%). This is necessary because higher concentrations of cholate will cause inactivation of Na$^+$,K$^+$-ATPase. It should also be noted that the concentration of cholate giving maximal stimulation of ATPase activity is dependent on the concentration of reconstituted vesicles: the higher the vesicle concentration, the higher the optimal cholate concentration. With a vesicle phospholipid concentration of 1 mg/ml in the assay and a 20:1 ratio of lipid:protein, the optimal cholate concentration was found to be 1 mg/ml, under which conditions a 30–40% stimulation of ATP hydrolysis was observed. In order to eliminate the possibility that cholate is stimulating ATP hydrolysis as a result of dissipating either ion gradients or an opposing membrane potential, it is advisable to carry out the experiments in the presence of an ionophore, such as nigericin, added as a 1/200 dilution from a 0.2 mg/ml ethanolic stock solution. The ionophore will dissipate the ion gradients prior to the addition of cholate.[19]

Density Gradient Purification of the Na,K-pump Incorporated into Sealed Vesicles

Separation of Sealed from Unsealed Lipid Vesicles

The following procedure can be employed to purify the sealed fraction of vesicles from the unsealed and partially sealed material.[10] The vesicles are prepared in a solubilization buffer of 30 mM imidazole (pH 6.8) containing 175 mM CsCl (Alfa, ultrapure, Cat. #87640), 175 mM NaCl, 10 mM 2-mercaptoethanol, 1 mM EDTA, and are dialyzed against this buffer by a procedure otherwise identical to that described above. The relatively high concentration of CsCl renders sealed vesicles denser than leaky or unsealed vesicles. It also provides for a sufficiently large change in density upon activation of inside-out-oriented Na,K-pump by addition of external ATP, because Cs^+ can be transported in place of K^+. The increase in NaCl concentration maintains adequate enzymatic activity, which is 2- to 3-fold less in this high ionic strength buffer as compared to the original 20 mM KCl/30 mM NaCl buffer.

The following solutions are required.

Solution A: 30 mM imidazole (pH 6.8) containing 292 mM sucrose (10%, w/v), 102 mM KCl, 10 mM 2-mercaptoethanol, and 1 mM EDTA
Solution B: 30 mM imidazole (pH 6.8) containing 175 mM KCl, 175 mM NaCl, 10 mM 2-mercaptoethanol, and 1 mM EDTA
Solution C: 1 vol solution A + 4 vol solution B.

Note that these have an osmolarity approximately equal to that of the high ionic strength CsCl buffer.

The vesicles are exchanged into solution B prior to loading them on the sucrose gradient by passage over a 1-ml Sephadex G-50 (fine) column, formed in a Pasteur pipet and previously equilibrated with solution B. A sample of 0.2 ml results in an eluate of 0.24 ml containing 75% of the initial lipid phosphate (measured by the method of Ames[16]), and essentially none of the original external medium. This rapid gel filtration has a 2-fold purpose. It makes it possible to gently layer the vesicles on the top of the gradient with minimal disruption, since the 2% (w/v) difference in sucrose

---

[19] J. F. Dixon and L. E. Hokin, *J. Biol. Chem.* **255**, 10681 (1980).

concentration generates a sharp interface. It also removes external small molecules, such as Cs and ATP, that interfere with subsequent assays (Cs interferes with the Lowry protein determination, and ATP interferes with the phosphate measurement). After effecting exchange, the vesicles should be kept at 0–4° to slow collapse of the concentration gradients.

A linear gradient of 2–10% sucrose is prepared from equal volumes of solutions A and C. The size of the gradient can be varied from 4.3 to 4.8 ml to accommodate a variation in sample size of 0.2 to 0.7 ml. The relative shallowness of the gradient necessitates careful handling and slow acceleration and deceleration. The gradients are centrifuged in a Beckman SW 50.1 rotor at 48,000 rpm (275,000 g) at 3° for 320 min, during which time the vesicles closely approach their equilibrium density. Fractions are collected by puncturing the bottom of the tube. If Beckman Ultra-Clear tubes (Cat. #344057) are used, the bands of turbidity corresponding to phospholipid vesicles are easily visualized against a black background.

Sealed vesicles band midway in the gradient, at about 6% sucrose. Leaky vesicles are distributed in lighter regions while unsealed vesicles just enter the gradient at 2–3% sucrose.

Resolution of Sealed Vesicles Containing Inside-Out from Right-Side-Out Reconstituted Na,K-Pump

The population of sealed vesicles can be further fractionated into right-side-out and inside-out populations after incubation in the presence of $MgCl_2$ (7 mM) and Tris–ATP (6 mM) for 2 hr at 23°. At this point the vesicles incorporating an inside-out-oriented Na,K-pump will have pumped out most of their internal Cs^+. Application of this mixture of sealed vesicles to a second round of sucrose density gradient sedimentation results in two well-separated regions of turbidity. The upper band appears at the same position as did the bulk of the unsealed vesicles in the first gradient. These vesicles contain inside-out-oriented Na,K-pump.

The proportions of sealed to unsealed vesicles and right-side-out- to inside-out-oriented Na,K-pump have been found to vary with the protocol used for cholate dialysis. This is probably a function of the rate of removal of detergent (see earlier section). The Bio-Fibers generate a 50 : 50 mix of right-side-out : inside-out-oriented enzyme. The Spectra/Por hollow fiber bundles and the Amicon 3P10 cartridge, for unknown reasons, favor right-side-out-oriented Na,K-pump by a factor of 3–10, and thus yield relatively small quantities of inside-out reconstituted Na,K-pump.

[14] Reconstitution of the Na,K-Pump by Freeze–Thaw Sonication: Estimation of Coupling Ratio and Electrogenicity

By LOWELL E. HOKIN and JOHN F. DIXON

Introduction

Initial studies on reconstitution of purified Na^+,K^+-ATPase were carried out by the cholate dialysis technique of Racker.[1] Reconstitution of Na^+ transport only was first reported[2,3] using the purified enzymes from dog kidney[4] and the rectal gland of *Squalus acanthias*.[5] Reconstitution of Na^+ transport by sonication without the freeze–thaw step using microsomes of the electroplax of *Electrophorus electricus* was also reported.[6] Later, the coupled Na,K-pump was reconstituted.[7–9] In all these studies, maximal rates of cation transport were of the order of 100–400 nmol/min/mg protein at 25°. The cholate dialysis technique suffers from several disadvantages. It is tedious, requiring 2–3 days for reconstitution, and the concentrations of cholate required for reconstitution markedly denature the Na^+,K^+-ATPase. For example, no reconstitution of the purified[10,11] Na^+,K^+-ATPase from *Electrophorus electricus* could be obtained (Hilden and Hokin, unpublished observations). The current method for reconstitution by the cholate dialysis technique has been published in detail.[12,13] The reader interested in various aspects of reconstitution of Na^+,K^+-ATPase should consult the recent extensive reviews of Anner[14,15] and earlier reviews.[16–19]

---

[1] E. Racker, *J. Biol. Chem.* **247**, 8198 (1972).
[2] S. M. Goldin and S. W. Tong, *J. Biol. Chem.* **249**, 5907 (1974).
[3] S. Hilden, H. M. Rhee, and L. E. Hokin, *J. Biol. Chem.* **249**, 7432 (1974).
[4] J. Kyte, *J. Biol. Chem.* **246**, 4157 (1971).
[5] L. E. Hokin, J. L. Dahl, J. D. Deupree, J. F. Dixon, J. F. Hackney, and J. F. Perdue, *J. Biol. Chem.* **248**, 2593 (1973).
[6] E. Racker and L. W. Fisher, *Biochem. Biophys. Res. Commun.* **67**, 1144 (1975).
[7] S. Hilden and L. E. Hokin, *J. Biol. Chem.* **250**, 6296 (1975).
[8] B. M. Anner, L. K. Lane, A. Schwartz, and B. J. R. Pitts, *Biochim. Biophys. Acta* **467**, 340 (1977).
[9] S. M. Goldin, *J. Biol. Chem.* **252**, 5630 (1977).
[10] J. F. Dixon and L. E. Hokin, *Arch. Biochem. Biophys.* **163**, 749 (1974).
[11] J. F. Dixon and L. E. Hokin, *Anal. Biochem.* **86**, 378 (1978).
[12] B. M. Anner and M. Moosmayer, *J. Biochem. Biophys. Methods* **5**, 299 (1981).
[13] B. M. Anner, M. M. Marcus, and M. Moosmayer, in "Enzymes, Receptors, and Carriers of Biological Membranes" (A. Azzi, U. Brodbeck, and P. Zahler, eds.), pp. 81–96. Springer-Verlag, Berlin, 1984.
[14] B. M. Anner, *Biochim. Biophys. Acta* **822**, 319 (1985).

Because of the above limitations of the cholate dialysis technique, some years ago we decided that it would be desirable to find an alternative method for reconstitution.[20] In 1977, Kasahara and Hinkle[21] developed the technique of freeze–thaw sonication for reconstituting facilitated glucose transport with the glucose transport protein from band 4.5 of the erythrocyte membrane. This technique consisted of adding the glucose transport protein to preformed liposomes, rapidly freezing in a dry-ice bath, thawing slowly at room temperature, followed by brief sonication.

Our laboratory applied this freeze–thaw sonication technique to the purified Na^+,K^+-ATPase from both the electroplax[10,11] and the rectal gland[5] with highly successful results.[20] The technique for reconstitution of Na,K-transport from the electroplax enzyme by freeze–thaw sonication is described here in detail. A variety of parameters for reconstitution were optimized as described below so as to obtain maximal transport rates.

Materials

The phospholipid used for preparing liposomes is asolectin, provided by Associated Concentrates, Woodside, Long Island, New York. It has a peroxide content as measured by the thiobarbituric acid assay[22] of 15.3 nmol/mg as compared to 27.2 and 135 nmol/mg for "EM" egg lecithin (Merck) and "type V" egg lecithin (Sigma), respectively. Purification of the asolectin by the method of Kasahara and Hinkle[21] does not reduce its peroxide content, and the color of its chloroform solution darkens. Asolectin is therefore used without further purification. Vanadium-free Tris–ATP and Tris–ADP are obtained from Sigma. $^{22}NaCl$ is obtained from Amersham-Searle. Dowex 50 (8% cross-linked, 20–50 mesh) is obtained from Sigma, and $^{86}RbCl$ is obtained from New England Nuclear Corp. The purified Na^+,K^+-ATPases from *Electrophorus electricus* and *Squalus*

---

[15] B. M. Anner, *Biochim. Biophys. Acta* **822,** 335 (1985).
[16] L. E. Hokin, in "Receptors and Hormone Action" (B. O'Malley, ed.), Vol. 1, pp. 447–462. Academic Press, New York, 1977.
[17] L. E. Hokin, in "Membrane Bioenergetics" (C. P. Lee, G. Schatz, and L. Ernster, eds.), pp. 281–295. Addison-Wesley, Reading, Massachusetts, 1979.
[18] L. E. Hokin, in "Function and Molecular Aspects of Biomembrane Transport" (E. M. Klingenberg, F. Palmieri, and E. Quagliariello, eds.), pp. 495–503. Elsevier/North-Holland, Amsterdam, 1979.
[19] L. E. Hokin, *Membr. Transp.* **1,** 547 (1982).
[20] L. E. Hokin and J. F. Dixon, in "Na,K-ATPase Structure and Kinetics" (J. C. Skou and J. G. Nørby, eds.), pp. 47–67. Academic Press, London, 1979.
[21] M. Kasahara and P. C. Hinkle, *J. Biol. Chem.* **252,** 7384 (1977).
[22] L. Warren, *J. Biol. Chem.* **234,** 1971 (1959).

acanthias are prepared by methods developed in our laboratory.[5,10,11] Chemicals in which the source is not specified are reagent grade.

Procedures for Preparation of Na^+,K^+-ATPase Liposomes

Preparation of Liposomes

A chloroform solution of 100 mg/ml asolectin and 2 mg/ml butylated hydroxytoluene is stored at −20°. It shows no loss of activity over a period of 1–2 months. A 0.45-ml aliquot is dried in a 15-ml Corex culture tube in a 40° water bath under a stream of nitrogen with a flow rate of 5600 ml/min. The lipid is redissolved in 1–2 ml of ether and dried twice with N_2 at the same flow rate until the end point is reached, as indicated by a sudden solidification of the lipid to a light yellow waxlike material. The stoppered tubes are left in the water bath until all have been dried in this manner. Each tube is then further dried with an N_2 flow rate of 7800 ml/min for 100 sec and then stoppered. These drying conditions are crucial since it was found that with less drying the transport rates were less, due presumably to excess residual ether which made the proteoliposomes very leaky.[20] On the other hand, excessive drying reduced transport. At optimal drying, there was a faint odor of ether. It would appear that residual ether stimulated reconstituted active Na^+,K^+-transport. No stimulatory effect of ether was seen if it was added to Na^+,K^+-ATPase liposomes over a wide concentration range from noninhibitory to inhibitory. It would appear that ether must be "dissolved" in the asolectin prior to formation of the liposomes for it to be stimulatory, and the concentrations are fairly critical. After drying, 1 ml of buffer containing 30 mM imidazole–HCl (pH 7.0) and 1 mM Na$_2$EDTA (pH 7.0) is added, and the tubes are again gassed with N_2 and stoppered. A bath sonicator (Laboratory Supplies Co., Hicksville, NY) is filled with distilled water to a height of 10.2 cm, and two drops of Triton X-100 are added to the bath. After gassing with N_2 and stoppering, the tubes are vortexed for 1 min and then placed at an angle in the sonicator with the contents of the tube just below the surface of the bath water. After sonication for 3–6 min, the contents of the tubes are translucent, indicating that liposomes have formed. This was confirmed by sonicating a 1-ml suspension of liposomes with $^{36}Cl^-$, [^{3}H]inulin, or [^{3}H]glucose and passing it over a Sephadex G-50 column.[23] Maximum entrapment had occurred, and the liposomal volume ranged from 10 to 20 μl/ml. The liposomes from separate tubes are pooled, and appropriate aliquots are taken for reconstitution.

[23] J. F. Dixon and L. E. Hokin, *J. Biol. Chem.* **255**, 10681 (1980).

Incorporation of Purified Na^+,K^+-ATPase into Liposomes

Unless otherwise indicated, 1 mg of purified Na^+,K^+-ATPase from the electroplax of *Electrophorus electricus*[10,11] or the rectal gland of *Squalus acanthias*[5] in 50 μl of 1 mM Na-EDTA, pH 7.4, is mixed with 1 ml of liposomes in 30 mM imidazole · HCl (pH 7.0) and 1 mM Na$_2$EDTA (pH 7.0). After gassing with N$_2$, the mixture is rapidly frozen in an ethanol–dry-ice bath, removed, allowed to thaw at room temperature, vortexed for 10 sec, and stored in ice. Reproducible conditions for sonication are crucial. Chipped ice is added to the sonicator bath to bring the height of the bath to 10.2 cm. The sonicator is allowed to run for 20 min. Ice is removed, and the height is adjusted to 10.2 ml with ice water. The temperature of the sonicator bath is 0–2°. One by one the tubes are submerged with the surface of the proteoliposome suspension about 1–2 mm below the bath surface, and sonication is carried out for 30 sec. The temperature of the sonication bath should not rise above 5°. Small amounts of ice can be added when necessary (sonication of proteoliposomes is recontinued when the ice has just melted). In the standard procedure, proteoliposomes are diluted in half with buffer containing 60 mM NaCl and 60 mM KCl. If K^+ transport is to be measured, 2.5 μCi of $^{86}Rb^+$ (specific activity, 1–10 mCi) is added immediately after the 2-fold dilution. Previous studies[7] showed that K^+ and Rb^+ behave in an identical manner. The Na^+,K^+-ATPase liposomes are stored for 24 hr to equilibrate the ions on both sides of the membrane.

Assay of Na^+ and K^+ Transport in Na^+,K^+-ATPase Liposomes

In the standard procedure, 100 μl of Na^+,K^+-ATPase liposomes is preincubated at 27° for 60 sec. Transport is initiated by addition of 20 μl of a solution containing 75 mM Tris–ATP (pH 7.0), 75 mM MgCl$_2$, 6×10^{-4} M ouabain, 1 mM Tris-EDTA, 30 mM imidazole–HCl (pH 7.0). When Na^+ transport is measured, $^{22}Na^+$ (100 mCi/mg) is added to give 2.5 μCi/ml. To measure K^+ transport, the proteoliposomes are preequilibrated for 24 hr, as described above. $^{86}Rb^+$ is added to the starting solution to make it equivalent in concentration to that in the equilibrated proteoliposomes.

Passive transport (non-ATP-dependent transport) is determined by substituting Tris–ADP for Tris–ATP in the above system. To stop the transport reaction and remove extravesicular isotope, 100 μl of the incubation medium is applied to a 0.7 (i.d.) × 17 cm Dowex 50-X8 column (Tris form) preequilibrated with 3 ml of 0.3 M sucrose containing 50 mM Tris chloride (pH 7.0), and pretreated with 10 mg bovine serum albumin dissolved in the same equilibration solution. The equilibration solution is isotonic with the Na^+,K^+-ATPase liposome suspension. The chilled

Na$^+$,K$^+$-ATPase liposome suspension is layered between the equilibration solution and the elution buffer. The elution buffer is 5.0 ml of 50 mM Tris–HCl (pH 7.0). Although the elution buffer is not isotonic with the Na$^+$,K$^+$-ATPase liposomes, measurements made with isotonic and nonisotonic elution buffer gave similar results and the use of 50 mM Tris elution buffer makes sample layering possible, which in turn makes sample application faster and more reproducible.

After addition of the elution buffer, the stopcock is opened and the 5 ml of the elution buffer is rapidly passed through the column. This requires 30 sec. One milliliter of column eluant is counted in a liquid scintillation spectrometer. Total counts in the incubation medium are determined by withdrawing 5-μl aliquots before application to the column. When incubations are carried out at 37°, the incubation tube is plunged into ice water and swirled for 30 sec. Column chromatography over Dowex 50 is carried out in a cold room. Under these conditions, it is necessary to make a correction for transport which occurs during chilling of the sample and possibly during the column run. This is done by plunging a sample into ice immediately after addition of the starting solution. Transport under these conditions is taken as "zero time transport." It is a small percentage of transport measured at the earliest times at 37°. When transport was carried out at 27°, it was found that a zero-time control was not necessary and samples could be run on columns at room temperature. The time for transfer of 100 μl of incubation sample to the column before opening the stopcock (5 sec) is included as part of the incubation time.

Effects of Various Parameters on Reconstitution by
  Freeze–Thaw Sonication

Sonication Time

Figure 1 shows the effect of increasing sonication time after the freeze–thaw step on reconstitution of Na$^+$ transport with 1 ml of electroplax Na$^+$,K$^+$-ATPase liposomes. Before sonication, transport of Na$^+$ into the Na$^+$,K$^+$-ATPase liposomes in the absence of ATP was very high and there was very little additional transport in the presence of ATP. This can best be accounted for by very leaky proteoliposomes after the freeze–thaw step, so that the leak essentially equalled active transport. However, on very brief sonication the proteoliposomes tightened, as evidenced by the drop in passive uptake of ^{22}Na$^+$ and the increase in ATP-dependent uptake. Maximum ATP-dependent uptake was reached after a 30-sec sonication and then fell slowly. ATP-dependent and -independent ^{22}Na$^+$ transport for the rectal gland enzyme is shown for a 60-sec sonication only

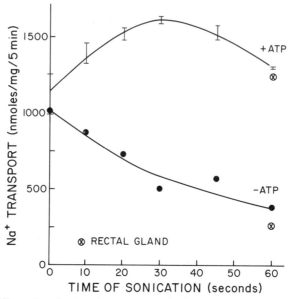

FIG. 1. Effect of sonication time on reconstitution. Na^+,K^+-ATPase liposomes were prepared as described under "Procedures" except the post freeze–thaw sonication time was as indicated, the asolectin concentration was 22.5 mg/ml, and 30 mM KCl and 30 mM NaCl were present throughout. $^{22}Na^+$ uptake was measured at 27° as described under the Procedures section. The solid circles are for the electoplax enzyme and the crossed open circles are for the rectal gland enzyme. Reproduced by permission of Academic Press.[20]

and is essentially the same as that for the electroplax enzyme. Other experiments (data not shown) showed the same shape for the sonication time course for the rectal gland enzyme. With volumes of proteoliposomes of 0.25 and 0.5 ml, the optimum sonication time was 10 and 20 sec, respectively.

Asolectin Concentration

Figure 2 shows the effects of increasing asolectin concentration on reconstitution of Na^+ transport. There was a sharp increase in ATP-dependent $^{22}Na^+$ accumulative capacity with increasing asolectin concentration, reaching a maximum at 45 mg/ml asolectin and then falling off sharply at 60 mg/ml asolectin. The $^{22}Na^+$ accumulation in the absence of ATP also increased to a maximum at 45 mg/ml and fell off slightly at 60 mg/ml. The 6-fold increase in ATP-dependent transport on increasing asolectin from 15 to 45 mg/ml may be partly explained by an increased vesicular volume (increased number and/or size of the vesicles), but since

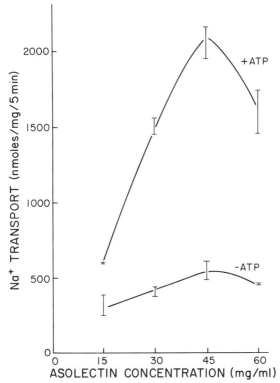

FIG. 2. Effect of asolectin concentration on reconstitution. Na^+,K^+-ATPase liposomes were prepared as described under the Procedures section except the asolectin concentration was varied as indicated and 30 mM KCl and 30 mM NaCl were present throughout. $^{22}Na^+$ uptake was measured at 27° as described under the Procedures section. The incubation time was 5 min. The vertical bars show correct duplicate values. Reproduced by permission of Academic Press.[20]

$^{22}Na^+$ accumulation in the absence of ATP (a function of vesicular volume) was only twice as high at 45 as at 15 mg/ml asolectin, the major cause of the stimulation of ATP-dependent accumulation would seem to be more functional Na^+,K^+-ATPase molecules incorporated into the liposomes.

Sodium Concentration during Reconstitution

If proteoliposomes were prepared at various concentrations of Na^+ and the post freeze–thaw sonication and assay for Na^+ transport were carried out in the usual way, results such as are shown in Fig. 3 were obtained. There was a sharp increase in transport from 10 to 30 mM Na^+,

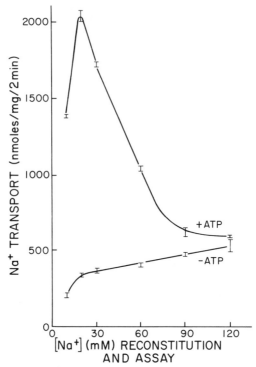

FIG. 3. Effect of increasing sodium concentrations on reconstitution. Na$^+$,K$^+$-ATPase liposomes were prepared as described under the Procedures section except the NaCl concentration was as indicated from the point of preparation of liposomes. KCl (30 mM) was present throughout. ^{22}Na$^+$ uptake was measured at 27° as described under the Procedures section with an incubation time of 2 min. Reproduced by permission of Academic Press.[20]

as would be expected from the effects of Na$^+$ on enzyme activity. However, above 30 mM Na$^+$ there was a marked inhibition of reconstituted Na$^+$ transport, so that active transport was essentially abolished between 90 and 120 mM Na$^+$. This inhibition was due to an effect of Na$^+$ on the actual reconstitution process itself because if reconstitution was carried out at 30 mM NaCl and the NaCl concentration was then adjusted upward or downward by dilution followed by appropriate additions of NaCl and storage for 24 hr, no inhibition by NaCl was seen.[20] Rather, ATP-dependent transport increased up to 30 mM and then remained constant up to 90 mM. ATP-independent transport increased with increasing Na$^+$ concentration, suggesting an effect of Na$^+$ on passive uptake of ^{22}Na, as to be expected. It was found that reconstitution was maximal if no Na$^+$ was present during preparation of the Na$^+$,K$^+$-ATPase liposomes prior to the 24-hr equilibration with 30 mM Na$^+$. It is unlikely that the Na$^+$ effect is on the lipid bilayer since preparation of liposomes in 90 mM Na$^+$, followed by dilution to 45 mM Na$^+$ and reconstitution, gave the same transport as liposomes prepared in 45 mM Na$^+$ and reconstituted in 45 mM Na$^+$ (White and Hokin, unpublished observations). Preparation of the liposomes in 45 mM Na$^+$ followed by reconstitution in 90 mM NaCl gave the usual strong inhibition of reconstitution by 90 mM Na$^+$. It would also seem unlikely that an effect on the bilayer would show such high selectivity for Na$^+$ vs K$^+$ (see below).

Potassium Concentration during Reconstitution

Increasing the K$^+$ concentration from 0 to 96 mM and adjusting to 36 mM, followed by soaking for 24 hr, had no effect on reconstitution. This is in contrast to the requirement of 50–100 mM K$^+$ for reconstitution by the cholate dialysis procedure in general.[1-3] Thus, with the freeze–thaw sonication procedure, neither Na$^+$ nor K$^+$ need be present during preparation of the Na$^+$,K$^+$-ATPase liposomes, as long as these ions are added after reconstitution and sufficiently in advance of transport assay to permit equilibration of the ions across the proteoliposomal membrane (24 hr or longer). In the case of (^{86}Rb$^+$)K$^+$ transport, it is important to add the ^{86}Rb$^+$ and KCl at the same time so that the ratio of ^{86}Rb$^+$ to K$^+$ will be the same inside and outside the proteoliposome, since calculations are based on the specific activity outside the proteoliposome, which would be the same inside the proteoliposome under these conditions. This assumption might not hold if the ^{86}Rb$^+$ and the K$^+$ were added at different times and one or the other did not equilibrate completely during the soaking period. This consideration is not critical for ^{22}Na$^+$, since it is transported inwardly rather than outwardly and its specific activity is that which is actually determined on adding ^{22}Na$^+$ at the beginning of the transport assay.

Effect of Purity of Na$^+$, K$^+$-ATPase on Reconstituted Na$^+$ Transport

Since the incorporation of protein into the liposome is probably nonspecific, one would anticipate that proteoliposomes reconstituted with pure Na$^+$,K$^+$-ATPase would have a higher transport activity than proteoliposomes reconstituted with impure Na$^+$,K$^+$-ATPase. This was found to be the case.[20] Purity of Na$^+$,K$^+$-ATPase and reconstituted transport paralleled each other closely.

Stability of the Na$^+$, K$^+$-ATPase and Reconstituted Transport to Sonication after the Freeze–Thaw Step

Figure 4 shows the stability of the Na$^+$,K$^+$-ATPase to sonication after the freeze–thaw step (assayed under standard conditions for Na$^+$,K$^+$-ATPase).[5] The enzyme which had not been exposed to sonication after

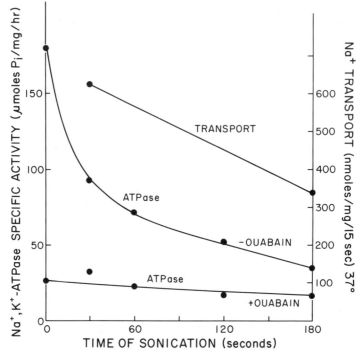

FIG. 4. Effect of sonication on ouabain-sensitive and ouabain-insensitive Na^+,K^+-ATPase and on active Na^+ transport. Na^+,K^+-ATPase liposomes were prepared as described under the Procedures section except the post freeze–thaw sonication time was varied as indicated. Na^+,K^+-ATPase was assayed under transport conditions. $^{22}Na^+$ uptake was measured for 15 sec at 37° as described under the Procedures section. Reproduced by permission of Academic Press.[20]

the freeze–thaw step had an ouabain-sensitive Na^+,K^+-ATPase activity about 60% of that of the enzyme assayed without liposomes. Although prolonged sonication virtually inactivated all of the Na^+,K^+-ATPase, with the usual sonication time (30 sec) after the freeze–thaw step, there was minimal inactivation. It should be pointed out that there is no Mg^{2+}-ATPase in this preparation.[10]

Since the ATP site in the Na^+,K^+-ATPase liposomes is utilized only on the outside of the liposome and the reconstituted transport is inhibited only by internal ouabain,[2,3,7] addition of ouabain externally will inhibit all unincorporated Na^+,K^+-ATPase but not Na^+,K^+-ATPase incorporated into the liposome with its ATP site on the outer surface and its ouabain site on the inner surface. (Na^+,K^+-ATPase with its substrate site on the inner surface would be silent, since no ATP would be available to it.)

About 15% of the total Na^+,K^+-ATPase prior to post freeze–thaw sonication was ouabain insensitive, indicating that about 85% of the measurable Na^+,K^+-ATPase was unincorporated.[20] Sonication inactivated the unincorporated Na^+,K^+-ATPase much more rapidly than the incorporated Na^+,K^+-ATPase. After a 180-sec sonication, about 80% of the unincorporated Na^+,K^+-ATPase had been inactivated, while only 30–40% of the ouabain-insensitive ATPase had been inactivated. Evidently, incorporation of the enzyme into the liposomes affords partial protection from the inactivating effects of sonication. Reconstituted Na^+ transport was measured in Na^+,K^+-ATPase liposomes sonicated for 30 and 180 sec.[20] The percentage inactivation of reconstituted transport was the same as the percentage inactivation of the ouabain-insensitive Na^+,K^+-ATPase, providing further evidence that the ouabain-insensitive Na^+,K^+-ATPase is the actively transporting Na^+,K^+-ATPase.

Effects of Various Parameters on Na,K-Transport after Reconstitution

Effect of Increasing Temperature on Reconstituted Na^+ Transport

With the cholate dialysis technique, Na^+ transport reached an optimum at 25° and then fell off sharply.[3] The simplest explanation for this observation is that after preparation of Na^+,K^+-ATPase liposomes by the cholate dialysis technique either the enzyme or the proteoliposome is unstable above 25°. Figure 5 shows that with Na^+,K^+-ATPase liposomes prepared by the freeze–thaw sonication technique, reconstituted transport increased linearly up to about 35° and then plateaued. ATP-independent $^{22}Na^+$ accumulation also increased with increasing temperature.

Time Course of $^{22}Na^+$ Accumulation and $(^{86}Rb^+)K^+$ Exit at Different Temperatures

At 27°, the transport of $^{22}Na^+$ and $(^{86}Rb^+)K^+$ were linear at least for 30 sec.[20] Sodium transport was linear for 2 min before falling off. At 37°, the ATP-dependent transport of Na^+ was linear for 30 sec.[20] However, K^+ transport began falling off slightly before 15 sec and fell off markedly after 15 sec. By comparing transport rates at 27 and 37°, it was found that approximately the same amount of K^+ still remained in the proteoliposomes. Since under these conditions Na^+ accumulation was still continuing at a linear rate, the data suggest that those proteoliposomes which had incorporated functionally active Na^+,K^+-ATPase liposomes had been essentially depleted of their K^+ at the time of the rapid fall-off. This interpretation is fortified by the observation that transport increased linearly

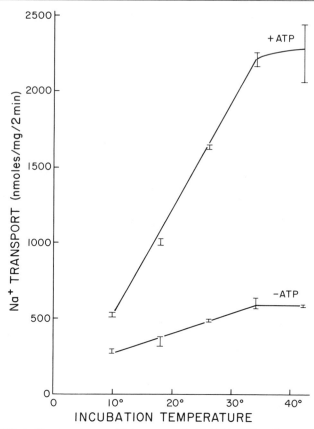

FIG. 5. Effect of temperature on Na^+ transport in Na^+,K^+-ATPase liposomes. Na^+,K^+-ATPase liposomes were prepared as described under the Procedures section. $^{22}Na^+$ transport was measured as described under the Procedures section except the incubation temperature was varied as indicated and the transport was stopped by application of the incubation solution to the Dowex 50 column at room temperature, as described for incubation at 27° under the Procedures section. NaCl (30 mM) and 30 mM KCl were present throughout. Reproduced by permission of Academic Press.[20]

with increasing protein concentration up to 10 mg/ml, indicating that the liposomes were not saturated with enzyme at 1 mg/ml protein and suggests that a sizeable population of liposomes may have been devoid or nearly devoid of enzyme. Na^+,K^+-ATPase liposomes prepared by the cholate dialysis technique showed that $^{42}K^+$ or $^{86}Rb^+$ concentrations in the proteoliposomes at the steady state were as low as one-tenth of that of the incubation medium.[7] With Na^+,K^+-ATPase liposomes prepared by freeze–thaw sonication, the gradients might even be higher because of the

TABLE I
COUPLING RATIO OF RECONSTITUTED
Na,K-TRANSPORT[a]

Temperature (°C)	Stoichiometry ± SEM ($n = 6$)
27	1.51 ± 0.12
37	1.75 ± 0.15

[a] Na^+,K^+-ATPase liposomes were prepared as described under the Procedures section. At 27°, the initial rate was measured for 1 min, and at 37°, the initial rate was measured for 15 sec. The slightly higher $Na^+:K^+$ coupling ratio at 37° was due to a 7.7% fall-off in K^+ transport by a 15-sec incubation.

much greater pumping capacity of the latter. The problem of depleting a pumped ion from a compartment would not apply to Na^+ since the Na^+ concentration in the medium, which is essentially an infinite volume, would not fall as compared to K^+ in the very small volume of the intravesicular space. The fact that the rate of K^+ transport could fall while the rate of Na^+ transport remained constant shows that the transports of Na^+ and K^+ are not tightly coupled. Uncoupling can also be produced by the lectin, wheat germ agglutinin[24] and substitution of SO_4^{2-} for Cl^- (see below).[21] Using the initial rates of 15 sec at 37°, the rates of Na^+ and K^+ transport were calculated to be approximately 3000 and 2000 nmol/mg/min, respectively.[20] These rates are at least one order of magnitude higher than the highest values reported with Na^+,K^+-ATPase liposomes prepared by the cholate dialysis technique. The rates of Na^+ and K^+ transport were approximately half as great at 27°, which is also considerably higher than rates with Na^+,K^+-ATPase liposomes prepared by the cholate dialysis technique and incubated at that temperature.[7-9]

Coupling Ratio of Na,K-Transport

Earlier studies[7-9] with Na^+,K^+-ATPase liposomes prepared by the cholate dialysis technique gave a coupling ratio of $3Na^+:2K^+$, which is the same as in the erythrocyte.[25-27] Table I shows that at both 27 and 37°,

[24] J. Pennington and L. E. Hokin, *J. Biol. Chem.* **254**, 9754 (1979).
[25] I. M. Glynn, *J. Physiol. (London)* **160**, 18P (1962).
[26] A. K. Sen and R. L. Post, *J. Biol. Chem.* **239**, 345 (1964).
[27] J. F. Hoffman, in "Organization of Energy-Transducing Membranes" (M. Nakao and L. Packer, eds.), pp. 9–21. University Park Press, Baltimore, 1973.

the coupling ratio in Na^+,K^+-ATPase liposomes prepared by freeze–thaw sonication was also $3Na^+:2K^+$. The somewhat higher coupling ratio at 37° is due to an 8% fall-off in K^+ transport by a 15-sec incubation.[20] Titration of binding sites in Na^+,K^+-ATPase shows $3Na^+:2K^+$ per phosphorylation site (see Anner[14,15]). As pointed out by Lauger,[28] the stoichiometric ratio is dependent on the number of ion-binding sites involved in ion transport, and this should have a fixed value which is determined by the molecular transport mechanism. The stoichiometric ratio should be distinguished from the coupling ratio, which is variable and which depends on the driving forces.

Electrogenicity of Reconstituted Na,K-Transport

Studies with proteoliposomes prepared by freeze–thaw sonication showed with the lipid-permeant anion thio[^{14}C]cyanate that the Na,K-pump is electrogenic,[23] as would be expected with a stoichiometric ratio of $3Na^+:2K^+$. Forgac and Chin[29,30] came to a similar conclusion from studies with [^{3}H]triphenylmethylphosphonium, which is lipid permeant. Recently, in a very thorough study with the fluorescent dye, 1,3,3,1',3',3'-hexamethylindodicarbocyanine, electrogenicity of Na^+,K^+-ATPase liposomes reconstituted by the cholate dialysis technique has been confirmed and extended.[31]

In the present case with proteoliposomes reconstituted by freeze–thaw sonication, the inward movement of $3Na^+$ with respect to the outward movement of $2K^+$ was balanced by the inward movement of Cl^-, thus maintaining electroneutrality. In fact, when Cl^- was replaced by the impermeant anion, SO_4^{2-}, the coupling ratio became $1Na^+:1K^+$, so as to maintain electroneutrality. These observations also indicate that no endogenous ion was present in the system (H^+, HCO_3^-, etc.) which could take the place of Cl^- to maintain the coupling ratio of $3Na^+:2K^+$. Anner[14,15] has reviewed in detail studies on the coupling ratio of Na^+ and K^+ transport and electrogenicity in Na^+,K^+-ATPase liposomes and factors which may influence the coupling ratio.

Conclusions

The technique of reconstitution by "freeze–thaw sonication" of coupled Na^+ and K^+ transport in asolectin liposomes using the purified Na^+,K^+-ATPase of the electroplax of *Electrophorus electricus* or the

---

[28] P. Lauger, *Biochim. Biophys. Acta* **779**, 307 (1984).
[29] M. Fogac and G. Chin, *J. Biol. Chem.* **256**, 3645 (1981).
[30] M. Forgac and G. Chin, *J. Biol. Chem.* **257**, 5652 (1982).
[31] H.-J. Apell, M. M. Marcus, B. M. Anner, H. Oetliker, and P. Lauger, *J. Membr. Biol.* **85**, 49 (1985).

rectal gland of *Squalus acanthias* is highly effective. Transport rates are obtained which are at least one order of magnitude higher than the highest rates reported with the cholate dialysis technique for reconstitution. Parameters for optimizing reconstitution have been studied in considerable detail. Optimal post freeze-thaw sonication time and asolectin concentration have been determined. Traces of residual ether in the asolectin increase reconstitution while no ether or excessive ether inhibit it. Concentrations of NaCl above 10 mM progressively inhibit reconstitution so that by 90 mM NaCl reconstitution is essentially abolished. There is no effect of K^+ on reconstitution. It appears that the effect of Na^+ is on the Na^+,K^+-ATPase and not on the phospholipid bilayer. Possible explanations for this effect are that the Na^+ conformation of the Na^+,K^+-ATPase is either not inserted into the liposome, is inserted in the wrong direction, or is inserted in a nonfunctional form. Reconstituted Na^+ transport rates parallel the purity of the enzyme used for reconstitution. Unlike transport in Na^+,K^+-ATPase liposomes prepared by the cholate dialysis technique, which peaks out at 25°, transport in Na^+,K^+-ATPase liposomes prepared by the freeze-thaw sonication technique reaches a maximum at about 35° and then plateaus. The unincorporated Na^+,K^+-ATPase and the incorporated Na^+,K^+-ATPase can be distinguished by the sensitivity of the former to externally added ouabain. The unincorporated Na^+,K^+-ATPase is also more sensitive to sonication so that it may be selectively denatured. Sonication produces a parallel inactivation of the ouabain-insensitive Na^+,K^+-ATPase (ATP site on outer surface and ouabain site on inner surface) and reconstituted transport. At 37°, Na^+ transport is linear for 30 sec and K^+ transport is linear for about 15 sec. The data suggest that the rapid fall-off in K^+ transport is due to depletion of intravesicular K^+ in liposomes containing Na^+,K^+-ATPase. The coupling ratio of $Na^+:K^+$ transport at both 27 and 37° is 3:2. The same ratio has been reported for Na^+,K^+-ATPase liposomes prepared by the cholate dialysis technique and incubated at 27°. However, the transports of Na^+ and K^+ are not tightly coupled since a variety of factors such as depletion of the proteoliposomes of K^+,[20] and substitution of sulfate for chloride,[23] can alter the coupling ratio. Chemical modifications such as addition of wheat germ agglutinin,[24] or partial proteolytic cleavage,[32] change the coupling ratio, which in this case is presumably due to a change in stoichiometric ratio.

Acknowledgments

We wish to thank Karen Wipperfurth for her dedication and skill in the preparation of this manuscript. This work was supported by grants from the National Institutes of Health (HL 16318) and the National Science Foundation (PCM 76-20602).

[32] P. L. Jørgensen and B. M. Anner, *Biochim. Biophys. Acta* **555**, 485 (1979).

[15] Incorporation of $C_{12}E_8$-Solubilized Na^+,K^+-ATPase into Liposomes: Determination of Sidedness and Orientation

By FLEMMING CORNELIUS

Introduction

Over the past 10 years reconstitution has become widely used to study the transport properties of Na^+,K^+-ATPase.[1-5] With purified membrane-bound or detergent-solubilized Na^+,K^+-ATPase in the test tube the catalytic properties can be investigated. However, extracellular and cytoplasmic sides are exposed simultaneously to the same ligands and with such "unsided" preparations ion transport, which is an essential characteristic of this enzyme, cannot be studied.

Upon reconstitution of purified detergent-solubilized Na^+,K^+-ATPase into phospholipid vesicles the "sidedness" of the Na^+,K^+-ATPase is reestablished and vectorial transport processes between the internal and external compartments can be studied.

In order to conclude that all components of the Na,K-pump as characterized in the intact cell have been reconstituted, full recovery of both catalytic activity and transport capacity is required.

To establish this it is necessary to (1) determine both the orientation of the inserted enzyme molecules and the total amount of enzyme inserted,[6] and (2) ensure optimal intravesicular ligand concentrations, which constitutes a difficulty owing to the small volume of the proteoliposomes.[7] In

---

[1] S. M. Goldin and S. W. Tong, *J. Biol. Chem.* **249**, 5907 (1974).
[2] S. Hilden and L. E. Hokin, *J. Biol. Chem.* **250**, 6296 (1975).
[3] E. Racker and L. W. Fischer, *Biochem. Biophys. Res. Commun.* **67**, 1144 (1975).
[4] S. M. Goldin, *J. Biol. Chem.* **252**, 5630 (1977).
[5] B. M. Anner, L. K. Lane, A. Schwartz, and B. J. R. Pitts, *Biochim. Biophys. Acta* **467**, 340 (1977).
[6] The enzyme can be inserted inside out, i.e., with the cytoplasmic side exposed (i : o), right side out, i.e., with the extracellular side exposed (r : o), or with both sides simultaneously exposed [nonoriented (n–o)]. Therefore it is important to measure the sidedness in order to determine how many of the incorporated enzyme molecules participate in the reaction. Moreover, a way to reopen the vesicles without loss of activity is needed in order to gain access to the internal directed substrate sites of r : o-oriented enzyme molecules when assaying total enzyme activity.
[7] When ATP is added externally, there is a danger of i : o-oriented enzyme molecules rapidly emptying the intravesicular pool of K^+. In order to avoid this, K^+ has to be replenished intravesicularly. This can be accomplished by adding the K^+–ionophore valinomycin in combination with the protonophore carbonyl cyanide *m*-chlorophenylhydrazone (CCCP). The latter will provide a pathway for an H^+ ion to be transported in exchange for the K^+-transported electrogenically with valinomycin.

order to accomplish successful reconstitution it is essential to use (1) a detergent which is not deleterious to the enzyme in concentrations necessary to ensure solubilization of both the enzyme and the lipid, (2) a proper lipid composition as well as the use of highly purified phospholipids, and (3) a proper protein-to-lipid concentration ratio.[8]

This chapter describes a method to incorporate purified and solubilized Na^+,K^+-ATPase from rectal glands of the spiny dogfish into large unilamellar phospholipid vesicles without loss of enzyme activity or transport capacity. Furthermore, a method to determine the orientation of the inserted enzyme molecules is described.

Formation of Phospholipid Vesicles Containing Na^+,K^+-ATPase

Preparation of Solubilized Na^+,K^+-ATPase

The principles and methods of preparing the solubilized Na^+,K^+-ATPase from shark rectal gland are described elsewhere in this volume.[9] In essence, the detergent octaethylene glycol dodecyl monoether ($C_{12}E_8$), the purified membrane-bound Na^+,K^+-ATPase,[10] and NaCl are gently mixed at 0° to a final concentration of 4 mg $C_{12}E_8$/mg Na^+,K^+-ATPase and 150 mM NaCl. After a 1-hr centrifugation at 280,000 g at 10° the solubilized enzyme is collected from the supernatant. Usually a protein concentration of about 3.0 mg/ml is used. The catalytic activity of the solubilized enzyme is determined as described elsewhere in this volume.[11] The specific enzyme activity is typically 600–900 μmol P_i/mg·hr at 22°, pH 7.0, and the *p*-nitrophenylphosphatase activity (*p*NPPase) 100–150 μmol/mg·hr.

Reconstitution of Vesicles with Na^+,K^+-ATPase

The phospholipids required in the reconstitution are a mixture of highly purified phosphatidylcholine (PC), phosphatidylethanolamine (PE), phosphatidylinositol (PI), and cholesterol in a weight ratio of 60:14:2:24. The bovine lipids are obtained from Avanti Polar Lipids, Alabama. The lipids are dissolved in chloroform (e.g., 10 mg/ml) and stored at −20°.

The method describes the preparation of about 2.0 ml proteoliposomes (reconstituted liposomes) with a protein-to-lipid weight ratio of 1:20 and

[8] F. Cornelius and J. C. Skou, *Biochim. Biophys. Acta* **772,** 357 (1984).
[9] M. Esmann, this volume [7].
[10] J. C. Skou and M. Esmann, *Biochim. Biophys. Acta* **567,** 436 (1979).
[11] M. Esmann, this volume [10].

containing 130 mM NaCl, 4 mM MgCl$_2$, and 30 mM histidine, pH 7.0. These are used for determining the sidedness. In order to measure Na:K exchange the liposomes are prepared in 130 mM NaCl, 20 mM KCl, 4 mM MgCl$_2$, and the same buffer is used. Usually multiple samples are prepared, pooled, and stored in small vials at $-70°$ where the proteoliposomes maintain their properties for several months.

PC (500 μl), 125 μl PE, 15 μl PI (all 10 mg/ml CHCl$_3$), and 20 μl cholesterol (100 mg/ml CHCl$_3$) are transferred to the bottom of a 100 × 15.5 mm round-bottom glass test vial. Carefully thaw the frozen lipid stocks before pipetting. The chloroform is completely evaporated by gassing with N$_2$ using a rotary evaporator (Büchi). A thin, even lipid film is deposited inside the test tube. The test tube is transferred to a desiccator, dried under a vacuum for 1 hr at room temperature, filled with N$_2$, and stoppered.

One thousand microliters C$_{12}$E$_8$ solution (10 mg C$_{12}$E$_8$/ml in 130 mM NaCl, 4 mM MgCl$_2$, and 30 mM histidine, pH 7.0) is added and the lipids are dissolved by sonic oscillations in a bath sonicator (Metason 200, Struers, Denmark) at room temperature. After sonication for 3–6 min the contents of the tube become translucent and opaque and the test tube is placed on ice.

Solubilized Na$^+$,K$^+$-ATPase (0.42 mg) (140 μl 3.0 mg/ml) is added to the dissolved lipids and the contents gently mixed. The volume is adjusted to 2.5 ml with cold buffer containing 130 mM NaCl, 4 mM MgCl$_2$, 30 mM histidine, pH 7.0. The mixed protein:lipid:detergent micelle solution contains 8.4 mg lipids, 10 mg detergent, and 0.42 mg protein.

In order to form closed vesicles with reconstituted Na$^+$,K$^+$-ATPase the detergent concentration must be lowered from the initial 4 mg/ml to below 1 mg/ml (Fig. 1), by adsorption to polystyrene beads (Bio-beads Sm-2, Bio-Rad, CA): 300 mg of Bio-beads prepared according to Holloway[12] and washed with the appropriate buffer solution is added and the tube is gently rocked overnight at 4° on a tissue incubater with 10 oscillations/min followed by 1 hr at room temperature with new Bio-beads. Usually about 2 ml proteoliposomes with a protein concentration of about 0.14 mg/ml is collected from the supernatant after sedimentation of the beads by a 15-min centrifugation at 20,000 g at 4°.

Protein

The protein determination after reconstitution is difficult due to interference from the lipids using the Lowry method.[13] Moreover, histidine

---
[12] P. W. Holloway, *Anal. Biochem.* **53**, 304 (1973).
[13] O. H. Lowry, N. J. Rosebrough, A. L. Farr, and R. J. Randall, *J. Biol. Chem.* **193**, 265 (1951).

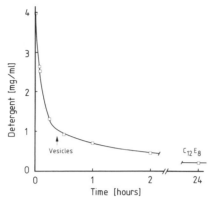

FIG. 1. Time course of $C_{12}E_8$ elimination from a mixed protein:lipid:detergent micelle suspension at 4°; 300 mg Bio-beads was added to 2.5 ml mixed micelle suspension initially containing 0.17 mg/ml protein, 3.3 mg/ml lipid, and 4 mg/ml detergent. The arrow indicates the time of appearance of Na$^+$,K$^+$-ATPase activity, which is insensitive to external ouabain (i.e., the formation of closed vesicles). The $C_{12}E_8$ level reached after 24 hr corresponds to 0.73 g/g protein.

interferes with the Lowry method. Therefore the modified method of Peterson[14] is employed. In this the protein is quantitatively precipitated with sodium deoxycholate and trichloroacetic acid. (TCA), followed by resuspension in H_2O, and sodium dodecyl sulfate (SDS) is included in the copper tartrate solution in order to leave the lipids transparent and noninterfering. Proteoliposome samples containing 10–100 μg protein are used and a standard curve using bovine serum albumin is run in parallel. The recovery of protein after Bio-bead treatment is greater than 80%.

ATPase Activity

The enzyme activity of reconstituted Na$^+$,K$^+$-ATPase is determined as for solubilized enzyme[11] with the exception that $C_{12}E_8$ is omitted in the assay solution. For samples preincubated in the presence of 0.2 mg/ml $C_{12}E_8$ in order to reopen closed vesicles this concentration of $C_{12}E_8$ is also included in the assay medium (see below). The assay medium contained (mM): 130 Na$^+$, 20 K$^+$, 4 Mg^{2+}, 3 ATP, 0.2 EGTA, 20 histidine (pH 7.0 at 22°), and 0.66 mg/ml albumin. The reaction is started by adding 50 μl proteoliposomes to 450 μl assay medium and incubating at 22°. After a given time interval the reaction is terminated by the addition of TCA. P_i is determined according to Baginsky *et al.*[15] with addition of 5% SDS to the arsenite–citrate reagent.

[14] G. L. Peterson, *Anal. Biochem.* **83,** 346 (1977).
[15] E. S. Baginski, P. P. Foa, and B. Zak, *Clin. Chim. Acta* **13,** 326 (1967).

Determination of Sidedness

The Ouabain/Ionophore Method

Assay Principle. This method determines the hydrolytic activities of Na^+,K^+-ATPase oriented either n–o, i:o, or r:o as a fraction of the total catalytic activity of reopened proteoliposomes. Since the turnover rates of the enzyme molecules in the different orientations are identical[8] these fractional activities reflect the fractional amount of enzyme molecules inserted in the different orientations.

Essentially, three proteoliposome samples are prepared as follows: proteoliposomes with 130 mM Na^+, 4 mM Mg^{2+}, and 30 mM histidine, pH 7.0, are either preincubated with K^+ (final concentration 20 mM) in the presence of valinomycin and CCCP (step 1 in Table I) in order to equilibrate the proteoliposomes with K^+ intravesicularly, or preincubated with $C_{12}E_8$ (step 3 in Table I) in order to reopen closed proteoliposomes, or finally preincubated without additions (step 2 in Table I).

Using these three proteoliposome preparations the Na^+,K^+-ATPase activity is determined with or without ouabain in the assay medium. Ouabain does not penetrate the proteoliposomes. Table I shows which orientations of the inserted Na^+,K^+-ATPase participate in the hydrolytic reaction under the different experimental conditions.

Assay Procedure. The three types of proteoliposome samples (1, 2, and 3), which are all prepared from the same batch of proteoliposomes with 130 mM Na^+ and no K^+ (Na^+ proteoliposomes), are as follows: (1) proteoliposomes which have been equilibrated with K^+, (2) Na^+ proteo-

TABLE I
HYDROLYTIC ACTIVITY OF PROTEOLIPOSOMES

Preincubation step	Internal/external ligands (Na, K)	Configurations which participate in hydrolysis[a]		
		−ouabain	+ouabain	+digitoxigenin
1. K^+ and valinomycin/CCCP	Na,K/Na,K	i:o + n–o (A)	i:o + ε(n–o) (C)	—
2. No addition	Na/Na,K	n–o (B)	ε(n–o) (D)	None (E)
3. $C_{12}E_8$	Na,K/Na,K	r:o + i:o + n–o (F)	None[b] (G)	None (G)

[a] (A–G) indicate the specific hydrolytic activities measured under the specified conditions.
[b] Usually digitoxigenin rather than ouabain is used here (see text).

liposomes, and (3) reopened (unsided) liposomes. The three samples are prepared as follows:

1. 275 μl Na$^+$ proteoliposomes
   30 μl Na$^+$ (130 mM), K$^+$ (200 mM), Mg^{2+} (4 mM), histidine (30 mM), pH 7.0
   5 μl valinomycin and CCCP (0.5 mg/ml and 1 mg/ml ethanol)

Incubate at 22° for 2 min and add 460 μl ice-cold buffer with 130 mM Na$^+$, 20 mM K$^+$, 4 mM Mg^{2+}, and 30 mM histidine, pH 7.0. Place on ice.

2. 400 μl Na$^+$ proteoliposomes

Incubate 2 min at 22° and add 720 μl ice-cold buffer with 130 mM Na$^+$, 4 mM Mg^{2+}, and 30 mM histidine, pH 7.0. Place on ice.

3. 100 μl Na$^+$ proteoliposomes
   1100 μl C$_{12}$E$_8$ solution containing 0.22 mg C$_{12}$E$_8$/ml, 130 mM Na$^+$, 20 mM K$^+$, 4 mM Mg^{2+}, and 30 mM histidine, pH 7.0
   5 μl valinomycin and CCCP (0.5 mg/ml and 1 mg/ml ethanol)

Incubate 2 min and place on ice.

From each of these three preparations the ATPase activity is determined in duplicate on 50-μl samples by incubation for, e.g., 1/2, 1, and 3 min in 450 μl of test solution (see above). Figure 2 shows a typical experiment.

The fraction of enzyme molecules incorporated in the i:o orientation is calculated as (see Table I)

$$f(i:o) = (A - B)/(F - G) \quad \text{or} \quad (C - D)/(F - G)$$

the fraction of n–o as

$$f(n-o) = (B - E)/(F - G)$$

and the fraction of r:o as

$$f(r:o) = (F - A)/(F - G)$$

As indicated in Table I the activity is determined on the three preparations (1, 2, and 3) both in the absence of ouabain (column 1) and in the presence of ouabain (column 2). In the presence of ouabain a fraction of hydrolytic activity due to n–o-oriented enzyme is included [ε(n–o)] which is caused by the delay in ouabain inhibition. $f(i:o)$ calculated from measurements in both the absence $(A - B)/(F - G)$, and in the presence of

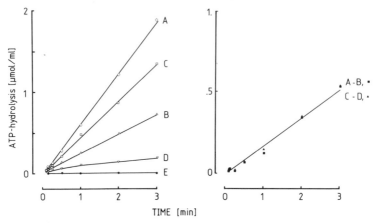

FIG. 2. *Left:* ATP hydrolysis at 22° as a function of time using proteoliposomes in the absence or presence of ouabain and in the absence or presence of intravesicular K^+ (see Table I). A and B: Proteoliposomes with intravesicular K^+ either in the absence (A) or presence (B) of ouabain. C (−ouabain) and D (+ouabain) represent proteoliposomes without intravesicular K^+. E: Proteoliposomes tested with digitoxigenin in the assay medium (blank). *Right:* The difference between A and B (■) and between C and D (▲). The slopes, which are identical, are proportional to the fraction of i : o-oriented enzyme molecules, as explained in the text.

ouabain (C − D)/(F − G) is identical, which indicates that ouabain and ATP cannot penetrate the liposomes. In order to calculate the total activity of reopened liposomes and f(n–o), blanks (E, G) are included. Here digitoxigenin (10^{-4} M) is preferable to ouabain because it is lipid soluble, readily penetrates the liposomes, and the inhibition is almost instantaneous.

The reason for using valinomycin and CCCP in combination in order to equilibrate the Na^+ proteoliposomes with K^+ is to short circuit the K^+ diffusion potential which is generated by the K^+ ionophore valinomycin, and which opposes equilibration.[8]

Determined in this way and using 1 : 20 protein : lipid liposomes, about 15% of the enzyme is oriented i : o, about 65% shows the opposite orientation (r : o), while 20% is unoriented (n–o).

Vanadate-Binding Method

Radioactive vanadate can be used as a marker of enzyme sidedness[8] since it has been found to bind exclusively to the cytoplasmic side of Na^+,K^+-ATPase. The presence of Mg^{2+} is necessary for binding of vanadate, Na^+ opposes its binding, but extracellular K^+ can abolish this Na^+ inhibition.[16]

Since vanadate binding is taking place at or near the ATP-binding site an almost identical methodology for detecting the enzyme sidedness can be employed, as described above using Na^+ or $Na^+ + K^+$ proteoliposomes; only now vanadate binding instead of ATP hydrolysis is being measured.

Assay Principle. The two different kinds of proteoliposomes (Na^+ vs. $Na^+ + K^+$) are incubated with [^{48}V]vanadate in the presence or absence of intravesicular K^+. After incubation the bound vanadate is measured after passage through a Sephadex column. The number of vanadate-binding sites calculated in each case is compared with the number calculated in experiments where the total number of vanadate-binding sites is measured on reopened proteoliposomes. In essence, the following four labeling experiments are performed. (1) Na^+ proteoliposomes are incubated with vanadate, (2) Na^+ proteoliposomes are incubated with vanadate and K^+, (3) $Na^+ + K^+$ proteoliposomes are incubated with vanadate, and (4) $Na^+ + K^+$ proteoliposomes are incubated with vanadate after reopening with $C_{12}E_8$.

From (1) the unspecific vanadate binding is calculated; it is practically zero. In (2) only enzyme molecules which have both the vanadate-binding site and the site for extracellular K^+ exposed are labeled, i.e., the n–o-oriented enzyme molecules. In (3) both n–o and i:o enzyme molecules can bind external vanadate since K^+ is present at both sides of the liposomes. In (4) all possible enzyme orientations can bind vanadate. As for the "ouabain/ionophore" method the fractions of i:o, r:o, and n–o can be calculated. The fraction of i:o-oriented enzyme molecules is the difference of calculated binding sites in (3) and (2) relative to the calculated total binding sites (4), the fraction of n–o-oriented enzyme molecules is the calculated number of binding sites in (2) relative to the total number of binding sites (4). Finally, the fraction of r:o-oriented enzyme molecules is the difference of calculated binding sites in (3) and (4) relative to the calculated total binding sites (4).

Assay Procedure. Before use the vanadyl ion VO^{2+}(IV) has to be oxidized to the active vanadium ion, VO_3^-(V). This is easily achieved by exposing the stock [^{48}V]vanadyl chloride (in 1 N HCl) to air after neutralization with 1 N NaOH. After incubation for 2 hr. Tris buffer is added and pH adjusted to 8.5. Before use the vanadium is diluted with an appropriate buffer to about 4 μM.

---

[16] L. C. Cantley, M. Resh, and G. Guidotti, *Nature (London)* **272**, 552 (1978).

Proteoliposomes with ($Na^+ + K^+$ proteoliposomes) or without (Na^+ proteoliposomes) intravesicular K^+ (20 mM) are incubated for 30 min at 22° as follows:

1. 400 μl Na^+ proteoliposomes
   100 μl 4 μM [^{48}V]vanadate in a buffer containing 130 mM Na^+, 4 mM Mg^{2+}, and 30 mM histidine, pH 7.0
   5 μl H_2O

2. 400 μl Na^+ proteoliposomes
   100 μl 4 μM [^{48}V]vanadium in a buffer containing 130 mM Na^+, 4 mM Mg^{2+}, and 30 mM histidine, pH 7.0
   5 μl 3 M KCl

3. 400 μl $Na^+ + K^+$ proteoliposomes
   100 μl 4 μM [^{48}V]vanadate in a buffer containing 130 mM Na^+, 20 mM K^+, 4 mM Mg^{2+}, and 30 mM histidine, pH 7.0
   5 μl H_2O

4. 400 μl $Na^+ + K^+$ proteoliposomes
   100 μl 4 μM [^{48}V]vanadate in a buffer containing 130 mM Na^+, 20 mM K^+, 4 mM Mg^{2+}, and 30 mM histidine, pH 7.0
   5 μl $C_{12}E_8$ (20 mg/ml)

After incubation 400 μl of each is eluted through a 1 × 10 cm Sephadex G-50 column at 4°. The columns are equilibrated and eluted with buffer containing either Na^+,[1] $Na^+ + K^+$,[2,3] or $Na^+ + K^+$ + 0.2 mg/ml $C_{12}E_8$.[4] Forty fractions, each 0.5 ml, are collected and the activity determined by counting in a gamma counter. Two 25-μl samples are collected before elution and counted directly (standards). From the activity eluting in the void column the number of binding sites is calculated in each experiment as

$$\frac{\text{activity in void (cpm/ml)}}{\text{specific activity of } VO_3^- \text{ (cpm/nmol)} \times \text{protein concentration (mg/ml)}}$$

which has unit nmol/mg protein. The specific activity of VO_3^- is determined from the standards. Figure 3 shows a typical experiment, corresponding to incubations 1, 2, and 3. Figure 4 shows the time course of [^{48}V]vanadate binding to solubilized enzyme and to reopened proteoliposomes (incubation 4).

The two methods, the ouabain ionophore and the vanadate-binding methods, compare satisfactorily in determining the orientation after reconstitution.[8]

Protein : Lipid Ratio. The protein : lipid weight ratio of the proteolipo-

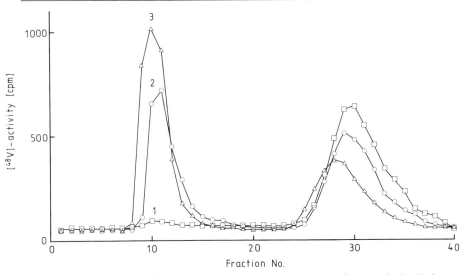

FIG. 3. [^{48}V]Vanadate binding to proteoliposomes in the presence of extravesicular K$^+$ in the absence (○–○) or presence (△–△) of intravesicular K$^+$ and without K$^+$ (□–□). Proteoliposomes are prepared in 130 mM Na$^+$, 4 mM Mg^{2+}, and 30 mM histidine, pH 7.0, and incubated with 4 μM [^{48}V]vanadate, 130 mM Na$^+$, 20 mM K$^+$, 4 mM Mg^{2+}, and 30 mM histidine, pH 7.0 (○–○), or without K$^+$ (□–□). Alternatively, the incubation is carried out with proteoliposomes prepared to contain 130 mM Na$^+$, 20 mM K$^+$, 4 mM Mg^{2+}, and 30 mM histidine, pH 7.0 (△–△). After a 30-min incubation at 22° the elution pattern is detected after passage through a Sephadex G-50 column at 4°. (△–△): [^{48}V]vanadate binding to n–o + i:o; (○–○): [^{48}V]vanadate binding to i:o; (□–□): unspecifiic vanadate binding. In the experiment shown the calculated VO$_3^-$-binding sites are as follows: 1.53 nmol/mg (n–o + i:o), 0.944 nmol/mg (n–o), and 3.4 nmol/mg total binding sites (not shown) giving 28% n–o, 17% i:o, and 55% r:o.

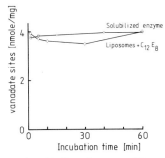

FIG. 4. The number of vanadate binding sites per milligram protein using either solubilized shark enzyme or reopened liposomes with reconstituted shark enzyme. The number of binding sites is in both cases about 4 nmol/mg protein. The binding of [^{48}V]vanadate is determined as described in the text and in Fig. 3 after various times of incubation at 22°, followed by gel chromatography.

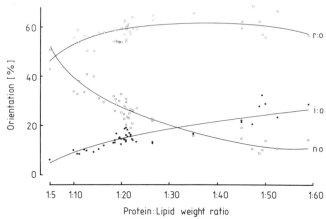

FIG. 5. The sidedness of proteoliposomes reconstituted with different protein:lipid weight ratios. The concentration of lipid is 3.3 mg/ml in all cases and the amount of protein varied between 0.66 mg/ml (1:5, protein:lipid) and 0.055 mg/ml (1:60, protein:lipid).

somes is important both for the orientation after reconstitution and for the recovered total activity. Figure 5 shows results from about 50 preparations of proteoliposomes using different protein:lipid ratio. As seen, the fraction of r:o orientations stays rather constant while the fraction of i:o

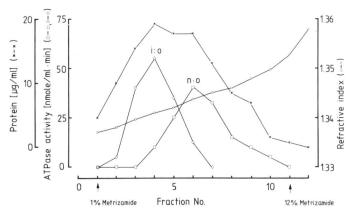

FIG. 6. Isoosmotic metrizamide gradient centrifugation of proteoliposomes (1:20, protein:lipid). Proteoliposomes (1 ml) were centrifugated for 3 hr at 280,000 g on a 1–15% (w/v) isoosmotic metrizamide gradient. The metrizamide concentration was increased by isoosmotic replacement of NaCl. ×–×, protein determined as in Ref. 14; ○–○, refractive index of fractions measured with an Abbe refractometer; ◇–◇ and □–□, ATPase activity originating from either i:o- or n:o-oriented enzyme molecules as determined using the ouabain/ionophore method.

increases and the fraction of n–o decreases when the protein–lipid weight ratio decreases.

The recovered specific enzyme activity and transport capacity is about 100% using a protein:lipid ratio greater than about 1:20. Decreasing the protein concentration below this value results in a decrease in recovered specific enzyme activity and a parallel decrease in transport capacity (Fig. 10 in Ref. 8).

It is not quite clear what n–o-oriented enzyme represents. Is it adsorbed enzyme, enzyme incorporated into leaky liposomes, or unincorporated enzyme, all of which will hydrolyze external ATP and show inhibition by external ouabain? A definite answer cannot be given at the moment but it seems not to represent unincorporated enzyme, since all protein is found together with the lipid on sucrose density gradients.[8] Using isoosmolar metrizamide density gradients[17] the i:o-orientated and n–o-orientated proteoliposomes are separated as expected if n–o represents activity from leaky proteoliposomes (Fig. 6). As shown in Fig. 6, n–o activity is found at a greater density than i:o activity, in agreement with i:o activity being associated with closed proteoliposomes that float at low densities due to a trapped volume of less dense medium.

[17] K. Shortman, this series, Vol. 108 [10].

Section IV

Analysis of the Pump Cycle

[16] Measurement of Na^+ and K^+ Transport and Na^+,K^+-ATPase Activity in Inside-Out Vesicles from Mammalian Erythrocytes

By RHODA BLOSTEIN

Inside-out membrane vesicles prepared from human red blood cells have special advantages for studies of the behavior of the Na,K-pump. Thus, the catalytic site of Na^+,K^+-ATPase is directly accessible to substrates and interactions of the enzyme with cations and other ligands can be controlled separately at either membrane surface. Furthermore, their relatively low passive permeability, large size, and low pump density eliminate the problem of rapid changes in intravesicular cations inherent in experiments with enzyme reconstituted into small liposomes and thereby facilitate accurate initial rate measurements of ion translocation and ATP hydrolysis. The preparations can be stored for at least 1 week at 0° and are amenable to studies carried out over a wide range of conditions of temperature and pH.

The method used to prepare inside-out vesicles is essentially that described in 1970 by Steck *et al.*[1] with slight modifications, as described below. Using several criteria of membrane protein sidedness,[2-4] earlier studies verified that these membrane preparations are predominantly inside-out vesicles.[4] In particular, ouabain can neither bind to the external surface nor inhibit Na^+,K^+-ATPase of inside-out vesicles.[3] For practical purposes, cardiac glycoside-sensitive activity can be quantitated by using the lipophilic cardiac glycoside, strophanthidin, rather than ouabain.

Most of the well-documented properties of the Na,K-pump described in intact red cells and red cell membranes are readily observed in inside-out vesicles and, because of their direct accessibility to ATP and ADP, concomitant measurements of hydrolysis and pump activity can be carried out. Examples of the activities and properties observed in inside-out vesicles are shown in Table I and are discussed below.

1. *Na^+-K^+ exchange:* The ATP-dependent (strophanthidin-sensitive) $^{22}Na^+$ uptake (normal efflux) is associated with Na^+,K^+-ATPase, the latter measured as strophanthidin-sensitive hydrolysis of ATP. Vesicles

---

[1] T. L. Steck, R. S. Weinstein, J. H. Strauss, and D. F. Wallach, *Science* **168**, 255 (1970).
[2] J. A. Kant and T. L. Steck, *Nature (London) New Biol.* **240**, 26 (1972).
[3] J. R. Perronne and R. Blostein, *Biochim. Biophys. Acta* **291**, 680 (1973).
[4] E. Reichstein and R. Blostein, *J. Biol. Chem.* **250**, 6256 (1975).

TABLE I
ACTIVITIES TYPICAL OF INSIDE-OUT MEMBRANE VESICLES DERIVED FROM HUMAN RED CELLS

Parametric measured	Alkali cations present (mM)		Substrate (mM)	Activity (nmol/mg/min)	Reference
	Intravesicular	Extravesicular			
Na^+ influx: Na^+–K^+ exchange	K (5)	Na (20)	ATP (1)	5.4	14
Na^+ influx					
Na^+–K^+ exchange	K (0.5)	Na (5)	ATP (0.2)	3.3	10
Na^+/0 flux	None	Na (5)	ATP (0.2)	0.3	
Na^+ influx					
ATP dependent	Na (50)	Na (5)	ATP (0.1)	0.3	10
ATP–ADP dependent	Na (50)	Na (5)	+ ADP (0.3)	0.5	
Na^+,K^+-ATPase	K (0.2)	K (0.2), Na (2)	ATP (0.2)	1.5^a	Harvey and Blostein (unpublished)
K^+-dependent p-nitrophenyl-phosphatase	None	K (10)	pNPP (4.5)		
	K (10)	None	pNPP (4.5)		12
	K (0.2)	K (0.2), Na (10)	+ ATP (0.1)	1.5^a	
Phosphorylated intermediate (steady state)	None	Na (1), K $(0.2)^b$	ATP (0.002)	0.5	11
	Na (1)	Na (0.01), K (0.2)	ATP (0.002)	0.0	

^{a} 10^{-6} M valinomycin present. All assays were carried out at 37°.
^{b} 0.2 mM KCl was added to counteract effects due to traces of extravesicular (cytoplasmic) Na^+.

are either loaded with K^+ (intravesicular K^+, i.e., normally extracellular K^+) by preincubation or by adding K^+ plus valinomycin to the assay medium.

2. *Na^+/O flux:* The ATP-dependent (strophanthidin-sensitive) $^{22}Na^+$ uptake (normal efflux) is associated with Na^+-activated ATP hydrolysis in the absence of intravesicular alkali cations. This transport mode was referred to originally as an uncoupled Na^+ efflux from human red cells[5] and has been shown recently to be anion coupled.[6]

3. *Na^+-Na^+ exchange:* (1) $^{22}Na^+$ uptake requires intravesicular (normally extracellular) Na^+, ADP, and ATP as observed in intact red cells[7] and associated, presumably, with Na^+-dependent [^{14}C]ADP–ATP exchange[8,9]; (2) Na^+ uptake requires intravesicular Na^+ and only ATP and is associated with Na^+-activated ATP hydrolysis.[10]

4. *Phosphoenzyme intermediate:* Na^+-activated phosphorylation of Na^+,K^+-ATPase requires extravesicular (normally cytoplasmic) Na^+.[11]

5. Distinct side-specific effects of K^+ on strophanthidin-sensitive *p*-nitrophenylphosphatase are observed.[12]

Preparation of Inside-Out Vesicles and Adjustment of Intravesicular Ionic Composition

Reagents

154 mM choline chloride (purchased from Syntex Agri Business, Springfield, MO)
5 mM Tris/phosphate, pH 8.2–8.3 (H_3PO_4 titrated with Tris base)
0.5 mM Tris/phosphate, pH 8.2–8.3
0.2 mM $MgSO_4$
Dextran T-70 solution (4.46 g/100 ml 0.5 mM Tris/phosphate, pH 8.2)
40, 20, and 10 mM Tris/glycylglycine, pH 7.4, containing 0.2 mM $MgSO_4$

Isotonic NaCl or choline chloride-washed red cells obtained from fresh blood collected into 0.1 vol 0.1 M EDTA, pH 7.4, are lysed with 30–40 vol ice-cold (0°) 5 mM Tris/phosphate, pH 8.2–8.3, prepared by titrat-

[5] I. M. Glynn and S. J. D. Karlish, *J. Physiol. (London)* **256**, 465 (1976).
[6] S. Dissing and J. F. Hoffman, *Biophys. J.* **41**, 188a (1983).
[7] I. M. Glynn and J. F. Hoffman, *J. Physiol. (London)* **218**, 239 (1971).
[8] R. Blostein, *J. Biol. Chem.* **245**, 270 (1970).
[9] J. H. Kaplan and R. J. Hollis, *Nature (London)* **288**, 587 (1980).
[10] R. Blostein, *J. Biol. Chem.* **258**, 7948 (1983).
[11] R. Blostein, *J. Biol. Chem.* **254**, 6673 (1979).
[12] P. Drapeau and R. Blostein, *J. Biol. Chem.* **255**, 7827 (1980).

ing phosphoric acid with Tris base. The suspension is stirred thoroughly and allowed to stand on ice for 20 min. After centrifugation for 20 min at 30,000 g, using a fixed-angle rotor (e.g., rotor SS34 of the Sorvall RC-5 centrifuge), the clear red supernatant is removed and the pellet is resuspended in the same volume of 5 mM Tris/phosphate and after 20 min on ice the suspension is centrifuged as above. Removal of the supernatant, suspension, and centrifugation are repeated two or three times and the resulting white pellet is then suspended in about 30 vol 0.5 mM Tris/phosphate, pH 8.2, and allowed to stand at least 1 hr or, preferably, overnight at 0°. Following centrifugation, the pellet, suspended in a few milliliters of the same 0.5 mM Tris/phosphate buffer, is passed five times through a 1½-in. 27 gauge hypodermic needle and 0.2 mM (final concentration) MgSO$_4$ is added. The usual procedure is to then gradually add increasing amounts of 10 mM Tris/glycylglycine, pH 7.4, containing 0.2 mM MgSO$_4$, and after centrifuging the pellet is suspended in 20 mM Tris/glycylglycine buffer. Centrifugation and suspension are repeated with either 20 mM Tris/glycylglycine or 40 mM Tris/glycylglycine, pH 7.4, with MgSO$_4$ (0.2 mM) included in the buffers. The vesicles are then carefully suspended in the "final" buffer solution (either 20 or 40 mM Tris/glycylglycine, pH 7.4, containing 0.2 mM MgSO$_4$) at a membrane protein concentration of 1–3 mg/ml. The percentage membranes which are inside-out vesicles is determined by the acetylthiocholine esterase assay described by Steck.[13] Acetylcholinesterase is exposed to the extracellular membrane surface so that detergent-stimulated activity is a measure of the percentage inside-out vesicles. In general, ≥60% of the preparation comprises inside-out vesicles; the rest are predominantly membrane fragments rather than right-side-out vesicles which can also be quantitated as the fraction of detergent-stimulated glyceraldehyde phosphate dehydrogenase activity[13] or valinomycin plus K$^+$-activated p-nitrophenylphosphatase (pNPPase) activity.[12] Inside-out vesicles can be further separated from membrane fragments by layering the needled preparation over a dextran cushion [about 3 ml layered above 8 ml of a dextran T-70 solution (1.115 g/25 ml 0.5 mM Tris phosphate, pH 8.0)] and centrifuged for 1–2 hr at 100,000 g in a swinging bucket rotor; a simpler method is described by Mercer and Dunham,[14] whereby 2 ml of needled vesicles is layered on top of 3 ml of the dextran T-70 solution in a polycarbonate tube (11 × 120 mM) and centrifuged for 40 min at 30,000 g. The vesicles can be stored at 0° for periods up to 1 week.

[13] T. L. Steck, *Methods Membr. Res.* **2**, 245 (1974).
[14] R. Mercer and P. B. Dunham, *J. Gen. Physiol.* **78**, 547 (1981).

Precautions

1. It is important to (1) remove completely, following the first lysis and centrifugation, the white button which clings to the bottom of the 50-ml centrifuge tube and (2) ascertain that all solutions and suspensions are maintained at 0°.

2. It is advisable to heat the dextran solution in a boiling water bath for 15–20 min to eliminate interference from inactivating components, possibly proteases, found in some lots of dextran T-70 (Pharmacia).

Assays

General Principle

Prior to use, the vesicles are equilibrated with a solution of the same tonicity as that used in the final assay medium. Typically, an aliquot of vesicles stored in 20 mM Tris/glycylglycine, pH 7.4, is concentrated by centrifugation to 4–8 mg/ml and diluted 1 : 1 with a solution of twice the final concentration of choline chloride (usually 100 mM) and MgSO$_4$ (usually 2 mM) and the assay is then initiated by adding an equivalent amount of the final desired chloride salt (50 mM), or mixture thereof, MgSO$_4$ (1 mM), and Tris/glycylglycine (10 mM). The suspension is then equilibrated overnight at 0° and then for 30 min at 37°. The assay is initiated by mixing the vesicles with the appropriate medium (see below) containing 50 mM chloride salt, 1 mM MgSO$_4$, and 10 mM Tris/glycylglycine, pH 7.4, and usually 0.2 mM EGTA, pH 7.4. For most assays, it is convenient to add a small aliquot (0.1 vol) of chilled equilibrated vesicles (1–4 mg protein/ml) to a large volume (0.9 vol) of prewarmed medium. Final reaction volumes of 0.1–0.2 ml are convenient provided that glass micropipets (e.g., Lang–Levy or a calibrated capillary type such as the SMI pipettors) are used for sampling the vesicles.

Transport Assays

Reagents

Stock solutions: 1 M choline chloride (concentration checked by osmometry)
1 M NaCl
1 M KCl
0.2 M MgSO$_4$
0.4 M Tris/glycylglycine, pH 7.4
0.1 M Tris–EDTA, pH 7.4

[22]NaCl, [86]RbCl

Na⁺–K⁺(Rb⁺) Exchange. Measurements of pump-mediated ^{22}Na$^+$ influx (or efflux) and K$^+$ efflux, measured with the congener ^{86}Rb$^+$, are carried out by first equilibrating the vesicles with the desired concentration of alkali cation replacing choline and then initiating the reaction by diluting the suspension with NaCl-containing medium of the same tonicity (see above and Ref. 15). One time course, or one set of replicate tubes incubated for a given period, are assayed without ATP (baseline) and another with ATP. Alternatively, ATP is included in both sets and for one set, vesicles are pretreated with the cardiac glycoside strophanthidin, i.e., a small aliquot (0.002 vol) of an ethanolic solution of 0.05 M strophanthidin is added during the last 5 min of vesicle preincubation. It is important to carry out assays with sufficient K$^+$ (or Rb$^+$) inside the vesicles, i.e., at the normally extracellular surface (K$_{ext}$ or Rb$_{ext}$) and for sufficiently short periods to assure minimal depletion of K$_{ext}$ or Rb$_{ext}$. In some experiments, it is convenient to add K$^+$ (or Rb$^+$) plus valinomycin (10^{-6}–10^{-5} M) to the assay medium in order to maintain a constant intravesicular K$^+$ or Rb$^+$ content. Either ^{22}Na$^+$ is added to the medium for the measurement of ATP-dependent (or strophanthidin-sensitive) ^{22}Na$^+$ influx or the vesicles are equilibrated with ^{86}Rb$^+$ for the measurement of ^{86}Rb$^+$ efflux. The reaction is terminated after a convenient interval, e.g., 4 min for ^{22}Na$^+$ influx into vesicles assayed with 5 mM ^{22}NaCl, 0.5 mM K$^+$ plus valinomycin, or for shorter intervals and/or lower temperatures depending on the ATP concentration. ^{22}Na$^+$ or ^{86}Rb$^+$ retention by the vesicles is determined by filtration (1.2-nm Millipore type RA filter presoaked in wash solution) after dilution of an aliquot (0.05–0.25 ml) of the vesicle suspension in 10 ml of ice-cold nonradioactive medium contained in a Millipore funnel (No. XX10-025-14) and identical with that used for the assay except that, for ^{86}Rb$^+$ efflux measurements into Rb$^+$-free medium, 0.1 mM RbCl is included in the wash medium to effect removal of nonspecific ^{86}Rb$^+$ binding. The filter is then washed with 15 ml of the same ice-cold solution after which the funnel is removed from the base and the filter edges are washed with 1–2 ml of wash solution. Radioactivity associated with the filter is measured in a liquid scintillation counter.

Na⁺–O Flux and Na⁺ Exchange. The procedure for measuring ^{22}Na$^+$ influx via the Na$^+$–O flux or Na$^+$–Na$^+$ exchange is similar to that described above. It is generally convenient to equilibrate the vesicles with either 50 mM choline chloride or 50 mM NaCl for measurements of either Na$^+$–O flux or Na$^+$–Na$^+$ exchange, respectively and to dilute them 10-fold into medium containing ^{22}Na$^+$ and a small amount of carrier NaCl (0.1 mM) so that the extravesicular Na$^+$ concentration, normally cyto-

[15] R. Blostein, *J. Biol. Chem.* **258**, 12228 (1983).

plasmic Na^+, is about 5 mM.[10] Assays of $^{22}Na^+$ efflux via Na^+–Na^+ exchange are carried out similarly except that the $^{22}Na^+$-loaded vesicles are diluted with 100 vol of assay medium containing 5 mM NaCl and the assay period is increased to 30 or 60 min.

Enzyme Assays

Reagents

Stock solutions as described for transport assays.
ATP (Tris form, vanadate free)
[γ-^{32}P]ATP prepared as described by Post and Sen[16]
5% trichloroacetatic acid
5% trichloroacetic acid containing 5 mM Na_2ATP and 2.5 mM NaH_2PO_4
Norit A charcoal
p-Nitrophenylphosphate (Tris form)
0.2 N NaOH containing 2.5% sodium dodecyl sulfate and 4 mM EDTA

ATP Hydrolysis. Vesicles are equilibrated and assays are initiated in the same manner as that described for transport measurements except that [γ-^{32}P]ATP prepared as described by Post and Sen[16] is used and $^{32}P_i$ released is measured after separation from [γ^{32}P]ATP bound to charcoal. The reaction volume is either 0.1 or 0.2 ml and the reaction is terminated by adding 0.9 or 0.8 ml, respectively, of an ice-cold solution (TCAI) comprised of 5% (w/v) trichloroacetic acid containing 5 mM Na_2ATP and 2.5 mM NaH_2PO_4. Following centrifugation, 0.9 ml of supernatant is removed and 0.45 ml 5% trichloroacetic acid containing Norit A charcoal (15 g/100 ml) is added. The suspension is kept at 0° for about 1 hr, mixed every 15–20 min, and then centrifuged. An aliquot of the clear supernatant is removed for the measurement of radioactivity ($^{32}P_i$) by liquid scintillation spectrometry. Assays are usually carried out in two sets of triplicates, one set with Na^+ and the other with either Na^+ replaced by K^+ or strophanthidin added, the difference being the Na^+-stimulated (or strophanthidin-sensitive) component.

Phosphoenzyme Intermediate. Phosphoenzyme can be quantitated following brief incubation of inside-out vesicles with very low concentrations of [γ-^{32}P]ATP (see Table I) in the presence of Na^+ and in the presence of K^+ (baseline phosphorylation) as described above for the ATP

[16] R. L. Post and A. K. Sen, this series, Vol. 10, p. 773.
[17] O. H. Lowry, N. J. Rosebrough, A. L. Farr, and R. J. Randall, *J. Biol. Chem.* **193**, 265 (1951).

hydrolysis assays. Following incubation (usually 15 sec at 37°), an aliquot of the TCAI-treated membranes is filtered on a glass fiber filter (Reeve Angel, 2.4 cm), fitted, and prewet on a perforated 2.4-cm porcelain funnel. The sample tube is rinsed with 1 ml of TCAI, the filter is washed with this solution, and then with 8–9 ml of TCAI added in portions of approximately 1.5 ml, allowing the funnel to fill and empty after each addition by applying vacuum. The filters are counted by liquid scintillation spectrometry. The vesicle protein added is determined[17] and the ^{32}P bound is expressed in terms of the amount of vesicle protein added. The phosphoenzyme intermediate is the ^{32}P bound in the presence of Na^+ after subtraction of the ^{32}P bound in the presence of K^+.

K^+-*Activated Phosphatase*. Vesicles are equilibrated and the reaction initiated as described above except that the final $MgCl_2$ concentration is 4 mM and 4.5 mM p-nitrophenylphosphate (Tris form, Tris/pNPP) is used as substrate. K^+ activation of pNPP hydrolysis is observed under either of the following conditions: (1) KCl is added to the reaction medium or (2) vesicles loaded with KCl as described above are added to medium containing 4.5 mM pNPP plus NaCl and ATP (See Table I and Ref. 12). Following incubation for 5 to 10 min at 37°, the reaction is stopped by addition of an equal volume of 0.2 N NaOH, 2.5% sodium dodecyl sulfate, and 4 mM EDTA as described by Ottolenghi.[18] Absorption of the product p-nitrophenol at 410 nm is determined. Specific K^+-activated phosphatase is the activity corrected for activity observed in the absence of KCl and presence of choline chloride.

Comments

The methods described here are probably generally applicable to mature and even immature red cells of various mammals. For example, inside-out (as well as right-side-out) vesicles have been prepared from sheep reticulocytes[19] and, as might be expected, the Na,K-pump activity is several-fold greater than that observed with vesicles from mature cells. Furthermore, even when the yield in terms of the percentage of membranes which are vesicles is relatively poor, transport measurements are not compromised since only the activity of the competent vesicles is observed.

[18] P. Ottolenghi, *Biochem. J.* **151**, 61 (1975).
[19] A. M. Weigensberg, R. M. Johnstone, and R. Blostein, *J. Bioenerg. Biomembr.* **14**, 335 (1982).

[17] Measurement of Active and Passive Na$^+$ and K$^+$ Fluxes in Reconstituted Vesicles

By S. J. D. Karlish

It is now recognized that in addition to the normal physiological function of active ATP-dependent Na$^+$–K$^+$ exchange, the Na,K-pump is capable of sustaining several abnormal modes of cation transport, including reversal of Na$^+$–K$^+$ exchange, uncoupled Na$^+$ efflux associated with ATP hydrolysis, (ATP + ADP) or ATP-dependent Na$^+$–Na$^+$ exchange, and (ATP + P$_i$)-dependent K$^+$–K$^+$ exchange. Study of the abnormal modes has been particularly informative for understanding the mechanism of transport.[1,2] In the past 5 years we have moved away from the classical red cell systems for studying Na,K-pump fluxes and developed the use of phospholipid vesicles reconstituted with partially purified pig kidney Na$^+$,K$^+$-ATPase.[3] The major advantages of this system for transport studies are (1) convenience, (2) simplicity in that it contains essentially only one protein, (3) the sidedness of ligand effects can be readily determined, and (4) the vesicles themselves show a very low passive permeability to cations. Exploitation of these features, in conjunction with sensitive transport assays, has led to the discovery of previously unknown modes of slow passive cation fluxes via the Na,K-pump,[4] and modulation of the Na,K-pump function by transmembrane allosteric effects of Na$^+$ ions.[5] In this chapter I describe the methods used for reconstituting vesicles and measuring the various flux modes of the Na,K-pump, utilizing particular examples to illustrate important features of experimental design and pitfalls. The original experiments and detailed interpretations are to be found in Refs. 3–7.

Reconstitution by Freeze–Thaw Sonication[3,8]

Na$^+$,K$^+$-ATPase is prepared from pig kidney red outer medulla by the procedures of Jørgensen.[9] The protein (3–4 mg/ml specific activity, 15–20

[1] I. M. Glynn and S. J. D. Karlish, *Annu. Rev. Physiol.* **37**, 13 (1975).
[2] I. M. Glynn, *Enzymes Biol. Membr.* 35 (1984).
[3] S. J. D. Karlish and U. Pick, *J. Physiol. (London)* **312**, 505 (1981).
[4] S. J. D. Karlish and W. D. Stein, *J. Physiol. (London)* **328**, 295 (1982).
[5] S. J. D. Karlish and W. D. Stein, *J. Physiol. (London)* **359**, 119 (1985).
[6] S. J. D. Karlish and W. D. Stein, *J. Physiol. (London)* **328**, 317 (1982).
[7] S. J. D. Karlish, W. R. Lieb, and W. D. Stein, *J. Physiol. (London)* **32**, 333 (1982).
[8] M. Kasahara and P. Hinkle, *J. Biol. Chem.* **252**, 7384 (1977).
[9] P. L. Jørgensen, this series, Vol. 32, p. 277.

U/mg) is stored at $-70°$ in a solution containing sucrose, 250 mM; imidazole, 25 mM, pH 7.5; EDTA, 1 mM; dithiothreitol (DTT), 1 mM. Prior to reconstitution the enzyme is dialyzed in the cold room overnight against 1000 vol of a solution containing histidine, 25 mM, or imidazole, 25 mM, pH 7.0; and EDTA (Tris), 1 mM. Soybean phospholipid (Sigma P5638), 50 mg/ml, is suspended by prolonged vortexing in a solution containing Tris–HCl, 500 mM, pH 7.0; imidazole, 25 mM, pH 7.0; EDTA (Tris), 1 mM, or choline chloride, 500 mM; and histidine, 25 mM, pH 7.0. The lipid is sonicated to near clarity using a Bransonic 12 bath sonicator, and then dialyzed overnight in the cold against 1000 vol of a solution containing histidine, 25 mM, pH 7.0; and EDTA (Tris), 1 mM (with one change of dialyzing solution). Dialysis reduces contaminating K^+ and Na^+ ions to less then 50 μM. The small degree of dilution occurring during dialysis is calculated by comparing the optical density (light scattering) at 420 nm before and after dialysis. Cholic acid, 50 mg/ml, neutralized with Tris (to pH 7.0) is purified from crude cholic acid by the method of Kagawa and Racker.[10]

For optimal reconstitution, dialyzed sonicated lipid (40 mg/ml) at 0° is placed in a hard glass tube and salts, ligands, isotopes, etc., to be incorporated into the vesicles are added. In a separate tube, cholate (50 mg/ml) is added to Na^+,K^+-ATPase (3 mg/ml) at 0° to dissolve the protein, using a ratio of cholate : protein of 6 mg : 1 mg. If desired, undissolved protein is removed by centrifugation on a Beckman airfuge for 5 min at 30 psi. The protein/cholate mixture is added immediately to the lipid suspension, mixed lightly, frozen in liquid nitrogen, and then allowed to thaw at room temperature. After mixing with the protein/cholate mixture the ratio of lipid : protein should be 40 mg : 1 mg. The turbid suspension is sonicated on the bath sonicator until partial clearing is observed (usually about 1 min). The exterior medium of the vesicles is then exchanged for one of choice by centrifugation at low speeds on short columns of Sephadex G-50 (fine, particle size 20–80 μm) equilibrated with the medium of choice (cf Penefsky[11]). An optimal ratio of vesicles to Sephadex is 1 : 10 (v/v). Two consecutive centrifugations of 500 μl of vesicles on 5-ml columns of Sephadex reduce the concentrations of external ligands to 1/1000 of that in the original suspension.

Transport Assays: General Design Considerations

In all isotope influx or efflux assays, vesicles will be warmed to the desired temperature (usually 20–22°), and mixed with the reaction me-

[10] M. Kagawa and E. Racker, *J. Biol. Chem.* **246**, 5477 (1971).
[11] H. S. Penefsky, *J. Biol. Chem.* **252**, 2891 (1977).

dium containing the appropriate activators, inhibitors, etc., and an isotope, such as ^{22}Na or ^{86}Rb (the ^{86}Rb being used as a potassium analog). Isotope taken up or remaining in the vesicles is estimated by applying the suspension to 5- or 6-cm columns of Dowex 50-X8, 50–100 mesh (Tris form),[12] poured in Pasteur pipets fitted with glass wool plugs. This stops the flux and the vesicles are then eluted with 1.5 ml of ice-cold sucrose solution, 250 mM, into counting vials. For isotope uptake over very short incubations (up to 30 sec) timing is more accurate if the isotope flux is stopped quickly by first mixing the aliquot of vesicles (say 40 μl) with an ice-cold solution (say 100 μl) containing NaCl, 200 mM (for ^{22}Na fluxes), or RbCl, 200 mM (for ^{86}Rb fluxes), and then transferring to the Dowex column. The columns will remove about 99.99% of external isotope from 100 μl of suspension of total salt concentration, 200 mM. It is unnecessary to wash the columns with BSA.[12] Most assays are run in duplicate, the reproducibility usually being equal or better than ±5%. ^{86}Rb is counted by Cerenkov radiation and ^{22}Na by scintillation counting. It is convenient to quantify absolute flux rates in terms of nanomoles Na$^+$ or Rb$^+$ uptake per minute per 10 μl of vesicle. Provided the lipid, protein, and cholate concentrations used for reconstitution are those given above, 10 μl of vesicle will contain 1.55 ± 0.19 μg of Na$^+$,K$^+$-ATPase.[3]

In all isotope flux assays it is essential to (1) ensure that the rate is linear with respect to time, (2) determine the fraction of isotope flux sustained by the Na,K-pump, and (3) ascertain the orientation of the pumps—right side out or inside out catalyzing the flux of interest, since in the process of reconstitution, the pumps are inserted into the vesicles in a random orientation. Points (2) and (3) are defined by effects of activators such as ATP added to the exterior medium or inhibitors such as ouabain or vanadate present at one or both sides of the vesicles. ATP and vanadate, of course, combine with the cytoplasmic surface and ouabain combines with the extracellular surface.[1,2]

We refer to the face of the pump to which ligands bind as cytoplasmic or extracellular, irrespective of the orientation of the pumps in the vesicles.

ATP-Dependent Na$^+$–K$^+$ Exchange[3,5]

Time Course of ATP-Dependent ^{22}Na Uptake (Fig. 1)

Experimental Protocol. Vesicles (550 μl) loaded with KCl, 200 mM, were prepared as described above using the supernatant obtained by cen-

---

[12] O. Gasko, A. F. Knowles, H. G. Shertzor, E. M. Soulina, and E. Racker, *Anal. Biochem.* **72**, 57 (1976).

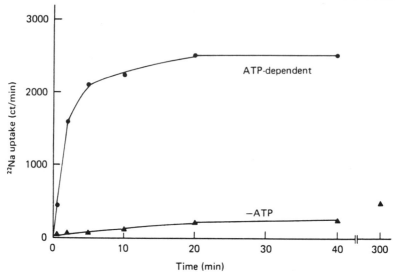

FIG. 1. Time course of ATP-dependent ^{22}Na uptake. Reproduced from Ref. 3.

trifuging the enzyme/cholate mixture in the airfuge. The vesicles were then centrifuged on a column of Sephadex G-50 equilibrated with Tris–HCl, 190 mM, pH 7.0; NaCl, 10 mM; and MgCl$_2$, 3 mM. For the assay 200 μl of vesicle suspension was added at room temperature to 300 μl of a reaction medium containing Tris–HCl, 190 mM; NaCl, 10 mM, plus ^{22}Na, 5 × 10^6 cpm; MgCl$_2$, 3 mM, with or without ATP (Tris), 6 mM. At the time indicated 40-μl samples (in duplicate) were transferred to Dowex columns.

Comments. (1) The experiment measures the fraction of ^{22}Na uptake dependent on ATP in the external medium, i.e., pumping of ^{22}Na from the cytoplasmic to extracellular face of inside-out-oriented pumps. This flux requires K$^+$ inside the vesicles.

(2) Because net cation movements are occurring, the initial rate of ^{22}Na uptake falls when K$^+$ is depleted from the vesicles. Linearity was maintained here for about 2 min. In general, if airfuge centrifugation is not performed, or if a high specific activity Na$^+$,K$^+$-ATPase is used for reconstitution, linearity will be maintained only for 0.5–1 min.

(3) The ^{22}Na uptake in the absence of ATP is the slow passive leak as demonstrated by (1) a linear dependence on Na$^+$ concentration (Fig. 2) and (2) in other experiments insensitivity to pump inhibitors.

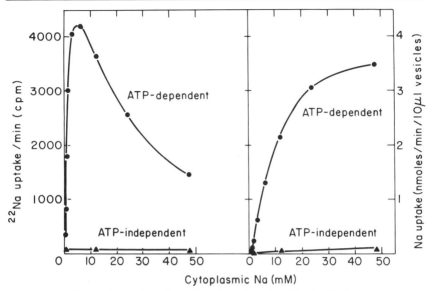

FIG. 2. Activation of cytoplasmic sodium (see Fig. 1 of Ref. 5).

Activation by Cytoplasmic Sodium (Fig. 2)

Experimental Protocol. Vesicles (600 μl) loaded with 200 mM KCl were prepared as described above and were centrifuged twice on Sephadex G-50 equilibrated with Tris–HCl, 200 mM, pH 7.0. For the assay 40 μl of vesicle suspension was mixed with 40 μl of reaction medium and, after a 1-min incubation, the suspension was removed to Dowex columns. The final concentrations of reaction components were NaCl, 0.19–47.6 mM; and Tris–HCl, pH 7.0, so that the sum of NaCl plus Tris–HCl was 200 mM; ^{22}Na, 7 × 10^5 cpm/tube; MgCl$_2$, 3 mM; with or without ATP (Tris), 3 mM.

Comments. Construction of cation activation curves: (1) The left-hand side of Fig. 2 shows the ATP-dependent ^{22}Na uptake and the right-hand side shows the conventional plot of rate against concentration. The procedure of using a fixed amount of radioactivity and a varying concentration of unlabeled ion is both convenient and accurate. With this design the radioactivity taken up is proportional to the absolute rate divided by the concentration of transported ion. For a hyperbolic activation (e.g., Fig. 3) unlabeled cation always competes with isotope for transport and hence the radioactivity falls monotonically. For a sigmoid activation in which more than one cation must be bound, isotope uptake will rise to a maximum before falling, as seen in Fig. 2. This occurs because at low concentrations unlabeled cations complete the binding requirements for transport and thus aid isotope uptake, while only at higher concentrations does competition prevail.

(2) The maximal rate of Na^+-K^+ exchange is 3–5 nmol Na/10 μl vesicles/min.

ATP-Dependent Na^+-K^+ Exchange in Vesicles Treated with Valinomycin Plus FCCP[3]

The following protocol allows measurement of ATP-dependent ^{22}Na uptake into vesicles, initially prepared without internal K^+, but suspended in a medium in which the ionophores cause K^+ to reenter the vesicles as fast as it is pumped out. Suspend vesicles, loaded with Tris–HCl, at room temperature in a medium containing Tris–HCl, 110 mM; NaCl, 30 mM; KCl, 10 mM; and MgCl$_2$, 3 mM. To 100 μl vesicles add 1 μl of valinomycin and 1 μl of p-trifluoromethoxycarbonyl cyanide phenylhydrazone (FCCP) (final concentration, 0.2 μM) and incubate 30 min. For assay add aliquots of vesicles to a reaction mixture containing NaCl, 30 mM; and KCl, 10 mM.

ATP-Dependent ^{86}Rb Efflux[3]

The protocol to use is essentially like that in Fig. 1 except that ^{86}Rb (1 μCi) is incorporated into the vesicles during formation and the reaction medium contains Mg^{2+}, unlabeled Na^+, and ATP. The Sephadex centrifugation step removes external ^{86}Rb. Using our procedure, isotope efflux measurements are intrinsically less accurate than uptake since one is looking at the isotope remaining in the vesicles. Also, about half of the vesicles do not contain pumps, so the ^{86}Rb trapped in this population cannot be pumped out and constitutes a large, nonvarying background. The latter problem can be overcome by selectively loading ^{86}Rb, by Rb^+-Rb^+ exchange, primarily into that fraction of vesicles with inside-out-oriented pumps. For this procedure (1) prepare vesicles with RbCl, 150 mM (but no ^{86}Rb), and centrifuge on Sephadex equilibrated with Tris–HCl, 150 mM. (2) Incubate the vesicles for 5 min at room temperature in a medium containing Tris–HCl, 140 mM, pH 7.0; RbCl, 2 mM, plus ^{86}Rb, 1 μCi; MgCl$_2$, 5 mM; phosphate/Tris, 10 mM; and ATP(Tris), 2 mM. (3) Recentrifuge twice, on Sephadex G-50 equilibrated with Tris–HCl, 150 mM, pH 7.0.

ATP-Dependent Na^+–Na^+ Exchange and Uncoupled Na^+ Flux[5]

See Fig. 3 of Ref. 5 for measurement of ATP-dependent ^{22}Na uptake into vesicles loaded with either 200 mM NaCl or 200 mM Tris–HCl, by protocols essentially like that in Fig. 2.

Rb^+–Rb^+ Exchange Activated by ATP and P_i[5,7] (Fig. 3)

Experimental Protocol

In the first experiment (see Fig. 3A), one set of vesicles was prepared to contain 150 mM RbCl and 1 mM $MgCl_2$. After reconstitution and centrifugation twice on Sephadex G-50 columns equilibrated with Tris–HCl, 150 mM, ouabain (0.8 mM) was added. Vanadate (Tris), 0.5 mM, plus $MgCl_2$, 1 mM, were added to one-half of the suspension. For the assay (3 min) 40 µl of vesicles (control or plus vanadate) was mixed with 40 µl of reaction medium. The final concentrations of components present

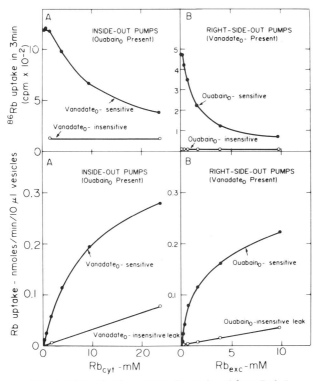

FIG. 3. Rb^+ activation curves. Reproduced from Ref. 5.

were as follows: RbCl, 0.23–23.9 mM; and Tris–HCl, pH 7.0, so that the sum of RbCl plus Tris–HCl was 150 mM; ^{86}Rb, 3 × 10^5 cpm/sample; MgCl$_2$, 2 mM; ATP(Tris), 1 mM; and phosphate/Tris, 10 mM.

In the second experiment (see Fig. 3B), a second set of vesicles was prepared to contain RbCl, 150 mM; MgCl$_2$, 2 mM; ATP(Tris), 1 mM; and phosphate/Tris, 10 mM. After reconstitution and centrifugation twice on Sephadex columns equilibrated with Tris–HCl, 150 mM, vanadate, 0.5 mM, plus MgCl$_2$, 1 mM was added. Ouabain (1 mM) was also added to one-half of the suspension. For the assay (3 min) 40 μl of vesicles (control or plus ouabain) was mixed with 40 μl of reaction medium. The final concentrations of components present were as follows: RbCl, 0.048–9.6 mM; Tris–HCl, pH 7.0, such that the sum of RbCl plus Tris–HCl was 150 mM; ^{86}Rb, 3 × 10^5 cpm/sample.

Comments. (1) Figure 3A measures classical (ATP + P$_i$)-activated Rb$^+$–Rb$^+$ (K$^+$–K$^+$) exchange on inside-out-oriented pumps. ATP and P$_i$ are required in the medium and ^{86}Rb moves from cytoplasmic to extracellular surface. Inhibition by exterior vanadate defines the size and direction of this flux. One might have measured the difference of ^{86}Rb uptake in the absence and presence of ATP and phosphate,[7] but this would underestimate the exchange on inside-out pumps since there is a measurable rate even in the absence of ATP and phosphate (see Fig. 4). Addition of ouabain to the medium improves sensitivity of the measurement by suppressing ^{86}Rb uptake through right-side-out-oriented pumps, without affecting that through inside-out pumps. The exchange shows a strictly hyperbolic dependence on the cytoplasmic Rb$^+$ concentration. The K_m in Fig. 3A is ~10 mM. The nonsaturating flux in the presence of both vanadate and ouabain is the passive Rb$^+$ leak.

(2) Figure 3B measures (ATP + P$_i$)-activated Rb$^+$–Rb$^+$ exchange on right-side-out-oriented pumps, i.e., ^{86}Rb flux from extracellular to cytoplasmic surface in vesicles preloaded with ATP and phosphate. Here

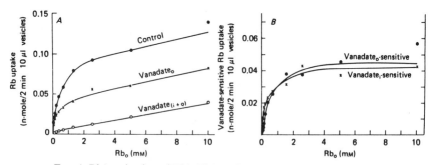

FIG. 4. Rb$^+$ activation of Rb$^+$–Rb$^+$ exchange. Reproduced from Ref. 4.

ouabain is used to define the size and orientation of the flux, and vanadate to suppress flux through inside-out-oriented pumps. Again Rb^+ is a hyperbolic activator. The K_m for extracellular Rb^+ is about 1.5 mM.

(3) Notice the relatively extended incubation time for the assay. Linearity of isotope uptake in exchange reactions is maintained for much longer then in fluxes involving net cation movements.

(4) The maximal rate of Rb^+–Rb^+ exchange is about 15% of active Na^+–K^+ exchange.

Rb^+–Rb^+ Exchange Activated by ATP or Phosphate Alone[6]

See Ref. 6 for protocols which are similar to that in Fig. 3.

Passive Rb^+ Fluxes in the Absence of ATP and Phosphate[4]

This section describes experiments showing that it is possible to detect and characterize slow vanadate- or ouabain-inhibited Rb^+ fluxes in the total absence of other pump ligands. These include Rb^+–Rb^+ exchange. Rb^+–congener exchange and net Rb^+ flux have maximal rates of 0.5–1.5% of active Na^+–K^+ exchange.

Rb^+ Activation of Rb^+–Rb^+ Exchange (Fig. 4).

Experimental Protocol. Two sets of vesicles were prepared, both containing RbCl, 150 mM, but one containing also vanadate (Tris), 50 μM, and $MgCl_2$, 1 mM. After centrifugation twice on Sephadex G-50 equilibrated with Tris–HCl, 150 mM, pH 7.0, the first set was divided and $MgCl_2$, 200 μM, and vanadate (Tris), 600 μM, were added to one-half (i.e., the + vanadate$_o$ sample) and also to all of the vesicles containing vanadate$_i$ (i.e., the vanadate$_{o+i}$ sample). The assay (2 min) was initiated by adding 40 μl of vesicles (control, + vanadate$_o$, or + vanadate$_{i+o}$) to 40 μl of reaction medium. Final concentrations of all components present were RbCl, 0.05–10 mM, plus Tris–HCl to make RbCl plus Tris–HCl, 150 mM; ^{86}Rb, 7 × 10^5 cpm/per sample, without or with vanadate, 300 μM, plus $MgCl_2$, 100 μM.

Comments. (1) In Fig. 4A the control flux shows both a saturating and linear component. Exterior vanadate blocks part of the ^{86}Rb movement, i.e., that from cytoplasmic to extracellular face of inside-out-oriented pumps. The difference between the fluxes with vanadate only outside or on both sides is the ^{86}Rb flux from extracellular to cytoplasmic surface of the right-side-out-oriented pumps. The linear flux with vanadate on both sides is the passive leak. The flux through the right-side-out-oriented pumps can also be determined in this experiment simply by subtracting, from the total ^{86}Rb uptake in the presence of external vanadate, the linear component. This could also be estimated by adding ouabain to the medium, i.e., as shown in Fig. 3B.

(2) The Rb^+–Rb^+ exchanges through inside-out and right-side-out pumps in Fig. 4B are of equal magnitudes, as expected for a random orientation. Notice that the apparent affinities for Rb^+, of 0.6 and 0.2 mM at the cytoplasmic or extracellular face, respectively, are roughly 10-fold higher than for Rb^+–Rb^+ exchange in the presence of ATP and phosphate (Fig. 3). Thus although the Rb^+–Rb^+ exchange without ATP and phosphate is very slow, the high affinity permits easy detection at low Rb^+ concentrations since the passive leak also is low.

Rb^+–Congener Exchange or Net Rb^+ Flux

Vanadate-sensitive or -insensitive ^{86}Rb uptake into vesicles prepared to contain KCl, CsCl, NaCl, LiCl, or Tris–HCl can be performed using protocols essentially like that in Fig. 4 and described in further detail.[4] The vanadate$_o$-sensitive ^{86}Rb uptake measures movement of Rb^+ on inside-out pumps, of Rb^+ from the cytoplasmic to extracellular face in exchange for congener (K, Cs, Na, Li) movement in the opposite direction, or a net Rb^+ flux in vesicles loaded with Tris–HCl. The vanadate$_o$-insensitive pump-mediated component measures movement of Rb^+ on right-side pumps from extracellular to cytoplasmic surface in exchange for congeners in the opposite direction or net Rb^+ flux from extracellular to cytoplasmic surface.

Acknowledgment

This work was supported by a grant from the U.S. P.H.S. GM 32286.

Section V

Measurement of Ligand Binding and Distance between Ligands

[18] Measurement of Binding of ATP and ADP to Na^+,K^+-ATPase

By JENS G. NØRBY and JØRGEN JENSEN*

This chapter describes the measurement of nucleotide binding to Na^+,K^+-ATPase by the rate of dialysis method[1,2] and the centrifugation method. The versatility and applicability of the former method has been demonstrated by Nørby and Jensen,[3] Hegyvary and Post,[4] and Jensen *et al.*[5] A centrifugation method has been used for measurement of ATP and ADP binding by, among others, Jørgensen[6] and Kaniike *et al.*[7,8]

Measurement of ligand binding at only one ligand concentration is generally of limited interest, whereas the determination of a complete binding isotherm can give information regarding the following parameters and phenomena. If a single type of binding site is present, the isotherm is rectilinear in a Scatchard plot and the binding capacity, E_t, as well as the dissociation constant, K_d, are easily obtained. More complex isotherms may reveal site heterogeneity, homotropic or heterotropic interactions, and changes in the properties of the isotherm might mirror different conformational states of the binding domain.

Principle

The Binding Experiment. In both methods the amount of ligand bound to the enzyme is determined as the difference between the known amount of ligand added and the measured amount of free ligand. Enzyme is mixed in a known volume with a given amount of radiolabeled nucleotide of known specific radioactivity. The conditions must be such that there is none or very little hydrolysis of the nucleotide, i.e., equilibrium binding is measured. This can be accomplished by working at low temperature and by adding EDTA to chelate contaminating Mg^{2+}. The equilibrium concentration of free ligand is then determined by monitoring the rate of dialysis

---

* We are indebted to the deceased Paul Ottolenghi for friendly collaboration.
[1] F. C. Womack and S. P. Colowick, this series, Vol. 27, p. 464.
[2] S. P. Colowick and F. C. Womack, *J. Biol. Chem.* **244**, 774 (1969).
[3] J. G. Nørby and J. Jensen, *Biochim. Biophys. Acta* **233**, 104 (1971).
[4] C. Hegyvary and R. L. Post, *J. Biol. Chem.* **246**, 5234 (1971).
[5] J. Jensen, J. G. Nørby, and P. Ottolenghi, *J. Physiol. (London)* **346**, 219 (1984).
[6] P. L. Jørgensen, *Biochim. Biophys. Acta* **356**, 53 (1974).
[7] K. Kaniike, E. Erdmann, and W. Schoner, *Biochim. Biophys. Acta* **298**, 901 (1973).
[8] K. Kaniike, E. Erdmann, and W. Schoner, *Biochim. Biophys. Acta* **352**, 275 (1974).

of radioactive nucleotide across a membrane separating the assay mixture and a perfusion system. In the centrifugation method, free ligand is measured as the radioactivity remaining in the supernatant after sedimentation of enzyme with bound ligand. This method also allows more direct estimation of amount of bound ligand since radioactivity in the pellet can be determined. The centrifugation method, of course, requires that the enzyme be sedimentable by centrifugation whereas the rate of dialysis method can be used on soluble or solubilized enzymes also.

The Binding Isotherm. In planning the determination of a binding isotherm one must realize that unequivocal interpretation requires that the data (1) show little experimental error, and (2) cover a wide range, e.g., that the occupancy of the binding sites is varied from about 0.1 to about 0.9, corresponding to a variation in the concentration of free ligand from well below to well above K_d. Since the concentration of bound ligand $[B]$ is determined as the difference between the concentration of total and free ligand:

$$[B] = [T] - [F] \tag{1}$$

the experimental error increases as $[F]$ approaches $[T]$. Let us for the sake of argument set an upper limit for $[F]/[T]$ of 0.95, corresponding to

$$[T]/[F] > 1.05 \tag{2}$$

Let us then consider the situation at the highest desired occupancy $[B]/[E_t] = 0.9$. The concentrations of unliganded and total binding sites are $[E]$ and $[E_t]$, respectively. For simple stoichiometric binding:

$$([E][F])/[B] = K_d \quad \text{or} \quad \{([E_t] - [B])[F]\}/[B] = K_d$$

which gives

$$[B]/[E_t] = 1 - \{(K_d/[E_t])([B]/[F])\} \tag{3}$$

At 90% saturation, $[B]/[E_t] = 0.9$, and from Eqs. (3) and (1)

$$0.9 = 1 - [(K_d/[E_t])\{([T]/[F]) - 1\}]$$

Combining this with the condition in Eq. (2) we obtain the minimum requirement:

$$[E_t]/K_d > 0.5 \tag{4}$$

As the dissociation constant for the enzyme–ATP or enzyme–ADP complex is of the order of 0.1 μM (or larger), it follows from the above considerations that the minimum concentration of enzyme sites in the assay must be around 0.05 μM.

Rate of Dialysis Method

Enzyme Preparations

Generally, in preparing Na^+,K^+-ATPase for nucleotide binding measurements, the following points should be kept in mind:

1. The determination of a binding isotherm takes about 2 hr. The enzyme preparation should not change properties during that period. This could, for example, be tested by activity measurements before and after an experiment.
2. The enzyme (site) concentration is important as described above [Eq. (4)], and the protein concentration also influences the assay (see below).
3. The enzyme solution usually constitutes the major part of the assay. This means that any contaminants will appear in the assay and some of them could interfere with nucleotide binding. Extensive washing of the preparation to remove these may therefore be necessary. See below.
4. Na^+,K^+-ATPase apparently displays a high affinity nucleotide site only in the so-called E_1 form, or sodium form.[5] The presence of Na^+, $Tris^+$, or certain other buffer cations such as those of histidine and imidazole induces this form.[9] In contrast, in the presence of K^+, even in quite small concentrations, the predominant form is the E_2 or potassium form, which has no *high* affinity nucleotide-binding site.[5] Removal of K^+ from the enzyme and addition of $Tris^+$ (or Na^+) to the assay thus favor nucleotide binding.

Dialysis Membrane

The membrane between the upper and lower chamber (Fig. 1) must be impermeable to the Na^+,K^+-ATPase protein so as to retain enzyme and bound ligand in the assay chamber. Furthermore, the permeability to free ligand should be large enough to allow a good counting accuracy in the effluent collected from the lower chamber, but nevertheless sufficiently low so as to assure equilibrium conditions in the upper chamber during the assay. A membrane cut from Visking seamless cellulose tubing No. 20/32, from Union Carbide, is suitable for the present purpose. It allows only 1–2% of the nucleotide to diffuse out of the upper chamber per assay (10 min).

Some hours before mounting a new membrane in the apparatus, the membrane is equilibrated with a perfusion buffer (see Procedure). Between experiments, drying out is prevented by the presence of perfusion

---

[9] J. C. Skou and M. Esmann, *Biochim. Biophys. Acta* **601**, 386 (1980).

buffer in the lower and upper chamber. Note that EDTA in the buffer prevents growth of bacteria that would eventually eat the membrane. The properties of even a frequently used membrane stay astonishingly constant for at least a year, unless accidentally punctured by a pipet.

Reagents

All Tris buffers are prepared from Tris base plus HCl and all pH values are obtained at 37° unless otherwise mentioned.

Na$_2$ATP and Na$_2$ADP from Boehringer, Mannheim

Uniformly labeled [U-^{14}C]ATP and [U-^{14}C]ADP, ammonium salts in H$_2$O containing 2% ethanol, from the Radiochemical Centre, Amersham. The specific radioactivity is about 500 Ci mol^{-1}, and 1 ml contains 50 μCi (concentration of nucleotide ca. 10^{-4} M)

The labeled and unlabeled nucleotides are purified and converted to their Tris salts as follows[3]: 5 mg Na$_2$ATP, Na$_2$ADP, or 1 ml of ^{14}C-labeled nucleotide solution is dissolved in 10 ml 100 mM Tris buffer, pH 8.3, and applied to a column of DEAE–Sephadex A-25 (Pharmacia). The length and cross-sectional area of the column is 1.5 cm and 0.7 cm^2, respectively. The elution flow is 20 ml/hr. After washing with 3 ml 100 mM Tris (pH 8.3), the compounds are stepwise eluted by 30 ml 100 mM Tris (pH 6.3) (removing P$_i$ and, in an overlapping peak, AMP), 30 ml 150 mM Tris (pH 6.3), giving an ADP peak, and finally 300 mM Tris (pH 6.3), which elutes ATP in a sharp peak. The elution is monitored by measurement of $A_{259\text{ nm}}$.

Following pooling of the appropriate fractions of the eluate, the nucleotide concentration, using $\varepsilon(259\text{ nm}) = 1.54 \times 10^4$ cm^{-1} M^{-1}, and radioactivity are measured. ^{14}C-labeled nucleotides are frozen in 500-μl portions, 20–40 μM, specific activity about 500 Ci mol^{-1}, one portion being used for each binding isotherm determination. The purity of the radiolabeled nucleotides can be checked by thin-layer chromatography on PEI cellulose plates (Polygram CEL 300 PEI No. 801-053, Macherey-Nagel, West Germany), using 1.2 M LiCl as the mobile phase and MgCl–Tris buffer for eluting the spots.[10] Solutions of the purified, unlabeled nucleotides, 1–2.5 mM, are likewise stored frozen.

The other stock solutions used in preparing the perfusion buffer and the assay medium are as follows:

ISE buffer: Imidazole (12.9 mM)–sucrose (250 mM)–EDTA (0.625 mM) buffer, pH 7.15

---

[10] K. Randerath and E. Randerath, this series, Vol. 12, p. 323. See especially pp. 336–338.

EDTA stock: EDTA (150 mM) acid form, buffered with imidazole base (1.2 M), pH 7.1–7.2
Tris (150 mM), pH 6.3
Tris (300 mM), pH 6.3
Tris (3 M), pH 6.3

Apparatus and Procedure

The dialysis cell is shown and described in Fig. 1. The lower chamber is perfused at a rate of 4 ml min^{-1} with buffer (see below) from a Mariotte bottle[11] above the cell. For experiments at 0–1° the Mariotte bottle and the major part of the tube feeding the lower chamber are kept in an ice bath, and the cell is thermostatted by pumping ice water through the jacket.

About 1 hr before the experiment the temperature control is initiated, and slow perfusion of the lower chamber is started, making sure that no air bubbles are trapped in any part of the perfusion system. The upper chamber is washed and filled with fresh, ice-cold perfusion buffer. The perfusion buffer (1 liter for an experiment comprising 8–12 assays) has the same composition as the binding assay except that it contains no protein or nucleotide (see below).

Immediately before an experiment the perfusion rate is adjusted to 4 ml min^{-1}, the upper chamber solution removed, and the chamber dried with a piece of Kleenex tissue. To the chamber are now added the following precooled solutions in the order mentioned.

1200 μl enzyme suspension (in ISE buffer)

100 μl "EDTA with Tris" consisting of 95 parts of EDTA stock and 5 parts of Tris (3 M)

200 μl nucleotide/Tris mixture consisting of X μl nucleotide solution and 200–X μl Tris stock (150 mM in ADP-binding or 300 mM in ATP-binding assays). The nucleotide solution could, for example, be made up of 475 μl ^{14}C stock (e.g., 30 μM, ca. 33,000 dpm/μl) and 25 μl unlabeled stock (e.g., 1 mM) and is in this case about 80 μM with ca. 30,000 dpm/μl. The amount X is then varied from 10 to 200 μl corresponding to about 0.5 to 10 μM nucleotide with constant specific activity in the assay.

Since pipetting of nucleotide into the upper chamber involves inserting the tip of the pipet into an enzyme-containing solution, cross-contamination problems must be avoided by using a fresh pipet for each assay.

The effluent from the lower chamber is sampled continuously and directly into empty polyethylene counting vials,[12] each sample covering a

---

[11] The simplest device for maintaining a constant perfusion pressure.

[12] ATP has been found to adsorb onto glass counting vials.[3] This results in erratic counting.

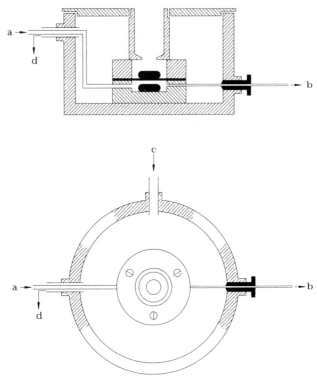

FIG. 1. Cross-sectional view and top view of the dialysis apparatus. The apparatus, which sits on a magnetic stirrer, is made of Plexiglas. It consists of two chambers, separated by a dialysis membrane, each containing a magnetic stirrer bar. The inner diameter of the cylindrical dialysis chambers is 19 mm and the heights of the upper and lower chambers are 10 and 6 mm, respectively. The volume of the magnetic bars is about 0.5 ml. The dialysis membrane is held tightly in place by three screws. The upper chamber is accessed via a cylindrical funnel glued onto the upper chamber. Perfusion buffer at 0–2° enters the lower dialysis chamber at (a) and exits at (b) through a polyethylene tube with internal diameter of 1 mm and a length of about 80 mm. Samples of dialysate dripping from the end of the latter tube are collected directly into counting vials. The contents of the dialysis chambers are maintained at constant temperature by the circulation of ice-cold water through a cooling jacket [outlet (d) and inlet (c), which enters the jacket at the same level as (b)]. To help maintain the perfusion buffer at near 0°, the tube carrying buffer to the apparatus is threaded through the outlet tube returning cooling water from the outer jacket of the apparatus to the ice-water circulation system.

period of exactly 30 sec. Usually eight samples are taken in the period 4–8 min after addition of nucleotide to the assay (Fig. 2). At the end, the assay solution is transferred from the upper chamber to a test tube, and for reasons that will become apparent, five aliquots of 100 μl are pipetted into

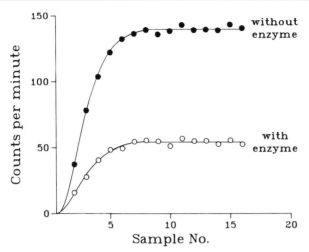

FIG. 2. The radioactivity in 2-ml samples of perfusate as a function of the number of samples collected after addition of ^{14}C-labeled nucleotide to the upper chamber. The ordinate values are calculated from four 10-min countings of each sample. The approach to steady state concentration of [^{14}C]ATP in the effluent from the lower chamber of the dialysis apparatus is shown for a typical ATP-binding experiment (○) and its buffer standard (●). The upper chamber contained 0.41 μM ATP, 50 mM Tris, 10 mM EDTA, and 0.38 μM ATP-binding sites at a protein concentration of 0.13 mg/ml. The standard experiment contained no enzyme (see text for further details). Steady state conditions are considered to have been reached after eight 30-sec samples (i.e., after 4 min of perfusion at a flow rate of 4 ml min^{-1}). The average radioactivity of samples 9 to 16 is used for the determination of the binding parameters of an experiment. In the experiments shown, the upper chamber contained radioactivity corresponding to 4.3×10^5 cpm/2 ml. The steady state concentration of nucleotide in the perfusate can thus be calculated to be 3.2×10^{-4} times the concentration of free nucleotide in the upper chamber.

counting vials for the determination of the concentration of ^{14}C in the upper chamber. The chamber is then made ready for the next experiment by rinsing several times with ice-cold perfusion buffer.

For every second or third binding experiment a standard assay is performed. The purpose of this is to determine the exact relationship between the concentration of free nucleotide in the assay chamber and the radioactivity in the samples from the lower chamber under the conditions of the binding experiment, i.e., the permeability of the membrane. There are three types of standard assays. One is called "buffer standard" since it contains 1200 μl ISE buffer instead of 1200 μl enzyme suspension and here all the added nucleotide is free to diffuse through the membrane. The second type of standard is a "protein standard." It uses heat-killed enzyme with the same nucleotide concentration as was used for the last preceding binding experiment. The third type of standard is also a "protein standard." It contains the same enzyme protein concentration as the assay and to ensure that essentially 100% of the ^{14}C-labeled nucleotide is unbound (free) a large excess of unlabeled nucleotide is added to the assay chambers so that the total nucleotide concentration is more than 500 times $[E_t]$. This can be achieved by using, as a source of unlabeled nucleotide, stock solutions of 30 mM ATP in 300 mM Tris or of 30 mM ADP in 150 mM Tris.

As one can see from Table I, when the protein concentration in the assay is sufficiently low, the membrane permeability measured by the three types of standards is essentially the same. When the amount of enzyme-containing material in the assay increases above 1 mg protein/ml one can expect to see differences due to a decrease in water content per milliliter of assay mixture with a consequent increase in concentration of free nucleotide per kilogram water. Under these latter conditions one of the protein standards must be used.

Apart from the just-mentioned special conditions, standard experiments are performed exactly as the binding experiments.

Counting of Radioactivity. To each counting vial is added a volume of any scintillator solution that will accommodate 2 ml of the sucrose-containing perfusion buffer and still give a high and constant counting efficiency. We use 10 ml of a solution consisting of 4 liters toluene, 2.5 liters

TABLE I
COMPARISON OF THE THREE TYPES OF
STANDARD EXPERIMENT[a]

Experiment	Relative $L(s)/U$
Buffer standard	1
Protein standard (heat-killed enzyme)	1.001 ± 0.006
Protein standard (live enzyme)	0.992 ± 0.0015

[a] See text for description. The ratio $L(s)/U$ for each type of standard is expressed relative to the value of this ratio for the buffer standard. The values given are averages of three determinations ± SEM. The protein concentration in the protein standards was 0.3 mg/ml. Where live enzyme was used, the nucleotide-binding site concentration was about 1 μM and the concentration of ATP was 1 mM.

Triton X-100, 1 liter ethanol (99%), and 10 g Permablend III, Packard. Blanks are prepared from 2 ml perfusion buffer and 10 ml scintillator solution. Before mixing on a vortex all samples are cooled to about 4° and the samples must be kept at that temperature during counting to avoid separation of phases and/or precipitation which reduces counting efficiency in a nonreproducible manner. Any scintillation counter with cooling may be used. Samples from the upper chamber are usually counted for 2 min whereas it may be necessary to count the effluent samples up to 4 × 10 min to ensure sufficient counting accuracy. Before calculations all results are appropriately corrected for blanks.

Calculation of [B] and [F]

A list of symbol definitions, where (i) refers to the number of the experiments, follows:

$L(i)$: Average cpm/sample of effluent from experiment (i)
$L(si)$: Calculated cpm/sample of effluent in a standard experiment corresponding to experiment (i)
$U(i)$: Average cpm/100 μl upper chamber solution
$[T](i)$: Concentration (μM) of nucleotide in the upper chamber
SA: Specific activity expressed as cpm/100 μl of the nucleotide solution used (see Procedure) divided by the nucleotide concentration (μM) in that solution

Let us as an example assume that experiment Nos. 2, 3, 4, 6, 7, and 8 are binding assays whereas Nos. 1, 5, and 9 are standard experiments. The concentration of free ligand $[F]$ in the binding assay is calculated as the fraction of free nucleotide times the total concentration. Using experiment 2 as an example:

$$[F](2) = [L(2)/L(s2)][T](2) \; \mu M$$

$L(2)$ is known from experiment 2 and $L(s2)$ and $[T](2)$ are determined as follows:

Since $L(i)$ is proportional to the concentration of free ^{14}C-labeled nucleotide in the upper chamber the relation $L(si) = gU(i)$, where g is a constant, holds for the standard experiments. $L(s2)$ is therefore calculated as

$$L(s2) = gU(2)$$

where g is obtained from the standard experiments as

$$g = [L(s1) + L(s5) + L(s9)]/[U(1) + U(5) + U(9)]$$

Furthermore, the specific activity of the nucleotide solution, SA, is known and $[T](2)$ is therefore

$$[T](2) = U(2)/SA \ \mu M$$

The concentration of bound nucleotide is

$$[B](2) = [T](2) - [F](2) = \{[L(s2) - L(2)]/L(s2)\}[T](2)$$

Centrifugation Method

Whenever it is of interest to measure the binding of nucleotide to insoluble and easily sedimentable preparations of Na^+,K^+-ATPase, it will usually be easier to use a centrifugation method.

The principle of the procedure used is the following. Enzyme is suspended in assay mixtures identical to those already described, each one containing one of a series of concentrations of nucleotide. Aliquots of 100 μl are pipetted into counting vials for the determination of total radioactivity and the assay mixtures are then centrifuged for a time sufficient to sediment the enzyme (and the enzyme–nucleotide complex), for example, 40 min at 10^5 g. Subsequently, 100-μl aliquots of the supernatants are withdrawn for determination of free (unbound) nucleotide and $[F]$ and $[B]$ are then calculated as described in the previous section. Note that this method demands lower specific radioactivity of nucleotide than necessary for the rate dialysis method, since a 100-μl aliquot of the supernatant will contain about 150 times as much nucleotide as a sample of perfusate from the lower chamber (see Fig. 2).

At $0°$ and even in the presence of 10 mM EDTA, Na^+,K^+-ATPase preparations from various sources are still able to hydrolyze ATP. The rate of hydrolysis is too low to be detectable in the rate dialysis procedure described above. Centrifugation procedures, however, are sufficiently prolonged for appreciable hydrolysis of ATP to occur. Such is not the case with ADP. Although ADP binds to Na^+,K^+-ATPase with lower affinity than ATP, measurements of ADP binding by the centrifugation method are a valid and useful alternative in qualitative and quantitative studies of the nucleotide-binding site of Na^+,K^+-ATPase.

Processing of the Data

Having determined corresponding values of $[B]$ and $[F]$, a binding isotherm, or part of it, can be constructed. For Na^+,K^+-ATPase the Scatchard plot, where $[B]$ is plotted as a function of $[B]/[F]$, has been shown to be useful as a diagnostic tool. In the presence of Na^+ and/or $Tris^+$, nucleotide binding to Na^+,K^+-ATPase shows a simple behavior[5] as the Scatchard isotherm is linear and corresponds to the equation (see Principle):

$$[B] = [E_t] - K_d([B]/[F])$$

The parameters $[E_t]$ and K_d can be obtained graphically or by linear regression analysis.

In contrast, the presence of K^+ introduces apparent negative cooperativity, which makes the evaluation of the binding isotherm much more complicated.[5,13]

A phenomenon which would complicate determination of the relevant parameters is unspecific binding. In our experience, this has not been encountered: first, Scatchard plots (from data with Na^+ or $Tris^+$ in the assay) are strictly rectilinear and second, both types of protein standards are identical to the buffer standard (e.g., Table I), all in all indicating an absence of unspecific binding of nucleotide.

[13] Various practical and theoretical problems of the more complex systems are treated by, for example, J. E. Fletcher, A. A. Spector, and J. D. Ashbrook, *Biochemistry* **9,** 4580 (1970); I. M. Klotz and D. L. Hunston, *J. Biol. Chem.* **250,** 3001 (1975); and J. G. Nørby, P. Ottolenghi, and J. Jensen, *Anal. Biochem.* **102,** 318 (1980).

[19] Interaction of Cardiac Glycosides with Na^+,K^+-ATPase

By EARL T. WALLICK and ARNOLD SCHWARTZ

Introduction

Cardiac glycosides bind to and inhibit the Na^+,K^+-ATPase (the sodium pump) present in the cell membrane of all eukaryotic cells. Matsui and Schwartz[1] were the first to measure specific [^{3}H]digoxin binding to membrane preparations of Na^+,K^+-ATPase and to demonstrate that physiological ligands regulated the binding. Since that time, many binding studies, using primarily [^{3}H]ouabain, have been carried out on a wide variety of preparations including intact tissue,[2] isolated cells,[3] re-

[1] H. Matsui and A. Schwartz, *Biochim. Biophys. Acta* **151,** 655 (1968).
[2] R. C. Deth and C. J. Lynch, *Pharmacology* **21,** 29 (1980).
[3] K. Werdan, B. Wagenknecht, B. Zwibler, L. Brown, W. Krawietz, and E. Erdmann, *Biochem. Pharmacol.* **3,** 55 (1984).

formed vesicles,[4] crude homogenates,[5] and membrane preparations,[6] with varying degrees of enrichment in Na^+,K^+-ATPase. Cardiac glycosides inhibit the hydrolysis of ATP and other substrates such as *p*-nitrophenyl phosphate. They prevent the phosphorylation of the enzyme by ATP and inhibit the ATP–ADP exchange reaction. In sided preparations, they inhibit the Na^+–K^+ transport function of the sodium pump.

Although Na^+,K^+-ATPases isolated from different sources have similar affinities for most ligands such as Na^+, K^+, Mg^{2+}, and ATP, which are involved in the transport process, the affinity for cardiac glycosides varies widely. For example, the I_{50} for inhibition of enzyme prepared from beef and rat heart are 33 nM and 60 μM, respectively.[6] Even within a single species, the affinity can vary. Erdmann[7] has shown that the K_D for [^{3}H]ouabain binding to enzyme isolated from brain, skeletal muscle, and kidney of rat are 16 nM, 125 nM, and 10 μM, respectively.

The second factor that determines the affinity of Na^+,K^+-ATPase for a given inhibitor is the chemical structure of the inhibitor. A wide variety of compounds from plant (primarily cardenolides) and animal (bufodienolides) sources as well as semisynthetic compounds have been studied. Because of its aqueous solubility, ouabain is by far the most studied compound. The high water solubility of ouabain allows the preparation of stock solutions (10 mM at 37°, 1 mM at 4°) without using organic solvents, which often have their own effect on enzyme activity, especially with cruder preparations. These stock solutions are adequate for inhibiting even insensitive preparations from, e.g., rat kidney. The high aqueous solubility of ouabain also makes it an excellent radioligand because of its low nonspecific binding.

The third factor that influences the affinity of Na^+,K^+-ATPase is the presence of ligands such as Mg^{2+}, ATP, Na^+, and K^+ in the binding medium. In presence of buffer alone, the K_D of Na^+,K^+-ATPase isolated from sheep kidney for ouabain is 1–2 μM. Addition of magnesium increases the affinity by 200- to 300-fold. The conditions most commonly used (because they are similar to physiological conditions) are MgATPNa or MgP$_i$. Under these conditions, the enzyme has its highest affinity for ouabain (K_D = 1–2 nM for enzyme from sheep kidney[6]).

[4] G. E. Lindenmayer and N. V. Wellsmith, *Circ. Res.* **47,** 710 (1980).
[5] L. H. Michael, A. Schwartz, and E. T. Wallick, *Mol. Pharmacol.* **16,** 135 (1979).
[6] E. T. Wallick, B. J. R. Pitts, L. K. Lane, and A. Schwartz, *Arch. Biochem. Biophys.* **202,** 442 (1980).
[7] E. Erdmann, *Handb. Exp. Pharmacol.* **56/I,** 337 (1981).

Methods

Model

The majority of evidence suggests that the interaction of cardiac glycosides (I) with purified homogeneous forms of Na^+,K^+-ATPase (E) follows the following simple scheme[6,7]:

$$E + I \underset{k_{-1}}{\overset{k_1}{\rightleftharpoons}} E \cdot I$$

where $E \cdot I$ represents the enzyme–cardiac glycoside complex. The number of cardiac glycoside-binding sites is equal to the number of phosphorylation sites and ATP-binding sites. Although the question of monomer versus dimer is not yet settled, no convincing evidence for cooperativity exists. The few experiments designed to test cooperativity, in fact, indicate that there is none. For example, the dissociation of ouabain from purified preparations follows first-order kinetics when the [^{3}H]ouabain is removed by centrifugation.[8] There is, however, ample evidence that isoforms of Na^+,K^+-ATPase exist in brain[9] and indirect evidence that isoenzymes are present in rat ventricle.[10] Therefore, the possibility of heterogeneity of binding sites in the membrane preparations must be considered.

Estimation of Affinities from Measurement of Enzyme Activity

With the current interest in endogenous substances which might have "digitalis-like" activity, it has become important to characterize fully the nature of the inhibition. A convenient and rapid method of measuring enzyme activity is the spectrophotometric linked enzyme assay.[11] The advantages of the method are severalfold: (1) the continuous nature of the assay enables one to determine linearity of the reaction; (2) inhibitors such as azide, ouabain, and EGTA can easily be added in small volumes to estimate, e.g., mitochondrial ATPase and calcium ATPase as well as Na^+,K^+-ATPase; and (3) the concentration of ligands such as Mg^{2+}, ATP, and Na^+ can easily be varied.

---

[8] E. T. Wallick, G. E. Lindenmayer, L. K. Lane, J. C. Allen, B. J. R. Pitts, and A. Schwartz, *Fed. Proc., Fed. Am. Soc. Exp. Biol.* **36,** 2214 (1977).
[9] K. J. Sweadner, R. Gilekson, and C. Chapman, *J. Biol. Chem.* **260,** 9016 (1985).
[10] R. J. Adams, A. Schwartz, G. Grupp, I. Grupp, S.-W. Lee, E. T. Wallick, T. Powell, V. W. Twist, and P. Gathiram, *Nature (London)* **296,** 167 (1982).
[11] A. Schwartz, K. Nagano, M. Nakao, G. E. Lindenmayer, J. C. Allen, and H. Matsui, *Methods Pharmacol.* **1,** 369 (1971).

Maximum activity at 37° is obtained in the presence of 25 mM histidine (pH 7.2), 5 mM Na$_2$ATP (Boehringer Mannheim), 5 mM MgCl$_2$, 100 mM NaCl, 10 mM KCl, 0.4 mM NADH, 1 mM phosphoenolpyruvate, and 20 μl of pyruvate kinase/lactate dehydrogenase suspension, PK/LDH (Sigma), in a volume of 2.5 ml. The absorbance at 340 nm is monitored continuously. The amount of enzyme needed will vary according to specific activity. With enzyme purified from sheep kidney (see [9] in this volume) with an activity of ~1 mmol P$_i$/hr/mg protein, 3–6 μg is sufficient. A potential disadvantage is that a compound could interfere with PK or LDH, but they are present in large excess, and we have never encountered any problems. We have had compounds, however, that interfered with the absorbance at 340 nm. Another disadvantage is that the Sigma enzymes are suspended in ammonium sulfate, and under these conditions, variation of potassium has no effect on activity. To determine the effect of potassium, the Sigma mixed enzyme must be dialyzed free of the ammonium sulfate or a colorimetric assay of liberated phosphate such as the Fiske-SubbaRow must be used. For activity to be expressed fully, the membrane preparation must be "leaky" to both ATP and the inhibitor. Detergent or freeze–thaw treatment might be necessary for certain sarcolemma and sealed microsome preparations.

Another criterion for extracting meaningful quantitative data from inhibition studies is that equilibrium must be reached. Binding of ouabain to enzyme isolated from sensitive tissue is incredibly slow at the low concentrations necessary to define an equilibrium I_{50}. Figure 1 graphically illustrates this phenomenon. The apparent I_{50} at 15 min is only 150 nM, whereas at 2 hr it is 28 nM. Lack of attention to this fact has led, especially in the older literature, to erroneous concepts such as "the I_{50} for inhibition of enzyme activity does not correlate with the K_D estimated from radioligand binding." If sufficient time is given for equilibrium to be reached, the two values agree very closely. For example, under the same conditions used for measurement of enzyme activity, the ratio of the dissociation rate constant to association rate constant yields a K_D of 16 nM for enzyme from sheep kidney. For these reasons it is preferable to use rapidly reversible inhibitors such as ouabagenin for these types of experiments with enzyme that has a high affinity for ouabain. If equilibrium is established, then the data can be fit to the following equation:

$$v_i = v/(1 + I/I_{50}) \tag{1}$$

where v and v_i is the velocity in the absence and presence of inhibitor, respectively; I is the concentration of inhibitor, and I_{50} is the inhibitory constant which should be equivalent to the K_D derived from radioligand binding. Although Eq. (1) can be linearized, it is recommended that a

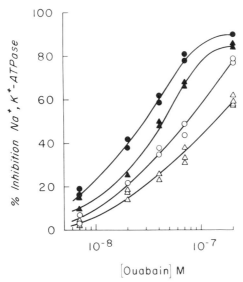

FIG. 1. Time dependence of ouabain inhibition of Na^+,K^+-ATPase purified from lamb kidney. Activity was measured continuously as described above over a 2-hr period. Activity at each concentration of ouabain, relative to that in the absence of ouabain, was determined at 15 min (△), 30 min (○), 60 min (▲), and 120 min (●). At 60 min, additional phosphoenolpyruvate (2 mM) and NADH (0.4 mM) were added to the cuvette. Reproduced with permission from Wallick et al.[6]

direct fit be carried out using nonlinear techniques. This is especially important if one is trying to determine if the data best fit a one-site or two-site model. If one suspects that there might be more than one form of ATPase, it is recommended that the following equation be fit:

$$v_T = v_0 + v_1/(1 + I/K_1) + v_2/(1 + I/K_2) \quad (2)$$

where v_T is the measured total ATPase activity and I is the concentration of inhibitor. The constant parameters to be obtained are v_0, the cardiac glycoside-insensitive ATPase activity; v_1 and v_2, the activities of the two forms of Na^+,K^+-ATPase in the absence of inhibitor; and K_1 and K_2, the I_{50} values for the two forms. The data can be analyzed using any nonlinear general purpose fitting program. Figure 2 shows such a fit to data corrected for v_0 from a sarcolemmal preparation[4] from rat heart. Nonlinear, least-squares analysis (CET Research Group) predicts an I_{50} of 44 ± 5 μM for the one-site model. The small number of runs (the number of times the actual data points cross the theoretical curve) indicates that this is not a very good fit. The fit to the two-site model was better ($p < 0.005$) and yielded I_{50} values of 165 ± 41 μM and 8.9 ± 2.5 μM with 54 ± 8% of the

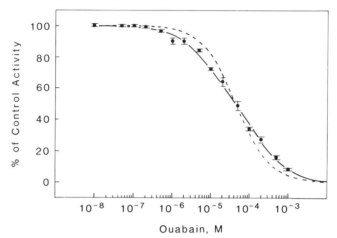

FIG. 2. Inhibition by ouabain of SDS-treated (0.25 mg/ml for 30 min at room temperature) rat heart sarcolemma. Dotted line represents best fit to a single-site model ($v_2 = 0$) and the solid line, best fit to a two-site model. Activity was measured at 37° as described in text over a 30-min time period. The assay medium also contained in final concentrations 5 mM NaN$_3$ and 0.1 mM Tris/EGTA, pH 7.2, to inhibit, respectively, mitochondrial ATPase and Ca^{2+}-ATPase. Total ATPase activity was 78 ± 6 and ouabain-sensitive ATPase 62 ± 9 μmol/mg/hr. Each data point represents the mean ± SEM of at least three determinations.

sites being low affinity. The data ideally should not be corrected for v_0 prior to fitting as this "infinite dose" data point can be treated like any other data point. This approach is statistically superior if the fit is able to converge.

The concentrations of ATP, Mg^{2+}, Na$^+$, and K$^+$ in the assay medium can be varied. Kinetic analysis revealed that ouabagenin is a noncompetitive inhibitor with respect to sodium and ATP but competitive with respect to potassium (KCl <10 mM).[12] Any putative "digitalis-like" inhibitor should have these same characteristics.[13]

Radioligand Binding

General. For the remainder of this chapter, the following shorthand notation for ligand conditions will be used: Mg is MgCl$_2$ alone; MgATP is MgCl$_2$ plus ATP; MgP$_i$ is MgCl$_2$ plus Tris phosphate; MgATPNa is MgCl$_2$ plus ATP plus NaCl; MgATPNaK is MgCl$_2$ plus ATP plus NaCl plus KCl;

---

[12] E. T. Wallick, F. Dowd, J. C. Allen, and A. Schwartz, *J. Pharmacol. Exp. Ther.* **189,** 434 (1974).
[13] M. Wehling, A. Schwartz, K. Whitmer, G. Grupp, I. L. Grupp, and E. T. Wallick, *Mol. Pharmacol.* **20,** 551 (1981).

when present, $MgCl_2$ = Tris ATP = $TrisP_i$ = 5 mM; NaCl = 100 mM; and KCl 1 or 10 mM. All media contain 50 mM Tris–HCl (pH 7.4).

As mentioned earlier, [^{3}H]ouabain is an excellent radioligand. Typically, commercial suppliers claim 97–99% purity as assessed by paper and thin-layer chromatography and by reverse isotopic dilution. Under ideal conditions, the rate of decomposition is said to be 1% for the first 6 months after preparation. It should be noted, however, that radioimmunoassays with anti-ouabain antibodies indicate that from 10 to 15% of the radioactivity in ouabain solutions is not absorbed by the antibodies. Consistent with this, we find that the majority (93%) of the radioactivity injected into an HPLC column is recovered in a single UV-absorbing peak. The missing radioactivity does not appear to be volatile since, consistent with earlier reports,[14] evaporation of the ethanol/benzene mixture (9:1, v/v) supplied by New England Nuclear Corporation to dryness does not result in loss of radioactivity. Radioligand-binding studies to purified homogeneous preparations from kidney do not suggest any heterogeneity of the radioligand. Thus, for most purposes, it is safe to assume that the radioactivity is primarily in the form of ouabain. Column or high-pressure liquid chromatography may be used if a ligand with purity greater than 90% is required.

Of greater concern is the possibility that the specific radioactivity (and hence the ouabain concentration) may be different from that specified by the supplier. Akera and Cheng[14] reported that the ouabain concentration of one stock solution of [^{3}H]ouabain was only 54% of the specified value. We have noticed the same problem. These same investigators give a fairly complex procedure for obtaining the true value. Our laboratory uses a somewhat simpler method. The ethanol/benzene solution, ostensibly 1 mCi/ml, is diluted 50-fold in distilled water to 20 μCi/ml (solution A). This will yield, depending on specific radioactivity, a ouabain concentration of 0.9–1.7 μM. This stock is further diluted 100-fold with a carefully prepared 1.00 μM unlabeled ouabain solution to yield a solution ca. 0.2 μCi/ml (solution B). The concentrations of ouabain solutions greater than 10 μM can be verified by spectroscopy (molar extinction coefficient at 220 nm is 18,800). Alternatively, solution A may be prepared by evaporating the ethanol/benzene stock to dryness and adding the requisite amount of distilled water if ethanol is deemed a problem. In our hands, the small amount of ethanol does not affect binding. Aliquots of solutions A and B are counted and the dpm/ml determined. The specific radioactivity of solution B can be calculated by dividing the measured μCi/liter by the known ouabain concentration of 1.0 μmol/liter. The specific radioactivity

---

[14] T. Akera and V.-J. K. Cheng, *Biochim. Biophys. Acta* **470**, 412 (1977).

should be close to 200 Ci/mol. To estimate the specific radioactivity of solution A, a ouabain-binding experiment is carried out at 37° using both ouabain stocks to a homogeneous Na^+,K^+-ATPase preparation using MgP_i-binding conditions. Under these conditions, the enzyme purified from lamb kidney has a K_D of approximately 2 nM and a capacity of 2000 pmol/mg. To keep radioligand in excess of receptor, we use a protein concentration of 5 µg/ml (ca. 10 nM in binding sites) and a final [^{3}H]ouabain concentration (diluted 10-fold from solution A or B) of 100 nM. The reaction is initiated by addition of enzyme and allowed to proceed for 30 min when appropriate aliquots (in this case 1.0 ml) are filtered, washed, and counted. After correction for nonspecific binding, measured in solutions as above but containing, in addition, 10^{-4} M unlabeled ouabain, the dpm due to specific binding is determined. Under these conditions, the picomoles of ouabain bound should be the same for both solutions A and B although the amount of radioactivity will be different. The specific radioactivity of solution A, SA(A) can be calculated from the specific radioactivity of solution B, SA(B), and the ratio of the amount of radioactivity bound from solutions A and B, dpm(A)/dpm(B).

$$SA(A) = SA(B) \cdot dpm(A)/dpm(B)$$

The concentration of ouabain in solution A is calculated by dividing the measured amount of radioactivity per unit volume of solution A by the specific radioactivity of solution A. For example, suppose that solutions A and B contained 16,570 and 172 µCi/liter, respectively, and that the enzyme specifically bound 521,834 and 2476 dpm when using solutions A and B, respectively. The specific activity of solution B would be (172 µCi/liter)/(1.0 µmol/liter) or 172 Ci/mol. Specific activity of solution A would be 36,300 Ci/mol (specification sheet stated 20,000 Ci/mol) and the concentration of ouabain in A would be 0.457 µM. Having this estimate, the binding experiment should now be repeated so that a final concentration of 100 nM [^{3}H]ouabain can be more closely obtained.

Using [^{3}H]ouabain of the highest specific radioactivity, as little as 0.1 pmol of binding sites can reliably be detected if the enzyme is moderately sensitive. Using the purified enzyme from sheep kidney, binding is linear with respect to protein over a range of 0.5–25 µg.

Binding measured in the presence of a 100- to 1000-fold excess of unlabeled ouabain over the concentration of [^{3}H]ouabain has been shown to be an adequate measurement of nonspecific binding. If the source of the enzyme preparation is new or if the density of receptors is very low, additional controls are recommended. Since Mg^{2+} is required for binding and K^+ retards binding, binding in the presence of buffer and 10 mM EDTA (pH 7.4) and in buffer plus 100 mM KCl serve as additional estimates of nonspecific binding.

Binding of [³H]ouabain to fragmented leaky preparations of Na^+,K^+-ATPase can be carried out under a variety of ligand conditions. For other preparations, however, certain precautions should be taken. Crude homogenates, for example, contain a large amount of ouabain-insensitive ATPase activity and can rapidly hydrolyze the ATP if binding is carried out in the presence of MgATPNa. Fortunately, the $K_{0.5}$ for ATP for stimulation of ouabain binding is in the micromolar range and maximum binding can usually be obtained without problems. Also detergent treatment of crude preparations is often needed to expose latent binding sites. For preparations of intact cells, one must realize that ATP and ouabain must have access to opposite sides of the cell or vesicle. If the cells contain sufficient internal ATP, only external sodium is required for binding.

Binding of [³H]ouabain can be carried out on intact tissue, although special attention should be given to separation of specific from nonspecific binding. Deth and Lynch[2] describe an excellent procedure which includes a zero-degree wash and inhibition of binding by external potassium and by decreasing internal ATP. Sarcolemmal preparations from heart and microsomal preparations from kidney can contain a mixture of leaky and sealed vesicles. Binding in the presence of MgATPNa occurs only to the fraction leaky to ATP and [³H]ouabain. The amount of binding in this medium can be stimulated by detergent or freeze–thaw treatment. Binding in the presence of MgP_i, however, evidently occurs to both leaky and intact vesicles, and this medium is recommended for this type of membrane preparation.

A variety of methods such as centrifugation, column chromatography, and dialysis may be used to separate the bound [³H]ouabain from the free [³H]ouabain, but by far the most common is filtration. Because of the high water solubility, almost any type filter may be used as long as the membrane preparation does not clog the filter. Three washes of 5 ml each are sufficient to reduce the nonspecific binding to a minimum for the various purified preparations, but this should be confirmed for cruder preparations.

Kinetics of Binding. The interaction of ouabain with Na^+,K^+-ATPase depicted earlier is a bimolecular process and the binding follows second-order kinetics. The analysis of radioligand-binding data is greatly simplified, however, if the concentration of the [³H]ouabain is in large excess of the concentration of Na^+,K^+-ATPase. In this case, the interaction follows pseudo-first-order kinetics and can be analyzed according to the following equation:

$$B = B_e(1 - e^{-kt}) \qquad (3)$$

where B and B_e are the amount of [³H]ouabain bound to the enzyme at time, t, and at equilibrium, respectively; and k is a rate constant representing a first-order approach to equilibrium. Although Eq. (3) can be linearized, it is recommended that the data be fit directly to Eq. (3) using nonlinear least-squares procedures (see Ref. 6).

The overall rate constant, k, is related to the association rate constant, k_1, and the dissociation rate constant, k_{-1}, by the following equation:

$$k = k_1 L + k_{-1} \tag{4}$$

where L is the concentration of the [^{3}H]ouabain. Thus if k is determined at several different [^{3}H]ouabain concentrations, k_1 and k_{-1} can be determined from the slope and intercept of a plot of k vs L. For enzyme prepared from the more sensitive species, k_{-1} can be very small (0.01 min^{-1} or less) and thus the intercept may not be very accurately determined. In any event, an independent measurement of k_{-1} is preferred.

To measure k_{-1}, the rate of dissociation, a chase experiment is carried out. The enzyme–[^{3}H]ouabain complex is formed as described in the previous section. At an appropriate time, unlabeled ouabain (200- to 1000-fold greater than the [^{3}H]ouabain) is added and aliquots removed at appropriate times. The disappearance of radioactivity follows the following equation:

$$B = B_0 e^{-k_{-1} t} \tag{5}$$

where B is the amount bound at time, t; B_0 is the amount bound at time, zero, and k_{-1} is the dissociation rate constant. Table I lists some representative association and dissociation rate constants and the calculated K_D under various ligand conditions. As can be seen from Table I, the effect of ligands on the affinity is mediated primarily through the rate of association.

TABLE I
KINETIC CONSTANTS FOR [^{3}H]OUABAIN BINDING[a]

Ligand condition	k_1 (sec^{-1} M^{-1} × 10^{-3})	k_{-1} (sec^{-1} × 10^3)	K_D (nM)
Mg	2.77	0.0566	23.2
MgATP	10.3	0.0700	6.23
MgATPNa	30.8	0.0642	2.08
MgP$_i$	45.4	0.0642	1.44
MgATPNaK^1	13.6	0.0642	4.72
MgATPNaK^{10}	4.14	0.0535	15.5

[a] Data compiled from Ref. 6. Concentration of ligands are as stated in text with potassium being 1 or 10 mM.

If there is significant deviation from linearity in either association or dissociation, then multiple sites should be suspected. The data for both types of experiments should then be fit using multiexponential fitting. The following equation for a three-compartment, noninteracting model is an example.

$$y = Ae^{-k_a t} + Be^{-k_b t} + C \tag{6}$$

For a dissociation experiment, A and B are the amounts bound to compartments A and B at time zero; C is the nonspecific binding, and y is the total amount bound at time, t. The dissociation rate constants from compartments A and B are k_a and k_b. For an association experiment, A is $-A_e$ and B is $-B_e$, the amount bound to compartments A and B at equilibrium, respectively. C is the sum of A_e and B_e and the nonspecific binding. The rate constants k_a and k_b represent the overall first-order approach to equilibrium [Eq. (4)] for each compartment.

Equilibrium Experiments. As can be seen from Table I, the dissociation of ouabain from Na^+,K^+-ATPase can be very slow. This fact leads to the major error which occurs in equilibrium (Scatchard or displacement) assays) studies of [^{3}H]ouabain binding. Since $K_D = k_{-1}/k_1$, Eq. (4) can be rearranged as shown:

$$k = k_{-1}(1 + L/K_D) \tag{7}$$

To obtain a good Scatchard plot, one would like to obtain values of B_e/B_{max} from 0.10 to 0.90, i.e., from 10 to 90% occupation of receptors. To obtain 10% occupancy L/K_D will equal 0.11 and $k = 1.11 k_{-1}$. For enzyme prepared from sensitive species, the off-rate has a $t_{1/2}$ of ca. 120 min. Thus to reach equilibrium (five half-times), 10 hr is required. Over this time

TABLE II
DISSOCIATION OF OUABAIN FROM Na^+,K^+-ATPase

Source	$t_{1/2}$ (min)[a]				
	Beef	Dog	Sheep	Guinea pig	Rat
Heart	112	144	96	1.1	1.2[b]
Kidney	128	120	180	4.5	—
Brain	90	—	—	45	39
Skeletal muscle	—	6.2	—	4.2	2

[a] Data compiled from Refs. 6 and 7.
[b] Rat heart contains two forms of Na^+,K^+-ATPase. This value probably applies to the more sensitive form. The off-rate from the other form in rat heart and the single form in rat kidney is extremely fast.

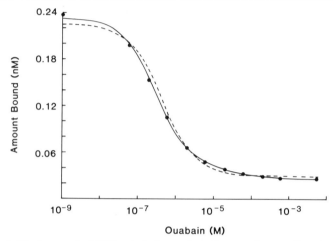

FIG. 3. Displacement of [³H]ouabain from rat cardiac sarcolemma. Sarcolemma, 30 μg/ml, was incubated at 37° under MgATPNaK conditions plus 0.5 mM CaCl$_2$ and 100 nM [³H]ouabain (19.5 Ci/mmol) and indicated concentrations of unlabeled ouabain for 60–90 min. Dotted line represents best fit to a single-site model and the solid line, the best fit to a two-site model.

period, one has to worry about the stability of the enzyme–ouabain complex. For enzyme prepared from lamb kidney, the denaturation process is fairly slow ($t_{1/2}$ = 15 hr). If consideration is paid to these aspects, then satisfactory equilibrium saturation or displacement experiments can be carried out. Within each set of ligand conditions, k_1 does not vary very much from species to species or organ to organ. In the presence of MgP$_i$ it is 1–5 × 10^4 M^{-1} sec^{-1}. The sensitivity of the enzyme to ouabain is determined primarily by the off-rate. Table II lists some typical off-rates.

Most purified preparations contain a single class of [³H]ouabain-binding sites, but multiple forms clearly exist in some membrane preparations. Recently, much attention has been paid to the necessity for using weighted nonlinear least-squares fitting procedures for analyzing radioligand-binding data.[15] This is especially important if one is trying to decide between a one-site or multiple-site model in, e.g., a heterogeneous preparation. An excellent computer program for this purpose is LIGAND,[16] which can be used for both saturation and displacement assays and is available for both Apple (Biomedical Computing Technology Information Center, Vanderbilt Medical Center, Nashville, TN) and IBM PC.[17] Figure

[15] P. J. Munson, in "Brain Receptor Methodologies" (P. J. Marangos, I. C. Campbell, and R. M. Cohen, eds.), Part A. Academic Press, New York, 1984.
[16] P. J. Munson and D. Robbard, Ann. Biochem. **107**, 220 (1980).
[17] G. A. McPherson, J. Pharmacol. Methods **14**, 213 (1985).

3 shows such a fit to a displacement assay using the Apple version. This sarcolemmal preparation from rat heart is best fit ($F = 57.4, p < 0.005$) by the two-site model representing two forms of Na^+,K^+-ATPase. The K_D and B_{max} for the two forms are 0.21 μM, 10.4 pmol/mg, and 55 μM, 122 pmol/mg. Because of the relative insensitivity of the low-affinity form, there is a large error associated with the constants. To obtain more quantitative data, the experiment should be repeated using many more concentrations of ouabain.

Akera and Cheng[14] have described a convenient graphical method for determination of affinity and binding site from displacement-type assays, which is sufficient for some purposes. For a more quantitative analysis, a program such as LIGAND is recommended.

[20] Estimation of Na,K-Pump Numbers and Turnover in Intact Cells with [^{3}H]Ouabain

By SETH R. HOOTMAN and STEPHEN A. ERNST

The asymmetric distribution of Na^+ and K^+ ions across the plasma membranes of most eukaryotic cells is maintained by the activity of the Na,K-pump (Na^+,K^+-ATPase; EC 3.6.1.3). In most instances, hydrolysis of one molecule of ATP by the Na,K-pump results in the cellular uptake of two K^+ in exchange for expulsion of three Na^+ ions. Aside from its general role in maintenance of cellular ionic homeostasis, the enzyme also plays a central role in the transepithelial transfer of solutes and water. Exposure of epithelia to specific inhibitors of the Na,K-pump such as ouabain and related steroid glycosides rapidly abolishes transport activity.

Rates of electrolyte transport in many tissues are modulated by neurohormonal agents. It is now well established that a number of hormones including aldosterone,[1] corticosterone,[2] triiodothyronine,[1,3] and insulin[4] regulate the numbers of Na,K-pumps in target cells, thereby establishing upper limits of transport activity. In addition to these long-term effects, which are manifested over several hours to days and likely involve altera-

---

[1] K. Geering, M. Girardet, C. Bron, J.-P. Kraehenbuhl, and B. C. Rossier, *J. Biol. Chem.* **257**, 10338 (1983).
[2] L. C. Garg, M. A. Knepper, and M. B. Burg, *Am. J. Physiol.* **240**, F536 (1981).
[3] C. S. Lo and I. S. Edelman, *J. Biol. Chem.* **251**, 7834 (1976).
[4] M. Fehlman and P. Freychet, *J. Biol. Chem.* **256**, 7449 (1981).

tions in biosynthesis of the enzyme, rapid changes in cellular Na,K-pump activity also can be elicited by hormones and neurotransmitters within seconds. Rather than an increase or decrease in the cellular Na,K-pump population, these effects appear to involve a change in the rate of cycling or turnover of pump units. Alternatively, the possibility of acute activation of quiescent pumps or of modification of pump affinities for activating ligands by secretogogues must be considered, although there is at present little direct evidence to support such mechanisms. An increase in pump activity not only serves to reestablish cytoplasmic Na^+ and K^+ activities that are perturbed by hormonally induced changes in permeability of the plasma membrane to these cations, but in epithelia also may potentiate routing of Na^+ and possibly K^+ into specific tissue subcompartments (i.e., paracellular channels) and thereby augment vectorial solute transport along selected pathways. Agents that have been shown to modulate acute changes in Na,K-pump activity in intact cells include catecholamines,[5] cholinomimetics,[6] and hormones such as islet hormones,[7] and cholecystokinin (CCK), secretin, and vasoactive intestinal peptide (VIP).[8]

Because changes in Na,K-pump turnover elicited by hormones or neurotransmitters in most cases appear to represent a secondary response, monitoring pump activity potentially can give information about more proximal transport events which closely follow the initial activation of membrane receptors. For example, a hormonally-induced increase in the passive Na^+ conductance of the apical plasma membrane in an absorptive epithelial cell might be expected to trigger a compensatory increase in Na,K-pump turnover at the basolateral cell surface. Thus measurement of the change in pump activity elicited will be reflective of the relationship between occupancy of receptors for the hormone of interest and opening of the conductance pathway. Similar relationships may be drawn for transport events in secretory epithelia as well. Thus in a secretory epithelium like the acinus of the exocrine pancreas, which secretes both digestive enzymes and fluid, release of amylase can be monitored as an indicator of enzyme secretion while changes in pump activity reflect fluid secretion.

Several techniques have been developed to assess Na,K-pump activity in intact cells. These include measurement of ouabain-sensitive oxygen consumption,[9] ouabain-sensitive $^{86}Rb^+$ uptake,[7] and [^3H]ouabain binding.[6,8,10] This last technique, which was developed recently in studies

[5] H. Furukawa, J. P. Bilezekian, and J. N. Loeb, *Biochim. Biophys. Acta* **598,** 345 (1980).
[6] S. R. Hootman and S. A. Ernst, *Am. J. Physiol.* **241,** R77 (1981).
[7] M. D. Resh, *Biochemistry* **22,** 2781 (1983).
[8] S. R. Hootman, S. A. Ernst, and J. A. Williams, *Am. J. Physiol.* **245,** G339 (1983).
[9] L. J. Mandel and R. S. Balaban, *Am. J. Physiol.* **240,** F357 (1981).
[10] S. R. Hootman and J. A. Williams, *J. Physiol. (London)* **360,** 121 (1985).

on secretory epithelial cells from duck salt glands and guinea pig parotid gland and pancreas, will constitute the primary focus of the following discussion.

Specificity of [^{3}H]Ouabain Binding to Na,K-Pumps

The Na$^+$,K$^+$-ATPase of plasma membranes is inhibited by ouabain and related cardiotonic steroids. This inhibition is quite specific, since the activities of other membrane nucleotide di- and triphosphatases are unaffected by these glycosides. The interaction of ouabain with partially purified Na$^+$,K$^+$-ATPase from tissues of most species is of high affinity, with equilibrium dissociation constants (K_d) in the low micromolar to nanomolar range. Binding of ouabain to membrane preparations enriched in Na$^+$,K$^+$-ATPase is antagonized by K$^+$, but supported by Na$^+$, Mg^{2+}, and ATP.[11] These data indicate that ouabain binds primarily to the phosphorylated intermediate or E$_2$P configuration of the Na,K-pump, a supposition that recently has received confirmation.[12] In intact cells, conditions that lower intracellular ATP levels such as anoxia, low temperature, or metabolic inhibitors (e.g., cyanide, 2,4-dinitrophenol) inhibit ouabain binding,[13–15] as does a reduction in cell Na$^+$ activity.[13,16] These observations suggest that enzyme turnover is a prerequisite for binding under normal physiological conditions and that in intact cells the rate-limiting step in the ATPase reaction cycle is the association of Na$^+$ and ATP with the enzyme. This reaction mechanism is illustrated in Fig. 1, as is the interaction of the enzyme with ouabain. When cells become ATP or Na$^+$ depleted, their Na,K-pumps "stall" in the E$_1$ configuration and the ouabain-binding site is not in a configuration amenable to binding. Conversely, if chilled or anoxic cells are warmed or oxygenated, ATP hydrolysis by Na,K-pumps may be resumed with consequent formation of phosphorylated intermediates and ouabain binding.[15]

Rationale

Several assumptions underlie the use of [^{3}H]ouabain to monitor Na,K-pump activity. (1) Binding of ouabain to the Na,K-pump is reversible and the binding reaction obeys the laws of mass action. (2) The reaction

---

[11] R. W. Albers, G. J. Koval, and G. J. Siegel, *Mol. Pharmacol.* **4**, 324 (1968).
[12] A. Yoda and S. Yoda, *Mol. Pharmacol.* **22**, 700 (1982).
[13] P. F. Baker and J. S. Willis, *J. Physiol. (London)* **224**, 441 (1972).
[14] J. Shaver and C. E. Stirling, *J. Cell Biol.* **76**, 278 (1978).
[15] J. W. Mills, A. D. C. Macknight, J. A. Jarrell, J. M. Dayer, and D. A. Ausiello, *J. Cell Biol.* **88**, 637 (1981).
[16] J. D. Gardner and C. Frantz, *J. Membr. Biol.* **16**, 43 (1974).

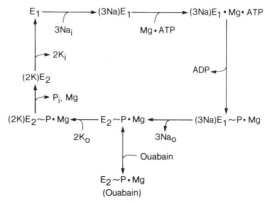

FIG. 1. Schematic representation of the reaction cycle of Na,K-pumps in intact cells and their interaction with ouabain. E_1 and E_2 designate different configurations of the pump enzyme. Binding and release of cytoplasmic Na^+ and extracellular K^+ are sequential, with the respective ions translocated across the plasma membrane in a ratio of 3 : 2 during a single reaction cycle. Formation of the phosphorylated intermediate of the enzyme is depicted by ~P. Ouabain interacts primarily with the E_2P configuration of the Na,K-pump. Modified from L. C. Cantley, *Curr. Top. Bioenerg.* **11**, 201 (1981).

mechanism of the Na,K-pump is cyclic, with a succession of configurations displayed (Fig. 1). (3) Ouabain binds primarily if not exclusively to only one configuration (E_2P) of the pump enzyme in intact cells. (4) There is a one-to-one correspondence between specific ouabain-binding sites and functioning Na,K-pumps. (5) A single cell has a number of Na,K-pumps which turn over or cycle independently of each other and whose rates of turnover are regulated primarily by the cytoplasmic Na^+/K^+ ratio and ATP level in their cytoplasmic microenvironments. From these assumptions, one can predict that at low turnover rates, Na,K-pumps in intact cells will spend proportionately more time in the E_1 configuration and less in succeeding states. If, however, the availability of the rate-limiting ligand (usually Na^+_i) is increased, the lag time between cycles and thus the time spent in the E_1 configuration will decrease and the frequency of presentation of the phosphorylated intermediate, E_2P, will rise to a new steady state level. Additionally, although the time per cycle occupied by E_2P may not be altered, an increase in turnover will result in a corresponding increase in the amount of time spent by the pump in E_2P over a finite interval. In a cell suspension where many millions of functional Na,K-pumps are present, at any given moment some percentage of the total pump population will be in the E_2P configuration. If ouabain is added to such a cell suspension at a saturating concentration, all pumps in the

E_2P configuration will be bound and their cycling will cease, at least temporarily. Pumps not in E_2P will be bound as soon as they cycle into this configuration. The initial rate of ouabain binding therefore will be directly proportional to the overall frequency at which the phosphorylated pump intermediate is presented; that is, the rate of pump turnover averaged across the entire pump population. However, since binding is reversible, each individual Na,K-pump is arrested only momentarily and turnover can resume once the glycoside dissociates from the enzyme. At saturating concentrations of the ligand only one or at most a few reaction cycles will occur before the pump is rebound by another ouabain molecule. Thus when equilibrium is attained, at any given moment virtually all Na,K-pumps are in a bound and thus inactive state. Equilibrium binding then approaches a maximal level indicative of the number of Na,K-pumps present.

If a concentration of ouabain is chosen, however, that is far below that needed to saturate all Na,K-pumps in a cell suspension, only a few percentages of the pump population will be bound at any time. Since each individual Na,K-pump may then turn over many times between successive interactions with ouabain molecules, the equilibrium binding level achieved will reflect the frequency of presentation of the E_2P configuration. Any circumstance that increases the rate of pump turnover likewise will potentiate chance encounters between free ouabain molecules and the receptive pump configuration. Practically speaking, this means that a relative measure of the turnover rate of Na,K-pumps in a suspension of intact cells can be obtained from [^{3}H]ouabain binding, provided that the concentration of the glycoside is maintained at a very low level. At such minimal concentrations (see below), the ligand will then have negligible effects on the cells' electrolyte balance and can be thought of not as an inhibitor, but as a specific probe for sensing alterations in Na,K-pump activity. We present here a protocol for measuring both pump numbers and activity in intact cells and illustrate with results from recent studies in our laboratories how insights into the effects of hormones and neurotransmitters on electrolyte transport processes in three secretory epithelia were obtained.

Preparation of Cell Suspensions

In order to accurately quantitate cellular Na,K-pumps and to assess the effects of neurohumoral agents on pump activity, diffusional barriers to the cell surface must be minimized. In most instances, this necessitates the preparation of suspensions of single cells. The following protocol, which will be described briefly, has been used in our laboratories to prepare isolated epithelial cells from duck salt gland[17] and guinea pig parotid gland[10] and pancreas.[8]

In these three tissues, the secretory parenchyma was dissociated in a similar fashion. In each case, the tissue was reduced after dissection to a fine mince and these pieces were transferred to a flask containing a HEPES-buffered Ringer's solution (HR). This solution contained in addition to a balanced complement of inorganic salts, the enzymes hyaluronidase, α-chymotrypsin, and collagenase. The concentration of each enzyme was adjusted to give the lowest level that resulted in adequate dissociation of the tissue. The presence of each enzyme was necessary in order to obtain good yields of dispersed cells. The tissue fragments were incubated at 37° for 15–45 min with vigorous shaking (90 cpm). At this and all subsequent steps in the dispersion protocol, the incubation medium was gassed with 100% O_2. At the end of this digestion, the medium was removed and replaced with an equal volume of Ca^{2+},Mg^{2+}-free HR containing 1.0 mM EDTA. Chelation of divalent cations at this step is essential, as it causes adjacent epithelial cells to pull away from each other and disrupts desmosomes and occluding junctions. Incubation in this medium was for 15–20 min. After this step, tissue fragments were rinsed with HR and incubated again at 37° for 15–45 min with the mixture of enzymes. Afterward, the fragments were rinsed with Ca^{2+},Mg^{2+}-free HR without EDTA, but containing bovine serum albumin, soybean trypsin inhibitor, and DNase. Final dispersion was obtained by pipetting of the tissue pieces through a siliconized Pasteur pipet. The number of passes through the bore of the pipet needed to effect complete disruption of tissue integrity varied. With pancreas, the tissue usually was completely dispersed after only a few passes. With duck salt gland many passes were needed to liberate sufficient numbers of cells. In this instance, after a few pipettings, the aliquot of medium was drawn off to protect the liberated cells from further mechanical damage, a fresh volume of medium was added to the fragments, and the process was repeated. In this tissue, a few small fragments invariably did not dissociate entirely and were discarded.

After dispersion, the medium which now contained subcellular debris, intact cells, and small tissue fragments was filtered through 25- to 50-μm nylon mesh and centrifuged for 5 min at 50 g. The resulting pellet which contained primarily single cells was resuspended in HR. Cells were checked for viability by mixing a drop of the suspension with a drop of 0.5% trypan blue in 0.9% NaCl on a glass slide and counting in a microscope the percentage of stained nuclei. In suspensions of pancreas and parotid gland, staining by the vital dye routinely was less than 5%. In salt gland cell suspensions, viability ranged from 75 to 95%. Suspensions

---

[17] S. R. Hootman and S. A. Ernst, *Am. J. Physiol.* **238**, C184 (1980).

where more than 20% of the cells were permeable to the dye were discarded. After resuspension in HR, cells were preincubated either at room temperature or 37° for 45–60 min before binding assays were initiated to allow reestablishment of normal cytoplasmic Na^+ and K^+ activities, which are markedly perturbed immediately after isolation.[6]

Quantitation of Na,K-Pumps in Intact Cells and Determination of the Equilibrium Binding Constant

To use [^{3}H]ouabain as a probe for assessing changes in Na,K-pump activity in intact cells, the concentration of free ligand chosen must be low enough to avoid perturbing the electrolyte balance of the cells being studied. In practice, this means using a concentration that is less than one-tenth that of the equilibrium binding constant (K_d) for the ouabain–Na,K-pump interaction. Since the K_d varies from tissue to tissue even within the same animal, it must be determined empirically by saturation binding analysis, which has the added advantage of simultaneously providing an estimate of the size of the cellular Na,K-pump population. The theoretical basis for such analysis was recently reviewed by Weiland and Molinoff,[18] and will be discussed only briefly here when applicable. Since association of ouabain with cellular Na,K-pumps is reversible, the amount bound at equilibrium will depend on the concentrations of both reactants and on the affinity of ouabain for the pump, the latter being reflected by the K_d value. Thus if the number of pump sites is held constant and the concentration of labeled ouabain is increased in steps, a series of equilibria will be established that will approach a maximal or saturation level (B_{max}). Scatchard analysis[19] of these values will yield an estimate of both K_d and B_{max}. For such studies, a range of ouabain concentrations is chosen that will include the K_d value. The choice of this range initially must be made by trial and error. In our studies with guinea pig parotid gland and pancreatic acinar cells, we found 0.1 to 10 μM to be a suitable range. After a pilot experiment or two, this range should be adjusted so that several ouabain concentrations will lie both above and below the K_d value, a factor of importance in obtaining an accurate estimate of B_{max}.[20]

Determination of Equilibrium Binding Conditions

It is essential that a time interval be chosen that will allow attainment of equilibrium binding conditions at the lowest [^{3}H]ouabain concentration

---

[18] G. A. Weiland and P. B. Molinoff, *Life Sci.* **29**, 313 (1981).
[19] G. Scatchard, *Ann. N.Y. Acad. Sci.* **51**, 660 (1949).
[20] I. M. Klotz, *Science* **217**, 1247 (1982).

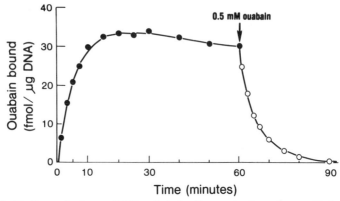

FIG. 2. Binding and release of [³H]ouabain by dispersed guinea pig parotid gland acinar cells incubated at 37° in HEPES-buffered Ringer's. At $t = 0$, 0.1 μM [³H]ouabain was added to the cell suspension. Association of [³H]ouabain with acinar cells was rapidly reversed by addition of a 5000-fold excess of unlabeled ouabain. Reprinted from Hootman and Williams[10] by permission of the *Journal of Physiology*.

employed. To determine an appropriate interval, a simple time course such as is shown in Fig. 2 is carried out. In this experiment, guinea pig parotid acinar cells (1–2 × 10⁶ cells/ml) were incubated in HR at 37°, and 0.1 μM [³H]ouabain (1.7 μCi/ml) and 0.2 μCi/ml [¹⁴C]sucrose were added to the suspension. [¹⁴C]Sucrose, which in our experience is not readily taken up by epithelial cells, is always included as an extracellular space marker. It is preferable to labeled inulin or other higher molecular weight markers, since being similar in size to ouabain it has a similar diffusion coefficient. At intervals after this addition, 0.5-ml samples of the suspension were removed, diluted with 3.0 ml of chilled HR or 0.9% NaCl, and poured over GF/A glass fiber filters (Whatman) mounted in a vacuum filtering manifold. Each filter was then quickly rinsed three times with 3.0 ml of chilled HR or 0.9% NaCl and placed directly into a scintillation vial where it was extracted for several hours in a cocktail consisting of 1.0 ml of Protosol and 10 ml of Omnifluor (New England Nuclear). GF/A glass fiber filters retain 100% of acinar cells, although we have noted that mammalian erythrocytes will pass through them. They also have a high loading capacity and flow rate and are readily rendered transparent in the scintillation cocktail. Counting of radioactivity on filters is carried out in a liquid scintillation spectrometer by the external standard channels ratio method. Appropriate quench standards are included and ³H and ¹⁴C are counted simultaneously by standard double-isotope procedures. The total [³H]ouabain associated with each filter then is corrected using the [¹⁴C]sucrose values for residual medium contamination, the difference representing ouabain that is cell associated. [^{3}H]Ouabain bound to cells may be expressed in convenient units, such as fmol/10^6 cells or fmol/μg DNA, which are obtained from the specific activity of the labeled ligand and by either counting cells directly using a hemacytometer or by measuring DNA by the diphenylamine[21] or other procedure.

At 37°, cell-associated [^{3}H]ouabain increases with time until equilibrium conditions are established and remains relatively constant thereafter (Fig. 2), as long as cell viability is maintained. Reversibility of binding can be demonstrated by adding an excess of unlabeled ouabain to the cell suspension. In Fig. 2, when reassociation of labeled ouabain with Na,K-pumps was thereby blocked, cell-associated radioactivity declined rapidly. At submicromolar concentrations, virtually all [^{3}H]ouabain was released, indicating that it was associated specifically with Na,K-pumps and that little if any of the labeled ligand had been internalized or degraded.

Determination of K_d and B_{max}

Having established the length of time needed to achieve equilibrium at the low end of the range of concentrations that will be used, saturation analysis can be carried out. Usually, two 1.0-ml aliquots of cells in HR are incubated in culture tubes at 37°. This medium contains 0.2 μCi/ml [^{14}C]sucrose and one of several concentrations of [^{3}H]ouabain. At ouabain concentrations above 1 μM, the labeled compound is diluted with unlabeled ouabain to obtain the desired final concentration and specific activity is recalculated accordingly. A duplicate set of tubes which contains in addition 0.5 mM unlabeled ouabain is coincubated. Binding is allowed to proceed until equilibrium is attained at all ouabain concentrations. At this time each cell suspension is diluted with 3.0 ml of chilled HR, filtered, rinsed, extracted, and counted as described above. In each cell type that we have examined, ouabain binding can be resolved into two components, as illustrated in Fig. 3. In the presence of a large excess of unlabeled ouabain, [^{3}H]ouabain binding is linearly dependent on concentration over the entire range. When this component is subtracted at each concentration from binding in the absence of unlabeled inhibitor (i.e., total binding), a saturable binding component is revealed. This component represents binding to the finite cellular population of Na,K-pumps. Scatchard analysis of the saturation binding curve invariably generates a straight line, as illustrated in the inset to Fig. 3. From the Scatchard plot, K_d and B_{max} are determined. Both values are listed in Table I for the three cell types we have studied. There is a large variation in numbers of cellu-

[21] D. N. Croft and M. Lubran, *Biochem. J.* **95**, 612 (1965).

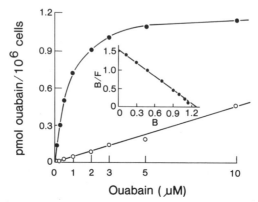

FIG. 3. Saturation analysis of [³H]ouabain binding to dispersed guinea pig pancreatic acinar cells in HEPES-buffered Ringer's at 37°. Cell-associated [³H]ouabain in the presence of an excess of unlabeled inhibitor (○) was linearly related to concentration. On subtraction of this binding component from total binding, a saturable component (●) representing binding to cell Na,K-pumps was observed. Inset: Scatchard analysis of the saturable component of [³H]ouabain binding. Maximal binding level (B_{max}) = 1.29 pmol/10⁶ cells and the equilibrium dissociation constant (K_d) = 0.86 μM. Reprinted from Hootman et al.[8] by permission of the American Journal of Physiology.

lar Na,K-pumps among these three tissues, which correlates with their established levels of electrolyte secretory activity. There is also a smaller, but appreciable, variation in the affinity of pumps for ouabain.

Effect of K⁺ on [³H]Ouabain Binding

It is important to note that in the binding experiments described here, cells were incubated in Ringer's solutions that contained physiological concentrations of K⁺ (i.e., 4–6 mM). Potassium competes with ouabain at

TABLE I
BINDING OF [³H]OUABAIN TO DISPERSED CELLS FROM THREE SECRETORY EPITHELIA[a]

Organ	K_d (μM)	B_{max} (ouabain-binding sites/cell)
Duck salt gland	0.23	2.6×10^7
Guinea pig parotid gland	2.05	2.9×10^6
Guinea pig pancreas	0.86	7.8×10^5

[a] Data are taken from Refs. 6, 8, and 10.

a site on the Na,K-pump[22] and thus reduces the rate of approach to equilibrium in the ouabain-pump binding reaction. At saturating concentrations of ouabain, changes in extracellular K^+ will not substantially alter the amount of labeled ligand bound. However, at low ouabain concentrations, changes in K^+ concentration can affect both the rate and equilibrium level of binding. Eliminating K^+ from the medium will increase the slope of the Scatchard line, thus resulting in a smaller K_d and an apparent increase in the affinity of Na,K-pumps for ouabain. Raising the extracellular K^+ concentration will have the opposite effect, an apparent decrease in binding affinity. Thus it is prudent to standardize the K^+ concentration of the binding medium at an early stage in studies with [^{3}H]ouabain and to utilize a cytocrit that is low enough that changes in cytoplasmic K^+ activity will not substantially alter medium K^+.

Effect of Ca^{2+} on [^{3}H]Ouabain Binding

Another cation that may potentially alter Na,K-pump turnover, and therefore [^{3}H]ouabain binding, is Ca^{2+}. Calcium inhibits the Na,K-pump by competing with Na^+ at the inner aspect of the plasma membrane and thus reduces the rate of formation of the phosphorylated intermediate.[23] However, this inhibition is not significant at submicromolar levels of free cytoplasmic Ca^{2+}.[23,24] We recently measured $Ca^{2+}{}_i$ in guinea pig pancreatic acinar cells using the fluorescent Ca^{2+} chelator Quin-2, and found it to be 161 ± 13 nM.[25] Thus it is not likely that free Ca^{2+} in intact epithelial cells plays an important role in regulating Na,K-pump turnover by a simple competitive interaction with Na^+. However, the ability of Ca^{2+} to reduce pump turnover at higher concentrations must be considered when investigating the effects of agents such as the Ca^{2+} ionophore A23187, which can at high concentrations induce a massive influx of Ca^{2+} across the plasma membrane.

Use of [^{3}H]Ouabain Binding to Assess Changes in Na,K-Pump Activity in Intact Cells

If a concentration of ouabain is added to a cell suspension such that less than 10% of Na,K-pumps are bound, the ability of the cells to maintain homeostatic ion balance will not be compromised appreciably and equilibrium binding will reflect pump turnover. From this reasoning, hor-

[22] H. Furukawa, J. P. Bilezekian, and J. N. Loeb, *J. Gen. Physiol.* **76**, 499 (1980).
[23] T. Tobin, T. Akera, S. I. Baskin, and T. M. Brody, *Mol. Pharmacol.* **9**, 336 (1973).
[24] M. J. Dunn and R. Grant, *Biochim. Biophys. Acta* **352**, 97 (1974).
[25] S. R. Hootman, D. L. Ochs, and J. A. Williams, *Am. J. Physiol.* **249**, G470 (1985).

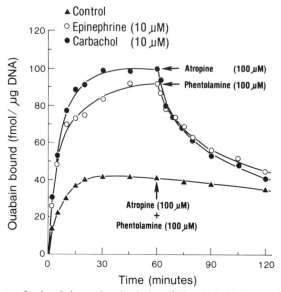

FIG. 4. Effects of epinephrine and carbachol on [^{3}H]ouabain binding to dispersed guinea pig parotid gland acinar cells incubated in HEPES-buffered Ringer's at 37°. Stimulation of binding evoked by carbachol or epinephrine was reversed by atropine and phentolamine, respectively. Neither antagonist, however, affected [^{3}H]ouabain binding in the absence of agonists. Reprinted from Hootman and Williams[10] by permission of the *Journal of Physiology*.

mones or neurotransmitters that stimulate Na,K-pump activity by increasing plasma membrane Na^+ and K^+ conductances may be expected likewise to elevate [^{3}H]ouabain binding. That this is in fact observed is shown in Fig. 4. In these experiments, three suspensions of guinea pig parotid acinar cells were incubated in HR at 37° as described above for determining the time course of binding. Carbachol and epinephrine were added separately to two suspensions 5 min before initiation of [^{3}H]ouabain binding. Both agonists markedly stimulated both the initial rate of binding and the equilibrium level of binding attained. The ratio of stimulated to control binding in each case was the same at all time intervals to 60 min. The increase in [^{3}H]ouabain bound evoked by carbachol or epinephrine could be abolished by atropine or phentolamine, respectively, demonstrating that maintenance of elevated levels of binding depended in this instance on agonist occupancy of muscarinic cholinergic and α-adrenergic receptors on the secretory cell surface.

The concentration dependence of stimulation of [^{3}H]ouabain binding by agonists also can be determined. Results of such experiments with

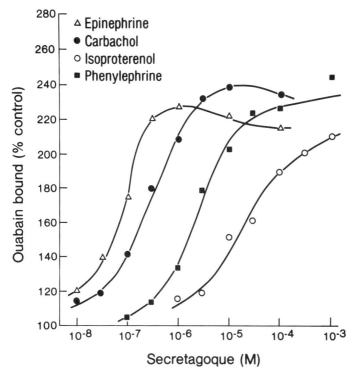

FIG. 5. Concentration dependence of agonist stimulation of [^{3}H]ouabain binding to Na,K-pumps in dispersed guinea pig parotid gland acinar cells. Incubation was for 60 min at 37° in HEPES-buffered Ringer's containing 0.1 μM [^{3}H]ouabain. Reprinted from Hootman and Williams[10] by permission of the *Journal of Physiology*.

guinea pig parotid gland acinar cells are shown in Fig. 5. Similar results also were obtained in our studies with pancreatic acinar cells,[8] although here we found, in addition to carbachol, that the gastrointestinal hormones cholecystokinin (CCK) and secretin were effective stimulators of [^{3}H]ouabain binding. In this earlier work, dose–response curves for stimulation of binding to pancreatic acinar cells by carbachol, CCK, and secretin were calculated from the increase in initial rate of binding of 0.1 μM [^{3}H]ouabain at room temperature. More recently we found that, as with parotid acinar cells, when the concentration of the labeled probe was decreased (in this case to 0.05 μM), equilibrium binding at 37° was enhanced by agonists to a degree identical to that seen for the initial rate of binding.[25]

The intracellular messengers that translate receptor occupancy into changes in Na,K-pump activity also can be elucidated utilizing the

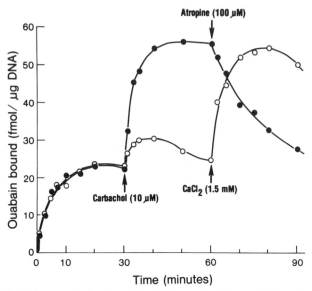

FIG. 6. Effect of removal of extracellular Ca^{2+} on stimulation of [^{3}H]ouabain binding to guinea pig parotid gland acinar cells by carbachol. Cells were incubated at 37° in HEPES-buffered Ringer's (●, 1.3 mM Ca^{2+}) or in Ca^{2+}-free HEPES-buffered Ringer's (○, 0.2 mM EGTA), each with 0.1 μM [^{3}H]ouabain. Reprinted from Hootman and Williams[10] by permission of the *Journal of Physiology*.

[^{3}H]ouabain-binding technique. It is now well established that Ca^{2+} plays a central role in mediation of the stimulatory effects of many agonists,[26] including carbachol, CCK, and epinephrine. Parotid and pancreatic acinar cells recently have been shown to possess Ca^{2+}-activated cation channels in the plasma membrane.[27,28] Channel opening is modulated by occupancy of muscarinic, CCK, and α-adrenergic receptors. Thus it seemed likely that changes in cellular Na,K-pump activity might reflect a response to changes in the rate of Na^+ influx across the cell surface induced by these agonists. To assess the Ca^{2+} dependency of Na,K-pump stimulation in guinea pig parotid acinar cells, we carried out experiments as shown in Fig. 6. Acinar cells were incubated at 37° in standard HR (1.3 mM Ca^{2+}) or in Ca^{2+}-free HR with 0.2 mM EGTA. [^{3}H]Ouabain was added to each suspension and binding was allowed to proceed to equilibrium before addition of carbachol. In standard HR, the agonist caused an immediate increase in cell-associated [^{3}H]ouabain. Subsequent addition

[26] H. Rasmussen and P. Q. Barrett, *Physiol. Rev.* **64**, 938 (1984).
[27] Y. Maruyama and O. H. Petersen, *Nature (London)* **299**, 159 (1982).
[28] Y. Maruyama, D. V. Gallacher, and O. H. Petersen, *Nature (London)* **302**, 827 (1983).

TABLE II
EFFECT OF CHLORIDE DEPLETION ON AGONIST STIMULATION OF [³H]OUABAIN
BINDING TO SECRETORY EPITHELIAL CELLS[a]

		[³H]Ouabain bound (% control)	
Cell type	Agonist	Standard Ringer's	Cl⁻-free Ringer's
Duck salt gland	None	100.0	95.9 ± 11.1
	0.5 mM MCh	156.7 ± 16.5	103.0 ± 9.6
Guinea pig pancreas	None	100.0	104.1 ± 3.3
	30 μM CCh	161.8 ± 7.4	166.1 ± 11.1
	300 pM CCK$_8$	158.6 ± 9.2	157.8 ± 8.0
	30 nM secretin	142.5 ± 10.4	136.9 ± 5.6

[a] [³H]Ouabain bound was measured as the initial rate of binding at 23°. In standard Ringer's, Cl⁻ concentrations was 126–128 mM. In Cl⁻-free Ringer's, Cl⁻ was replaced by SO_4^{2-} and osmolarity was adjusted with mannitol. MCh, methacholine. CCh, carbachol. CCK$_8$, cholecystokinin octapeptide. Data adapted from Refs. 6 and 8.

of atropine abolished this stimulation. By contrast, when carbachol was added to acinar cells incubated in Ca^{2+}-free HR, only a small and transient increase in binding was observed. However, readmission of Ca^{2+} into the incubation medium resulted in a rapid increase in binding to a level equivalent to that seen in suspensions of carbachol-stimulated cells in standard HR. When epinephrine was substituted for carbachol in this experiment, a similar pattern of response was seen.[10] These results can be interpreted as representing the Ca^{2+} requirements of membrane channels that mediate Na^+ and K^+ fluxes. The transient increase in [³H]ouabain binding seen after carbachol or epinephrine addition to acinar cells in Ca^{2+}-free HR with 0.2 mM EGTA may reflect the Na,K-pump response to the release of Ca^{2+} from intracellular stores following either muscarinic or α-adrenergic receptor occupation. This supposition is supported by the observation that when carbachol is added to suspensions of parotid acinar cells several minutes after the transient response to epinephrine, no increase in binding is seen.[10]

Experiments of this nature show clearly how monitoring Na,K-pump activity by [³H]ouabain binding can reveal characteristics of transport events more proximal to receptor activation. Another illustration of this strategy is shown in Table II. In many epithelia, a transport component has been described that mediates the coupled uptake of Na^+ and Cl^- (or Na^+, K^+, and Cl^-).[29] We asked in our studies with duck salt gland cells

[29] R. A. Frizzell, M. Field, and S. G. Schultz, Am. J. Physiol. **226**, F1 (1979).

and guinea pig pancreatic acinar cells whether this mechanism might be involved in regulating the influx of Na^+ and thus increased Na,K-pump activity elicited by secretagogues. Dispersed cells from both tissues were incubated in Cl^--free Ringer's and the Na,K-pump response to agonists was determined. In suspensions of duck salt gland cells, the increase in [^{3}H]ouabain binding evoked by the cholinomimetic methacholine was clearly dependent on extracellular Cl^-. Additional experiments showed that the Cl^--dependent stimulatory response in salt gland cells also was blocked by furosemide, an inhibitor of the Na^+/Cl^- transporter.[6] By contrast, in suspensions of guinea pig pancreatic acinar cells, removal of Cl^- from the medium bathing the cells had no effect on Na,K-pump stimulation by carbachol, CCK_8, or secretin. These data reveal a fundamental difference in the mechanisms of anion transport in two chloride-secreting epithelia following activation by secretagogues.

Corroboration of [^{3}H]Ouabain-Binding Data with Other Techniques for Measuring Na,K-Pump Activity

While the [^{3}H]ouabain-binding technique provides a sensitive and convenient method for assessing the effects of hormones, neurotransmitters, and other exogenous compounds on Na,K-pump numbers and activity in

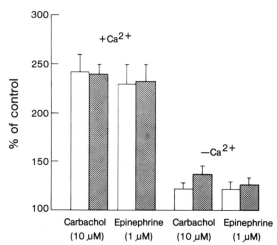

FIG. 7. Effect of extracellular Ca^{2+} on carbachol and epinephrine stimulation of [^{3}H]ouabain binding (open bars) and ouabain-sensitive oxygen consumption (shaded bars) by guinea pig parotid gland acinar cells. Cells were incubated in HEPES-buffered Ringer's (1.3 mM Ca^{2+}) or Ca^{2+}-free HEPES-buffered Ringer's (0.2 mM EGTA). Results represent the means ± SEM of three to five experiments.

intact cells, it is not without limitations. Absolute rates of pump-mediated Na$^+$ and K$^+$ transport are not obtainable. In addition, in some common laboratory animals, most notably rats and mice, the K_d for ouabain binding is on the order of 100 μM, a circumstance that makes quantitation of binding using the filter assay impractical due to the extremely rapid rate of ouabain dissociation from cellular pump sites. Thus one must turn to other methods. As we developed the [^{3}H]ouabain-binding technique, we corroborated binding data with measurements of ouabain-sensitive oxygen and ^{86}Rb$^+$ uptake. The excellent agreement between methods is illustrated in Fig. 7. When the effects of carbachol and epinephrine on [^{3}H]ouabain binding and ouabain-sensitive oxygen uptake by guinea pig parotid acinar cells were compared, no significant difference was noted. Similarly, in prancreatic acinar cells, ouabain-sensitive ^{86}Rb$^+$ uptake was stimulated to the same extent as [^{3}H]ouabain binding by carbachol, CCK$_8$, and secretin.[25] Since each of the three methods give data in a form that is not directly obtainable from the other two, the investigator may wish to utilize more than one technique to quantitate different parameters of Na,K-pump activity.

Acknowledgments

This research was supported by National Institutes of Health Grants AM32708 and AM32994 and a grant from the Cystic Fibrosis Foundation to S. R. Hootman and NIH Grant AM27559 to S. A. Ernst. The authors thank Dr. John A. Williams for critical reading of this manuscript.

[21] Measurement of Binding of Na$^+$, K$^+$, and Rb$^+$ to Na$^+$,K$^+$-ATPase by Centrifugation Methods

By HIDEO MATSUI and HARUO HOMAREDA

Direct measurement of Na$^+$ and K$^+$ binding to Na$^+$,K$^+$-ATPase is important for an understanding of the molecular mechanism of the Na,K-pump. The measurement, however, is difficult primarily because of the relatively low affinity of the specific binding and the coexistence of a high level of nonspecific binding. The determination of the specific binding, therefore, requires the use of an enzyme preparation with high specific activity as well as the employment of a reliable criterion to discriminate the specific from the nonspecific binding and of an accurate assay method with high precision, and these requirements constitute practical difficulties in the measurement.[1] Recent progress in enzyme purification has made it possible to use the high specific activity preparation, which increases the specific binding while decreasing the nonspecific one relatively. Consequently, several reports have appeared with different assay methods and different discriminating criteria.[2-5] As for the sensitivity and precision of the measurement, the centrifugation method is the best among those hitherto employed.[1] Here we introduce the assay system of cation binding with the centrifugation method in combination with ouabain sensitivity for discriminating specific from nonspecific binding.[1]

Assay Method

Principle

Since Na^+,K^+-ATPase is usually purified as a membrane-bound enzyme, radioactive Na^+ or K^+ bound to the enzyme is concentrated into a pellet by centrifugation and its radioactivity can be assayed. The specific binding of cation is designated as an ouabain-sensitive binding, i.e., the difference in cation in the absence and presence of ouabain, since the enzyme to which ouabain was previously bound has a reduced affinity for the cation.

Reagents

Triethanolamine buffer, 0.5 M, pH 7.2 (at 0°), contains 0.52 M triethanolamine, 0.46 M HCl, and 5 mM EDTA
Ouabain, 1 mM
Oligomycin, 0.5 mg/ml ethanol solution
^{42}KCl (^{43}KCl, ^{86}RbCl, or ^{204}TlCl), 10 μM–2 mM, $\sim\mu$Ci/μmol
^{22}NaCl, 0.1–10 mM, $\sim\mu$Ci/μmol
Other reagents, e.g., EDTA (recrystallized to eliminate contamination by sodium), CDTA (recrystallized), ATP, MgCl$_2$, choline, and sucrose, and enzyme preparation should be tested for contamination by Na^+ and K^+.[1]

[1] H. Matsui and H. Homareda, *J. Biochem. (Tokyo)* **92**, 193 (1982).
[2] K. Kaniike, G. E. Lindenmayer, E. T. Wallick, L. K. Lane, and A. Schwartz, *J. Biol. Chem.* **251**, 4794 (1976).
[3] H. Matsui, Y. Hayashi, H. Homareda, and M. Kimimura, *Biochem. Biophys. Res. Commun.* **75**, 373 (1977).
[4] D. F. Hastings, *Anal. Biochem.* **83**, 416 (1977).
[5] M. Yamaguchi and Y. Tonomura, *J. Biochem. (Tokyo)* **86**, 509 (1979).

Standard Procedure

One-half milligram of membrane-bound enzyme is preincubated in a push-on, capped polycarbonate tube (for Beckman rotor type 40) containing 0.5 ml of 1 mM EDTA(triethanolamine)$_4$ with or without 0.2 mM ouabain at 0° (pH ~ 6.5) for 1 hr to complete ouabain binding to the enzyme. Then the buffer solution (final concentration, 50 mM), with any necessary additions, and the radioactive cation are added to make up the final reaction mixture of 1 ml in an ice bath. After adding the labeled cation, the tubes are immediately centrifuged at 40,000 rpm and 2° ± 1° for 20 min. The supernatants are taken off by aspiration and wiped off with a Kimwipe. The centrifuged pellets are dissolved in 0.5 ml of 1 M NaOH by warming at 45° for 30 min. The entire solution in each tube is transferred into counting vials and neutralized with 0.5 ml of 1 M HCl with which the tube walls were rinsed. After mixing the samples with 10 ml of liquid scintillation cocktail (emulsifier type), the radioactivities of cation in the pellets are counted in a liquid scintillation spectrometer.

The ouabain-sensitive bound cation is calculated by subtracting the amount of cation in the presence of ouabain (nonspecific bound cation plus unbound cation) from that in the absence of ouabain (total cation in the pellet). An example of measurement for the ouabain-sensitive K$^+$ binding is shown in Fig. 1, which illustrates the concentration dependency of binding.

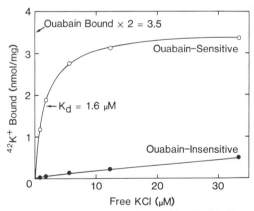

FIG. 1. Concentration dependency of ouabain-sensitive K$^+$ binding under standard condition of measurement. ○, Ouabain-sensitive binding, which is the difference in total K$^+$ in the pellets in the absence and the presence of ouabain. ●, Ouabain-insensitive binding, which is calculated from total K$^+$ in the pellet in the presence of ouabain by subtracting unbound K$^+$. Unbound K$^+$ in the pellet, which is at a similar level as the ouabain-insensitive binding under the standard condition, is obtained by multiplication of free K$^+$ concentration by water volume estimated as [^{3}H]sucrose space (13–15 μl/mg protein). Values are means of duplicate tubes. Specific activity of the enzyme is 16 U/mg.

Comments on Assay Conditions and Procedure

Enzyme. The specific cation bound per milligram of protein of the enzyme is proportional to the specific enzyme activity, while the nonspecific cation bound remains almost constant. Accordingly, it is difficult to detect the specific cation binding to a crude enzyme preparation with low specific activity. It is desirable to use membrane-bound enzyme with specific activity of more than 15 μmol/min · mg protein, purified from microsomes of enzyme-rich organs (e.g., kidney) by treatment with deoxycholate or sodium dodecyl sulfate.[6,7] To reduce the Na^+ and K^+ contamination of the enzyme preparation, washing of the enzyme (once with 5 mM $MgCl_2$, then with 1 mM CDTA) is recommended.[1]

Temperature. The experiment is carried out at low temperature for the following reason. Increasing the reaction temperature not only decreases the affinity for the specific binding of Na^+ and K^+, but also increases nonspecific binding of the cations.[8] A large decrease in K^+ affinity at high temperature has been suggested by a large negative enthalpy change in the K^+ effect on ouabain binding.[9]

Buffer. Buffers usually decrease the affinities of the enzyme for cations. Inhibition depends on the kind of organic base of the buffer and buffer concentrations. Imidazole and tris(hydroxymethyl)aminomethane inhibit strongly, but triethanolamine and histidine weakly.[10] One disadvantage of triethanolamine is an interference with the protein assay by the Lowry method because of a heavy color development in control tubes.

pH. The affinity for K^+ is constant at pH values higher than 7, but decreases in the acidic pH range.[10,11] The affinity for Na^+ increases in the alkaline pH range.[11]

EDTA. EDTA chelation is used to reduce a possible contamination by inhibitory divalent cations. When Tl^+ binding is to be measured, EDTA should be omitted from the reaction mixture because of the significant chelation by EDTA of Tl^+.[10]

Oligomycin. Oligomycin (5 μg/ml) increases the affinity for Na^+ by about one order, but has no effect on the K^+ binding. So, the addition of oligomycin is recommended to detect the saturation level of Na^+ binding.[1]

---

[6] P. L. Jørgensen, this series, Vol. 32 [26].

[7] Y. Hayashi, M. Kimimura, H. Homareda, and H. Matsui, *Biochim. Biophys. Acta* **482**, 185 (1977).

[8] H. Homareda, Y. Hayashi, and H. Matsui, *Proc. Meet. Jpn. Bioenerg. Group* **3**, 41 (1977) (in Japanese).

[9] E. T. Wallick, B. M. Anner, M. V. Ray, and A. Schwartz, *J. Biol. Chem.* **253**, 8773 (1978).

[10] H. Homareda and H. Matsui, *J. Biochem. (Tokyo)* **92**, 219 (1982).

[11] H. Homareda and H. Matsui, in "The Sodium Pump" (I. M. Glynn and J. C. Ellory, eds.). Company of Biologists, Cambridge, England, 1985.

Cation Concentration. If the concentration of cation is higher than 0.2 mM (at this concentration the unbound cation is comparable to the specific bound cation), the increased experimental error results mainly from variation in the unbound cation proportional to water volume in the centrifuged pellet. To decrease this error it is useful to subtract the unbound cation from the total cation in the pellet by measuring the water volume in each pellet as a sucrose space with [^{3}H]sucrose added to the reaction mixture.[1]

Ouabain Binding. The amount of the enzyme is usually determined from the amount of ouabain bound to the enzyme. The ouabain-binding level, if necessary, can be measured simultaneously with cation binding by using [^{3}H]ouabain in place of unlabeled ouabain. To measure the unbound ouabain and the nonspecific binding, control tubes have to contain either 0.1 M NaCl to prevent the specific binding of ouabain or excess unlabeled ouabain to dilute the ^{3}H label before the addition of [^{3}H]ouabain.[1]

Simultaneous Measurement of Na$^+$ and K$^+$ Binding. Simultaneous measurement is possible in a double-labeling experiment with ^{22}Na and ^{42}K because of the large difference between half-times of ^{22}Na and ^{42}K. The ^{42}K counts are calculated from the initial counts of ^{22}Na plus ^{42}K by subtracting the ^{22}Na counts measured at the tenth day after decay of ^{42}K.[1]

Micromodification. To decrease the amount of enzyme used, the total volume of the reaction mixture can be reduced to one-fifth by using microtubes and a Beckman TL-100 tabletop ultracentrifuge.[11a]

Criteria for Specific Binding

Selection of adequate criteria is essential for the determination of the true specific binding of cations to the enzyme. Various criteria for the specific binding are listed in Table I together with the assay methods employed by different investigators.

High-affinity binding, which is thought to be specific, is not always easily and accurately separated from low-affinity binding by Scatchard plot,[2,12] because the affinity difference between the specific and nonspecific sites is not so large for Na$^+$ binding and a positive cooperativity observed at the specific sites makes the Scatchard plot nonlinear particularly for K$^+$ binding.[1] Heat treatment of the enzyme not only inactivates the specific binding activity, but also alters the nonspecific binding.[1] The use of heat-denatured enzyme, as a control in the study of the native enzyme,[4,13] therefore, would not be suited for accurate measurement.

[11a] H. Homareda and H. Matsui, *J. Biochem. (Tokyo)* **101**, 789 (1987).
[12] L. C. Cantley, Jr., L. G. Cantley, and L. Josephson, *J. Biol. Chem.* **253**, 7361 (1978).
[13] D. Hastings and J. C. Skou, *Biochim. Biophys. Acta* **601**, 380 (1980).

TABLE I
ASSAY METHODS AND CRITERIA FOR SPECIFIC BINDING OF CATIONS

Method	Criterion	Radioisotopes and ions determined	References
Centrifugation	High-affinity sites by Scatchard plot	$^{22}Na^+$	2
Centrifugation	Ouabain sensitivity	$^{42}K^+$, $^{43}K^+$, $^{22}Na^+$	1, 3, 10, 11, 11a, 17
Centrifugation	Inhibition by 2 mM Rb$^+$	$^{86}Rb^+$	16
Filtration	High-affinity sites by Scatchard plot	$^{86}Rb^+$	12
Filtration	Inhibition by 0.1 M Mg^{2+}, Na$^+$, or K$^+$	$^{42}K^+$, $^{86}Rb^+$, $^{22}Na^+$	5, 15
Electrode	Heat inactivation	K$^+$	4, 13
Electrode	ATP dependency in the presence of K$^+$	Na$^+$	14

ATP-dependent Na$^+$ binding in the presence of K$^+$ seem to have a high specificity, but its application is limited to a system containing only K$^+$ and ATP.[14] The use of a high concentration of competitive cations (e.g., 0.1 M of Na$^+$, K$^+$, or Mg^{2+}) is inadequate for the definition of specific binding,[5,15] because these cations at high concentrations displace not only specific but also nonspecific binding.

The use of ouabain sensitivity of the criteria listed in Table I has the least effect on nonspecific binding, because ouabain has no electrical charge and is used at concentrations less than 0.1 mM. Thus the ouabain-sensitive binding of Na$^+$ and K$^+$ exhibits high specificity, but it may not be a perfect criterion for specific binding. Since the enzyme to which ouabain was once bound does not lose its binding ability but is reduced in its binding affinity for Na$^+$ and K$^+$ at the specific sites, the simple difference in the cations in the absence and presence of ouabain becomes smaller than the true value of specific binding at relatively high concentrations of Na$^+$ or K$^+$ (e.g., 1 mM). To avoid this disadvantage, high-affinity competition by K$^+$ (2 mM) is useful for the measurement of Na$^+$ binding at millimolar concentrations of Na$^+$.[11] This high-affinity competition by nonradioactive K$^+$ (or Rb$^+$) can be used for the measurement of the specific binding of radioactive K$^+$ (or Rb$^+$) itself.[16]

[14] Y. Hara and M. Nakao, in "Cation Flux across Biomembranes" (Y. Mukohata and L. Packer, eds.), p. 21. Academic Press, New York, 1979.
[15] M. Yamaguchi and Y. Tonomura, J. Biochem. (Tokyo) **88,** 1365 (1980).
[16] P. L. Jørgensen and J. Petersen, Biochim. Biophys. Acta **705,** 38 (1982).

Characteristics of Na^+ and K^+ Binding

Studies on the direct measurement of Na^+ and K^+ binding have revealed that high-affinity sites of the enzyme for Na^+ are different from those for K^+. Basic properties of Na^+ and K^+ binding are discussed below.[1,10,11,17]

K^+ Binding

The affinity of the enzyme for K^+ is observed to be very high by direct measurement. The dissociation constant (K_d) under favorable buffer conditions decreases to 2 μM, too low to be observed by kinetic methods, whereas K_d for nonspecific (ouabain-insensitive) binding is more than 10 mM (Fig. 1). The saturation level of bound ouabain-sensitive K^+ is 2 mol/mol of ouabain bound.[1,17]

The K^+ binding is specifically inhibited by Na^+ at a concentration where the nonspecific binding is not affected, but is not inhibited by choline at the same concentration as that of Na^+.[1] Monovalent cations other than Na^+ compete with K^+ binding at high-affinity K^+ sites; the order of efficiency is $Tl^+ > Rb^+ (\geq K^+) > NH_4^+ > Cs^+ (> Na^+) > Li^+$.[10] Divalent cations also inhibit K^+ binding; the inhibitory order is $Ba^{2+} \simeq Ca^{2+} > Zn^{2+} \simeq Mn^{2+} > Sr^{2+} > Co^{2+} > Ni^{2+} > Mg^{2+}$ and the ratio of K_d for Mg^{2+} to that for K^+ is about 120:1.[10] ATP, ADP, AMPPNP, and AMPPCP inhibit K^+ binding, whereas AMP, P_i, and p-nitrophenyl phosphate have little effect.[10]

Although the K^+-binding curve as a function of K^+ concentration shows a nearly hyperbolic shape but not a sigmoidal one under standard conditions (Fig. 1), positive coooperativity between the two K^+-binding sites is increased in accordance with the increase in K_d for K^+ by adding imidazole buffer, Na^+, and/or Mg^{2+}.[1,11] Two individual K_d values for cooperative binding can be obtained graphically.[1]

Na^+ Binding

The K_d for specific Na^+ sites is about 0.3–0.2 mM at pH 7–8 and the affinity ratio of specific K^+ vs specific Na^+ is about 100, whereas the K_d for the nonspecific Na^+ sites is almost the same as that for the nonspecific K^+ sites (>10 mM).[1,10,11] Oligomycin increases the Na^+ affinity by nearly one order of magnitude. The saturation level of the specific Na^+ binding is difficult to demonstrate at Na^+ concentration less than 1 mM because of the relatively high value of K_d for Na^+. In the presence of oligomycin,

---

[17] H. Matsui, Y. Hayashi, H. Homareda, and M. Taguchi, *Curr. Top. Membr. Transp.* **19**, 145 (1983).

however, the saturation binding of ouabain-sensitive Na^+ is observed to be about 3 mol/mol ouabain bound to the enzyme.[1,17] Na^+ binding is specifically inhibited by K^+ but not by choline.[1] The binding is inhibited by Mg^{2+} but not by ATP, contrasting with K^+ binding.[1] Heat treatment of the enzyme (65°, 10 min) abolishes the specific binding capacity for both Na^+ and K^+ as well as ouabain.[1]

Positive cooperativity between the Na^+-binding sites is weak under standard experimental conditions, but is increased by the addition of oligomycin in accordance with the increase in affinity for Na^+.[1] The effect of pH on Na^+ and K^+ binding is stated above in the comments on assay conditions.

Competition between the ouabain-sensitive binding of $3Na^+$ and $2K^+$ and the different effects on Na^+ and K^+ binding by ATP, oligomycin, and pH do not imply real competition at identical binding sites, but rather alternative binding of Na^+ or K^+ at high-affinity Na^+ sites existing on the sodium form of the enzyme or at high-affinity K^+ sites on the potassium form, respectively.[1]

A stoichiometry of $3Na^+$, $2K^+$, and one ouabain is always observed with the enzyme having specific activity of 15–25 U/mg. An experiment with an enzyme having high specific activity (45 U/mg) also confirms that the ligand-binding stoichiometry of $Na^+ : K^+ : ATP : ouabain : (\alpha\beta)$ protomer of enzyme is $3 : 2 : 1 : 1 : 1$.[17]

[22] Estimating Affinities for Physiological Ligands and Inhibitors by Kinetic Studies on Na^+,K^+-ATPase and Its Partial Activities

By JOSEPH D. ROBINSON

Kinetic studies on Na^+,K^+-ATPase can provide valuable information on ligand interactions with the enzyme as well as affording mechanistic insights.[1,2] A preeminent virtue of kinetic approaches is the direct linkage observed between enzymatic activity and added ligand: thus problems of ligand binding either to catalytically irrelevant ("silent") sites or to inactivated enzyme or to impurities in the preparation are avoided. Only occupancy of sites that matter to the enzyme are revealed. On the other hand,

---

[1] J. D. Robinson and M. S. Flashner, *Biochim. Biophys. Acta* **549**, 145 (1979).
[2] L. C. Cantley, *Curr. Top. Bioenerg.* **11**, 201 (1981).

the obligatory requirement for such kinetic approaches is that the catalytic activity observed does indeed result from the specific enzyme being studied and only from that enzyme. This stricture may be met most satisfactorily by removing during purification all other enzymes using the substrate being studied. Alternatively, inhibitors may be used to abolish the studied or the extraneous activities, but this approach requires great care that the necessary specificity and extent of inhibition is obtained under all the experimental conditions.

General Considerations

In this chapter steady state kinetic studies of the Na^+,K^+-ATPase will be considered with reference chiefly to three hydrolytic activities.

1. Na^+,K^+-ATPase activity is assayed in the presence of ATP (usually as the Tris salt), Mg^{2+}, Na^+, and K^+ (usually as the chloride salts). Activity may be measured in terms of P_i release using colorimetric assays; the Lowry–Lopez method[3] is sensitive enough for ATP concentrations as low as 0.3 mM, allowing 10% hydrolysis of substrate within initial velocity conditions. For substrate concentrations below this level, $[^{32}P]P_i$ release from $[\gamma\text{-}^{32}P]$ATP may be measured,[4] using an isobutyl alcohol–toluene extraction[5] of the $[^{32}P]$phosphomolybdate complex. Coupled assays using pyruvate kinase[6] allow continuous spectrophotometric monitoring of reaction progress, but suffer from the sensitivity of pyruvate kinase to changes in K^+ concentration. In the past, ATP preparations from muscle contained an unrecognized contaminant, the vanadate anion, that is a potent inhibitor of the ATPase[7]; ATP free of significant levels of vanadate is commercially available and obviously should be used.

The Na^+,K^+-ATPase reaction will be considered here in terms of a modified Albers–Post scheme (Fig. 1),[8] that includes phosphorylation of the enzyme by ATP to form E_1–P and dephosphorylation of E_2–P, with these two changes in covalent bonding separated by two conformational changes, E_1–P to E_2–P and E_2 to E_1.

2. Na^+-ATPase activity is assayed similarly, but in the absence of K^+. In this reaction dephosphorylation of E_2–P proceeds either in the absence of monovalent cations at the K^+ sites or with Na^+ activating dephosphor-

[3] O. H. Lowry and J. A. Lopez, *J. Biol. Chem.* **162**, 421 (1946).
[4] A. H. Neufeld and H. M. Levy, *J. Biol. Chem.* **244**, 6493 (1969).
[5] E. Schacter, *Anal. Biochem.* **138**, 416 (1984).
[6] R. E. Barnett, *Biochemistry* **9**, 4644 (1970).
[7] L. Josephson and L. C. Cantley, Jr., *Biochemistry* **16**, 4572 (1977).
[8] J. D. Robinson, *Curr. Top. Membr. Transp.* **19**, 485 (1983).

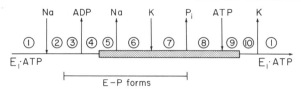

FIG. 1. Reaction scheme for Na$^+$,K$^+$-ATPase. The vertical arrows show addition and release of substrates, activators, and products; the thin horizontal line represents the E_1 enzyme conformations, and the stippled horizontal bar the E_2 conformations. The successive enzyme forms are (1) $E_1 \cdot$ ATP, (2) $E_1 \cdot$ ATP $\cdot$ Na, (3) E_1–P $\cdot$ ADP $\cdot$ Na, (4) E_1–P $\cdot$ Na, (5) E_2–P $\cdot$ Na, (6) E_2–P, (7) E_2–P $\cdot$ K, (8) $E_2 \cdot$ K, (9) $E_2 \cdot$ ATP $\cdot$ K, and (10) $E_1 \cdot$ ATP $\cdot$ K. From Robinson.[8]

ylation through occupancy of those sites; ATP then binds to the E_1 enzyme conformation.

3. K$^+$-phosphatase activity, which parallels the terminal hydrolytic steps of the Na$^+$,K$^+$-ATPase reaction, is assayed using a variety of acid anhydride phosphatase substrates (as the Tris salts), including p-nitrophenyl phosphate[9] (p-nitrophenol measured spectrophotometrically) and 3-O-methylfluorescein phosphate[10] (3-O-methylfluorescein measured spectrofluorometrically). In addition to substrate the reaction requires Mg^{2+} and K$^+$.

In these assays the usual precautions for kinetic studies must be observed.[11] Most approaches require measurements of initial velocity, and care must be taken that under all experimental conditions this stricture is satisfied. The most straightforward measurements of initial velocity are made when product increases linearly with time throughout the chosen incubation period; less easy are estimates of initial velocity derived from tangents to the curve of product formation drawn through the origin. In addition, the extent of inhibition caused by the reaction products should be determined and compared with the amounts of products formed during the assay incubations. Most kinetic analyses also assume a sufficient excess of substrates or other ligands so that their concentrations do not change appreciably during the incubation period. Related concerns are the extent to which the concentration of ligand added represents the free concentration available to the enzyme, beyond any loss due to chelation or complexing, and the possibility that the pertinent form of the ligand is in fact not the free form but a complex, such as MgATP. Nucleotide-bivalent cation complexing is particularly relevant to kinetic studies on

[9] K. Ahmed and J. D. Judah, *Biochim. Biophys. Acta* **93**, 603 (1964).
[10] W. Huang and A. Askari, *Anal. Biochem.* **66**, 265 (1975).
[11] R. D. Allison and D. L. Purich, this series, Vol. 63, p. 3.

the Na^+,K^+-ATPase, and since the K_d for the complex varies with temperature, pH, buffer composition, ionic strength, and presence of other ligands, the K_d value should be determined for the specific experimental conditions used[12]; a relatively simple spectrophotometric method for measuring K_d uses 8-hydroxyquinoline.[13]

Another potential problem in analysis of the kinetic studies is the possibility of competition between ligands for specific sites, notably between Na^+ and K^+ for their respective activating sites. Since *in vivo* the Na^+ sites for activating enzyme phosphorylation are accessible from the cytoplasm whereas the K^+ sites for activating enzyme dephosphorylation are accessible from the extracellular milieu, one way to minimize such competition is to use enzyme preparations where sidedness of exposure to these cations can be maintained: right-side-out vesicular preparations from kidney,[14] inside-out vesicles from erythrocytes,[15] or the purified enzyme reconstituted into lipid vesicles[16] (after reconstitution the enzyme is oriented randomly, but restricted access of substrate or inhibitor can allow only enzyme oriented in one particular sense to be assayed).

Finally, it should be emphasized that an important feature of proposed reaction schemes for the Na^+,K^+-ATPase is the transition between two major conformational families, E_1 and E_2, which are depicted as having different affinities for ATP, Na^+, and K^+. Thus an apparent affinity observed for some ligand may represent an average between the actual affinities of the alternative conformations, weighted by the distribution of the total enzyme between these conformations. Correspondingly, apparent affinities for ligands may vary with conditions that influence not the actual microscopic dissociation constants for each conformation, but the relative population sizes of the conformations. Such conditions include temperature, pH, and buffer composition, as well as the presence of specific ligands.[17,18]

Substrate Interactions

For the ATPase reactions the essential reactants include nucleotide triphosphate and bivalent cation, of which ATP and Mg^{2+} are most effective. Although free ATP binds to the enzyme with high affinity, shown in

[12] J. F. Morrison, this series, Vol. 63, p. 257.
[13] W. J. O'Sullivan and G. W. Smithers, this series, Vol. 63, p. 294.
[14] B. Forbush, *J. Biol. Chem.* **257,** 2678 (1982).
[15] T. L. Steck, *Methods Membr. Res.* **2,** 245 (1974).
[16] S. J. D. Karlish and W. Stein, *J. Physiol. (London)* **328,** 295 (1982).
[17] J. C. Skou and M. Esmann, *Biochim. Biophys. Acta* **601,** 386 (1980).
[18] A. C. Swann, *Arch. Biochem. Biophys.* **221,** 148 (1983).

direct measurements to have a K_d on the order of 0.1 μM,[19,20] such studies cannot measure binding in the presence of Mg^{2+} due to hydrolysis of the ATP that then occurs; however, with a nonhydrolyzable analog of ATP, the β,γ-imido derivative, Mg^{2+} decreased the K_d at most 2-fold.[21] Direct measurement of Mg^{2+} binding has not been reported, but indirect measurements through competition with Mn^{2+} indicate a K_d on the order of 1 mM.[22] Since Mg^{2+} is required for enzymatic activity and through the range of ATP concentrations optimally at roughly a 1:1 molar ratio with ATP, the effective substrate would thus seem to be MgATP. Moreover, both free ATP and free Mg^{2+} inhibit ATPase activity.[23,24] Determining the K_m for substrate thus requires attention to the concentrations of all three variables, MgATP, free Mg^{2+}, and free ATP. As noted above, the actual concentrations should be calculated from the K_d for MgATP evaluated for the specific reaction conditions. Since the three species inevitably vary as any one is changed, a reasonable compromise is to vary total ATP concentration with a small but constant excess of $MgCl_2$, such as 0.5 mM. Under the experimental conditions chosen in Fig. 2 the K_d for MgATP was 32 μM[25] and thus the free ATP concentration ranged from 0.02 to 140 μM when ATP was varied from 0.3 to 3000 μM with a constant 0.5 mM excess of $MgCl_2$, and the free Mg^{2+} concentration thus ranged from 500 to 640 μM. It should also be remembered that varying the concentration of two interacting ligands at a 1:1 molar ratio will not result in a constant ratio of complex to free species.

For the Na^+,K^+-ATPase reaction, plots of MgATP concentration are biphasic, with double-reciprocal Lineweaver–Burk plots approximating two straight lines (Fig. 2).[26] Such a plot is interpretable as demonstrating "high-affinity" and "low-affinity" sites, but such kinetic studies cannot determine whether such sites must be spatially distinct and temporally coexistant, or represent negative cooperativity between sites, or a cyclical interconversion of one class of sites during the reaction sequence. The last possibility is the most economical hypothesis and is the interpretation incorporated into usual reaction schemes such as that in Fig. 1, which shows ATP binding to a low-affinity substrate site of the E_2 conformation that is subsequently transformed into the high-affinity substrate site of the

[19] C. Hegyvary and R. L. Post, *J. Biol. Chem.* **246**, 5234 (1971).
[20] J. G. Nørby and J. Jensen, *Biochim. Biophys. Acta* **233**, 104 (1971).
[21] J. D. Robinson, *J. Bioenerg. Biomembr.* **12**, 165 (1980).
[22] C. M. Grisham and A. S. Mildvan, *J. Biol. Chem.* **249**, 3187 (1974).
[23] T. Hexum, F. E. Samson, Jr., and R. H. Himes, *Biochim. Biophys. Acta* **212**, 322 (1970).
[24] J. D. Robinson, *Biochim. Biophys. Acta* **341**, 232 (1974).
[25] J. D. Robinson, *FEBS Lett.* **47**, 352 (1974).
[26] J. D. Robinson, *Biochim. Biophys. Acta* **429**, 1006 (1976).

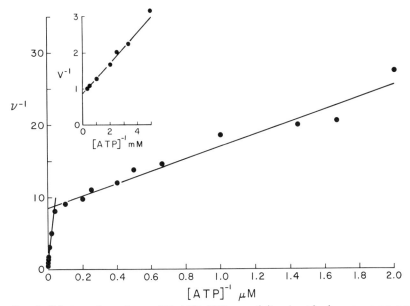

FIG. 2. Substrate dependence of Na$^+$,K$^+$-ATPase activity. A rat brain enzyme preparation was incubated at 37° with 30 mM histidine · HCl/Tris (pH 7.8), 90 mM NaCl, 10 mM KCl, the concentrations of [γ-^{32}P]ATP shown, and with the MgCl$_2$ concentrations 0.5 mM greater than the ATP concentrations. Data are plotted in double-reciprocal form; in the inset data with the higher ATP concentrations are replotted on an expanded scale. From Robinson.[26]

E_1 conformation. In this scheme the enzyme is phosphorylated at the high-affinity site of E_1 and P_i is released from the low-affinity site of E_2.

Direct binding studies of free ATP in the absence of K$^+$, as cited above,[19,20] report a K_d near 0.1 μM, consistent with the K_m for the high-affinity site in the presence of K$^+$ an order of magnitude higher, 1 μM, considering the demonstrated antagonism[19,20] of ATP binding by K$^+$ and the frequent deviation of the K_m derived from steady state kinetics from the K_d [reflecting, in the simplest case, the difference between $(k_{-1} + k_2)/k_1$ and k_{-1}/k_1].

Direct measurements of binding to the low-affinity site are technically far more difficult, but K_d values near the K_m value have been reported.[27] These studies, however, indicate binding of two ATP molecules, one at a high-affinity and one at a low-affinity site, per enzyme monomer (i.e., per

[27] F. M. A. H. Schuurmans Stekhoven, H. G. P. Swarts, J. J. H. H. M. DePont, and S. L. Bonting, *Biochim. Biophys. Acta* **649**, 533 (1981).

enzyme phosphorylation site), and thus do not correspond to the reaction scheme shown in Fig. 1.

In addition to factors affecting the congruence of K_d and K_m, such as differences in ligands and experimental conditions, and the potential difference between steady state values for K_m and the actual K_d, evaluating K_m values from biphasic plots depends on the reaction model proposed. A reaction scheme with two independent sites requires that the contribution to velocity from occupancy of one class of sites be subtracted from the total velocity in order to evaluate the actual velocity resulting from occupancy of the other class of sites. With the Na^+,K^+-ATPase, however, occupancy of the low-affinity sites increases total velocity by an order of magnitude, so the error introduced by merely fitting two straight lines to the uncorrected total velocity plots is minimal. An alternative, and mechanistically more specific approach, is to fit the velocity data to a steady state model that explicitly formulates the interconversion of sites,[28,29] rather than merely fitting two straight lines in accord with a kinetic scheme for two independent catalytic sites.

Alternative substrates and competitive inhibitions toward ATP may be studied at these two classes of sites, in terms of the usual kinetic parameters K_m and K_i, and such studies reveal differing specificities at the high-affinity and low-affinity sites.[26] These experiments obviously require consideration of bivalent cation–substrate and cation–inhibitor complexes. Again, the dissociation constants should be determined for each complex and the values in the incubation medium then calculated.

A fourth relationship between ligands and substrate should be emphasized, the antagonism between K^+ and ATP. As noted above, K^+ reduces the apparent K_d for ATP measured in binding studies.[19] Increasing concentrations of K^+ also increase the apparent K_m calculated from both limbs of the biphasic double-reciprocal plot. However, the double-reciprocal plots corresponding to occupancy of the low-affinity substrate sites are displaced in parallel by successively higher K^+ concentrations (Fig. 3).[30] This response is consistent with the intervention of an irreversible step between the addition of K^+ and the addition of ATP, such as the release of a product, P_i, in Fig. 1. Adding P_i to the reaction mixture, which converts that step from an irreversible to a reversible one, transforms the plots from a parallel to an intersecting pattern.[31] These observations thus favor the sequence: K^+ adds, P_i is released, and ATP adds (Fig. 1).

[28] R. L. Smith, K. Zinn, and L. C. Cantley, *J. Biol. Chem.* **255**, 9852 (1980).
[29] E. G. Moczydlowski and P. A. G. Fortes, *J. Biol. Chem.* **256**, 2357 (1981).
[30] J. D. Robinson, *Biochemistry* **6**, 3250 (1967).
[31] D. A. Eisner and D. E. Richards, *J. Physiol. (London)* **326**, 1 (1982).

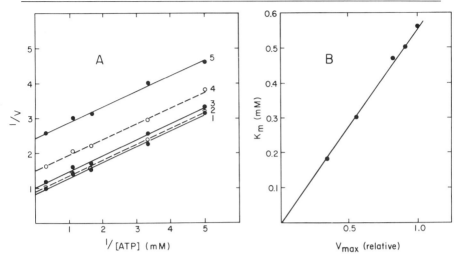

FIG. 3. Effect of KCl concentration on substrate dependence of Na^+,K^+-ATPase activity. A rat brain enzyme preparation was incubated at 30° with 50 mM Tris · HCl (pH 7.3), 90 mM NaCl, the concentrations of ATP indicated, equimolar concentrations of $MgCl_2$, and the following concentrations of KCl: (1) 10 mM, (2) 4 mM, (3) 2 mM, (4) 1 mM, and (5) 0.5 mM. In panel A the data are plotted in double-reciprocal form; in panel B the various K_m values for ATP are plotted against the corresponding V_{max} values derived from the plot in panel A. From Robinson.[30]

When Na^+-ATPase activity is examined over the same range of MgATP concentrations as the Na^+,K^+-ATPase activity, only a monophasic response is found, with a linear double-reciprocal plot having a K_m comparable to that for the high-affinity substrate site of the Na^+,K^+-ATPase activity.[26] Similarly, there have been no reports of nonlinear double-reciprocal plots for substrates of the K^+-phosphatase activity.

Patterns of product inhibition can be related also to the reaction sequence[32]: for an ordered release of two products, the first produces noncompetitive inhibition and the second competitive inhibition toward the substrate. In accord with this formulation, nitrophenol, a noncompetitive inhibitor of K^+-p-nitrophenylphosphatase activity, would be the first product released, whereas P_i, a competitive inhibitor toward p-nitrophenyl phosphate, would be the second.[33]

With p-nitrophenyl phosphate as substrate the K_d for the Mg–substrate complex is sufficiently large that, with the usual millimolar concentrations of $MgCl_2$ and p-nitrophenyl phosphate, negligible change in free

---

[32] W. W. Cleland, *Biochim. Biophys. Acta* **67**, 104 (1963).
[33] J. D. Robinson, *Biochim. Biophys. Acta* **212**, 509 (1970).

concentrations of these ligands occurs.[34] The K_m for Mg^{2+} is on the order of 1 mM,[35] corresponding to the K_i for Mg^{2+} evaluated as a competitor toward Mn^{2+} binding.[22]

Mg^{2+} is also required for phosphorylation of the enzyme by P_i, an activity demonstrable in the presence of millimolar $MgCl_2$ and phosphate (as the Tris salt).[36] These similar values of apparent affinity for Mg^{2+} suggest activity through occupying a single site, presumably part of the substrate site. However, experiments with CrATP indicate a site for Mg^{2+} as well as for Cr^{2+},[37] and kinetic studies imply a positive role for free Mg^{2+} even when MgATP saturates the high-affinity substrate sites.[38] On the other hand, millimolar Mg^{2+} can inhibit dephosphorylation of the enzyme by forming a "K-insensitive phosphoenzyme,"[36] but the nature of the sites through which Mg^{2+} exerts such an effect has not been identified definitively.

Free Mg^{2+} can also inhibit the K^+-phosphatase reaction when present at concentrations equivalent to that of K^+, acting as a competitor toward K^+; conversely, K^+ acts as a competitor toward Mg^{2+}.[39] Analogously, free Mg^{2+} can inhibit both the Na^+,K^+-ATPase and Na^+-ATPase reactions, acting, at least in part, as a competitor toward Na^+.[40]

Enzyme phosphorylation by ATP, as discussed above, requires Mg^{2+}, and Mg^{2+} remains bound to the phosphorylated enzyme after dissociation of free ADP, at least through hydrolysis of the acyl phosphate.[41] Whether it must dissociate from the enzyme when P_i is released has not been determined.

Monovalent Cation Interactions

Kinetic approaches to studying monovalent cation interactions with the enzyme follow similar procedures. The enzyme is incubated with fixed concentrations of buffer, substrate, and bivalent cation, and the monovalent cation is varied (Na^+ for the ATPases, and K^+ for the Na^+,K^+-ATPase and the K^+-phosphatase). Although complexing of the monovalent cations with other reagents is much less of a problem than

---

[34] J. D. Robinson, *Biochemistry* **8**, 3348 (1969).
[35] B. J. R. Pitts and A. Askari, *Biochim. Biophys. Acta* **227**, 453 (1971).
[36] R. L. Post, G. Toda, and F. N. Rogers, *J. Biol. Chem.* **250**, 691 (1975).
[37] S. E. O'Connor and C. M. Grisham, *FEBS Lett.* **118**, 303 (1980).
[38] J. D. Robinson, *Biochim. Biophys. Acta* **727**, 63 (1983).
[39] J. D. Robinson, *Biochim. Biophys. Acta* **384**, 250 (1975).
[40] M. S. Flashner and J. D. Robinson, *Arch. Biochem. Biophys.* **192**, 584 (1979).
[41] Y. Fukushima and R. L. Post, *J. Biol. Chem.* **253**, 6853 (1978).

complexing of bivalent cations, a number of other considerations require attention.

1. Changes in ionic strength with variation in cation concentrations may affect the enzyme.[17,42] Ideally, ionic strength should be maintained over the range of varied cation concentrations by substituting complementary concentrations of a "neutral" cation having no effect on the system beyond maintaining ionic strength. Choline chloride has been used for this purpose, but it seems to have a weak sodium-like effect on enzyme conformational transitions[43]; other organic ions, including histidine–HCl and Tris–HCl, also affect the enzyme,[17,42] and there may be no absolute solution to this problem. Comparison of data in the absence and presence of the minimally necessary compensatory cation (e.g., choline) should, nevertheless, be made. In assaying the Na^+,K^+-ATPase reaction Na^+ and K^+ can be varied reciprocally to maintain ionic strength, but kinetic analysis is then complicated not only by changing simultaneously both activating species but also by shifting the ratio of these two antagonistic species (see below).

2. Competition among the cations for their respective sites seems to occur: the measured $K_{0.5}$ for one cation varies with the concentration of the others.[44,45] This would be expected on several grounds. (1) Actual competition for occupancy seems probable, representing the lack of absolute specificity at any site. This response may be complicated by the competitor having a qualitatively if not quantitatively similar effect through occupying the site: thus, Na^+ can occupy with low affinity the extracellularly accessible K^+ sites and thereby activate E_2–P hydrolysis, although not as effectively as K^+.[46] The most straightforward solution to this problem is to segregate the cations and their appropriate sites, as suggested above, by using preparations in which sidedness is preserved, such as oriented membrane vesicles or liposomes in which the enzyme is reconstituted. This approach would allow Na^+ and K^+ each to be varied in the absence of the other; however, it obviously will not prevent competition between cations that act at the same enzyme face, such as Na^+ and Mg^{2+}. The second and more general approach to this problem is to vary one cation at successive fixed levels of the others, and then to extrapolate the series of $K_{0.5}$ values for the varied cation to the zero value of the

---

[42] G. Rempeters and W. Schoner, *Biochim. Biophys. Acta* **727,** 13 (1983).
[43] P. L. Jørgensen, *Biochim. Biophys. Acta* **401,** 399 (1975).
[44] J. D. Robinson, *Arch. Biochem. Biophys.* **139,** 17 (1970).
[45] J. D. Robinson, *Biochim. Biophys. Acta* **482,** 427 (1977).
[46] R. Blostein, *J. Biol. Chem.* **258,** 7948 (1983).

competitors.[45] (2) For multisubstrate reactions the apparent K_m for one substrate can vary with the concentration of the other as a reflection of the reaction sequence, in the absence of any actual competition for the same sites.[47] In particular, any two reactants that bind to the same enzyme form will act as competitors to each other. From the reaction scheme in Fig. 1 it will be seen that Na^+ dissociates from the same enzyme form to which K^+ binds and K^+ from that to which Na^+ binds: on this basis alone Na^+ and K^+ should act as kinetic competitors toward each other.[8] (3) Insofar as the various cations favor different enzyme conformations, which in turn have different intrinsic affinities for the cations, then one cation can alter the apparent affinity for another cation. Na^+, which favors E_1 enzyme conformations, would therefore diminish the apparent affinity for K^+ since the E_2 conformations have higher affinity sites for K^+ than do the E_1 conformations.[48]

3. The kinetic model selected to analyze the data will also affect the evaluation of the affinity. A notable feature of many but not all studies of monovalent cation activation of Na^+,K^+-ATPase, Na^+-ATPase, and K^+-phosphatase reactions is that the double-reciprocal plots are not linear but concave upward (Fig. 4).[8,30,34,44]

A standard enzymological method to obtain linear plots utilizes the Hill equation[49]:

$$v/V_{\max} = 1/[1 + (K/S^n)] \tag{1a}$$

or

$$v/V_{\max} = 1/[1 + (K_{0.5}/S)^n] \tag{1b}$$

In the latter formulation $K_{0.5}$ is equal to the concentration of substrate at which $v = 0.5 V_{\max}$, and $K_{0.5} = K^{1/n}$. For reactions satisfying this relationship, plots of $\log v/(V_{MAX} - v)$ against $\log S$ are linear with a slope equal to n. Alternatively, but less easy to plot for nonintegral values of n, double-reciprocal plots of $1/v$ against $1/S^n$ are also linear.

The Hill equation is frequently interpreted in terms of allosteric enzymes with cooperative interactions between multiple substrate (or activator) sites, with n reflecting the strength of interaction between these sites, and, when n is rounded to the next higher integer, the minimal number of such interacting sites.[49] Nevertheless, other although rarer mechanisms can approximate the response of the Hill equation in the absence of multiple interacting sites. For allosteric enzymes and the Hill

[47] I. H. Segal, "Enzyme Kinetics," p. 274. Wiley (Interscience), New York, 1975.
[48] S. J. D. Karlish, D. W. Yeats, and I. M. Glynn, *Biochim. Biophys. Acta* **525**, 252 (1978).
[49] I. H. Segal, "Enzyme Kinetics," p. 346. Wiley (Interscience), New York, 1975.

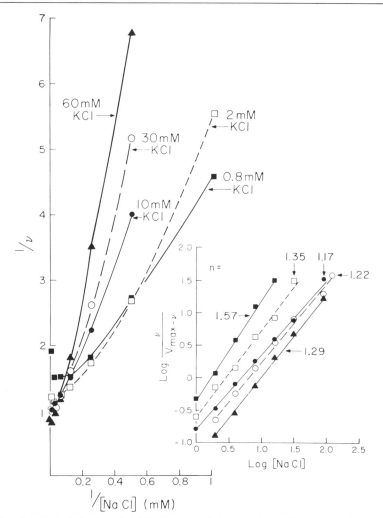

FIG. 4. Effect of NaCl on Na$^+$,K$^+$-ATPase activity at various KCl concentrations. A rat brain enzyme preparation was incubated at 30° with 50 mM Tris · HCl (pH 7.8), 3 mM ATP, 3 mM MgCl$_2$, the concentrations of NaCl shown on the abscissa, and the following KCl concentrations: 60 mM (▲), 30 mM (○), 10 mM (●), 2 mM (□), and 0.8 mM (■). Data are presented in double-reciprocal form on the left, and, on the right, as a Hill plot with the slopes, n, labeled: both $K_{0.5}$ and n vary with KCl concentration. From Robinson.[44]

equation, however, it should be recognized that finite enzyme velocities can occur in the absence of all substrate sites filled.

On the other hand, studies on the pump activity of this enzyme reveal stoichiometries for the ratio of Na$^+$ transport to K$^+$ transport to ATP hydrolyzed close to $3:2:1$,[50] often interpreted as an obligatory requirement that 3 Na^+ sites and 2 K^+ sites must be occupied for catalytic activity. In such a case, when no ATP hydrolysis occurs without all cation sites being occupied, the kinetic equation would be as follows:

$$v/V_{max} = 1/[1 + (K'/S)]^{n'} \qquad (2)$$

where n' here equals the number of sites for the substrate (or activator) in question, and must have an integral value. (This equation also stipulates that the affinity is the same for each of the n' ligands bound.) Equation (2) will result in a linear double-reciprocal plot in terms of $v^{1/n'}$ (i.e., $1/v^{1/n'}$ against $1/S$), as opposed to double-reciprocal plots of the Hill Eq. (1a) which are linear in terms of $1/S^n$. Moreover, Eq. (2) does not result in a linear Hill plot, and the slope of the tangent where $v = 0.5\ V_{max}$ is less than n'. A further difference between these two kinetic equations is the relationship between $K_{0.5}$ and K or K': for example, when $n = 2$ then $K_{0.5} = K^{1/2}$ with Eq. (1), but when $n' = 2$ then the $K_{0.5} = K'/(\sqrt{2} - 1)$ with Eq. (2).

Despite these differences, kinetic data for cation activation have been fitted to Eq. (1) (Fig. 4) and to Eq. (2) with integral values of n' for both transport[51] and enzymatic experiments.[52] Nevertheless, the applicability of Eq. (2) may seem less appropriate in light of competition between Na^+ and K^+ for their respective sites, the demonstrated efficacy of Na^+ at the K^+ sites, and the existence of alternative enzymatic and transport pathways that could affect the stoichiometry (i.e., Na^+-ATPase occurring as well as Na^+,K^+-ATPase activity; uncoupled Na^+ efflux occurring as well as Na^+–K^+-exchange).

Effects of Inhibitors and Modifiers

Previous sections have referred to the use of inhibitors to investigate the reaction sequence, to compare the properties of various classes of sites, and to examine the relative responses to alternative substrates and activators, as well as noting the mutual interactions among various ligands that compete for certain sites. Studies with reversible inhibitors follow the usual kinetic procedures, although the analysis of inhibition in multisite systems can be complex: interpretation of the response depends on the specific model of inhibition, and includes such considerations as whether the extent of cooperativity is altered when the inhibitor occupies

---

[50] I. M. Glynn and S. J. D. Karlish, *Annu. Rev. Physiol.* **37**, 13 (1975).
[51] R. P. Garay and P. J. Garrahan, *J. Physiol. (London)* **231**, 297 (1973).
[52] J. S. Charnock, C. L. Bashford, and J. C. Ellory, *Biochim. Biophys. Acta* **436**, 413 (1976).

one or more interacting sites.[49] Unfortunately, there is no means of determining a priori what model is applicable.

Just as K_m need not equal K_d for the substrate under steady state kinetic conditions, so K_i need not equal K_d for the inhibitor. Moreover, modifiers that shift the distribution of enzyme conformations between E_1 and E_2 may alter the observed $K_{0.5}$ for a ligand without actually altering the K_d, as recently calculated, using a rapid-equilibrium model of the K^+-phosphatase reaction, for effects of oligomycin and dimethyl sulfoxide on the $K_{0.5}$ for K^+ and for Na^+.[53]

With irreversible inhibitors (or inactivators) a further opportunity for assessing the affinity for ligands becomes available if the ligands affect measurably the rate of inactivation. Furthermore, this approach has a major advantage over studies on inhibition of enzymatic catalysis in that effects of the ligand on rate of inactivation need not be measured in the presence of the full set of ligands necessary for enzymatic activity, and thus certain potential interactions and competitions can be avoided. Specificity for the enzyme is, nevertheless, still assured by measuring the residual enzyme activity to assess the extent of inactivation.

The simplest case for analysis is inactivation that requires certain ligands, as with K^+-dependent inactivation of the enzymes by F^-.[39] In this case the rate of inactivation depends, hyperbolically, on the concentration of inhibitor F^-, and the extent of inactivation after incubation for a given time with F^- is readily measured by dilution and rapid assay of residual activity. The pseudo-first-order rate constant for inactivation, k', is determined as a function of the modifying ligand, K^+, from measurements of residual activity after incubation of enzyme with fixed levels of F^- and Mg^{2+} and a series of K^+ concentrations (Fig. 5). The model[54]

$$E + L \rightleftharpoons EL \rightarrow E_{inactive}$$

proposes that only the enzyme–ligand complex (in the example chosen, EK^+) undergoes irreversible inactivation at an appreciable rate. However, if this rate is slow compared to the rate of association and dissociation of L, so that EL is in equilibrium with E and L (the validity of this assumption should be assessed for specific cases, e.g., in terms of the reported rates of dissociation of K^+ from the occluded sites), then:

$$k'/k'_{max} = 1/[1 + (K_d/L)] \qquad (3)$$

where k'_{max} is the maximal pseudo-first-order rate constant for inactivation at saturating L. This expression is formally equivalent to the Michaelis–

---

[53] J. D. Robinson, L. J. Robinson, and N. J. Martin, *Biochim. Biophys. Acta* **772**, 295 (1984).
[54] J. D. Robinson, *Arch. Biochem. Biophys.* **156**, 232 (1973).

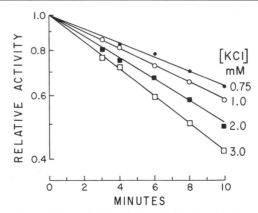

FIG. 5. K^+ dependence of enzyme inactivation by F^-. A rat brain enzyme preparation was first incubated at 37° with 30 mM histidine · HCl/Tris (pH 7.8), 1.5 mM LiF, and 0.5 mM $MgCl_2$, in the absence or presence of the KCl concentrations shown. This inactivating incubation was terminated at the times indicated by adding 4 vol of medium in which the residual Na^+,K^+-ATPase activity was then measured. This residual activity is plotted on the ordinate, on a logarithmic scale, against the duration of the inactivating incubation. From the slopes of the plots the pseudo-first-order rate constants for inactivation, k', can be calculated, to be plotted as a function of KCl concentration. From Robinson.[39]

Menten equation, so that plots of $1/k'$ against $1/L$ should be linear; in the case of multiple sites for the ligand, Eq. (3) can be modified to equivalence with Eqs. (1) or (2).

Analogous approaches are applicable for ligands that slow the rate of inactivation. For the simplest case, in which free enzyme, E, is inactivated but the enzyme–ligand complex, EL, is not susceptible to inactivation,[55]

$$EL \rightleftharpoons L + E \rightarrow E_{inactive}$$

a corresponding relationship can be derived, again assuming that the rate of association and dissociation of L is rapid relative to the rate of inactivation, and that the extent of inactivation is observable (e.g., by rapid enzyme assay). With these conditions met, then:

$$[1 - (k'_1/k'_0)] = 1/[1 + (K_d/L)] \qquad (4)$$

where k'_0 is the pseudo-first-order rate constant for inactivation in the absence of the ligand L, and k'_1 is the pseudo-first-order rate constant observed in the presence of a specific concentration of the ligand L. Again, this expression is formally equivalent to the Michaelis–Menten equation, so that plots of $1/(1 - k'_1/k'_0)$ against $1/L$ should be linear; in the

[55] J. D. Robinson, *FEBS Lett.* **38**, 325 (1974).

case of multiple sites for the ligand, Eq. (4) can be modified to equivalence with Eqs. (1) or (2).

These approaches may also be applied to ligand-modified inactivation due to such other processes as tryptic digestion of the enzyme, photoinactivation, and thermal denaturation. For all such studies the standard precautions pertain, including the potential for competition between ligands, the complexing of ligands, and the possibility that the K_d calculated by Eq. (3) reflects a weighted average of the K_d values of different enzyme conformations.

[23] Inhibition of Translocation Reactions by Vanadate

By LUIS BEAUGÉ

Vanadium is a rather complex element: it is highly reactive, can form polymers, and exists in several oxidation states (see Ref. 1 for references). The oxidized form, vanadate, is a powerful inhibitor of the Na,K-pump; the reduced form, vanadyl, is almost powerless in that effect.[2] Because it is possible to couple the vanadyl–vanadate transformation to oxidation–reduction systems present in cells and because of the K^+–vanadate interactions in the development of inhibition,[3] vanadium appears to be a potential candidate for *in vivo* regulation of the Na,K-pump.[2-4] The ligands related to the function of the Na,K-pump which are relevant to vanadate inhibition are Mg^{2+}, K^+, Na^+, and ATP. Some of their relationships and sidedness of interactions will become evident in the experiments described below. Nevertheless, to help one understand the results of these experiments the following facts are important: (1) Mg^{2+} is essential for vanadate binding to the enzyme[5-7] and for inhibition[8,9]; the inhibitory ternary complex is then of the MgEV type where the

[1] B. R. Nechay, *Annu. Rev. Pharmacol. Toxicol.* **24**, 501 (1984).
[2] J. J. Grantham and I. M. Glynn, *Am. J. Physiol.* **236**, F530 (1979).
[3] L. A. Beaugé and I. M. Glynn, *Nature (London)* **268**, 355 (1977).
[4] L. C. Cantley and P. Aisen, *J. Biol. Chem.* **254**, 1781 (1979).
[5] R. L. Smith, K. Zinn, and L. C. Cantley, *J. Biol. Chem.* **255**, 9852 (1980).
[6] J. D. Robinson and R. W. Mercer, *J. Bioenerg. Biomembr.* **13**, 205 (1981).
[7] B. R. Nechay and J. P. Saunders, *J. Environ. Pathol. Toxicol.* **2**, 247 (1978).
[8] L. C. Cantley, L. G. Cantley, and L. Josephson, *J. Biol. Chem.* **253**, 7361 (1978).
[9] G. H. Bond and P. M. Hudgins, *Biochemistry* **18**, 325 (1979).

Mg^{2+} effect takes place at intracellular sites[10]; (2) K^+ acts as a cofactor for inhibition (see Ref. 11 for references), increasing the affinity of the enzyme for the inhibitor[5,6]; (3) sodium and ATP antagonize inhibition[3,5–8]; they also increase the rate of vanadate debinding for the enzyme[8,12]; (4) because of the structural similarities between vanadate and phosphate[13] it has been suggested that the phosphate site is the binding site for vanadate[5]; on the other hand, a high- and a low-affinity site for vanadate have been reported,[5] but the relevance of the latter in the inhibition process remains uncertain.

Some key experiments on vanadate inhibition are based on its effects on translocation reactions of the Na-pump. The way these results fit with those obtained from biochemical studies and the way they help in understanding the operation of the Na-pump will also be examined. It seems particularly important to stress the usefulness of transport experiments using entire (intact or not) cells. On the one hand they allow the establishment of sidedness of interactions; on the other, if the substance under investigation is to have a physiological or pharmacological role, its effects are studied under the closest physiological conditions possible, something that cannot be done in most biochemical experiments.

Dialyzed Squid Giant Axons: Sodium Efflux (Na^+, K^+-Coupled Active Transport; $Na^+–Na^+$ Exchange)[14,15]

General Procedure

Internal dialysis consists of passing a solution through a tube (glass or plastic) longitudinally steered into an axon mounted in a special chamber. The tube has a selective porous region (1.0- to 1.5-cm length); the size of the pores is such that all solutes with a molecular weight of 1000 or less can be equilibrated across the capillary wall. In this way the internal solute composition (ions, nucleotides, metabolites, etc.) can be accurately controlled. The extracellular composition is controlled by the solution bathing the axons (artificial seawater). The flow rates are 1 μl/min for the dialysis solution (using a syringe pump and a gas-tight Hamilton sy-

---

[10] L. A. Beaugé, J. S. Cavieres, I. M. Glynn, and J. J. Grantham, *J. Physiol. (London)* **301**, 7 (1980).
[11] L. Beaugé, in "Na,K-ATPase: Structure and Kinetics" (J. C. Skou and J. C. Nørby, eds.), p. 373. Academic Press, New York, 1979.
[12] W. Huang and A. Askari, *J. Biol. Chem.* **259**, 13287 (1984).
[13] V. J. Lopez, T. Stevens, and R. N. Lindquist, *Arch. Biochem. Biophys.* **175**, 31 (1974).
[14] F. J. Brinley and L. J. Mullins, *J. Gen. Physiol.* **50**, 2303 (1967).
[15] L. Beaugé and R. DiPolo, *Biochim. Biophys. Acta* **551**, 220 (1979).

ringe) and 1 ml/min for the external solution (using an LKB peristaltic pump). Complete washout of intracellular ATP can be accomplished in about 90 min, whereas isotope equilibration for ^{22}Na takes about 20 min. Additional advantages of this preparation are that (1) all intracellular structures normally present in the axon are preserved, and (2) several changes in internal as well as external solutions can be made, with the same axon serving as its own control.

The composition of a typical dialysis solution (Fig. 1) is the following (mM): Na$^+$, 70; K$^+$, 310; Mg^{2+}, 4 in excess to ATP concentration; Tris$^+$, 30; Cl$^-$, 108; aspartate, 310; glycine, 300; EGTA, 1; ATP (vanadium free), 3; and phosphoarginine (PA), 5; osmolarity is 1000 mOsm, pH 7.3 at 17°. Phosphoarginine is included in order to buffer the ATP concentration and eliminate all intracellular ADP through the reaction

$$ADP + PA \rightleftharpoons ATP + A$$

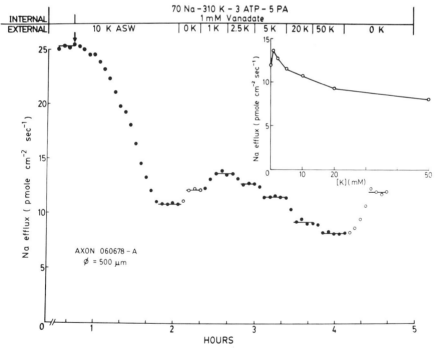

FIG. 1. Effect of different external potassium concentrations of Na$^+$ efflux in a vanadate-poisoned dialyzed squid giant axon. Insert: Na$^+$ efflux plotted as a function of the external potassium concentration. All solute concentrations are given in millimoles. Taken from Ref. 16.

catalyzed by the arginine phosphokinase present in the axoplasm. The amount of ^{22}Na is usually 100 μCi/ml. Vanadate is added from a 200 mM stock solution and adjusted to pH 7.3 with HCl.

On the other hand, the composition of the artificial seawater (ASW) is as follows (mM): Na$^+$, 400; K$^+$, 10; Mg^{2+}, 50; Ca^{2+}, 10; Tris$^+$, 10; Cl$^-$, 580; EDTA, 0.1. The osmolarity is 1000 mOsm, pH 7.8 at 17°. Removal or addition of KCl is usually done together with the addition or removal of equal amounts of NaCl.

Results and Comments

Figure 1 is an experiment where the effects of 1 mM vanadate on Na$^+$ efflux were explored in a dialyzed squid giant axon.[16] The important points that emerge are the following: (1) Previous observations in red blood cells[17] are confirmed in the sense that the inhibitor acts from the inside of the cell. (2) A time lag of about 1 hr is required to acquire full inhibition. This is extremely interesting because first, this is much longer than the time taken to reach steady state distribution for molecules of that size and charge[14] and, second, the $K_{0.5}$ for vanadate inhibition of Na$^+$,K$^+$-ATPase activity in membrane fragments from squid nerves is about 10^{-8} M.[16] Plausible explanations for this behavior are the presence of intracellular structures that bind the inhibitor and/or oxidation–reduction systems that render vanadium ineffective.[2] At any rate this calls for caution in a straightforward extrapolation from *in vitro* to *in vivo* responses to the inhibitor. (3) The K_o^+-dependent Na$^+$ efflux was inhibited (partially or fully depending on the K_o^+ concentration), but Na$^+$ efflux in the absence of K_o^+ was not at all affected. However, although practically all Na$^+$ efflux in dialyzed squid axons is ATP-dependent, how much of the fraction remaining in K_o^+-free Na$^+$ solutions is due to the Na-pump (uncoupled or Na$_o^+$-stimulated) is not certain. This is important because it has been shown that the ATP-dependent strophanthidin-insensitive Na$^+$ efflux is via a Na$^+$–Na$^+$ exchange and not affected by 1 mM vanadate concentration.[18] (4) In the presence of vanadate, increasing [K$^+$]$_o$ resulted in a *biphasic* response of Na$^+$ efflux (insert on Fig. 1): an increase at low K_o^+ followed by a decline at high K_o^+ concentrations. A similar effect of extracellular K$^+$ has been described for the ouabain-sensitive ATPase activity[19] and also for the ouabain-sensitive phosphatase activity provided Na$_i^+$, Na$_o^+$, and ATP are present.[20] On the other hand, the actual values of Na$^+$

---

[16] L. Beaugé, *Curr. Top. Membr. Transp.* **22,** 131 (1984).
[17] L. C. Cantley, M. D. Resh, and G. Guidotti, *Nature (London)* **272,** 252 (1978).
[18] L. Beaugé and R. DiPolo, *J. Physiol. (London)* **315,** 447 (1981).
[19] L. Beaugé and I. M. Glynn, *Nature (London)* **272,** 551 (1978).
[20] L. Beaugé and G. Berberian, *Biochim. Biophys. Acta* **727,** 336 (1983).

efflux into solutions containing 10 mM K_o^+ or more are *lower* than those seen in K^+-free conditions. A *reverse K^+-free effect* on ouabain-sensitive ATPase activity has been reported in resealed human red cell ghosts without ADP when ATP concentration was in the micromolar range.[21] This is ascribed to a shift in the rate-limiting step from K_o^+-promoted dephosphorylation to an ATP-stimulated $E_2(K)-E_1K$ transformation. For that reason a possible explanation for the reverse K^+-free effect in vanadate-poisoned cells (see also below) is that the inhibitor actually interferes with the $E_2(K)-E_1K$ conformational change, even at high ATP concentrations.[15] However, as it was pointed out,[15] a simple positive cooperative effect between K_o^+ and vanadate in promoting inhibition could also account for the data. Actually, the need for an external regulatory site with which K_o^+ interacts favoring inhibition seems inescapable from the biphasic curve in the insert of Fig. 1 (see also later). The changes in fluxes observed in Fig. 1 once the full effects of vanadate have developed reach true steady state values; that is, no artifacts due to time-dependent phenomena are introduced.

Intact Human Red Blood Cells: Rubidium (Potassium) Influx (Na^+,K^+-Coupled Active Transport)[10,20]

General Procedure

Blood is drawn in heparinized syringes and centrifuged (at room temperature or at about 4°) at 1500 g for 10 min. After plasma and buffy coat (containing white elements) are removed by aspiration the red cells are washed and subjected to different treatment depending on the type of experiment to be performed. Therefore, this part of the procedure applies to all experiments using red blood cells and will not be repeated under the following headings.

The red cells are washed three to five times at room temperature with a solution of the following composition (mM): NaCl, 150; $MgCl_2$, 1; Tris–HCl, pH 7.4 (37°), 10; and EGTA, 0.1. The cells are then resuspended at 20% hematocrit in the same solution containing 14 mM glucose with and without different concentrations of sodium orthovanadate; the inhibitor is taken from a 100 mM stock solution neutralized to pH 7.4 with HCl. In order to allow for vanadate equilibration the suspension is incubated for 0.5 to 2 hr at 37°, then cooled to 0° and washed three times in a solution of the same composition except that NaCl has been replaced by an equal amount of choline chloride. After the final wash the cells are resuspended in the same choline solution at about 20% hematocrit. Aliquots of 0.1 ml

[21] I. M. Glynn and S. J. D. Karlish, *J. Physiol. (London)* **256**, 465 (1976).

are mixed in 1.4-ml Eppendorf tubes with 0.9 ml of solutions containing variable NaCl, KCl, or choline chloride and the other constituents already mentioned in the presence or absence of 2×10^{-4} M ouabain and the tubes transferred to a 37° water bath. After 30 min the tubes are returned to the ice-cold bath (for about 10 min), centrifuged, and the cells washed three times with the same incubation solution and then lysed with 0.9 ml of distilled water. A fraction of 0.1 ml of the lysate is used for hemoglobin determination (estimated by the absorbance at 541 nm). Taking the optical density of packed cells (when only red cells without extracellular space are present) equal to 284, the hemoglobin content of the lysate permits an estimation of the volume of cells present in each tube. For example, if 10 ml of lysate gives an optical density of 0.5 it means the sample contains $(0.5/284) \times 10$ ml = 0.0176 ml of red cells. The remaining lysate is precipitated with trichloroacetic acid (5% final concentration) and centrifuged. The ^{86}Rb activity in the supernatant is determined by scintillation counting. Control experiments show that under all these conditions Rb fluxes are linear with time at least up to 60 min.

Results and Comments

The ouabain-sensitive uptake of K^+ (or its congeners like Rb^+) is considered the counterpart of Na^+ extrusion in the coupled active Na^+–K^+ transport mechanism. The experiments described in Fig. 2 indicate that the biphasic effect of external K^+ already observed for Na^+ efflux in vanadate-poisoned axons (Fig. 1) is also seen for Rb^+ influx in human red blood cells subjected to the same treatment; i.e., Na^+ efflux and K^+ influx, coupled in normal conditions, remain coupled under inhibition by vanadate. In addition to the positive cooperativity between K_o^+ and vanadate the results also show antagonism between K_o^+ and Na_o^+ in promoting inhibition. This is clearly seen in Table I where $K_{0.5}$ for Rb_o^+-promoted inhibition decreases when the concentrations of intracellular vanadate increases or when the extracellular Na^+ concentration is lowered. Under certain conditions (rows 1 to 3 in Table I) the apparent affinity for Rb_o^+ as coinhibitor is lower than as Na-pump activator, and in some cases (row 1 in Table I) inhibition begins to develop at Rb_o^+ concentrations at which the activation sites are already saturated. This suggests that both effects take place at different sites. Furthermore, the action of Rb_o^+ as cofactor for inhibition seems to depend on the presence of external Na^+. When Rb^+ influx is followed in Na_o^+-free choline solutions the biphasic response disappears and the fractional vanadate inhibition of pumped Rb becomes independent of Rb_o^+ concentration (Fig. 3). An explanation consistent with these results is that external K^+ removes Na^+ from extracellular sites at which binding of Na^+ prevents vanadate from acting.

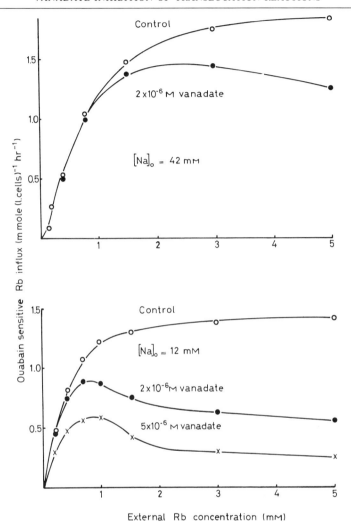

FIG. 2. The relation between ouabain-sensitive rubidium influx and the external rubidium concentration in human red blood cells. Top: In the absence (open circles) and presence (filled circles) of 2×10^{-6} M vanadate in media containing 42 mM Na$^+$. Bottom: In the absence (open circles) and presence of 2×10^{-6} M (filled circles) and 5×10^{-6} M (crosses) vanadate in media containing 12 mM Na$^+$. Each point is the mean of duplicate determinations. Taken from Ref. 10.

In purified Na$^+$,K$^+$-ATPase K$^+$ ions, in the absence of Na$^+$, increase the binding affinity for vanadate without changing the maximal amounts of inhibitors bound to the enzyme[5]; likewise, vanadate increases the equilibrium binding affinity for Rb$^+$.[5] On the other hand, in the same prepara-

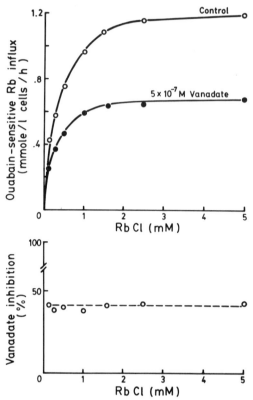

FIG. 3. The relation between ouabain-sensitive rubidium influx and external rubidium concentration in human red blood cells. Top: Rubidium influx in the absence (open circles) and presence (filled circles) or 5×10^{-7} M vanadate in sodium-free choline media. Bottom: Percentage of vanadate inhibition at different extracellular rubidium concentrations. Each point is the mean of duplicate determinations. Taken from Ref. 20.

tion the rate of vanadate dissociation from the MgEV complex is accelerated by Na^+ [5,12]; K^+ antagonizes Na^+ but in itself has no effect.[12] It is not easy to reconcile observations in which a nonturning enzyme has both "surfaces" simultaneously exposed to all ligands with experiments on transport (or ATPase activity), under turnover conditions in a sided preparation. The results described here point unambiguously to interactions of Na^+ and K^+ with extracellular sites, almost surely regulatory ones. It is likely that the observations with nonturning purified enzyme (in which the K^+ sites exposed are mostly "intracellular") are consequences of ligand interactions with intracellular sites. In this regard there are two important points to consider. (1) Vanadate binding brings the enzyme to the E_2

TABLE I
EFFECTS OF EXTERNAL Na AND VANADATE CONCENTRATIONS
ON THE $K_{0.5}$ FOR ACTIVATION AND INHIBITION OF Rb INFLUX
BY EXTERNAL Rb IN HUMAN RED BLOOD CELLS[a]

		Rb$^+$ on ouabain-sensitive Rb$^+$ influx	
Na$_o^+$ (mM)	Vanadate (M)	$K_{0.5}$ for activation (mM)	$K_{0.5}$ for inhibition (mM)
130	2×10^{-6}	1.10	10
45	2×10^{-6}	0.68	4.1
12	2×10^{-6}	0.35	1.1
12	5×10^{-6}	0.35	0.4

[a] The data were obtained from experiments like those described in Fig. 2. To estimate the $K_{0.5}$ for inhibition the magnitude of the inhibited Rb$^+$ influx (control minus vanadate) was plotted against Rb$^+$ concentration. The $K_{0.5}$ for activation refers to Rb$^+$ influx in the absence of vanadate. Taken from Ref. 11.

conformation[22] and all ligands favoring that conformation (except Mg^{2+}, Mn^{2+}, and Ca^{2+} that stabilize the E$_1$ state[23]) also favor binding of the inhibitor, and (2) in vanadate-poisoned Na$^+$,K$^+$-ATPase *external* K$^+$ shows biphasic effects only in the presence of external Na$^+$ and when there is simultaneous phosphorylation at the catalytic ATP site.[20]

Sodium Efflux (Na$^+$,K$^+$-Coupled Active Transport and Na$^+$–Na$^+$ Exchange)[10,11]

General Procedure

Red cells are washed five times at room temperature with a solution of the following composition (mM): NaCl, 150; MgCl$_2$, 1; Tris–HCl (pH 7.4 at 37°) 10; EGTA, 0.1; and glucose, 11. The washed cells, resuspended at about 12% hematocrit in the same media with the addition of different concentrations of sodium orthovanadate (0, 10^{-5}, 10^{-4}, and 10^{-3} M) and with 5 μCi/ml of ^{22}Na, are incubated for 2 hr at 37°. After that time they are washed seven times at 0° with the same solutions without radioactive Na$^+$ and resuspended in the washing media at 20% hematocrit. Aliquots

---

[22] S. J. D. Karlish, L. Beaugé, and I. M. Glynn, *Nature (London)* **282**, 333 (1979).
[23] G. Berberian and L. Beaugé, *PAABS Int. Meet., 4th* Abstr. 505 (1984).

of 0.05 ml are then mixed in 1.4-ml Eppendorf tubes with 0.95 ml of either K^+-free 150 mM Na^+ or 11 mM K^+–139 mM Na^+ solutions with the appropriate vanadate concentrations with or without 10^{-4} M ouabain. The tubes are incubated for 1 hr at 37°, cooled in an ice-cold bath, and centrifuged. The ^{22}Na present in the supernatant is assayed by liquid scintillation counting. The ^{22}Na originally present in the cells is determined in the same way after precipitating the cells with trichloroacetic acid (5% final concentration) followed by centrifugation.

Results and Comments

Figure 4 illustrates an experiment on the effects of different vanadate concentrations on ouabain-sensitive Na^+ efflux into K^+-free and 10 mM K^+ media containing 140 mM sodium. Because with 10 mM K^+ the external K^+ sites of the Na^+ pump are practically saturated, the ouabain-sensitive Na^+ efflux into that solution is taken as an expression of

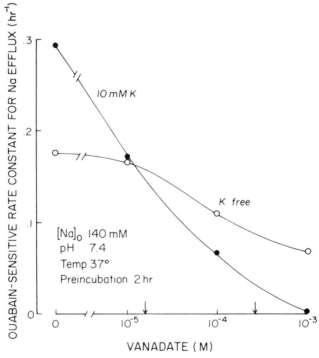

FIG. 4. The effects of different vanadate concentrations on ouabain-sensitive Na^+ efflux from human red blood cells incubated in K^+-free (open circles) and 10 mM K^+ (filled circles) full sodium media. Each point is the mean of two determinations. Taken from Ref. 11.

Na^+,K^+-coupled active transport. These cells contain ATP as well as ADP and are incubated in Na^+ media; therefore, in the absence of external K^+ the ouabain-sensitive Na^+ efflux through the Na-pump could theoretically have three components: Na^+-Na^+ exchange linked to ATP–ADP exchange[24] and Na^+ efflux associated to ATP hydrolysis, either uncoupled[21] or in exchange for external Na^+.[25] What fraction corresponds to each one cannot be determined, although it is likely that the uncoupled fraction, if it exists, would be minimal.[25] The coupled Na^+-K^+ transport appears more sensitive to vanadate than the Na^+-Na^+ exchange (about 20-fold difference in $K_{0.5}$ for the inhibitor). In addition, at vanadate concentrations above 10^{-5} M a reverse K^+-free effect is observed. Although this preparation is not biochemically equal to the dialyzed axon in Fig. 1 the results in both cases can be considered in fairly good agreement. The fact that in the absence of K_o^+ sodium efflux was not affected by 1 mM vanadate in the axon could be due to difference in the "active state" of vanadate or in the reactivity of both preparations to the inhibitor. The latter possibility cannot be neglected considering the different intra- and extracellular ionic environments in the two cases. On the other hand, the results in Fig. 4, including inhibition of the Na^+-Na^+ exchange in the presence of ATP and ADP, were reproduced in inside-out vesicles from human red cell membranes.[25]

Potassium Efflux (K^+-K^+ Exchange and Pump Reversal)[10]

General Procedure

Red cells are washed three times at room temperature with a solution of the following composition (mM): choline chloride, 140; $MgCl_2$, 1.5; imidazole (pH 7.4, 37°), 10. To load the cells with ^{42}K they are incubated for 3.5 hr at 37° and 30% hematocrit in a solution containing (mM) ^{42}KCl, 62; choline chloride, 82; $MgCl_2$, 1; EGTA, 0.1; K_2HPO_4, 2.5; and HEPES–Tris, 5. The pH is kept at 7.4 (20°). Sodium orthovanadate (final concentration 4×10^{-6} M) is added to half of the cell suspension and the incubation continues for 30 more minutes. The suspension is cooled at 0° and the cells are washed four times in an ice-cold solution containing (mM): NaCl, 140; $MgCl_2$, 1; EGTA, 0.1; Na_2HPO_4, 2.5; HEPES–Tris, 5 (pH 7.4 at 20°) with and without 4×10^{-6} M vanadate. The hematocrit is taken to 6% with the same solutions and 1 ml of cell suspension is mixed in 1.4-ml Eppendorf tubes with 0.2 ml of similar media where NaCl is

[24] I. M. Glynn and J. F. Hoffman, *J. Physiol. (London)* **218**, 239 (1971).
[25] R. Blostein, *J. Biol. Chem.* **258**, 7948 (1983).

replaced with KCl to give final K$^+$ concentrations of 0, 5, and 14 mM. One-half of the tubes also contain ouabain (10^{-4} M final concentration). The tubes are then transferred into a 37° water bath; after 5 min and after 65 min tubes are removed and transferred back into the ice-cold bath, allowed to stand for 10 min and centrifuged. Radioactivity is assayed in the supernatant by Cerenkov radiation. In each case K$^+$ losses due to hemolysis are checked measuring the absorbance of the supernatant at 541 nm. The ^{42}K present in the cells before efflux measurements is determined after precipitating the cells with 5% trichloroacetic acid (final concentration). The difference in ^{42}K lost between 5 and 65 min is used to calculate the rate constant for K$^+$ efflux. The ouabain-sensitive fraction of K$^+$ efflux into K$^+$-free Na$^+$ solutions is considered equivalent to pump reversal; on the other hand, the ouabain-sensitive fraction lost in K$^+$-containing media is taken as an expression of K$^+$–K$^+$ exchange through the pump.

Results and Comments

Table II summarizes two experiments on the effects of 3.3×10^{-6} M vanadate on K$^+$ efflux through two modes of cation translocation by the Na-pump: reversal (K$_i^+$–Na$_o^+$ exchange) and K$^+$–K$^+$ exchange. At the

TABLE II
Effects of Vanadate (3.3 μM) on the Ouabain-Sensitive Efflux of Potassium from Red Cells[a]

Sample	Rate constants for ouabain-sensitive potassium efflux[b]		
	Potassium-free medium (hr^{-1})	5 mM potassium medium (hr^{-1})	14 mM potassium medium (hr^{-1})
Experiment 1			
Control	0.0078 ± 0.0009	0.0112 ± 0.0011	0.0110 ± 0.0011
Vanadate	0.0073 ± 0.0006	0.0096 ± 0.0006	0.0048 ± 0.0007
Experiment 2			
Control	0.0071 ± 0.0008	0.0129 ± 0.0006	0.0118 ± 0.0007
Vanadate	0.0083 ± 0.0004	0.0108 ± 0.0008	0.0050 ± 0.0007

[a] Incubated in high sodium, potassium-free media (pump reversal), or high-sodium media containing 5 or 14 mM potassium (potassium–potassium exchange). The figures represent the differences in the means of rate constants for potassium efflux in the presence and absence of ouabain ± the standard error of the differences of means. Taken from Ref. 10.

[b] In these experiments, vanadate had no significant effect on the potassium efflux in the presence of ouabain under any condition.

concentrations used the inhibitor had no effect on pump reversal. On the other hand, K^+-K^+ exchange was reduced by 15% (only one experiment gave statistical significance) with 5 mM K_o^+ and by 57% with 14 mM K_o^+; considering the absolute values, K^+-K^+ exchange in vanadate-poisoned cells also showed a biphasic response to external K^+ concentration. These results could mark intrinsic differences between cation translocation linked to ATP synthesis and to reversible binding of P_i to the enzyme.[10] However, there is another explanation that seems more coherent with vanadate inhibition of the Na-pump: The concentration of vanadate used in these experiments was 3.3×10^{-6} M; with 2×10^{-6} M concentration and in cells incubated in 112 mM Na_o^+ inhibition of Na^+,K^+-activated transported was observed only when external K^+ was 30 mM.[10] Thus, a protective effect of Na_o^+ antagonized by K_o^+ [10,20] can explain why reversal (followed in K^+-free Na^+ media) was resistant to inhibition, whereas K^+-K^+ exchange was sensitive and became even more so as K_o^+ was increased.

Resealed Human Red Cell Ghosts: Sodium Efflux in the Absence of External Na^+ and K^+ (Uncoupled Na^+ Efflux)[10]

General Procedure

By definition, uncoupled Na^+ efflux is a mode of Na^+ extrusion associated with ATP hydrolysis in cells incubated in the absence of external K^+ (or its congeners) and Na^+. To avoid stimulation of the Na-pump by recapture of K^+ leaking out of the cells it is advisable to have the intracellular K^+ at its minimum level. One way to accomplish this is to use resealed ghosts.

The red cells are washed three times at room temperature with the following solution (mM): choline chloride, 150; $MgCl_2$, 1; HEPES-Tris, pH 7.4, 20°, 2; and EGTA-Tris, 0.1. The cells are then lysed at 0° under constant stirring in 50 vol of a solution of the following composition (mM): $ATPNa_2$, 1; $^{24}NaCl$, 6; $MgCl_2$, 1.5; HEPES-Tris, pH 7.3, 20°, 5; and EGTA-Tris, 0.2. The 1:50 dilution of all diffusible intracellular components means that K^+ concentration in the lysate, and inside the cells after resealing, will be around 3 mM. After 5 min tonicity is taken to 107 mOsm/liter, adding enough choline chloride from a 3 M stock solution, and resealing is brought about by incubating the suspension for 20 min at 37°. The resealed ghosts are washed five times in the following solution (mM): choline chloride, 50; $MgCl_2$, 1; HEPES-Tris, pH 7.4, 20°, 2; EGTA-Tris, 0.2. The first, second, and fourth washes are performed with ice-cold solutions; the third and fifth at 37°. After the third and final wash the ghost suspension is incubated at 37° for 5 min; it was found that this reduces the initial nonspecific Na^+ losses during the actual efflux period. The final incubation takes place in the same washing solution supplemented, when appropriate, with the following ligands: 5 mM KCl, 3.3 × 10^{-6} M vanadate, and 10^{-4} M ouabain. One-half of the tubes remain 15 min in the incubation bath, and the rest 60 min. They are then transferred to an ice-cold bath, cooled, and centrifuged. Radioactivity is measured by Cerenkov radiation in the supernatants and in the supernatant of a whole-cell suspension after precipitation with trichloroacetic acid (5% final concentration) and centrifugation. The rate constants for Na^+ efflux are calculated from the difference between the ^{24}Na lost in 60 and 15 min.

Results and Comments

Table III summarizes the results of an experiment where the effects of 3.3 × 10^{-6} M vanadate on Na_i^+-K_o^+-activated coupled exchange (in the absence of Na_o^+) and the uncoupled Na^+ efflux are compared. At that concentration the inhibitor did not affect Na^+-Na^+ exchange and pump reversal; likewise, its effects on active Na^+-K^+ exchange in Na^+ media would have been rather insignificant (see above). In this case active Na^+-K^+ transport was fully inhibited; the uncoupled Na^+ efflux was also inhibited but the low levels of fluxes do not allow precise establishment of the extent of that inhibition. In other experiments performed under similar conditions the inhibition of uncoupled Na^+ efflux did not reach statistical significance; however, this seems a consequence of the low levels of

TABLE III
EFFECTS OF 3.3 μM VANADATE ON THE OUABAIN-SENSITIVE EFFLUX OF Na^+ FROM RESEALED GHOSTS[a]

	Rate constants for ouabain-sensitive sodium efflux[b]	
	Potassium-free medium (hr^{-1})	5 mM potassium medium (hr^{-1})
Control	0.048 ± 0.011	0.242 ± 0.011
Vanadate	0.013 ± 0.009	0.010 ± 0.006

[a] Incubated in Na^+-free media containing no potassium (uncoupled Na^+ efflux) or 5 mM potassium (sodium–potassium exchange). The flux in each condition was measured in quintuplicate. Each entry is the mean ± SEM. Taken from Ref. 10.

[b] In this experiment, vanadate had no significant effect on the efflux of sodium in the presence of ouabain, either with no potassium or with 5 mM potassium in the medium.

fluxes and the natural dispersion of the results rather than an erratic response of this mode of pump operation to vanadate. This line of reasoning is supported by the following additional observations: (1) the values of "residual" Na^+ efflux are the same for both conditions in Table III; (2) the ATPase activity associated to this flux was fully inhibited by similar concentrations of vanadate[10]; and (3) in inside-out vesicles (IOV) from human red cells uncoupled Na^+ efflux was about 65% inhibited by $3 \times 10^{-6}\ M$ vanadate concentration.[25]

Concluding Remarks

The first conclusion that can be drawn from the results presented and discussed here is that all translocation reactions carried out by the Na,K-pump can eventually be inhibited by vanadate. However, their sensitivity to the inhibitor is conditioned by the intracellular and, especially in intact cells, the extracellular ionic composition, namely the K^+/Na^+ ratio.

The Na_o^+–K_o^+ interactions and the biphasic response of active Na^+–K^+ exchange to external K^+ in the presence of Na_o^+ and vanadate appear unexplainable, within the accepted scheme of transport and biochemical reactions, unless these interactions take place at regulatory pump sites. A simple effect of K_o^+ due to the formation of the $E_2(K)$ state after dephosphorylation as proposed[26] would result in a monotonic inhibition of the Na-pump as K_o^+ increases, but could never account for a biphasic response. The existence of regulatory interactions is reinforced by the fact that with no external Na^+ the magnitude of inhibition becomes independent of K_o^+ concentration (Fig. 2). The identification of the sites at which Na_o^+ protects against inhibition has been elusive so far[10,20,25]; likewise, it is not known if under certain conditions other cations (perhaps choline or Tris) could also show some Na^+-like effects.[25]

The idea that vanadate binds to a phosphate site[5] is possible, and even likely. This is not only supported by the fact that Mg^{2+} is an essential cofactor for binding and inhibition[5–10] but by the additional observations that (1) Mg^{2+} acts intracellularly[10] and (2) at low vanadate concentrations the levels of Mg^{2+} required to promote inhibition are higher than those needed to stimulate the pump.[10] What remains less clear is at which step(s) in the cycle does the inhibitor bind. Magnesium seems able to bind (allowing P_i binding to follow) at more than one step in the pump cycle, leading to "product" and "dead-end" type inhibitions.[27,28] There are two

[26] S. J. D. Karlish and U. Pick, *J. Physiol.* (*London*) **312,** 509 (1981).
[27] L. Plesner and I. W. Plesner, *Biochim. Biophys. Acta* **643,** 449 (1981).
[28] C. H. Pedemonte and L. Beaugé, *Biochim. Biophys. Acta* **748,** 245 (1983).

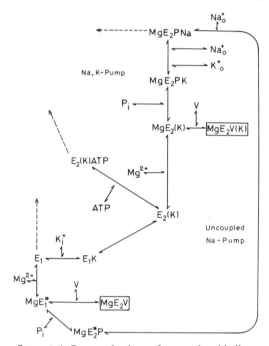

SCHEME 1. Proposed scheme for vanadate binding.

possible points at which this could happen: immediately after the release of P_i and at a later step, perhaps after the release of K^+ from E_1K.[28] In either case, the Mg^{2+}-promoted binding of P_i and vanadate have different effects on the enzyme, for although both ligands produce Mg^{2+} occlusion[5,29] and take the enzyme into an E_2 state[22] the binding of P_i induces K^+ disocclusion or prevents K^+ from binding to the enzyme,[30] whereas vanadate favors K^+ occlusion (or at least prevents its disocclusion[22,30,31]). The possible sites for vanadate binding during turnover given in Scheme 1 are taken from a scheme proposed by Pedemonte and Beaugé.[28] They explain the protective effect of ATP (acting with low apparent affinity[8]) as well as the need for Mg^{2+} as cofactor for inhibition at concentrations higher than those required to stimulate the pump. What is important, and puzzling, is that at whatever stage in the cycle vanadate enters into the enzyme, the Na_o^+–K_o^+ interactions demand that the regulatory external sites for Na^+

[29] I. M. Glynn, D. E. Richards, and Y. Hara, *Int. Conf. Na,K-ATPase, 4th* p. 27 (1984).
[30] K. Taniguchi and R. L. Post, *J. Biol. Chem.* **250,** 3010 (1975).
[31] I. M. Glynn and D. E. Richards, *J. Physiol. (London)* **330,** 17 (1982).

(or K^+) remain with one of these ions bound to them at the time of vanadate binding.

Both alternatives given in the scheme above result in the formation of a dead-end inhibitory complex with vanadate. The position of that complex in the cycle is such that it is difficult to visualize how it could slow the $E_2(K)-E_1K$ conformational change as proposed.[11,15,22] Again, the key point here is to reconcile observations in a system under turnover with those seen under nonturning conditions with only some of the ligands involved present. As pointed out before, the reversal K^+-free effect in vanadate-poisoned cells, seen even at high ATP concentrations, could be fully accounted for by the $Na_o^+-K_o^+$ antagonisms over regulatory sites without the need for slowing the conformational change of the unphosphorylated enzyme. A slowing effect has been reported in fluorescence studies under nonturning conditions.[22] However, this observation might be misleading. In these experiments the change from E_1 to E_2 in purified Na^+,K^+-ATPase was brought about by adding K^+, and the change back into the E_1 state by adding enough Na^+ to neutralize the effect of K^+; the rate of return to E_1 fluorescence levels was slow if, prior to Na^+, vanadate was added to the media. Under the conditions at which the fluorescence experiments were done it is known that Na^+ ions favor vanadate release from the enzyme, and that this effect is antagonized by K^+ [12] (if that takes place on "regulatory sites" or by favoring E_1 and E_2 states is not known at present). On the reasonable assumption that the enzyme will stay in the E_2 state as long as vanadate remains bound to it, there is another way to interpret these results: They show not a slow conformational change of the MgEV complex, but rather the rate at which vanadate debinds from the complex, which is then followed by the E_2-E_1 transition, at its usual rate, due to the high Na^+/K^+ concentration ratio.

Acknowledgment

Supported by the Consejo Nacional de Investigaciones Científicas y Técnicas of Argentina (CONICET) (Grant 0155). L.B. is an established investigator of CONICET.

Section VI

Measurements of Conformational States of Na^+,K^+-ATPase

[24] Use of Formycin Nucleotides, Intrinsic Protein Fluorescence, and Fluorescein Isothiocyanate-Labeled Enzymes for Measurement of Conformational States of Na^+,K^+-ATPase

By S. J. D. KARLISH

Hydrolysis of ATP and active cation transport by Na^+,K^+-ATPase involve oscillations between two major conformations, E_1P and E_2P, which are phosphorylated forms involved in Na-pumping, and E_1 and $E_2(K)_{occ}$, which are nonphosphorylated forms involved in K-pumping.[1,2] Fluorescent probes are particularly convenient tools for detecting nonphosphorylated conformations and measuring the rates of $E_1 \rightleftharpoons E_2(K)_{occ}$ transitions.[2-5] Both intrinsic and extrinsic probes have been developed, each probe having strengths and limitations. Initially the formycin nucleotides (FTP and FDP), which are fluorescent analogs of ATP and ADP, were shown to undergo an appreciable increase in fluorescence intensity on binding to Na^+,K^+-ATPase.[3] E_1 and $E_2(K)_{occ}$ states were distinguished by tight or weak binding of FTP or FDP, as is also the case for ATP or ADP, and the rates of $E_1 \rightleftharpoons E_2(K)_{occ}$ were measured for the first time. Disadvantages include a signal amplitude of only 5–10% compared to background fluorescence, light scattering, etc., the necessity of synthesizing FTP and FDP; the experiments also consume large quantities of enzyme. Subsequently the $E_1 \rightleftharpoons E_2(K)_{occ}$ transition was shown to be accompanied by 1–3% changes in intrinsic protein fluorescence.[4] This signal is important because it provides information on unmodified Na^+,K^+-ATPase, but its small size precludes many kinetic measurements. Finally it was found that fluorescein 5'-isothiocyanate (FITC) labels covalently and selectively the ATP binding region, 1 fluorescein/pump, and the bound fluorescein acts as a convenient reporter group.[5] At present fluorescein-labeled Na^+,K^+-ATPase is the system of choice for many purposes due to the large signal (25–35%), but this excludes studies on effects of nucleotides which do not bind.

[1] I. M. Glynn, *Enzymes Biol. Membr.* 35 (1984).
[2] S. J. D. Karlish, D. W. Yates, and I. M. Glynn, *Biochim. Biophys. Acta* **525**, 252 (1978).
[3] S. J. D. Karlish, D. W. Yates, and I. M. Glynn, *Biochim. Biophys. Acta* **525**, 230 (1978).
[4] S. J. D. Karlish and D. W. Yates, *Biochim. Biophys. Acta* **527**, 115 (1978).
[5] S. J. D. Karlish, *J. Bioenerg. Biomembr.* **12**, 111 (1980).

In media of high ionic strength the conformational changes reported by fluorescence can be understood as follows[2,5]:

$$E_1 3Na \rightleftharpoons 3Na + E_1 + 2K \rightleftharpoons E_1 2K \rightleftharpoons E_2(2K)_{occ} \quad (1)$$

E_1 forms bind Na^+ or K^+ competitively at the cytoplasmic surface. Binding of K^+ to E_1 induces a spontaneous conformational transition to $E_2(2K)_{occ}$ containing occluded K^+.[6] In media of low ionic strength the principal form in the absence of monovalent cations is an E_2 form, but in high ionic strength media E_1 predominates.

E_1 states can also be stabilized by $ATP^{7,8}$ or high pH^9 in the absence of Na^+, and E_2 by ouabain,[5] vanadate,[10] or phosphate,[5] or low pH^9 in the absence of K^+. These effects are detectable by the fluorescence probes, particularly fluorescein, and the experiments provide evidence for subconformations of E_1 and E_2 states. For lack of space they are not discussed in detail here. The interested reader is referred to the original papers.[5,7-10]

Labeling Na^+, K^+-ATPase with Fluorescein 5′-Isothiocyanate

Na^+, K^+-ATPase is prepared from pig or rabbit kidney.[11] It is dialyzed overnight in the cold against 1000 vol of a solution containing histidine, 25 mM, pH 7.0, and EDTA, 1 mM. For optimal labeling, FITC (Sigma F7250 isomer I), dissolved in dry dimethylformamide, is added at 5 μM (Stock 2 mM) to dialyzed enzyme, 1 mg/ml, to which Tris base has been added until the pH is 9. The suspension is incubated at room temperature in the dark for 4 hr. It is then diluted by about 15-fold with the histidine/EDTA buffer and incubated for a further 2 hr to allow dissociation of nonspecifically bound fluorescent products. The labeled membranes are centrifuged in the Ti-60 rotor at 50,000 rpm for 45 min, and are resuspended at 3-4 mg/ml in the histidine/EDTA buffer.

Preparation of Formycin Nucleotides

FTP, FDP, and FMP are prepared from formycin [3-(β-D-ribofuranosyl)pyrazolo[4,3-d]-6H-7-pyrimidone] by published chemical[12] or enzy-

---

[6] I. M. Glynn and D. E. Richards, *J. Physiol.* (*London*) **330**, 17 (1982).
[7] P. L. Jørgensen and S. J. D. Karlish, *Biochim. Biophys. Acta* **597**, 305 (1980).
[8] L. A. Beaugé and I. M. Glynn, *J. Physiol.* (*London*) **299**, 367 (1980).
[9] J. C. Skou and M. Esmann, *Biochim. Biophys. Acta* **601**, 386 (1980).
[10] S. J. D. Karlish, L. A. Beaugé and I. M. Glynn, *Nature* (*London*) **282**, 333 (1979).
[11] P. L. Jørgensen, this series, Vol. 32, p. 277.
[12] J. Eccleston and D. R. Trentham, *Biochem. J.* **163**, 15 (1977).

matic[3,13] procedures. FTP and FDP are separated from other compounds by elution from Dowex-1. Analysis of nucleotides is checked by thin-layer chromatography on PEI-cellulose sheets. Concentrations of formycin nucleotides are estimated from absorption measurements at 295 nm (A_{295} = 1.10^4 M^{-1} cm^{-1}). Originally formycin was obtained from Meiji Seika Kaisha, Ltd., Kawasaki, Japan, but it is now commercially available (Sigma, Calbiotech).

Equilibrium or Steady State Fluorescence Measurements

Measurements are performed on a Perkin-Elmer MFP 44A spectrofluorimeter modified to allow addition of solutions to the cuvette, during continuous stirring and recording. Excitation and emission wavelengths: FTP and FDP, excitation, 310 nm, emission 380 nm; intrinsic protein fluorescence, excitation, 295 nm, emission, 325 nm; fluorescein, excitation, 495 nm, emission, 520 nm. Slit widths are generally 10 nm, and the time constant is 0.3 sec except for noisy signals when it is 1.5 sec. Ligand-induced fluorescence responses can be titrated by adding small volumes of solutions, or preferably by injection of ligand at a constant slow rate into the well-stirred cuvette.

The $E_1 \rightleftharpoons E_2(K)_{occ}$ Equilibrium. Suspend enzyme or fluorescein–enzyme in 2.5 ml of a medium containing Tris–HCl, 100 mM, pH 7.0, EDTA (Tris), 1 mM. For formycin measurements use enzyme at 150–200 μg/ml (0.4–0.6 μM in sites) and add FTP or FDP, 4 μM. For intrinsic protein fluorescence, use enzyme at 10–20 μg/ml. For fluorescein fluorescence use labeled enzyme at 20 μg/ml. Add KCl to 2 mM, record the change, and then add NaCl to 10 mM. Conversion of the initial stage E_1 in Tris to $E_2(K)_{occ}$ upon addition of K^+ is accompanied, as the case may be, by (1) dissociation of FTP and a fall in fluorescence of 5–10%, (2) by a 1–3% rise in intrinsic fluorescence, and (3) 25–35% quenching of fluorescein fluorescence. Addition of NaCl reverses these changes. Figure 1 demonstrates both excitation and emission spectra for fluorescein-labeled Na^+,K^+-ATPase, and quenching of fluorescence by K^+ and reversal by Na^+. In the optimal conditions of labeling and experimental conditions given above one would observe larger fluorescence responses. Titration of the K^+ response gives hyperbolic curves with a $K_{0.5}$ of 50–100 μM for FTP dissociation or intrinsic protein fluorescence change, and $K_{0.5}$ = 200–300 μM for the fluorescein-labeled enzyme. Congeners of K^+ displace formycin nucleotides or quench fluorescein fluorescence with the following order of affinities: $Tl^+ > Rb^+ \geq K^+ > NH^+ > Cs^+$. Li^+ is ineffective. In a high Na^+

---

[13] D. C. Ward, A. Cerami, E. Reich, G. Acs, and L. Altweger, *J. Biol. Chem.* **244**, 3243 (1969).

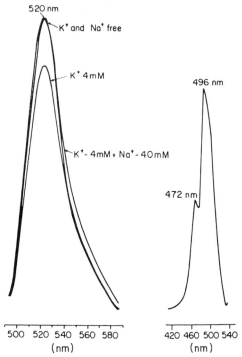

FIG. 1. Excitation and emission spectra of fluorescein-labeled Na$^+$,K$^+$-ATPase and the effects of K$^+$ and Na$^+$ ions. Enzyme was labeled at a protein-to-FITC concentration ratio of 1:1.5 in a 30-min incubation at 24°, and was then dialyzed for 48 hr. Labeled protein, 60 μg, was suspended in 2.5 ml of a solution containing 50 mM Tris–HCl, pH 7.7, at 20° and 1 mM EDTA (Tris). Emission spectra (left) were run with a fixed excitation wavelength of 485 nm, and for the excitation spectrum (right) the fixed emission wavelength was 540 nm. Slit widths on both monochromators were 5 nm; 2.5 μl of 4 M KCl or 25 μl of 4 M NaCl were added as indicated and the spectra were rescanned. Reproduced from Ref. 5.

medium (100 mM), competition between Na$^+$ and K$^+$ leads to a $K_{0.5}$ of 5–10 mM. Titrating Na$^+$ required to reverse the K$^+$ response is also hyperbolic, with a K_{Na} of 1–2 mM.

K$^+$-Nucleotide Antagonism.[7,8] ATP and other nucleotides stabilize E$_1$ forms due to a higher binding affinity than to E$_2$ forms. Therefore higher K$^+$ concentrations are required to stabilize E$_2$(K)$_{occ}$ in the presence of ATP or ADP (μM) than in their absence, and nucleotides can reverse K$^+$ responses. This antagonism is seen conveniently with intrinsic protein fluorescence.

Natural Poise of the E$_1$ ⇌ E$_2$ Equilibrium.[9] *Monitoring Intrinsic Protein Fluorescence:* Suspend enzyme in a medium of low ionic strength: histidine, 25 mM, pH 7.0, EDTA, 1 mM. In this condition E$_2$ predominates. Upon addition of NaCl, 2 mM, the fluorescence falls 1–3% due to

$E_2 \rightarrow E_1 \cdot Na$. Subsequent addition of KCl reverses the signal. The poise of $E_1 \rightleftharpoons E_2$ in native enzyme for different conditions (ionic strength, pH, temperature, etc.) is determined simply by measuring the amplitude of the response to Na^+ and to K^+ as a fraction of the maximal signal on transition from E_1 to $E_2(K)_{occ}$. The sum of the individual Na^+ or K^+ responses should equal the total signal for $E_1 \rightarrow E_2(K)_{occ}$. E_1 is favored by high ionic strength and pH (8–9) and E_2 by low ionic strength and pH (6–7).

Effects of Phosphate, Vanadate, Ouabain, and Divalent Cations[5,14]

Intrinsic Protein Fluorescence Signals Accompanying ATP Hydrolysis.[4,7] *Experimental Conditions:* Suspend enzyme (30 μg/ml) in a solution containing NaCl, 10–250 mM, plus Tris–HCl, pH 7.0, to a total salt concentration of 300 mM; add $MgCl_2$, 1 mM, with or without KCl, 2 mM, and ATP, 1 μM.

Observations and Comments: (1) In a K^+-containing medium, fluorescence rises 1–3%, remains in the high fluorescence steady state until ATP is exhausted, and then falls to the initial level. These changes reflect interconversions between the initial E_1 form and the steady state intermediate $E_2(K)_{occ}$. (2) In a K^+-free medium and low Na^+, 10 mM, a transient signal is observed, about 70% in amplitude of that in the K^+-medium, but as Na^+ is raised to 250 mM the signal falls to about 30%. (3) Assuming the fluorescence of E_2P and $E_2(K)_{occ}$ are the same, the ratio of the signal amplitudes (without K : with K) gives the ratio $E_2P : (E_1P + E_2P)$ in the absence of K^+.[8] (4) The signals are too small to permit measurement of the rates of approach to steady state using stopped flow. (See below for FTP experiments.)

Stopped Flow Fluorescence Measurements

The stopped flow machine used for most measurements has been described before.[15] Fluorescence is excited using a monochromator, while emitted light is detected after passage through cut-off filters; formycin, excitation, 310 nm, cut-off filters Schott U. G. 11 and K. V. 370; intrinsic protein fluorescence, excitation, 295 nm, cut-off filters Schott U. G. 11 plus W. G. 335; fluorescein, excitation, 495 nm, cut-off Wratten 16. Signals are recorded on a Tektronix type 564 storage oscilloscope and then photographed or more usually transferred to a Datalab D. L. 95 transients recorder. The time constant on the oscilloscope must be no greater than one-tenth of the half-time of the signal. Rate constants from exponential signals are obtained graphically.

[14] C. Hegyvary and P. L. Jørgensen, *J. Biol. Chem.* **256**, 6296 (1981).
[15] C. R. Bagshaw, J. F. Eccleston, D. R. Trentham, D. W. Yates, and T. Goody, *Cold Spring Harbor Symp. Quant. Biol.* **37**, 127 (1972).

Rate of $E_2(K)_{occ} \rightarrow E_1$.[2,4,5] Enzyme or fluorescein-labeled enzyme is premixed with K^+ (or congener) at a sufficient concentration to stabilize $E_2(K)_{occ}$ (2 mM K^+) and placed in syringe I. Use 0.4–0.5 mg/ml for formycin measurements, 0.05–0.1 mg/ml for intrinsic protein fluorescence, and about 30 μg/ml for fluorescein-labeled enzyme. Syringe II contains NaCl at a high concentration (100–150 mM) (and if relevant, FTP or FDP, 4 μM). After mixing in the stopped flow apparatus the initial $E_2(K)_{occ}$ form is converted to the $E_1 \cdot$ Na form, and the rates of fluorescence changes are monitored. Figure 2 (left) provides an example using fluorescein fluorescence.

Comments: The characteristic features are (1) the very slow rate of $E_2(K)_{occ} \rightarrow E_1$, 0.2–0.3 sec^{-1} in these conditions. This is observed equally well using intrinsic protein or formycin fluorescence; (2) a large acceleration by ATP acting with a low affinity. This is not observed with fluorescein–enzyme due to lack of binding of ATP (compare Fig. 2, left side). It is observed using intrinsic protein fluorescence or FTP. Acceleration of $E_2(K)_{occ} \rightarrow E_1$ by ATP bound with low affinity to $E_2(K)_{occ}$ reflects the

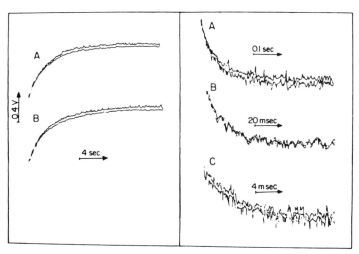

FIG. 2. Stopped flow traces of conformational transitions in fluorescein-labeled Na$^+$,K$^+$-ATPase. Left: transition $E_2(K)_{occ} \rightarrow E_1$Na. Syringe I contained, in 5 ml, 150 μg Na$^+$,K$^+$-ATPase labeled as in Fig. 1, 150 mM Tris–HCl, pH 7.7 at 20°, 1 mM EDTA (Tris), and 3 mM KCl. Syringe II contained (A) 150 mM NaCl and (B) 150 mM NaCl and 6 mM ATP(Na)$_2$. The time constant on the oscilloscope was 100 msec. Right: transition $E_1K \rightarrow E_2(K)_{occ}$. Syringe I contained, in 5 ml, 300 μg Na$^+$,K$^+$-ATPase labeled as in Fig. 1, 150 mM Tris–HCl, pH 7.7, and 1 mM EDTA (Tris). Syringe II contained KCl and Tris–HCl, pH 7.7, at a total concentration of 150 mM, 1 mM EDTA (Tris), and (A) KCl, 4 mM, time constant 5 msec; (B) KCl, 20 mM, time constant 1 msec; (C) KCl, 150 mM, time constant 0.1 msec. The final concentration of all materials not present in both syringes was half that given above. All experiments were conducted at room temperature, 22°. Reproduced from Ref. 5.

ATP-induced shift of the $E_1 \rightleftharpoons E_2$ equilibrium toward E_1; (3) the rate of $E_2(K)_{occ} \rightarrow E_1$ is raised about 5-fold by raising the pH from 7.4 to 8.4.[16]

Rate of $E_1 + K \rightarrow E_2(K)_{occ}$.[2,5] Syringe I contains enzyme or fluorescein–enzyme in the E_1 form, suspended in a solution containing Tris–HCl (and if relevant, FTP or FDP, 4 μM). Syringe II contains K^+ or congener at a concentration such that after mixing in the stopped flow apparatus the final state of the protein will be all $E_2(K)_{occ}$. Upon transition of E_1 to $E_2(K)_{occ}$ the opposite fluorescence changes will occur to those accompanying $E_2(K)_{occ} \rightarrow E_1$. Figure 2 (right side) shows an example of $E_1 + K^+ \rightarrow E_2(K)_{occ}$ using fluorescein fluorescence, at increasing concentrations of KCl (2, 10, and 75 mM). Similar results are obtained with FTP. Intrinsic protein fluorescence signals are too small to permit this measurement.

Comments: The characteristic features are (1) the rates of $E_1 + K \rightarrow E_2(K)_{occ}$ are much faster than $E_2(K)_{occ} \rightarrow E_1$; (2) the rate increases greatly over the range of KCl, 2–75 mM (Fig. 2, right side), even though 2 mM is sufficient to fully stabilize $E_2(K)_{occ}$ in an equilibrium titration. The maximal rate of $E_1 + K \rightarrow E_2(K)_{occ}$ was deduced to be 286 sec^{-1}, and the $K_{0.5}$ for K^+ is 74 mM; (3) the high $K_{0.5}$ for K^+ in the rate measurement reflects a low intrinsic affinity for E_1. The low $K_{0.5}$ for K^+ in equilibrium fluorescence titrations reflects the quotient of the low intrinsic affinity and the conformational equilibrium poised far in the direction of $E_2(K)_{occ}$ (see Scheme 1); (4) the rate of $E_1 + K \rightarrow E_2(K)_{occ}$ is reduced at least 4-fold upon raising the pH from 7.4 to 8.4.[16]

Transient Fluorescence Changes Accompanying Hydrolysis of FTP.[3] *Experimental Conditions:* Syringe I contains enzyme, 300 μg/ml, FTP, 4 μM, NaCl, 40 mM (with or without K^+, 1 mM), Tris–HCl, 60 mM, pH 7.7, and *trans*-1,2-diaminocyclohexanetetraacetic acid (CDTA) (Na) salt, 10 mM. Syringe II contains the same basic medium and in addition FTP, 4 μM, and MgCl$_2$, 22 mM.

Observations and Comments: Observed are three phases of fluorescence: (1) an initial fall due to dissociation of FDP from newly formed phosphoenzyme ($E_1FTP \rightarrow E_1P \cdot FDP \rightarrow E_2P + FDP$); (2) a low fluorescence steady state in which FTP or FDP are not bound (E_2P or $E_2(K)_{occ}$); and (3) a final rise in fluorescence due to rebinding of FDP to E_1 when FTP approaches zero. The characteristic features are that (1) the rate of FDP dissociation (30–40 sec^{-1}) in unaffected by the presence of K^+; (2) the duration of the steady state is greatly prolonged by K^+.

Acknowledgment

Parts of this work were supported by a grant from U.S. PHS GM 32286.

[16] J. C. Skou, *Biochim. Biophys. Acta* **688**, 369 (1982).

[25] Eosin as a Fluorescence Probe for Measurement of Conformational States of Na$^+$,K$^+$-ATPase

By J. C. Skou and Mikael Esmann

Introduction

Eosin Y (CI 45380) binds noncovalently to the E_1 conformation of the Na$^+$,K$^+$-ATPase with a high affinity and with an increase in fluorescence but to the E_2 conformation with a low affinity.[1] Eosin Y can thus be used as an extrinsic fluorescence probe to distinguish between the E_1 and E_2 conformations of the enzyme. Eosin seems to bind to the ATP site of the enzyme.[1,2]

Assay

Eosin Y is dissolved in a histidine–HCl buffer, pH 7.2–7.4. The excitation maximum is 518 nm and the emission maximum 538 nm (Fig. 1). Addition of enzyme in the same buffer or of enzyme + K$^+$ (the E_2 conformation) has no effect on the maxima and gives no or a very slight increase in fluorescence. With Na$^+$ instead of K$^+$ (the E_1 conformation) there is a pronounced increase in fluorescence, the excitation maximum is increased to 524 nm, the emission maximum to 542 nm, and a shoulder appears on the excitation curve around 490 nm. This effect on the fluorescence of eosin Y is similar to the effect of dissolving eosin Y in alcohol.[1]

The effect of Na$^+$ on the fluorescence saturates at about 20–30 mM.[3] As eosin is bound noncovalently to the enzyme, the difference in the fluorescence with 20 mM Na$^+$ and with 20 mM K$^+$ depends on the concentration of the enzyme as well as on the concentration of eosin. At a given protein concentration, for example, 100 μg/ml of an enzyme preparation which is about 50% pure, the difference in fluorescence with 20 mM Na$^+$ and with 20 mM K$^+$ approaches saturation at around 1–2 μM eosin.[1] At lower eosin concentrations, for example 0.1 μM, the fluorescence in the presence of Na$^+$ is about 130% higher than in the presence of K$^+$, when measured in a Perkin Elmer MPF 44A spectrofluorometer with an excitation at 530 nm and an emission at 560 nm and with a 10-nm slit for both excitation and emission. Excitation at 530 nm and emission at 560 nm are the conditions which give the highest fractional increase in fluores-

---

[1] J. C. Skou and M. Esmann, *Biochim. Biophys. Acta* **647**, 232 (1981).
[2] J. C. Skou and M. Esmann, *Biochim. Biophys. Acta* **727**, 101 (1983).
[3] J. C. Skou and M. Esmann, *Biochim. Biophys. Acta* **746**, 101 (1983).

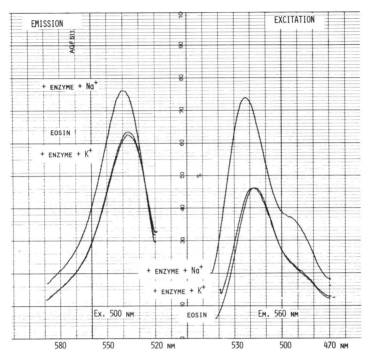

FIG. 1. Fluorescence excitation and emission spectra of 30 nM eosin in a 30 mM histidine–HCl buffer, pH 7.2, at 22° without and with 0.1 mg Na$^+$,K$^+$-ATPase/ml with 12 mM K$^+$ and with 20 mM Na$^+$, respectively. The enzyme is from rectal glands of shark; the specific activity is 1270 μmol P$_i$/mg protein/hr, and with 2.1 nmol ^{32}P-labeling sites/mg protein. Ten-nanometer slit for both excitation and emission.

cence. The fractional increase in fluorescence decreases as the eosin concentration is raised and is 70–80% at 1 μM eosin. At a given eosin concentration, the difference in fluorescence with enzyme in the E$_1$ and E$_2$ conformation increases with the protein concentration.

The rate constant for eosin binding to E$_1$Na$_n$ is $3.7 \times 10^7 \, M^{-1} \, \text{sec}^{-1}$ and for eosin release 11 sec^{-1} at 6°, pH 7.4.[3] This is comparable to the rate constants for binding and release of the ATP analog formycin triphosphate (FTP).[4] K_D for eosin in the presence of Na$^+$ is 0.3–0.45 μM[1,3] which is comparable to K_D for FTP, 1 μM,[4] and to K_D for ATP, 0.1–0.2 μM.[5,6] The extinction coefficient at 525 nm is $8.3 \times 10^4 \, M \, \text{cm}^{-1}$.[7]

[4] S. J. D. Karlish, D. W. Yates, and I. M. Glynn, *Biochim. Biophys. Acta* **525**, 230 (1978).
[5] J. G. Nørby and J. Jensen, *Biochim. Biophys. Acta* **233**, 104 (1971).
[6] C. Hegyvary and R. L. Post, *J. Biol. Chem.* **246**, 5234 (1971).
[7] S. S. Chan, D. J. Arndt-Jovin, and T. M. Jovin, *J. Histochem. Cytochem.* **27**, 56 (1979).

Eosin Derivates

The chemical composition of eosin Y is shown in Fig. 2. An ethyl group on the carboxyl group, eosin S (CI 45386), has no effect on the difference in fluorescence with the enzyme in the E_1 and in the E_2 conformation. Neither has the replacement of the ethyl group with a maleimide group (eosin 5′-maleimide) when the maleimide group is blocked by mercaptoethanol, dithiothreitol, or cysteine. Removal of two of the bromides in the molecule, i.e., eosin B (CI 45.400), also has no effect. With erythrosin B (CI 45430) or erythrosin Y (CI 45425), which have iodine in place of bromide (Fig. 2), Na^+ in the presence of enzyme has the same effect on fluorescence as that shown with eosin Y. However, K^+ gives an increase

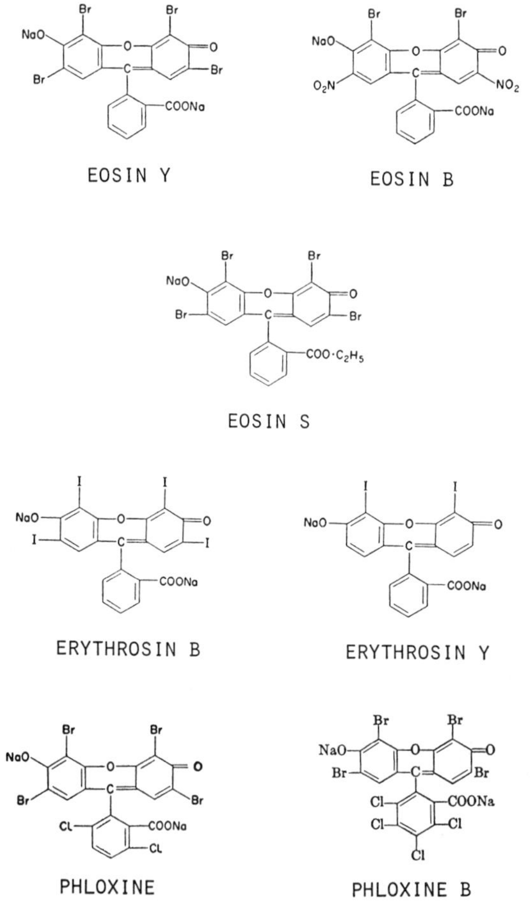

FIG. 2. Chemical composition of fluorescent compounds (see text).

in fluorescence like that seen with Na$^+$, but a higher concentration of K$^+$ than of Na$^+$ is needed to produce the same effect. The same effect is observed when eosin with two or four chlorides on the lower six-membered ring, i.e., phloxine (CI 45405) and phloxine B (CI 45410), respectively (Fig. 2), is used.[1] These latter probes can therefore not be used to distinguish between the two conformations of the enzyme.

Comments

ATP competes with eosin for binding to the enzyme,[1,2] and eosin can therefore not be used to test conformational transitions which involve ATP or other nucleotides.

A problem in measuring the conformational transition is that buffer cations like histidine and Tris have Na$^+$-like effects, i.e., turn the enzyme into the E$_1$ conformation.[2,8] At pH 7.4 the effect of histidine on the conformation is, however, very low and, with 10 mM histidine and with no K$^+$ in the medium, practically all the enzyme is in the E$_2$ conformation. $K_{0.5}$ for Tris at pH 7.4 for the transition of E$_2$ to E$_1$ is about 7 mM with 0.5 μM eosin. As eosin binds noncovalently with a high affinity to E$_1$ and a low affinity to E$_2$, $K_{0.5}$ for the ligand effect depends on the eosin concentration.

With no cations in the medium and under conditions where the buffer effect is eliminated the enzyme is in the E$_2$ conformation and the fluorescence of the enzyme in the E$_2$ conformation without K$^+$ and with K$^+$ is the same. Therefore, the effect of K$^+$ on the conformation can only be tested from the reversal by K$^+$ of the effect of ligand which turns the enzyme into the E$_1$ conformation.

[8] J. C. Skou and M. Esmann, *Biochim. Biophys. Acta* **601,** 386 (1980).

[26] Rapid Ion-Exchange Technique for Measuring Rates of Release of Occluded Ions

By I. M. Glynn, D. E. Richards, and L. A. Beaugé

Principle

The procedures described in this chapter measure the rate at which cations dissociate from a protein. They were designed, first, to test whether Na$^+$,K$^+$-ATPase in certain defined conditions contains occluded Rb$^+$ or Na$^+$ ions and, second, to measure the rates of release of the occluded ions, and to investigate the factors affecting those rates.[1-4] The principle of the method is extremely simple. Na^+,K^+-ATPase suspended in conditions such that it contains occluded (labeled) ions is forced through a cation-exchange column at a rate slow enough for the resin to remove virtually all of the free ions, yet fast enough for the enzyme to emerge from the resin before there is time for the conformational changes that would release the occluded ions. In practice, at room temperature, for adequate removal of free ions it is necessary that the enzyme spends at least 0.9 sec in contact with the resin. The method is therefore applicable only if the time constant for the conformational change leading to the release of the occluded ions is at least several seconds (or can be made so by suitable pretreatment of the enzyme). To study the release of ions under conditions in which the rate-limiting step has a smaller time constant, the filtration method introduced by Forbush[5] must be used.

Apparatus

In its simplest form, the apparatus consists of (1) a 0.5-ml column of Dowex 50W sulfonic resin (X8 cross-linked; 200–400 mesh) in the bottom half of a vertically mounted disposable 1-ml syringe (4.6-mm i.d.), and (2) a device for depressing the plunger of the syringe at an accurately controlled rate, in the range 0.01–0.3 ml/sec. A small air hole is drilled in the syringe barrel at what would be the 1.3-ml mark, if the scale were extended, so that the plunger can be inserted to this point without raising the pressure in the syringe. For the pushing device, a convenient arrangement is to use a Siemens variable speed stepping motor (1 AD 500 OB), geared down by a factor of 2.5 or 25 with a two-speed gear box, to drive a vertical screw of 5-mm pitch on which is mounted a movable horizontal platform. (A stepping motor of the kind recommended is found in Braun homogenizers, and an old-fashioned Palmer stand provides an ideal source of screw and platform.) A simple friction clutch is interposed between the gear box and the screw. The syringe may either be mounted on the movable platform, so that as it is raised the plunger is depressed by a fixed stop, or a projecting bar attached to the movable platform may be used to depress the plunger of a fixed syringe. In either case, it is convenient to include a reversing switch in the motor circuit to restore the movable platform to its original position. It is obviously important that the syringe

[1] L. A. Beaugé and I. M. Glynn, *Nature (London)* **280**, 510 (1979).
[2] I. M. Glynn and D. E. Richards, *J. Physiol. (London)* **330**, 17 (1982).
[3] I. M. Glynn, Y. Hara, and D. E. Richards, *J. Physiol. (London)* **351**, 531 (1984).
[4] I. M. Glynn, J. L. Howland, and D. E. Richards, *J. Physiol.* **368**, 453 (1985).
[5] B. Forbush, *Anal. Biochem.* **140**, 495 (1984).

plunger moves at a constant speed throughout its traverse and this is ensured by (1) arranging that at the start of each run there is a gap of at least 5 mm between the end of the syringe plunger and the pushing device, so that the platform is running at the proper speed by the time contact is made, and (2) stopping the platform abruptly by the use of a second fixed stop, placed to act just when the plunger tip is about to touch the surface of the resin—the friction clutch allows the motor to continue to run until it is switched off manually. By mounting an array of syringe barrels in a Perspex water bath so that the nozzles protrude through the bottom of the bath, it is possible to work at controlled temperatures.

As will be explained below, for some purposes it is necessary to mix the enzyme suspension with another solution immediately before it is forced through the resin. This can be achieved by placing the two fluids to be mixed in identical parallel syringes whose plungers can be pushed at the same rate by a bar attached to the movable platform, and inserting the nozzles of the two syringes into a purpose-made tangential flow mixer (Fig. 1). This consists of a block of Perspex containing a central cylindrical chamber (3.6 mm high and 2 mm in diameter) into which the fluids from the two syringes are forced through four fine channels of 400-μm bore. These tubes enter the cylindrical chamber tangentially and are arranged in such a way that fluids from the two syringes swirl anticlockwise in the top half of the chamber and clockwise in the bottom half. The mixed fluids escape from the chamber by passing, first, through two perforated Teflon disks separated by Teflon O rings, and then through a Teflon plug inserted into the bottom of the mixing chamber. A 20-mm-long polythene tube of internal diameter 1.5 mm conveys the mixed fluids from the Teflon plug to the top of the resin column. This column is formed in the bottom 0.5 ml of the sawn off barrel of a 1-ml plastic syringe, and the connection between the polythene tube and the column is made by inserting the tube in a hole made in the rubber tip taken from the discarded plunger of the syringe and pushed into the sawn off barrel.

Procedure

The precise procedure depends on (1) whether the occluded-ion form being studied is in a steady state in the initial conditions, and (2) whether one wishes to know the rate of release (i.e., the off-rate) under these initial conditions or following some perturbation to the system.

The simplest procedure (Fig. 2a) is that used when the form being studied is in a steady state under the initial conditions, and one wishes to measure only the off-rate under these conditions. An example would be the measurement of the rate of release of Rb^+ ions occluded within un-

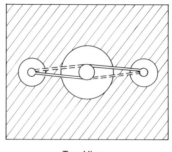

Top View

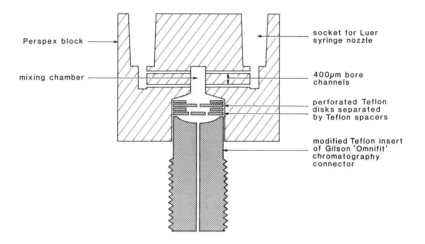

Vertical Section

FIG. 1. Tangential flow mixing device. For description, see text.

phosphorylated Na^+,K^+-ATPase, suspended in a Tris medium containing a low concentration of $^{86}RbCl$. Here, the procedure is simply to place 0.5-ml portions of enzyme suspension above resin that has been equilibrated with a similar medium lacking radioactivity, to force the enzyme through the resin at different rates, and to correlate the amounts of radioactivity in the effluents with the times spent by the enzyme in passing through the resin. (To avoid premature contact of the enzyme suspension with the resin, a layer of washed sand 1 mm thick is placed above the resin layer before the enzyme suspension is added.) One cannot of course assume that all radioactivity appearing in the effluents represents occluded Rb^+, so it is necessary to do control runs in which occlusion in the original suspension is prevented by the presence of, say, 15 mM Na^+ or 2 mM

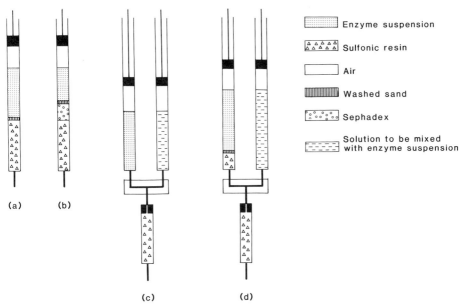

FIG. 2. (a) Arrangement for measuring the off-rate when the occluded-ion form is in the steady state. (b) Arrangement, using Sephadex as a mixing device, (1) for synthesizing and detecting transient occluded-ion forms, and (2) for demonstrating the release of occluded ions by an added reagent. (c) Arrangement, using tangential flow mixer, (1) for measuring the rate of release of occluded ions following the addition of some reagent, and (2) for synthesizing and detecting transient occluded ion forms [as an alternative to (b)]. For the measurement of rate of release, a variable delay is introduced by changing the length of the tube connecting the mixer to the resin column. (d) Arrangement, using tangential flow mixer, (1) for observing the effect of an added reagent on the release of occluded ions in the absence of the corresponding free ions, and (2) for detecting the occlusion of free K^+ congener ions that accompanies the loss of occluded K^+ congener ions following the addition of orthophosphate plus Mg^{2+}. For further details see text.

ADP. Because the amounts of protein emerging in the effluents vary somewhat from tube to tube, it is also necessary to measure the protein content of each effluent[6] and to relate the amount of ion occluded to the amount of protein present.

When the occluded-ion form being studied is not in a steady state but exists only transiently, it is necessary to generate it immediately before the enzyme is pushed through the resin. An example would be the demonstration that newly dephosphorylated Na^+,K^+-ATPase in a high-Na^+, low-Rb^+ medium contains occluded Rb^+, although the stable form of dephosphoenzyme in such a medium does not. Here, the procedure is to

[6] G. L. Peterson, *Anal. Biochem.* **83**, 346 (1977).

suspend the enzyme in a medium lacking the ATP necessary for phosphorylation, and then to add ATP shortly before the enzyme enters the resin. This can be done most simply by placing just above the resin 0.2 ml of Sephadex G-25 equilibrated with a medium containing 40 μM ATP (Fig. 2b) and reducing the volume of the enzyme suspension to 0.3 ml. At the flow rates employed there is no time for complete equilibration across the surface of the Sephadex particles, but the Sephadex does act as a rough-and-ready mixing device[2] and is convenient provided that the concentration of the reagent to be added is not critical. It is, of course, important that the fluid in the Sephadex does not contaminate the enzyme suspension above it; to prevent contamination the inside of the syringe above the Sephadex is carefully washed with water, and a 1-mm-thick layer of washed sand is placed above the Sephadex before the enzyme suspension is added. An alternative technique that gives a more precise control of concentration is the use of a tangential flow mixer with the outflow fed directly to the resin column (Fig. 2c). Because of the necessarily low flow rates employed, however, it is important to check that mixing is adequate—for example, by preliminary experiments in which a slight excess of HCl is used to neutralize a solution of NaOH containing phenolphthalein. The apparatus shown in Fig. 1 gives adequate mixing at flow rates (after mixing) of 0.2 ml/sec or faster.

To determine the rate of release of occluded ions following the addition of some reagent to the medium, it is best to use a tangential flow mixer to add the reagent, and to vary the duration of the period between mixing and the passage of the mixture into the resin by interposing appropriate lengths of Polythene tubing of 1.5 mm i.d. between the mixer and the resin column. With the lengths of tube necessary to achieve delays of over 4 sec, the "dead space" in the tube makes it necessary to use 2-ml syringes. (The simpler procedure of varying the delay by varying the flow rate is not satisfactory because mixing with the tangential flow mixer is incomplete at the slower flow rates.) This procedure has been used, for example, to investigate the release of occluded Rb^+ following the addition of orthophosphate plus Mg^{2+} ions, and also to investigate the effects of K^+ congener ions on the rate of that release. A simpler procedure for adding the reagent whose effect on ion release is to be tested is to use a layer of suitably loaded Sephadex G-25 (0.2 ml) above the resin and below the enzyme suspension (Fig. 2b). This technique has been used to show that ADP causes the rapid loss of occluded Na^+ from the E_1P form of phosphorylated $Na,^+K^+$-ATPase, but it is not satisfactory if the concentration of the reagent added needs to be known with any precision, and it cannot be used to measure the rate at which the occluded ions are released following the addition of the reagent.

Whether a tangential flow mixer or the Sephadex method is used to add the reagent to be tested, unless special precautions are taken the free ions in the original enzyme suspension will still be present at the moment of mixing. Since these ions may themselves affect the response to added reagents, it is desirable to be able to remove them before mixing occurs. With the tangential flow mixing technique, this can be done by placing 0.2 ml of suitably loaded resin at the bottom of the enzyme syringe (Fig. 2d). If the radioactive label is omitted from the enzyme suspension and labeled RbCl, say, is included in the right-hand syringe, the arrangement shown in Fig. 2d can also be used to demonstrate the occlusion of free Rb^+ (or other K^+ congener) ions that accompanies the rapid release of 50% of the occluded Rb^+ ions caused by the addition of orthophosphate and Mg^{2+} ions.

Sensitivity

With the techniques described, and using ^{22}Na, ^{86}Rb, ^{137}Cs, or ^{204}Tl, it is possible to measure 10-pmol quantities of occluded Na^+, Rb^+, Cs^+, or Tl^+ in 20-μg portions of $Na,^+K^+$-ATPase (specific activity 10–20 U/mg). Unfortunately, with such small quantities of enzyme, ^{42}K has too low a specific activity to be useful for labeling occluded K^+ ions and it is necessary to use the more expensive and less pure ^{43}K.

Section VII

Modification of Na^+,K^+-ATPase

[27] Proteolytic Cleavage as a Tool for Studying Structure and Conformation of Pure Membrane-Bound Na^+,K^+-ATPase

By PETER L. JØRGENSEN and ROBERT A. FARLEY

Introduction

Proteolytic enzymes provide effective tools for studying structure and conformation of Na^+,K^+-ATPase. This enzyme is composed of two polypeptide subunits, the α subunit with an M_r of about 100,000, and the β subunit with an M_r of 32,000–40,000, plus carbohydrate. In carefully controlled conditions, trypsin or chymotrypsin cleaves bonds in cytoplasmic loops of the α subunit in membrane-bound Na^+,K^+-ATPase. A characteristic pattern of proteolytic fragments is formed and the splits are associated with inhibition of Na^+,K^+-ATPase activity.

The sites of primary cleavage in α subunit of Na^+,K^+-ATPase depend on the conformational state of the protein. In the E_1 conformation cleavage is restricted to the N-terminal half of the α subunit, while in the E_2 form bonds near the middle and in the C-terminal half of the α subunit are exposed to cleavage. The splits are located in hydrophilic loops protruding at the cytoplasmic surface. The α subunit does not expose bonds that are sensitive to proteolysis at the extracellular surface. All hydrophilic residues of the β subunit are located at the extracellular surface,[1] but the β subunit is remarkably resistant to proteolysis. Combination of controlled proteolysis with specific chemical labeling has allowed identification of regions within a linear representation of the α subunit that participate in the binding of ATP, the formation of the phosphoenzyme intermediate during turnover, the penetration of the polypeptide into the cell membrane, and the binding of cardiac glycosides.[2-5] In addition, examination of the effects of controlled proteolysis on enzyme activity has demonstrated specific correlations between enzymatic partial reactions and conformational transitions and cation transport.[5]

---

[1] K. Dzandzhugazyan and P. L. Jørgensen, *Biochim. Biophys. Acta* **817**, 165 (1985).
[2] S. J. D. Karlish, P. L. Jørgensen, and C. Gitler, *Nature (London)* **269**, 715 (1977).
[3] J. Castro and R. A. Farley, *J. Biol. Chem.* **254**, 2221 (1979).
[4] P. L. Jørgensen, S. J. D. Karlish, and C. Gitler, *J. Biol. Chem.* **257**, 7435 (1982).
[5] P. L. Jørgensen, E. Skriver, H. Hebert, and A. B. Maunsbach, *Ann. N.Y. Acad. Sci.* **402**, 207 (1982).

Proteolytic Cleavage

The substrate for the proteolytic enzymes in the experiments described below is membrane-bound Na^+,K^+-ATPase that has been purified by isopycnic zonal centrifugation after incubation of crude membranes from outer renal medulla with sodium dodecyl sulfate (SDS) in the presence of ATP.[6,7] The preparation consists of the two Na^+,K^+-ATPase subunits embedded in membrane disks such that the polypeptides are accessible to the proteolytic enzymes at both surfaces of the membranes.

Cleavage of the E_1 Form of the α Subunit by α-Chymotrypsin A

In the E_1 conformation in NaCl (E_1Na) or after binding of ATP (E_1ATP) to the enzyme, chymotrypsin at low ionic strength cleaves a bond in a cytoplasmic loop about 250 residues away from the N-terminus of the α subunit.[5] The use of α-chymotrypsin A is important, since cleavage with β, γ, or δ-chymotrypsin A is slower and less specific than cleavage with α-chymotrypsin A. The rate constants for cleavage of Na^+,K^+-ATPase are about 50-fold lower at 170 mM salt than at 20 mM salt. Digestion with chymotrypsin is effectively stopped by dilution into buffer containing 150 mM NaCl, KCl, or Tris–HCl. As seen from Fig. 1 inactivation of both Na^+,K^+-ATPase and K^+-phosphatase is a linear function of time in semilogarithmic plots.[5] Sites for binding of cations and nucleotides are preserved after cleavage. The mechanism of inactivation appears to be that the split stabilizes both dephospho and phospho forms of the protein in E_1 forms.

Procedure. Chymotryptic digestion of pure membrane-bound Na^+,K^+-ATPase is started by mixing 1.25 μg α-chymotrypsin (from Sigma or Merck) with 25 μg Na^+,K^+-ATPase in a volume of 250 μl of 10 mM NaCl, 15 mM Tris, pH 7.5, at 37°. Digestion is stopped by transfer to the assay media containing 150 mM salt.

At the indicated time points, 10-μl aliquots containing 0.8 μg protein are transferred to tubes for assay of Na^+,K^+-ATPase. The tubes contain 1 ml of 130 mM NaCl, 20 mM KCl, 3 mM $MgCl_2$, 3 mM ATP, 25 mM imidazole, pH 7.5. After incubation for 1, 3, or 5 min at 37° the reaction is stopped with 1 ml ice-cold 0.5 M HCl containing 30 mg ascorbic acid, 5 mg ammonium heptamolybdate, and 10 mg SDS. The tubes are transferred to ice. For color development 1.5 ml containing 30 mg sodium m-arsenite, 30 mg sodium citrate, 30 μl acetic acid is added. The tubes are heated for 10 min at 37° and absorbance is read at 850 nm. The Na^+,K^+-

---

[6] P. L. Jørgensen, *Biochim. Biophys. Acta* **356**, 36 (1974).
[7] P. L. Jørgensen, this series, Vol. 32, p. 277.

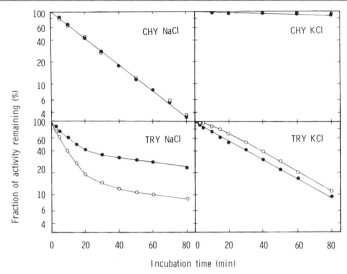

FIG. 1. Time course of inactivation of Na$^+$,K$^+$-ATPase (●) or K$^+$-phosphatase (○) by chymotrypsin or trypsin in NaCl or KCl media. From Refs. 5 and 10. Procedures are described in the text.

ATPase activity remaining at each time point is calculated from initial velocity curves.

For K$^+$-phosphatase assay aliquots are transferred to test tubes containing 1 ml of 150 mM KCl, 10 mM MgCl$_2$, 20 mM p-nitrophenyl phosphate, 25 mM imidazole, pH 7.5. The reaction is stopped by adding 2 ml 0.2 M NaOH, 0.05 M Na$_2$EDTA and absorbance is read at 410 nm.

Tryptic Cleavage of E_1 or E_2 Form of α Subunit

E_1 *Form.* In NaCl medium (E$_1$Na) or after binding of ATP (E$_1$ATP) tryptic inactivation of Na$^+$,K$^+$-ATPase and cleavage proceed in two phases, as seen from Fig. 1. The initial fast phase of inactivation is associated with removal of a small peptide with M_r about 2000 from the amino terminal of the α subunit. The second, slower phase of trypsin cleavage of the α subunit is associated with the generation of a large fragment with M_r about 77,000. At 20°, the difference in rate of inactivation in the two phases is larger than at 37°.[8] The ratio of the rate constants for inactivation in the initial faster phase and the secondary slower phase is 50–60 at 20° and 15–25 at 37°. Degradation rates depend strongly on ionic strength.

[8] S. J. D. Karlish and U. Pick, *J. Physiol. (London)* **312**, 505 (1981).

Salts protect toward cleavage of Na^+,K^+-ATPase through an effect on the membrane protein, since the activities of trypsin toward casein or synthetic substrates are unaffected by salt.[9]

E_2 *Form*. The E_2 form is stabilized by binding of K^+ (E_2K) or by phosphorylation at low concentration of Na^+ (E_2P). In The E_2 form inactivation of Na^+,K^+-ATPase is linear in semilogarithmic plots and correlates with cleavage of a bond near the middle of the α subunit resulting in 41- to 46-kDa and 58-kDa fragments. Inactivation of K^+-phosphatase is delayed, because consecutive cleavage of two bonds is required for inactivation. Cleavage of the bond near the middle of the α subunit precedes cleavage of the bond near the N-terminus of the α subunit.[10]

Procedure for Tryptic Digestion at 20° and Low Ionic Strength. Digestion is started by adding 0.2–1 μg TPCK trypsin and 30 μg Na^+,K^+-ATPase protein to 600 μl of 10 mM NaCl (E_1) or 10 mM KCl (E_2) in 15 mM Tris–HCl, pH 7.5, at 20°. At the indicated time 50 μl of the medium is mixed with 20 μl containing 2 μg soybean trypsin inhibitor in 25 mM imidazole, pH 7.5, and transferred to an ice bath. Aliquots containing 25 μl of this mixture are transferred to test tubes and assay of Na^+,K^+-ATPase or K^+-phosphatase activities is conducted as above.

Procedure for Tryptic Digestion at Physiological Ionic Strength and 37°. Mix 6 μg trypsin and 150 μg Na^+,K^+-ATPase in 1500 μl containing 150 mM NaCl (E_1) or KCl (E_2) in 25 mM imidazole, pH 7.5. At the indicated intervals 100 μl of the medium is mixed with 1 μg trypsin inhibitor in 40 μl 25 mM imidazole, pH 7.5. Aliquots containing 25 μl of this mixture are transferred to test tubes for assay of Na^+,K^+-ATPase or K^+-phosphatase as above.

Cleavage by Chymotrypsin of E_2 Form in the Presence of $MgCl_2$ and Ouabain

At high ratios of chymotrypsin to ATPase, in the presence of $MgCl_2$ and ouabain, chymotrypsin cleaves C(77) very rapidly to generate an amino-terminal M_r 35,000 fragment, C(35), and a carboxy-terminal M_r 40,000 fragment, C(40) (Fig. 1, lane 6). Procedure for cleavage is given in Table I. Cleavage at this site also occurs in the absence of ouabain, but at a slower rate. The β subunit is also cleaved under these conditions to an M_r 40,000 fragment and an M_r 18,000 fragment.[10a] The M_r 40,000 fragment is derived from the amino terminal of the β subunit.[10b]

---

[9] P. L. Jørgensen, *Biochim. Biophys. Acta* **401**, 399 (1975).
[10] P. L. Jørgensen, *Biochim. Biophys. Acta* **466**, 97 (1977).
[10a] R. A. Farley, R. P. Miller, and A. Kudrow, *Biochim. Biophys. Acta* **873**, 136 (1986).
[10b] G. Chin, *Biochemistry* **24**, 5943 (1985).

TABLE I
PROTEOLYTIC FRAGMENTATION REACTIONS FOR
Na^+,K^+-ATPase[a]

Compound present	Lane[b]					
	1	2	3	4	5	6
Na^+,K^+-ATPase (μg)	19	29	29	29	29	29
NaCl (mM)	—	90	90	—	—	—
KCl (mM)	—	—	—	90	—	—
$MgCl_2$ (mM)	—	—	—	—	—	5
Ouabain (mM)	—	—	—	—	—	1
Trypsin (μg)	—	0.33	0.33	1.65	—	—
Chymotrypsin (μg)	—	—	—	—	6	12
Reaction time (min)	—	5	20	10	20	15

[a] All reactions were performed at 37° for the indicated times. The buffer was 25 mM imidazole/HCl, 1 mM Na_2EDTA, pH 7.4, except for lanes 4–6, where the Na_2EDTA was omitted.

[b] The products of the cleavage reactions described in this table are shown in Fig. 1. The number of the lane in the table corresponds to the number of the lane in the figure.

Phosphorylation and Separation of Phosphopeptides after Proteolysis

Four peptides, i.e., the intact α subunit and the 77-kDa, 41- to 46-kDa, and 18-kDa fragments, can incorporate ^{32}P from [γ-^{32}P]ATP in an Na^+-dependent reaction. These peptides can be separated after sequential cleavage of the α subunit, first with chymotrypsin in NaCl and then with trypsin in KCl medium. The four phosphopeptides show different reactivities to ADP and K^+.[5]

Procedure for Sequential Cleavage. Cleavage is started by adding α-chymotrypsin to 25 μg/ml in 12 ml of 15 mM Tris–HCl/10 mM NaCl, pH 7.5, containing 0.5 mg/ml protein of membrane-bound Na^+,K^+-ATPase. After 15 min at 37° the reaction is stopped by adding KCl to 150 mM and cooling to 0°. The mixture is centrifuged for 90 min at 50,000 rpm. The pellet is resuspended and diluted to 0.5 mg protein/ml in 150 mM KCl, 25 mM imidazole, pH 7.5. Trypsin is added to 30 μg/ml. After 7 min at 37° the reaction is stopped by adding soybean trypsin inhibitor to 90 μg/ml. The mixture is centrifuged for 90 min at 50,000 rpm and resuspended in 25 mM Tris–HCl, 150 mM NaCl, and washed twice to remove KCl. The pellet is resuspended in 25 mM Tris–HCl, 150 mM NaCl, pH 7.5.

Procedure for Phosphorylation. For phosphorylation of α subunit and

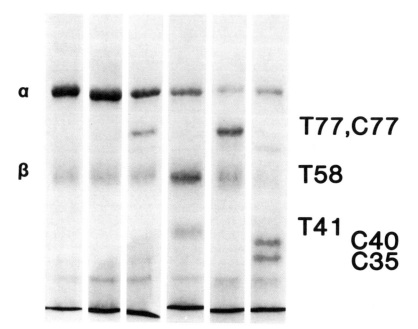

FIG. 2. Proteolytic fragmentation of Na$^+$,K$^+$-ATPase. Na$^+$,K$^+$-ATPase, purified from dog renal medulla[7] (lane 1), was cleaved as described in Table I by trypsin in the presence of NaCl (lanes 2 and 3), by trypsin in the presence of KCl (lane 4), by chymotrypsin (lane 5), or by chymotrypsin in the presence of MgCl and ouabain (lane 6). Trypsin reactions were stopped by a 2-fold weight excess of soybean trypsin inhibitor to trypsin, and chymotrypsin reactions were stopped by the addition of phenylmethylsulfonyl fluoride to 1 mM. Peptide fragments were separated on 7.5% SDS–polyacrylamide gels[11] and stained with Coomassie brilliant blue.

proteolytic fragments, 80–130 μg Na$^+$,K$^+$-ATPase protein is incubated at 0° for 6 sec in 3 ml 30 μM [γ-^{32}P]ATP (65,000 cpm/pmol), 3 mM MgCl$_2$, 0–20 mM NaCl, 20 mM Tris–HCl, pH 7.5, as before. The reaction is stopped with 3 ml 8% perchloric acid, 1.2 mM ATP, 1.2 mM P$_i$. The protein is separated by centrifugation and washed two times with 3 ml 4% perchloric acid, 0.6 mM P$_i$/0.6 mM ATP. The sediment is kept on an ice bath and dissolved in 80 μl 1% SDS, 10 mM dithiothreitol, 50 mM sodium phosphate, pH 2.4. One drop of 40% glycerol and 5 μl 360 μg/ml pyronin Y are added. The phosphoproteins are separated by electrophoresis at pH 2.4.

Gradient Slab Gels for Separation of Phosphopeptides at Low pH. Gradient slab gels are formed from 7.5 and 15% (T%) acrylamide in 50

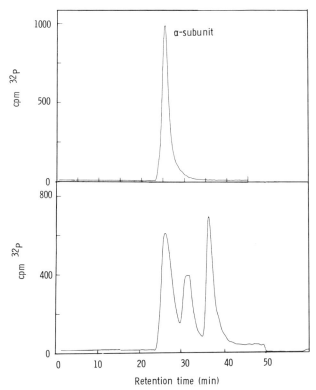

FIG. 3. Separation on TSK 3000 SW gel filtration column of phosphopeptides of Na^+,K^+-ATPase after sequential chymotryptic and tryptic cleavage. After cleavage and phosphorylation the protein is dissolved in SDS and injected into the column as described in the text. Elution with 1% SDS, 50 mM sodium phosphate, pH 2.8, at 0.5 ml/min. Fractions are collected at 30-sec intervals and counted for ^{32}P. Top frame: ^{32}P-Labeled α subunit after phosphorylation of native Na^+,K^+-ATPase. Bottom frame: elution of (from left) intact α subunit together with 78-kDa fragment followed by 46-kDa fragment and the 18-kDa fragment, the segment between bond 1 and bond 3 in Figs. 4 and 5.

mM sodium phosphate, pH 2.4, 1% SDS,[11] 1 mg/ml ascorbic acid, 2.5 μg/ml $FeSO_4$, 0.03 mg/ml H_2O_2, a modification of previous procedures.[12] Polymerization is initiated by adding H_2O_2. The solutions are pumped from the gradient mixer into the gel stand in less than 3 min at a rate of 18 ml/min using a Hughes Hiloflow pump (Surrey, England) to avoid premature polymerization.

[11] U. K. Laemmli, *Nature (London)* **227**, 680 (1970).
[12] J. Avruch and G. Fairbanks, *Proc. Natl. Acad. Sci. U.S.A.* **69**, 1216 (1972).

Separation of Phosphopeptides on TSK Gel Filtration Columns in SDS. Gel filtration on a TSK 3000 SW column allows separation of peptides with about a 2-fold difference in molecular weight. It is seen from Fig. 3 that the intact α subunit and the 77-kDa fragment appeared in the same peak, while the 41- to 46-kDa and 18-kDa fragments could be purified in phosphorylated form by this procedure.

Procedure: After phosphorylation the protein is dissolved with SDS as above and injected into a TSK 3000 SW column, 7.5 × 600 mm, that is equilibrated with 1% SDS, 50 mM sodium phosphate, pH 2.8, and operated at 0.5 ml/min with an HPLC pump. Absorbance is monitored at 277 nm and fractions are collected at $\frac{1}{2}$-min intervals.

Orientation of Fragments within the α Subunit Polypeptide

Table II summarizes results of several experiments in which some of the proteolytic fragments of the α subunit were purified and analyzed by amino-terminal and carboxy-terminal amino acid sequencing,[3] or experiments in which the distribution of specific chemical labels incorporated

TABLE II
END-TERMINAL ANALYSIS AND CHEMICAL LABELING PROCEDURES USED TO ALIGN MAJOR PROTEOLYTIC FRAGMENTS OF Na$^+$,K$^+$-ATPase

Fragment	Amino-terminal[a]	Carboxy-terminal[a]	Labels[b]
α	H$_2$N-G-R-D-	-(K,S,T)-Y-COOH	^{32}P, IAA, FITC
T(77)	H$_2$N-I-	-(K,S,T)-Y-COOH	^{32}P, IAA[c]
T(58)	ND	ND	IAA, FITC
T(41)	H$_2$N-A-	ND	^{32}P
C(77)	ND	-(K,S,T)-Y-COOH	^{32}P, IAA, FITC
C(35)	ND	ND	^{32}P, IAA[d]
C(40)	ND	ND	—[d]

[a] One-letter symbols are used for amino acids: G, Gly; R, Arg; D, Asp; K, Lys; S, Ser; T, Thr; Y, Tyr; I, Ile; A, Ala.

[b] ^{32}P, labeled by [γ-^{32}P]ATP; IAA, labeled by iodoacetate; FITC, labeled by fluorescein 5'-isothiocyanate.

[c] T(77) is not produced by trypsin cleavage after the Na$^+$,K$^+$-ATPase is labeled by FITC. T(58) and T(41) are produced by trypsin cleavage in the presence of either NaCl or KCl.

[d] Neither C(35) nor C(40) retains the FITC-modified amino acid after FITC-labeled Na$^+$,K$^+$-ATPase is cleaved by chymotrypsin in the presence of MgCl$_2$ and ouabain.

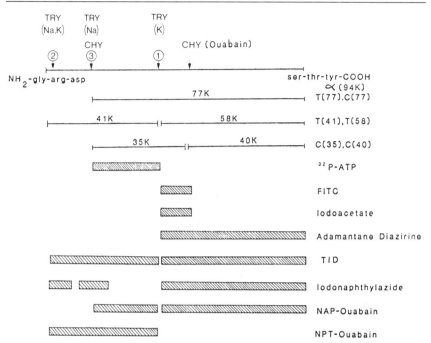

FIG. 4. Alignment of the major proteolytic fragments within the subunit of Na$^+$,K$^+$-ATPase. Locations of the proteolytic fragments were determined as described in the text. Fragments containing chemical labels were identified after cleavage of labeled Na$^+$,K$^+$-ATPase, also as described in the text.

into the α subunit was followed in each of the fragments.[2-5,13,14] From these data it can be determined that both T(77) and C(77) are generated from the carboxy-terminal region of the α subunit. Since T(77) or C(77) can be converted to T(58), but not T(41), T(58) must also arise from the carboxy-terminal region of the intact polypeptide and T(41) must be derived from the amino-terminal region. Also, since the aspartate residue that is phosphorylated by [γ-^{32}P]ATP is located within T(41), C(35) must overlap the site of trypsin cleavage between T(41) and T(58). This establishes that C(40) is also obtained from the carboxy-terminal region of the α subunit. A linear map of the locations of the proteolytic fragments can be inferred from this analysis, and is shown in Fig. 4. Also shown in Fig. 4

[13] S. J. D. Karlish, *J. Bioenerg. Biomembr.* **12**, 111 (1980).
[14] C. T. Carilli, R. A. Farley, D. M. Perlman, and L. C. Cantley, *J. Biol. Chem.* **257**, 5601 (1982).

are the locations within the α subunit of several chemical labels that have been used to identify functionally important sites in Na^+,K^+-ATPase.[4,5,13,15,16] These locations were determined by cleavage of the appropriately labeled ATPase with either trypsin or chymotrypsin, and identification of the labeled fragments on SDS–polyacrylamide gels.

Folding of the α Subunit through the Membrane and Identification of Functional Domains

Experiments on red blood cells,[17] reconstituted vesicles,[8] and right-side-out vesicles[18] show that the bonds exposed to tryptic cleavage in the E_1 and the E_2 forms are accessible only from the cytoplasmic surface of the cell membrane.[8] As indicated in Fig. 4, the photochemical probes iodonaphthylazide,[2,4] adamantane diazirine[15,19] and trifluoromethylphenyldiazirine[20] have been used to identify membrane-embedded regions of the α subunit, and the approximate locations of the ouabain-binding site and the phosphorylated site on the subunit have also been determined relative to the locations of the cleavage sites for trypsin and chymotrypsin. The binding site for ouabain is located on the extracellular surface of the membrane, and the enzyme is phosphorylated by ATP at the cytoplasmic surface. These data support the model in Fig. 5 for the folding of the α subunit through the cell membrane.[4,21] The combination of hydrophilic and hydrophobic labeling of the α subunit suggested that six transmembrane segments were distributed along the sequence of the α subunit. Four of these segments were located to the N-terminal side of bond 1 and two were located to the C-terminal half of the α subunit.[4] After hydrophobic labeling with adamantane diazirine,[19] six hydrophobic segments were isolated from the α subunit.

In models for the function of Na^+,K^+-ATPase,[5] sites for binding of cations, nucleotides, and phosphorylation are located in different domains of the α subunit. Consequently, cleavage of bond 1, bond 2, or bond 3 has widely different consequences for the function of the system.[5,10]

---

[15] R. A. Farley, D. W. Goldman, and H. Bayley, *J. Biol. Chem.* **255**, 860 (1980).
[16] B. Rossi, G. Ponzio, and M. Lazdunski, *EMBO J.* **1**, 859 (1982).
[17] G. Giotta, *J. Biol. Chem.* **250**, 5159 (1975).
[18] G. Chin and M. Forgac, *Biochemistry* **22**, 3405 (1983).
[19] R. A. Nicolas, *Biochemistry* **23**, 889 (1984).
[20] P. L. Jørgensen and J. Brunner, *Biochim. Biophys. Acta* **735**, 291 (1983).
[21] L. Cantley, C. T. Carilli, R. A. Farley, and D. M. Perlman, *Ann. N.Y. Acad. Sci.* **402**, 289 (1982).

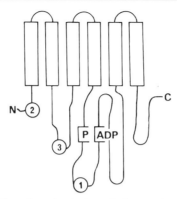

FIG. 5. Model for folding of α subunit through the membrane. Columns indicate transmembrane segments, consisting of predominantly hydrophobic residues arranged in helices. The encircled numbers mark the sites of primary tryptic cleavage in the E_1 form (bond 2 and 3) and in the E_2 form (bond 1 and 2). The model is based on localization of tryptic splits on the cytoplasmic surface, and the combination of proteolytic cleavage with labeling with hydrophobic compounds from the membrane bilayer and with specific ligands from the cytoplasmic and extracellular membrane surface as shown in Figs. 3 and 4.

Cleavage of bond 1 interferes with the proper orientation between the aspartyl phosphate residue in the 46-kDa fragment and a domain in the 58-kDa fragment that is engaged in nucleotide binding. Removal of the ionizable group at the amino-terminal end of the α subunit by cleavage of bond 2 shifts the equilibrium between E_1 and E_2 forms in direction of E_1. Cleavage of bond 3 effectively disrupts the coupling of phosphoryl transfer to cation exchange. Cation exchange, passive as well as active, is blocked and split 3 prevents the cation effects on the conformational equilibrium between the phosphoenzyme forms. A straightforward explanation for this could be that split 3 interrupts the association between the aspartyl phosphate residue and a structure involved in cation binding and translocation which may be formed by the part of the α subunit between bond 3 and bond 2.

[28] Irreversible and Reversible Modification of SH Groups and Effect on Catalytic Activity

By WILHELM SCHONER, MARION HASSELBERG, and RALF KISON

Introduction

Na^+,K^+-ATPase contains 11–14 sulfhydryl groups/mol of $\alpha\beta$ dimer, of which 8–9 SH groups reside on the α subunit and 1–2 on the β subunit.[1–4] Using radioactive N-ethylmaleimide it was determined that two to six sulfhydryl groups reside in the active center.[1,4,5] Alkylation of SH groups leads to the inactivation of Na^+,K^+-ATPase and its partial reactions, Na^+-dependent phosphorylation of the α subunit and K^+-activated phosphatase reaction.[4–10] This modification and inactivation is partly prevented by ATP[4,5,9] but not by ITP.[10] The number of alkylated sulfhydryl groups varies with the nature of the ligands present during the alkylation procedure.[4,5,11,12] This observation can be explained either by an ATP-induced alteration of the enzyme's conformation,[12,13] leading to a decrease of the reactive sulfhydryl groups, or by the assumption that a sulfhydryl group resides in or near the ATP-binding site, whereby ATP protects the SH group against modification. To resolve this dilemma it may be possible to use protein-reactive ATP analogs, which react with sulfhydryl groups.[10,14–17] Since ATP but not ITP protects Na^+,K^+-ATPase

[1] W. H. Peters, J. J. H. H. M. de Pont, A. Koppers, and S. L. Bonting, *Biochim. Biophys. Acta* **641**, 55 (1981).
[2] A. Askari, W. Huang, and G. R. Henderson, in "Na,K-ATPase: Structure and Kinetics" (J. C. Skou and J G. Nørby, eds.), p. 205. Academic Press, London, 1979.
[3] M. Kawamura, T. Ohta, and K. Nagano, *Curr. Top. Membr. Transp.* **19**, 153 (1983).
[4] A. J. Jesaitis and P. A. G. Fortes, *J. Biol. Chem.* **255**, 459 (1980).
[5] W. M. Hart and E. O. Titus, *J. Biol. Chem.* **248**, 4674 (1973).
[6] J. C. Skou and C. Hilberg, *Biochim. Biophys. Acta* **110**, 359 (1965).
[7] S. Fahn, G. J. Koval, and R. W. Albers, *J. Biol. Chem.* **243**, 1993 (1968).
[8] R. L. Post, S. Kume, T. Tobin, B. Orcutt, and A. K. Sen, *J. Gen. Physiol.* **54**, 200 S (1969).
[9] S. P. Banerjee, S. M. E. Wong, and A. K. Sen, *Mol. Pharmacol.* **8**, 8 (1972).
[10] R. Patzelt-Wenczler, H. Pauls, E. Erdmann, and W. Schoner, *Eur. J. Biochem.* **53**, 301 (1975).
[11] J. W. Winslow, *J. Biol. Chem.* **256**, 9522 (1981).
[12] F. Schuurmans Stekhoven and S. L. Bonting, *Physiol. Rev.* **61**, 1 (1981).
[13] P. L. Jørgensen, *Biochim. Biophys. Acta* **694**, 27 (1982).
[14] R. Patzelt-Wenczler and W. Schoner, *Biochim. Biophys. Acta* **403**, 538 (1975).
[15] R. Patzelt-Wenczler and W. Schoner, *Eur. J. Biochem.* **114**, 79 (1981).
[16] M. Hasselberg and W. Schoner, *Hoppe-Seyler's Z. Physiol. Chem.* **365**, 999 (1984).
[17] H. Koepsell, F. W. Hulla, and G. Fritzsch, *J. Biol. Chem.* **257**, 10733 (1982).

against the inactivation by 5,5′-dithiobis(2-nitrobenzoate), it is probable that the sulfhydryl group of the active site opposes the 6-amino group of the purine ring. Consequently derivatives of 6-mercaptopurine 5′-triphosphate (s^{6}ITP) were investigated which based on their chemical structure behaved as reversible [(s^{6}ITP)$_2$, CNs^{6}ITP] or irreversible inhibitors [Dnps^{6}ITP] of the enzyme.

Preparation of Derivatives of Thioinosine Triphosphate

Thioinosine triphosphate and its derivatives are synthesized from 6-mercapto-9-β-D-ribofuranosylpurine (thioinosine) by modification of previously described procedures.[18-26]

Synthesis of 2′,3′-O-Isopropylidene-6-Mercaptopurine. Copper p-toluene sulfonate (25.3 g) is prepared by refluxing a suspension of 5.6 g CuO (0.07 mol) and 22.8 g p-toluenesulfonic acid monohydrate in 200 ml water for 90 min and drying the filtrate coming through a Büchner funnel.

Dried thioinosine (4 g) is dissolved in 600 ml dry acetone under stirring in a stoppered flask and 26.8 g freshly dried p-toluenesulfonic acid monohydrate (141 mmol) and 3.6 g of the dried copper p-toluene sulfonate (8.2 mmol) are added. After 2 hr of stirring at room temperature, the dark solution is added to a vigorously stirring mixture of 200 ml 1 N ammonia containing about 150 g crushed ice while maintaining the pH above pH 7 (correct any fall by the addition of ammonia). Having added some drops of octanol, the foaming yellowish suspension is concentrated under reduced pressure to about 200 ml in a rotatory evaporator. The precipitating yellowish copper complex is collected by centrifugation at 12,500 g for 5 min. The precipitate is suspended in a mixture of 200 ml water and 30 ml 1 N ammonia. Subsequently 15 ml of 1 M ammonium sulfide (15 mmol) is added under vigorous stirring. Ten minutes later 4 g of Celite is added and 10 min thereafter the mixture is filtered. The filtrate is combined with the washing fluid (50 ml 0.01 N NH$_4$OH) of the Celite and concentrated to 50 ml by flash evaporation. The precipitating product is collected by filtration and dried. The product may be recrystallized from water after solubi-

[18] A. Hampton and M. H. Maguire, *J. Am. Chem. Soc.* **83**, 150 (1961).
[19] G. M. Tener, *J. Am. Chem. Soc.* **83**, 159 (1961).
[20] J. G. Moffat, *Can. J. Chem.* **42**, 599 (1964).
[21] A. J. Murphy, J. A. Duke, and L. Stowring, *Arch. Biochem. Biophys.* **137**, 297 (1970).
[22] J. L. Doerr, J. Wempen, D. A. Clarke, and J. J. Fox, *Org. Chem. (N.Y.)* **26**, 3401 (1961).
[23] R. G. Yount, J. S. Frye, and K. R. O'Keefe, *Cold Spring Harbor Symp. Quant. Biol.* **37**, 113 (1972).
[24] H. Fasold, F. W. Hulla, F. Ortanderl, and M. Rack, this series, Vol. 46, p. 289.
[25] J. G. Moffat and H. G. Khorana, *J. Am. Chem. Soc.* **83**, 649 (1961).
[26] M. Saneyoshi and G. Chihara, *Chem. Pharm. Bull.* **15**, 909 (1967).

lization of the product in an 800-fold excess of boiling water; mp, 252°; yield, 54%.

6-Mercaptopurine riboside 5′-monophosphate (thioinosine monophosphate) is prepared by phosphorylation of 2′,3′-O-isopropylidene-6-mercaptopurine with cyanoethyl phosphate and the subsequent hydrolysis of the isopropylidene residue.[19] The pyridinium salt of cyanoethyl phosphate is obtained by suspending 4 g (12.4 mmol) of the barium salt [BaPO$_4$-(CH$_2$)$_2$-CN · 2H$_2$O] in a few milliliters of water and stirring that mixture with 17 ml Dowex 50 (H$^+$ form) with a glass rod. After dissolution of the barium salt, the slurry is added to a column (1.6 × 10 cm) of Dowex 50 (H$^+$ form) and the substance is washed out with 250 ml of water. The eluate is collected in 15 ml pyridine (standing in ice) and is concentrated by flash evaporation to a few milliliters. A few milliliters of pyridine are added followed by 990 mg isopropylidenethioinosine dissolved in 30 ml 50% (v/v) pyridine–water mixture. The solution is dried in a rotatory evaporator under reduced pressure and the dry residue is dissolved again in 30 ml dry pyridine. This process is repeated three times. To the oily residue, 30 ml dry pyridine and 6 g N,N′-dicyclohexylcarbodiimide (32.4 mmol) are added and the mixture is stirred at room temperature in a stoppered flask. After 18 hr of reaction 3 ml H$_2$O is added. Pyridine is removed 30 min later by flash evaporation, dissolving the dry residue in 60 ml H$_2$O and removing the solution under reduced pressure. The cyanoethyl residue is split off by dissolving the dry residue in 120 ml 0.4 N LiOH and boiling it for 1 hr. The cooled solution is filtered using a Büchner funnel to remove lithium phosphate and dicyclohexylurea and the filtrate is added to a 3 × 12 cm column filled with Dowex 50 (H$^+$ form). The column is washed with 200 ml water. Fractionation of the eluate and measuring the absorption at 322 nm localizes the thioinosine 5′-monophosphate. The fractions containing s^{6}IMP are combined and freeze dried. The yield is 77%. The product has a ratio of thioinosine : phosphate of 1 and migrates on PEI cellulose thin-layer chromatography in 0.5 M NaCl with an R_f value of 0.35 and at pH 3.5 in 0.75 M phosphate solution with R_f 0.59.

6-Mercaptopurine riboside 5′-triphosphate (thioinosine triphosphate) is prepared in 60% yield by condensation of tetra(tributylammonium) pyrophosphate with the tributylamine salt of s^{6}IMP.[21] Glassy, syrupy tetra(tributylammonium) pyrophosphate is obtained by dissolving 5.57 g (12.5 mmol) Na$_2$P$_2$O$_7$ · 10H$_2$O in 120 ml water and pouring that solution on a Dowex 50 column (pyridine form) (3.2 × 10 cm). The pyridinium salt is eluted with 400 ml water and concentrated with a rotatory evaporator to about 80 ml. After addition of 120 ml pyridine and 9.5 ml tributylamine

(29.5 mmol) the mixture is evaporated to dryness under reduced pressure. The syrupy clear residue is dried by adding 30 ml dry pyridine and evaporating. This process is repeated three times. Pyridine is removed by addition of dry benzene which is evaporated. The residue is then taken up in dry dimethylformamide and immediately added to the tributylammonium salt of thioinosine monophosphate. Methylene diphosphonic acid or imido diphosphonate may be used instead of the diphosphate to obtain the β,γ-methylene or β,γ-imido derivatives of s^6ITP.

The tributylammonium salt of thioinosine monophosphate had been prepared by dissolving 2.5 mmol thioinosine monophosphate in 50 ml H_2O, adding 50 ml pyridine and 0.62 ml (2.6 mmol) tributylamine, and drying the mixture under reduced pressure to dryness in a rotatory evaporator. The residue is then solubilized in 30 ml dry pyridine and the solution is evaporated. That step is repeated three times and thereafter two times with 30 ml dry dimethylformamide. The resultant dry residue is dissolved in 25 ml dry dimethylformamide and 2.0 g (12.5 mmol) 1,1'-carbonyldiimidazole dissolved in 25 ml of the same solute is added. That mixture is allowed to react for 4 hr at room temperature after an initial 30-min reaction under stirring. Thereafter, the addition of 0.84 ml (20 mmol) of dry methanol is followed 30 min later by 12.5 mmol tributylammonium pyrophosphate in 150 ml DMF. A cloudy gray suspension, which has been formed, is centrifuged 24 hr later at 48,000 g for 5 min and the sediment is extracted with 60 ml dimethylformamide. The combined supernatants are added to 250 ml methanol and the solution is lyophilized in a rotatory evaporator. The residue is taken up in 150 ml water and added to a DEAE-Sephadex A-25 column (1.6 × 18 cm). Thioinosine triphosphate is eluted by applying a 0 to 0.6 M linear gradient of triethylammonium bicarbonate (pH 7.5). Thioinosine triphosphate is eluted at about 0.35 M buffer and is lyophilized. The pure substance with a ratio of thioinosine to total phosphorus of 1 : 3 migrates in a thin-layer chromatography on PEI-cellulose in 0.5 M NaCl with R_f 0.025 and in 0.75 M potassium phosphate (pH 3.5) with R_f 0.33.

The disulfide of thioinosine triphosphate is prepared by the oxidation of thioinosine triphosphate with iodine[14,22,23]: 16 μmol thioinosine triphosphate is dissolved in 0.9 ml Tris–HCl buffer (pH 7.5). To this solution 16 μl of 1 M iodine is added dropwise under stirring. After a few minutes of reaction, the mixture is added to a Sephadex G-10 column (1.5 × 20 cm) which had been swollen in distilled water. The disulfide of thioinosine triphosphate is eluted near the void volume showing an absorption maximum at 290 nm of $E_{290}^{1cm} = 30,000\ M^{-1}\ cm^{-1}$ at pH 7.5 instead of the absorption maximum of the thioinosine triphosphate of $E_{322}^{1cm} = 25\ 800\ M^{-1}$

cm^{-1} at pH 4.7. The disulfide is divided into 0.5-ml aliquots and quick frozen in Eppendorf vials. Due to the lability of the disulfide bridge the thawed material should be used within 1 day for the experiments.

The synthesis of S-(2,4-dinitrophenyl)-6-mercaptopurine riboside 5'-monophosphate and 5'-triphosphate described by Fasold et al.[24] was modified, because we had difficulties reproducing their procedure.[10]

For the synthesis of the S-(2,4-dinitrophenyl)-6-mercaptopurine riboside 5'-monophosphate 1 mmol of freshly prepared thioinosine monophosphoric acid was solubilized in a mixture of 14 ml ethanol and 6 ml water at room temperature and adjusted to pH 7 by the addition of 2 N NaOH. To this mixture, which was refluxed at 60–70°, 0.62 ml of a 2 M ethanolic solution of 2,4-dinitro-1-fluorobenzene was added and the pH was maintained at pH 7–7.5 by the dropwise addition of 0.5 ml 2 N NaOH. The formation of the dinitrophenyl thioether of s^{6}IMP was terminated within some minutes. After cooling the reaction mixture, precipitating NaF was filtered off and the filtrate was extracted twice with diethyl ether. The water phase containing Dnps^{6}ITP was poured onto a column (2 × 12 cm) of Dowex W-50 (H$^+$ form). The Dnps^{6}ITP was eluted with water after and well separated from other remaining compounds. The yield of the lyophilized product was 60%. The spectrum and the R_f value on cellulose thin-layer chromatography corresponded to that reported.[24]

The synthesis of S-(2,4-dinitrophenyl)-6-mercaptopurine riboside 5'-triphosphate proceeds via the 5'-phosphomorpholidate derivative, which is prepared in 67% yield according to Moffat and Khorana.[25] The dry morpholidate (0.4 mmol) is dissolved in 16 ml dimethyl sulfoxide and added to 1.6 mmol tributylammonium pyrophosphate. The solution was allowed to react in the dark for 2 days at room temperature. After addition of 120 ml water, the mixture is transferred to a column (1.5 × 10 cm) of DEAE-Sephadex A-25 (bicarbonate form). The elution of the triphosphate was performed by 300 ml of a linear gradient 0–0.8 M triethylammonium carbonate, pH 7.5, with a yield of 60%.

6-Thiocyanato-9-β-D-ribofuranosylpurine 5'-triphosphate (-6-thiocyanatoinosine triphosphate) was prepared by a modification of the reported procedure[26]: To a solution of 216 μmol of thioinosine triphosphate in 2.5 ml water standing in ice 0.22 ml 1 M NaOH (220 μmol) is added while stirring, followed immediately by the addition of 23.6 mg (223 μmol) of CNBr in 644 μl ethanol within 20 min. After a further 20 min of reaction the mixture was concentrated *in vacuo* to an oily residue which is added to a 1 × 18 cm column of Sephadex G-10 swollen in distilled water. 6-Thiocyanatoinosine triphosphate is eluted in water well separated from salts. The fractions with an absorbance maximum at 275 nm are collected and freeze dried.

[γ-^{32}P]Thioinosine triphosphate is prepared essentially as described by Glynn and Chappell[27] except that 60 μmol thioinosine triphosphate is included instead of ATP.

Synthesis of ^{14}CNs^{6}ITP. (s^{6}ITP)$_2$ (13.4 μmol) in 800 μl H$_2$O is added to 19.9 μmol K^{14}CN (50.2 mCi/mmol) in 100 μl and allowed to react at room temperature for 15 min. During that time the spectrum of (s^{6}ITP)$_2$, which has a maximum in absorbancy at 290 nm, gets a shoulder at 320 nm. After the addition of 26 μl of 1 N NaOH, 42.2 μmol BrCN in 5.6 μl acetonitrile is added, which changes the maximum of absorbancy to 275 nm. Chromatography of the radioactive product on Sephadex G-10 (1.2 × 24 cm) previously swollen in water gives a single radioactive peak, which migrates in a thin-layer chromatography on PEI-cellulose in 0.5 M NaCl with an R_f value of 0.08, which resembles that of authentic CNs^{6}ITP.

Inactivation and Reactivation Studies

Modification Procedure. An amount of either purified[28] or partially purified Na$^+$,K$^+$-ATPase[29] hydrolyzing between 1 and 2 mol ATP min^{-1} at 37° is incubated in a total volume of 0.5 ml in a test tube in a water bath at 37°. The test tube contains in 20–60 mM imidazole–HCl or Tris–HCl buffer (pH 7.2) 10 mM EDTA and varying concentrations of the above ATP analogs. Controls are run in parallel, where the ATP analog is missing or where additionally either 3.3 mM ATP or 20 mM KCl are present. Since K$^+$ lowers the affinity of ATP for its binding sites,[30] no affinity labeling should be possible in its presence. For preparative purposes the inactivation mixture is scaled up. That mixture is shaken gently every third minute.

When the reversibility of the modification is to be studied, dithiothreitol is added in a 20-fold excess over the nucleotide triphosphate to both the untreated and the ca. 50% inactivated Na$^+$,K$^+$-ATPase and the incubation at 37° is continued for at least 5 hr. During that time enzyme activity is measured by transferring 45-μl aliquots of the reaction medium to the coupled optical test.[29]

Inactivation and the Incorporation of [γ-^{32}P]s^{6}ITP into the Enzyme Protein Using ([γ-^{32}P]s^{6}ITP)$_2$. About 4 mg of enzyme protein is incubated at 37° in a total volume of 20 ml containing 60 mM imidazole buffer, pH 7.25, 10 mM EDTA. The blank contains 3.3 mM ATP in addition. The

[27] J. M. Glynn and J. B. Chappell, *Biochem. J.* **90**, 147 (1964).
[28] P. L. Jørgensen, this series, Vol. 32, p. 277.
[29] W. Schoner, C. von Ilberg, R. Kramer, and W. Seubert, *Eur. J. Biochem.* **1**, 334 (1967).
[30] M. Hasselberg, Ph.D. Thesis, Fachbereich Biochemie, Pharmazie und Lebensmittelchemie University of Frankfurt/M (1986).

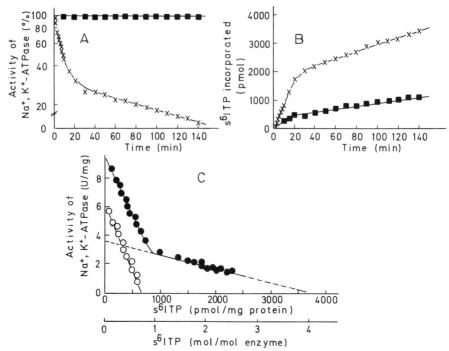

FIG. 1. Titration of active sites of 2.1 mg Na$^+$,K$^+$-ATPase and time course of inactivation with 21 μM ([γ-^{32}P]s^{6}ITP)$_2$: (A) Time course of inactivation with (×) and without (■) 3.3 mM ATP. (B) Incorporation of [γ-^{32}P]s^{6}ITP into the enzyme protein. (C) Inactivation of Na$^+$,K$^+$-ATPase as a function of the covalent modification of the enzyme protein. The radioactivity found in the presence of 3.3 mM ATP was subtracted. The dashed line extrapolates the slowly inactivating part to complete inactivation. The open symbols represent the values obtained by subtracting the experimental points from the dashed line [sites reacting fast with (s^{6}ITP)$_2$]. Reprinted with permission from Patzelt-Wenczler and Schoner.[15]

inactivation is started by the addition of 0.42 μmol radioactive (s^{6}ITP)$_2$. Aliquots of 0.04 ml are withdrawn at the times indicated in Fig. 1 and used for activity determinations in the optical assay.[29] At the same time 1.0-ml aliquots of the inactivation mixture are put into centrifuge tubes standing in ice, already holding 0.6 ml of a solution containing 6 mM dithioerythritol and 30 mM ATP [this treatment does not lead to a reactivation of Na$^+$,K$^+$-ATPase activity, but immediately destroys any free (s^{6}ITP)$_2$]. Shortly thereafter the enzyme protein is precipitated with 5 ml ice-cold 5% (w/v) trichloroacetic acid containing 1 mM ATP and 5 mM P$_i$. The denatured enzyme protein is centrifuged down for 30 min at 30,000 rpm in the 50 Ti rotor of the Beckman ultracentrifuge. The sediment is resuspended in 5 ml of 5% trichloroacetic acid. The suspension is centrifuged again at 80,000 g for 30 min, the supernatant is discarded, and the walls of the centrifuge tubes are cleaned carefully. The sediment is solubilized with 0.5 ml of 1 M NaOH, neutralized with some drops of HCl, and counted in 10 ml of a scintillation medium containing Triton X-100.[31]

Tryptic Map of the ([γ-^{32}P]s^{6}ITP)$_2$-Inactivated Enzyme. Enzyme protein (2.4 mg) is incubated in a total volume of 10.5 ml containing 60 mM imidazole, pH 7.25, 10 mM EDTA, and 20 μM ([γ-^{32}P]s^{6}ITP)$_2$ until the activity of Na$^+$,K$^+$-ATPase has been decreased to 20–30% of the initial activity. The reaction is quenched by the addition of 3.3 mM ATP. The denaturated enzyme is centrifuged down for 90 min in a 60 Ti rotor of the Beckman ultracentrifuge at 60,000 rpm. The sediment is taken up in 2 ml 0.1 mM NH$_4$HCO$_3$ containing 3 mg trypsin. The digestion proceeds for 90 min at room temperature. Thereafter any unsolubilized material is centrifuged off at 100,000 g for 30 min. The supernatant is subjected to a paper electrophoresis on Whatman 3 MM for 1.5 hr at 25 V/cm in a plate-cooled (5°) high-voltage electrophoresis Pherograph (model 64 of Vetter KG, Wiesloch, FRG) in 0.1 M pyridine acetate buffer, pH 5.3. The radioactive peptide is further chromatographed in a second direction in pyridine : n-butanol : water : acetic acid (10 : 15 : 12 : 3). The radioactivity is measured with a Berthold LB 2723 scanner II equipped with a strip counter of Laboratorium Prof. Dr. Berthold (Wildbad, FRG).

Assays

Na$^+$,K$^+$-ATPase activity during the inactivation/reactivation process is determined by transferring a 45-μl aliquot of the incubation mixture into 1.6 ml of an optical assay (37°), where the ADP formation is detected via the pyruvate kinase and lactate dehydrogenase reactions by the decrease of the optical density of NADH + H$^+$ at 366 nm.[29]

Binding Capacities of Sites. The capacity of the high-affinity ATP-binding site is calculated from a Scatchard plot of equilibrium binding experiments.[32,33]

The ouabain-binding capacity and the capacity of the phosphointermediate formed from [γ-^{32}P]ATP is measured according to previously published procedures.[10,15,28,34]

[31] J. G. Nørby and J. Jensen, *Biochim. Biophys. Acta* **233**, 104 (1971).
[32] K. Kaniike, E. Erdmann, and W. Schoner, *Biochim. Biophys. Acta* **298**, 901 (1973).
[33] G. Rempeters and W. Schoner, *Biochim. Biophys. Acta* **727**, 13 (1983).
[34] E. Erdmann and W. Schoner, *Biochim. Biophys. Acta* **307**, 386 (1973).

Derivatives of Thioinosine Triphosphate as Affinity Labels of Na+,K+-ATPase

Most of the derivatives of thioinosine triphosphate studied so far are substrates of Na+,K+-ATPase[15,17] and compete with ATP for its binding site (Table I).[10,14,15,17,30,31,35,36] However, the apparent dissociation constants of the enzyme–substrate complexes with these ATP analogs as determined at 0° by competition with radioactive ATP or ADP are about 100 times higher than that of ATP itself. All of these ATP analogs inactivate the enzyme by forming a covalent bond with either the nucleotide triphosphate.[15,17,30] The inactivation reaction is temperature dependent. The reaction with 72 μM disulfide of thioinosine triphosphate shows an activation energy of 95.5 kJ mol^{-1}.[15] The inactivation is abolished by the presence of ATP[10,14–17] in higher concentrations and by the presence of K+.[16,17,37] All these ATP analogs show therefore the properties of an ATP affinity label of this enzyme. The inactivation proceeds with two exponentials with CNs^{6}ITP,[30] Nbs^{6}ITP,[17] and (s^{6}ITP)$_2$[15] (Fig. 1). This phenomenon has been interpreted to indicate the existence of two ATP-binding sites with different affinities for ATP[15] or to indicate the slow formation of a reactive form from an unreactive ATP-binding site.[17] Presently it is impossible to determine which possibility is correct. As expected, 1 mol of site is modified per mole of α subunit in order to reach the full inactivation[15,17] (Fig. 1) using two different ATP analogs. Consistent with the labeling of one site only, one single spot is seen in a fingerprint of a tryptic digest of the ([γ-^{32}P]s^{6}ITP)$_2$-inactivated enzyme.[15]

It is suggested that all these ATP analogs react with a sulfhydryl group within or near the active site. This conclusion is in part analogous, since the S-(2,4-dinitrophenyl)-6-mercaptopurine and the S-(3-carboxy-4-nitrophenyl)-6-mercaptopurine derivatives have been shown to modify SH groups in forming new thio ethers by a transfer of the purine moiety.[24,38,39] Also cyano-6-thioinosine triphosphate does so,[30] although cyano-6-thioinosine has been described to cyanylate SH groups in myosin.[40] The validity of these analogous conclusions in the irreversible modifications of Na+,K+-ATPase, which due to their stability are rather suitable for the isolation of a tryptic peptide, haven't yet been proved. So far it cannot be excluded that a tyrosine residue is modified by Dnps^{6}ITP and

---

[35] C. Hegyvary and R. L. Post, *J. Biol. Chem.* **246**, 5234 (1971).
[36] R. Patzelt-Wenczler and W. Schoner, *Verh. Dtsch. Ges. Kreislaufforsch.* **41**, 311 (1975).
[37] R. Patzelt-Wenczler and W. Mertens, *Eur. J. Biochem.* **121**, 197 (1981).
[38] F. W. Hulla and H. Fasold, *Biochemistry* **11**, 1056 (1972).
[39] U. Faust, H. Fasold, and F. Ortanderl, *Eur. J. Biochem.* **43**, 273 (1974).
[40] H. Wiedner, R. Wetzel, and F. Eckstein, *J. Biol. Chem.* **253**, 2763 (1978).

TABLE I
RESULTS OBTAINED FROM INACTIVATION STUDIES WITH ANALOGS OF THIOINOSINE TRIPHOSPHATE

Nucleotide triphosphate	Dissociation constant (0°)	Substrate of Na$^+$,K$^+$-ATPase	Component transferred	Moles transferred per α subunit	Sensitivity of the inactivated Na$^+$,K$^+$-ATPase toward dithiothreitol
ATP	1.2–2.2×10^{-7} M[10,31,35]	$K_{0.5} = 2.8$ mM[17]	—	—	—
Dnps^{6}ITP	5.00×10^{-6} M[a]	Yes[10]	s^{6}ITP(?)	ND	No reactivation[10]
Nbs^{6}ITP	1.14×10^{-3} M[b]	$K_{0.5} = 1.4$ mM[17]	s^{6}ITP[c]	0.8[17]	ND
s^{6}ITP	2.00×10^{-5} M[a]	Yes[10]	—	—	—
(s^{6}ITP)$_2$	4.8×10^{-5} M[a]	$K_{0.5} = 0.83$ mM[15]	s^{6}ITP[c,d]	0.97–1.12[15]	Reactivation[36]
CNs^{6}ITP	7.4×10^{-5} M[a]	ND yes[30]	s^{6}ITP[d]	1.03[30]	Reactivation[30]

[a] From equilibrium binding experiments by competition with [^{14}C]ADP[10] or [α-^{32}P]ATP.[30]
[b] From kinetic analysis.[17]
[c] From inactivation with [^{32}P]Nbs^{6}ITP,[17] ([^{32}P]s^{6}ITP)$_2$,[14,15] and spectroscopic evidence.[15]
[d] From spectroscopic evidence and inactivation studies with [γ-^{32}P]CNs^{6}ITP and ^{14}CN-s^{6}ITP. ND, not determined.

Nbs⁶ITP.[17,41] Moreover, the transfer of the dinitrophenyl group from Dnps⁶IMP to a sulfhydryl group of the adenosine triphosphate phosphoribosyltransferase has been reported.[42] Such a transfer could also take place from Dnps⁶ITP and Nbs⁶ITP in Na$^+$,K$^+$-ATPase. It is therefore essential that the modification and inactivation of Na$^+$,K$^+$-ATPase by (s⁶ITP)$_2$ and CNs⁶ITP is reversible after the addition of dithiothreitol.[30,36] Due to that finding it can be concluded that a sulfhydryl group resides in or near the ATP-binding site and interacts with the 6-amino group of the purine moiety of ATP when this substrate is bound to Na$^+$,K$^+$-activated ATPase.

Acknowledgments

This work has been supported by the Deutsche Forschungsgemeinschaft, Bonn-Bad Godesberg and by the Fond der Chemischen Industrie, Frankfurt/Main.

[41] F. W. Hulla, M. Höckel, M. Rack, S. Risi, and K. Dose, *Biochemistry* **17**, 823 (1978).
[42] T. Dall-Larsen, H. Fasold, L. Klungsøyr, H. Kryri, C. Meyer, and F. Ortanderl, *Eur. J. Biochem.* **60**, 103 (1975).

[29] Photoaffinity Labeling with ATP Analogs

By WILHELM SCHONER and GEORGIOS SCHEINER-BOBIS

Introduction

Na$^+$,K$^+$-ATPase undergoes conformational changes during the active cation transport process.[1] Affinity labeling with ATP analogs is a means to localize the ATP-binding site on the subunits of the enzyme and to obtain information on the amino acids interacting with ATP.[2-7] In addition, the reactivity of the enzyme toward protein-reactive ATP analogs can be used as a probe to obtain information on the effects of the transport substrates and cosubstrates on the ATP-binding site.[3] Since Na$^+$,K$^+$-ATPase appears to contain specific domains for at least the acceptor site for the

[1] P. L. Jørgensen, *Biochim. Biophys. Acta* **694**, 27 (1982).
[2] B. E. Haley and J. F. Hoffman, *Proc. Natl. Acad. Sci. U.S.A.* **71**, 3367 (1974).
[3] G. Rempeters and W. Schoner, *Eur. J. Biochem.* **121**, 131 (1981).
[4] K. B. Munson, *Biochemistry* **22**, 2301 (1983).
[5] G. Ponzio, B. Rossi, and M. Lazdunski, *J. Biol. Chem.* **258**, 8201 (1983).
[6] R. Patzelt-Wenczler and W. Schoner, *Eur. J. Biochem.* **114**, (1981).
[7] H. Koepsell, F. Hulla, and G. Fritzsch, *J. Biol. Chem.* **257**, 10733 (1982).

terminal phosphate of ATP[8] and for the adenosine moiety,[3,5,9] we intend to demonstrate by photoaffinity labeling at the high-affinity ATP-binding site how the ribosyl and purine subsites behave toward their substrate in the presence and absence of univalent cations and Mg^{2+}.[3,9]

Synthesis of ATP Photoaffinity Labels

8-Azidoadenosine 5'-triphosphate (8-N_3ATP) is prepared according to Schäfer et al.[10] from 8-bromoadenosine 5'-triphosphate (8-BrATP), which is obtained essentially according to Ikehara and Uesugi.[11]

ATP-Na_2 (1 mmol) is placed in a 50-ml round-bottom flask and dissolved in 13.3 ml 1 M sodium acetate (pH 3.8). To this is added 13.3 ml bromine water containing 67 μl Br_2 and the flask is sealed with a ground-in stopper. The bromination is allowed to proceed at room temperature for at least 6 hr. The reaction may be followed by observing the shift of the λ_{max} of the ultraviolet spectrum from 259 to 265 nm or by thin-layer chromatography (TLC) on polyethyleneimine cellulose (PEI-cellulose) in 0.75 M KH_2PO_4 (pH 3.5), where 8-BrATP migrates with an R_f value of 0.29 and ATP with R_f 0.38.

After bromination of ATP the pH is adjusted to 6–8 with 4 N NH_4OH, which results in a color change from yellow to red. After 10-fold dilution with water the solution is added to a column (2 × 10 cm) filled with DEAE-Sephadex A-25 (HCO_3^- form) and a linear gradient of 2.8 liters of 0–0.6 M triethylammonium bicarbonate (pH 7.2) is applied to separate out degradation products of ATP and 8-BrATP and to free 8-BrATP from ATP. 8-BrATP (50–60% yield) elutes at approximately 0.3 M triethylammonium bicarbonate. The separation from ATP may be incomplete. In that case 8-BrATP should be localized by measuring the λ_{max} of the ultraviolet spectrum of each fraction. The fractions containing the λ_{max} at 265 nm are pooled and lyophilized.

8-N_3ATP in 40–50% yield is obtained by placing 0.6 mmol 8-BrATP in a round-bottom flask covered with aluminum foil. To it are added 4.8 mmol HN_3, 4.8 mmol triethylamine, and 18 ml freshly distilled dry dimethylformamide. This mixture is maintained for 8 hr at 75° (in the dark). To the yellowish mixture 400 ml of distilled water is added and the solution is transferred onto a column (2 × 10 cm) filled with DEAE-Sephadex A-25 (HCO_3^- form). The 8-N_3ATP and its degradation products 8-N_3ADP (20–30% yield) and 8-N_3AMP (ca. 14% yield) are eluted from the column

---

[8] J. Castro and R. A. Farley, J. Biol. Chem. 254, 2221 (1979).
[9] G. Scheiner-Bobis and W. Schoner, Eur. J. Biochem. 152, 739 (1985).
[10] H.-J. Schäfer, P. Scheurich, and K. Dose, Liebigs Ann. Chem. 1978, 1749 (1978).
[11] M. Ikehara and S. Uesugi, Chem. Pharm. Bull. 17, 348 (1969).

under the conditions described above with triethylammonium bicarbonate (pH 7.2). 8-N_3ATP with a λ_{max} of 281 nm is localized by UV photometry of the eluate, measuring its concentration in a coupled optical test[12] and testing the mobility of the substance in PEI-cellulose TLC in 0.75 M KH_2PO_4 (pH 3.5). 8-N_3ATP has an R_f value of 0.32, 8-N_3ADP of 0.5, and ADP of 0.57. 8-N_3ATP and 8-N_3ADP do not migrate in TLC, when the wet substance is photolyzed prior to chromatography. To check the purity of the 8-N_3ATP, the spectrum of the substance should be measured before and after photoinactivation with UV light of 254 nm. This procedure results in the loss of the λ_{max} of 281 nm.[10,13] The combined fractions containing 8-N_3ATP are lyophilized. The dry residue is dissolved in several milliliters of distilled water and stored in 2- to 3-ml aliquots in Eppendorf vials in the dark at $-20°$.

Hydrazoic acid, HN_3, is obtained according to von Braun[14] by adding to a three-necked bottle containing a stirrer, thermometer, dropping funnel and gas outlet tube (return condenser) equal amounts (w/w) of sodium azide and water. For 0.1 mol sodium azide, 40 ml benzene is added and the mixture is cooled down to $0°$. Then an equivalent amount of concentrated H_2SO_4 is added dropwise while stirring. Care is taken that the temperature does not exceed $10°$. Benzene is separated from the water phase at $0°$ in a funnel and the benzene phase is dried over Na_2SO_4. The solution is stored in a refrigerator at $4°$. For quantitation of HN_3 3 ml of the benzene mixture is shaken with 30 ml H_2O and titrated with 0.1 N NaOH. Care should be taken in handling the pure acid, because HN_3 is highly explosive.

Radioactive 8-N_3ATP can be obtained by exchanging enzymatically the terminal phosphate with ortho[^{32}P]phosphate using 8-N_3ATP instead of ATP by the method of Glynn and Chappell.[15] [α-^{32}P]ATP may also be used for the chemical synthesis of 8-N_3ATP. We were unable to use the commercially available triethylammonium salt of [α-^{32}P]ATP for the synthesis of [α-^{32}P]-8-BrATP. To convert it into the sodium form 1 mCi [α-^{32}P]ATP was added to a column (1.5 × 3 cm) of CM-Sephadex A-25 (sodium form) and was eluted with 100 ml distilled water. After lyophilization and appropriate dilution with nonradioactive ATP further synthesis could be performed as described above.

---

[12] D. Jaworek, W. Gruber, and H. U. Bergmeyer, *in* "Methods of Enzymatic Analysis" (H. U. Bergmeyer, ed.), p. 2097. Verlag Chemie, Weinheim, Federal Republic of Germany, 1974.

[13] J. Czarnecki, R. Geahlen, and B. Haley, this series, Vol. 56, p. 642.

[14] J. von Braun, *Liebigs Ann. Chem.* **490**, 125 (1931).

[15] I. M. Glynn and B. Chappell, *Biochem. J.* **90**, 147 (1964).

3′-O-[3-(4-Azido-2-nitrophenyl)propionyl]adenosine 5′-triphosphate (3′-N₃ATP) is synthesized according to Schäfer and Onur.[16] This ATP photoaffinity label has a higher stability against ester hydrolysis[16] than the 3′-esters derived from N-substituted amino acids.[17] The 3′-ester of ATP is synthesized by the method of Gottikh et al.[18] via the imidazolide of the 3-(4-azido-2-nitrophenyl)propionic acid which is prepared from 3-phenyl-propionic acid according to Gabriel and Zimmerman[19,20] and Ugi et al.[21]

Preparation of 3-(4-Azido-2-nitrophenyl)propionic Acid

2,4-Dinitrophenylpropionic acid is synthesized by adding 200 ml of fuming HNO_3 into a 2-liter two-necked bottle containing a magnetic bar. Through a dropping funnel 110 ml of concentrated H_2SO_4 is slowly added while stirring. Care is taken that the temperature is kept at −5°. After the addition of H_2SO_4 the mixture is cooled down to −10° and 0.2 mol of phenylpropionic acid added dropwise while stirring (while the temperature is kept at −5°). After completion of that addition the resulting mixture is refluxed for 1 hr at 80°. The cooled solution is then poured into a beaker containing 2.7 liter ice water. 2,4-Dinitrophenylpropionic acid is obtained as long yellow needles, which are collected on a Büchner funnel. This product is recrystallized from boiling water. 2,4-Dinitrophenyl-propionic acid is obtained in 74–80% yield; mp, 280°.

3-(4-Amino-2-nitrophenyl)propionic acid is obtained by partial reductive amination of the 2,4-dinitrophenylpropionic acid. 2,4-Dinitrophenylpropionic acid (0.145 mol) is added to a two-necked bottle equipped with a reflux and 128 ml 20% $(NH_4)S$ (0.37 mol) is slowly added while stirring. Heat is produced by the reaction (65–70°) and the mixture is continuously stirred for another 60 min. Water (145 ml) is then added and the pH is lowered to 1 by the addition of concentrated HCl. A bright yellow precipitate is removed by filtration. The remaining clear solution is brought to pH 2 by the addition of NH_3. An orange precipitate, 3-(4-amino-2-nitrophenyl)propionic acid, is obtained by filtration with a funnel. The substance is recrystallized from boiling water in 37–44% yield; mp, 138°.

3-(4-Azido-2-nitrophenyl)propionic acid is obtained in 37–52% yield by dissolving 53 mmol of 3-(4-amino-2-nitrophenyl)propionic acid in 54 ml

---

[16] G. Schäfer and G. Onur, *Eur. J. Biochem.* **97**, 415 (1979).
[17] J. R. Guillory and S. J. Jeng, this series, Vol. 46, p. 259.
[18] B. P. Gottikh, A. A. Krayevsky, N. B. Tarussova, P. P. Purygin, and T. L. Tsilevich, *Tetrahedron* **26**, 4419 (1970).
[19] S. Gabriel and J. Zimmermann, *Ber. Dtsch. Chem. Ges.* **12**, 600 (1879).
[20] S. Gabriel and J. Zimmermann, *Ber. Dtsch. Chem. Ges.* **13**, 1680 (1880).
[21] J. Ugi, H. Perlinger, and L. Behringer, *Ber. Dtsch. Chem. Ges.* **91**, 2330 (1958).

2 N HCl in a flask. Care is taken that the temperature does not exceed 5°. All subsequent preparatory steps are, if not otherwise stated, done in ice and protected from light. NaNO$_2$ (57 mmol) dissolved in 16.2 ml H$_2$O is then added dropwise with stirring. Subsequently the mixture is neutralized by the addition of solid CaCO$_3$. NaN$_3$ (60 mmol) (dissolved in 16.2 ml H$_2$O) is then added slowly. A creamlike reddish-orange foam resulting therefrom is placed in a funnel and extracted with boiling water until the filtrate no longer shows a yellow tint. After cooling in ice the pH of the combined filtrates is brought to pH 2 by the addition of 2 N HCl. The precipitating product is recrystallized from boiling water and dried. The product showing in UV light a λ_{max} of 250 nm in ethanol (molar extinction coefficient 21.4 × 10^3 liters mol^{-1} cm^{-1}) is destroyed by photolysis, which results in the disappearance of that maximum[15]; mp 125°; NMR (chemical shift is expressed in ppm, J in Hz, solvent = CD$_3$OD): 7.54 (aromatic H-3, J = 2), 7.44 (carboxylic proton, s), 7.26 (aromatic H-5, d, J = 2), 7.21 (aromatic H-6, d, J = 2), 3.04 (CH$_2$ β to the carboxylic proton, t, J = 7), 2.58 (CH$_2$ α to the carboxylic proton, t, J = 7).

Synthesis of the 3'-Ribosyl Ester and ATP with 3-(4-Azido-2-nitrophenyl)propionic Acid

3-(4-Azido-2-nitrophenyl)propionic acid (118 mg, 0.5 mmol) is dissolved in 1.0 ml dry dimethylformamide. Carbonyl-1,1'-diimidazole (270 mg, 1.67 mmol) is added and the mixture is stirred for 10 min. Thereafter 65 mg Na$_2$ATP (0.108 mmol) dissolved in 2 ml 1 M imidazole–HCl (pH 7) is added and the mixture is allowed to react for 4 hr at room temperature. The solvent is then evaporated. The dry residue is extracted with a mixture of ether/acetone (1/1), which extracts the free 3-(4-azido-2-nitrophenyl)propionic acid. The dried residue is dissolved in water and chromatographed on 18 × 16 cm Whatman 3MM sheets in n-butanol : water : acetic acid (5 : 3 : 2) at room temperature. From the resulting well-defined UV-absorbing bands at R_f 0.28 (3' isomer), a diffused band at R_f 0.45 (2' isomer), and a narrow band at R_f 0.67 (2',3' isomer),[17] the band with R_f 0.028 is eluted from the paper with water filtered through Selectron filters type BA 85 (Schleicher and Schüll, Dassel) and lyophilized. Fifty to 70 μmol is obtained.

By the inclusion of [α-^{32}P]ATP into the above procedure, [α-^{32}P]-3'-N$_3$ATP is obtained.[3] The identity of 3'-N$_{3p}$ATP is determined from the relationship of the acid labile to total phosphate of 2 : 3. The substance with a λ_{max} = 252 shows a molar absorption coefficient of 27 × 10^3 liters mol^{-1} cm^{-1}. Upon irradiation by ultraviolet light the spectrum of 3'-N$_{3p}$ATP changes drastically due to the photolysis of the azido group and a

λ_{max} of 260 nm is obtained.[3] The substance is stored in 1- to 2-ml aliquots at $-20°$ in the dark.

Synthesis of the Chromium(III) Complexes of 8-N_3ATP and 3'-N_{3p}ATP

The chromium complexes of the photoreactive ATP analogs are synthesized by a modification of the procedure reported by DePamphilis and Cleland.[22]

8-N_3ATP and $Cr(H_2O)_6(ClO_4)_3$ are dissolved in water to give a solution of 10 mM in each reactant. The pH of the solution is then adjusted to 3.0 with 1 M HCl and the mixture is heated for 10 min at 80°. After cooling to room temperature the mixture is placed on top of a column (1.5 × 10 cm) filled with Dowex AG-WX-2 (H^+ form), which had been washed well with water. After adsorption of the blue–green chromium complex of 8-N_3ATP and of the chromium salt, the column was washed with 1 liter water at 4°. The upper blue–green band, which had then developed, is removed. The second green band is removed and transferred to another column (0.7 × 1.5 cm), packed with the same material, and the chromium complex of 8-N_3ATP (Cr-8-N_3ATP) is eluted with 0.3 M aniline, pH 2.8. The aniline is extracted three times with 5 vol of diethyl ether. The product is analyzed for its spectral properties,[22] lyophilized, and stored at $-20°$ (yield, 17.6%).

The chromium complex of 3'-N_{3p}ATP is prepared in the same way as described above with the variation that the lower green top band taken from the first column is eluted from the second column with 0.3 M pyridinium acetate, pH 3.8 (0.3 M acetic acid adjusted with distilled pyridine). The fractions containing the product are pooled, immediately extracted three times with ether, and evaporated *in vacuo* to a small volume. The final product is rapidly frozen and stored at $-20°$ (yield, 37%).

Assays and Analytical Procedures

Photoinactivation and Photolabeling of Na^+,K^+-ATPase

Under standard conditions 1–2 U of purified Na^+,K^+-ATPase from pig or sheep[23] (1 enzyme unit hydrolyzes 1 μmol ATP at 37°/min under the conditions of the optical test[9]) is mixed to a total volume of 1 ml containing 10 mM Tris–HCl, pH 7.25, 5 mM $MgCl_2$, 10 mM NaCl, 50 μM photolabel, and 0.4 mM adenosine. Adenosine is included to maintain a relative constant absorbance at varying concentrations of the photolabel and to

---

[22] M. L. DePamphilis and W. W. Cleland, *Biochemistry* **12**, 3714 (1973).
[23] P. L. Jørgensen, this series, Vol. 32, p. 277.

protect the enzyme protein against the unspecific photoinactivation by the UV light.[2] Consequently the concentration of adenosine is raised or lowered, when the photolabel was decreased or increased. The mixture is placed in a watch glass of 4 mm depth and 50 mm diameter, which is inserted into a plastic plate and can be temperature controlled from the lower side with a water bath. The photoinactivation proceeds at room temperature with the photolabels 8-N_3ATP and the 3'-ribosyl ester derivatives of ATP. Since the chromium complex inactivates Na^+,K^+-ATPase slowly at room temperature by forming a stable phospho intermediate.[24] The photoinactivation with the chromium complexes of ATP has to be performed at 4°. The mixture is illuminated by placing the dish into an ultraviolet box (Fluotest, Hanau, type 4201, 80 W), which is normally used to visualize spots on TLC plates. The distance of the sample to the light source of 254 nm is 20 cm. Aliquots of 20 μl of the incubation medium are transferred after different time intervals of illumination to the coupled optical test[22] and the enzyme activity remaining is measured.

Digestion of Photolabeled Na^+,K^+-ATPase with TPCK-Trypsin

Radioactively labeled Na^+,K^+-ATPase (0.1 mg/ml) is partially digested at 37° with 6 g/ml TPCK-trypsin in 25 mM imidazole–HCl, pH 7.25, and 150 mM KCl for 10 min.[25] At that time a 4-fold excess (w/w) over trypsin of soybean trypsin inhibitor is added. The reaction mixture is sedimented for 30 min at 33,000 rpm in a Beckman 50 Ti rotor (100,000 g) and the resultant sediment is resuspended in sufficient sample buffer according to Laemmli[26] to give a final concentration of 0.5 mg/ml and the electrophoresis is carried out for 5 hr at 5 mA/gel in cylindrical 10% acrylamide [acrylamide-N,N'-methylenebis(acrylamide), 30:0.8] gels (0.5 × 10 mm) in 2% sodium dodecyl sulfate at room temperature. The gels are cut in slices of 2 mm length. The gel slices are transferred to vials containing 10 ml of 0.55% (w/w) Permablend III and 1% Soluene-350 (Packard Instruments) in toluene. After 24 hr standing at room temperature the radioactivity is counted in a scintillation counter. Alternatively slab gels of 10% polyacrylamide are run under the same conditions.[26] The localization of the radioactive protein and polypeptides is carried out by autoradiography of the dried gels by exposing Cronex NDT 75 X-ray film (Du Pont) for 2–5 days at $-20°$. Proteins are detected by staining with Coomassie brilliant blue.

[24] H. Pauls, B. Bredenbröcker, and W. Schoner, *Eur. J. Biochem.* **109**, 523 (1980).
[25] P. L. Jørgensen and R. Farley, this volume [27].
[26] U. K. Laemmli, *Nature (London)* **227**, 680 (1970).

Gel Chromatographic Separation of the Peptides Obtained after Complete Trypsinolysis of [α-^{32}P]-8-N$_3$ATP-Labeled Na$^+$,K$^+$-ATPase

Complete trypsinolysis of Na$^+$,K$^+$-ATPase is achieved by incubating 0.5 mg photoaffinity-labeled Na$^+$,K$^+$-ATPase in 1 ml 10 mM Tris–HCl, pH 7.25, with 10 μg TPCK-trypsin at 37° for 5 hr. The addition of trypsin was repeated three times in 1.5-hr intervals. The digestion mixture is added to a column (1.5 × 60 cm) filled with Sephadex G-50 fine. The gel had been swollen in 0.1 M sodium phosphate buffer, pH 7.0, containing 50 mM NaCl, 0.1% 2-mercaptoethanol (v/v), and 1% Triton X-100 (v/v) and the column had been packed at a flow rate of 40 ml cm^{-2} hr^{-1}. Chromatography proceeds in the same buffer mixture at 30 ml cm^{-2} hr^{-1}. This column is standardized with myoglobin (M_r 17,800), cytochrome c (M_r 12,600), and vitamin B$_{12}$ (M_r 1355).

The radioactive peptide elutes at a position equivalent to M_r 1800.

Photoreactive ATP Analogs as Affinity Labels and Conformational Probes of Na$^+$,K$^+$-ATPase

The ribosyl-esterified and purine-modified ATP analogs studied so far fulfill the requirements of affinity labels of the high-affinity ATP-binding site (Table I)[3,4,9,27–31]: (1) All ATP analogs have dissociation constants close to that of the enzyme–ATP complex despite the bulky substituent at the ribose moiety and despite the restriction of rotation of the purine moiety around the N-glycosidic linkage. The substitution at the purine C-8 position probably shifts conformation of the nucleotide from the anti [the favored conformation (Table I)] to the syn conformation.[32] (2) The interaction with the ATP-binding site is decreased or prevented by the additional presence of an excess of ATP.[3,4,9,31] (3) As is to be expected, the additional presence of K$^+$ hinders the photoinactivation.[3,9,27] (4) Consistent with the action at the active site, all ATP analogs are either substrates of the enzyme[3,9,33] or compete with ATP for its hydrolysis.[3,31] With inactivation photolabeling leads to a parallel decrease in the number of high-affinity ATP-binding sites.[9]

[27] J. G. Nørby and J. Jensen, *Biochim. Biophys. Acta* **233**, 104 (1971).
[28] C. Hegyvary and R. L. Post, *J. Biol. Chem.* **246**, 5234 (1971).
[29] W. Schoner, E. H. Serpersu, H. Pauls, R. Patzelt-Wenczler, H. Kreickmann, and G. Rempeters, *Z. Naturforsch. C: Biosci.* **37**, 692 (1982).
[30] G. Scheiner-Bobis and W. Schoner, *Eur. J. Biochem.* **152**, 739 (1985).
[31] K. Munson, *J. Biol. Chem.* **256**, 3223 (1981).
[32] J. J. Czarnecki, *Biochim. Biophys. Acta* **800**, 41 (1984).
[33] G. Bobis, G. Rempeters, and W. Schoner, *Hoppe-Seyler's Z. Physiol. Chem.* **364**, 1101 (1983).

TABLE I
RESULTS OBTAINED BY PHOTOAFFINITY LABELING OF Na^+,K^+-ATPase WITH ATP ANALOGS[a]

ATP analog	Enzyme source	Apparent dissociation constant (μM)	Effect of ions on photoinactivation	Subunit labeled	Tryptic fragments formed	Reference
ATP	Beef brain	0.1–0.2 (4°)	—	—	—	27, 28
	Guinea pig kidney					
8-Br-ATP	Beef brain	4.2 (4°)	—	—	—	29
8-N$_3$ATP	Pig kidney	3.1[b] (4°)	Mg^{2+} absolutely necessary $K_{0.5}^{Mg^{2+}} = 0.77$ mM K^+ inhibits $K_{0.5}^{K^+} = 0.55$ mM	α	56,000[c]	9
Cr-8-N$_3$ATP	Pig kidney	ND	Mg^{2+} enhances $K_{0.5}^{Mg^{2+}} = 0.71$ mM	ND		
3'-N$_{3p}$ATP	Pig kidney	19[e] (20°)	Mg^{2+} absolutely necessary $K_{0.5} = 0.53$ mM Na^+ enhances $K_{0.5} = 4.4$ mM	α	—	3
3'-N$_{3p}$ADP	Pig kidney	24[e] (20°)	ND	ND	—	3
3'-N$_{3p}$AMP-PCP	Pig kidney	26[e] (20°)	ND	ND	—	3
Cr-3'-N$_{3p}$ATP	Pig kidney	ND	Mg^{2+} enhances	ND	—	30
Cr-3'-N$_{3a}$ATP	Dog kidney	8–9[e,f] (4°)	ND	α	77,000[g] 58,000[c] 40,000[h]	4, 31

[a] Abbreviations: 8-BrATP, 8-Bromoadenosine 5'-triphosphate; 8-N$_3$ATP, 8-azidoadenosine 5'-triphosphate; Cr-8-N$_3$ATP, chromium complex of 8-N$_3$ATP; 3'-N$_{3p}$ATP, 3'-O-[3(4-azido-2-nitrophenyl)propionyl]adenosine 5'-triphosphate; 3'-N$_{3p}$ADP, 3'-O-[3(4-azido-nitrophenyl)propionyl]adenosine 5'-diphosphate; 3'-N$_{3p}$AMP-PCP, 3'-O-[3-(4-azido-2-nitrophenyl)propionyl]adenosine 5'-[β,γ-methylene triphosphate; Cr-3'-N$_{3p}$ATP, chromium complex of 3'-N$_{3p}$ATP; Cr-3'-N$_{3a}$ATP, chromium complex of 3'-O-{3-[N-(4-azido-2-nitrophenyl)amino]propionyl}adenosine 5'-triphosphate; ND, not determined.
[b] equilibrium binding with the radioactive ligand
[c] limited proteolysis in the presence of 150 mM K^+ with trypsin
[d] extensive proteolysis with trypsin
[e] from photoinactivation
[f] from competition studies with 2',3'-O-(2,4,6-trinitrocyclohexylidine)adenosine 5'-triphosphate

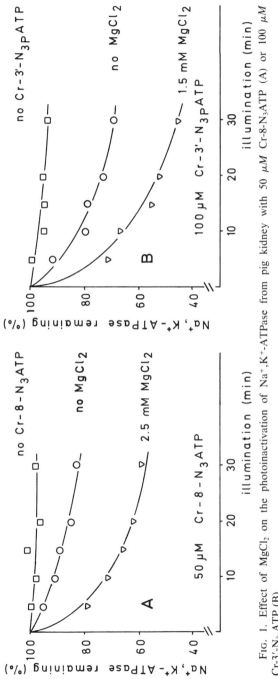

FIG. 1. Effect of MgCl$_2$ on the photoinactivation of Na$^+$,K$^+$-ATPase from pig kidney with 50 μM Cr-8-N$_3$ATP (A) or 100 μM Cr-3'-N$_{3p}$ATP (B).

ATP-sensitive photolabeling occurs at the α subunit only, although other impurities in the enzyme preparation (but not the β subunit) may be labeled to a minor extent.[3,4] The extent and the specificity of the photoinactivation seem to depend somewhat on the light source used.[31] The labeling pattern obtained after limited trypsinolysis under controlled conditions is consistent with the knowledge concerning the localization of the ATP-binding site on the α subunit.[1,5,8] The demonstration of a single peptide of M_r 1800 after extensive digestion[9] of the 8-N_3ATP-labeled enzyme indicates that only a single type of interaction with the binding site takes place. The tryptic peptide is adjacent to the N-terminal site of the FITC-fragment.[34]

8-N_3ATP binds to the high-affinity ATP-binding site of Na^+,K^+-ATPase in the absence of Mg^{2+} or in the presence of EDTA (Table I),[9,33] but it does not photoinactivate the enzyme under these conditions. Photoinactivation is only seen in the presence of Mg^{2+}, as is also the case with 3'-N_{3p}ATP.[3,9,33] But the stable MgATP–complex analog of these photoaffinity labels in the form of chromium complexes photoinactivate the enzyme in the absence of Mg^{2+} (Fig. 1).[9] Despite the binding at the active and high-affinity ATP-binding sites the formation of an MgATP–complex analog is necessary to get proper conformation for reaction with the photolabels. This might be a nondissociable MgATP complex.[3,35] Moreover, additional Mg^{2+} enhances considerably the photoinactivation by Cr-8-N_3ATP and Cr-3'-N_{3p}ATP (Fig. 1). This seems to indicate that Mg^{2+} changes by binding to a low Mg^{2+} affinity site ($K_{0.5}$ = 0.71 mM) the conformation of the active site and makes more reactants available for these ATP analogs by some sort of closing of the ATP-binding site at the purine and the ribose subsites.[30]

Acknowledgments

This work has been supported by the Deutsche Forschungsgemeinschaft, Bonn—Bad Godesberg and by the Fonds der Chemischen Industrie, Frankfurt/Main.

---

[34] C. M. Tran, G. Scheiner-Bobis, W. Schoner, and R. A. Farley, *J. Biol. Chem.*, submitted for publication.
[35] T. Kanazawa, M. Saito, and Y. Tonomura, *J. Biochem.* (*Tokyo*) **67**, 693 (1970).

[30] Affinity Labeling of the Digitalis-Binding Site

By BERNARD ROSSI and MICHEL LAZDUNSKI

Digitalis compounds are natural products extracted from plants. They have been used for centuries in the treatment of various cardiac disorders because of their positive inotropic effect. At the cellular level, the unique effect of this category of compounds is to specifically inhibit the active transport of Na^+ and K^+ mediated by Na^+,K^+-ATPase. This enzyme is an oligomer composed of two types of subunits. The α subunit (M_r 90,000–100,000) possesses the catalytic activity, the β subunit ($M_r = 45,000$–58,000) is a glycoprotein, the function of which still remains unknown. The interaction of cardiac glycosides with Na^+,K^+-ATPase occurs at the external face of the protein. The positive inotropic effect of digitalis is the indirect consequence of the inhibition of cardiac Na^+,K^+-ATPase.[1] Because of the usefulness of Na^+,K^+-ATPase to biochemists, cellular biologists, physiologists, and pharmacologists, it has become important to identify the domain of the enzyme which interacts with digitalis. For this purpose, several digitalis affinity labels have been designed. They belong to two categories: (1) alkylating derivatives which contain a highly reactive chemical group and (2) derivatives which contain a photoactivatable group which is converted into a very reactive intermediate upon photolysis.

A First Generation of Derivatives to Affinity Label the Digitalis-Binding Site

Alkylating derivatives of cardiac glycosides, like oxidized ouabain[2] or haloacetyl derivatives of strophanthidin[3] or hellebrigenin,[4] carry electrophilic groups that need to find a suitably positioned nucleophile at the binding site on Na^+,K^+-ATPase to be able to react with the protein moiety. These derivatives usually have poor stability in solution. These two features have strongly limited the use of such compounds which turn out to be poorly efficient covalent probes for the digitalis site.

---

[1] T. Kazazoglou, J. F. Renaud, B. Rossi, and M. Lazdunski, *J. Biol. Chem.* **258,** 12163 (1983).
[2] C. Hegyvary, *Mol. Pharmacol.* **11,** 588 (1975).
[3] L. E. Hokin, M. Mokotoff, and M. Kupchan, *Proc. Natl. Acad. Sci. U.S.A.* **55,** 797 (1966).
[4] A. E. Ruoho, L. E. Hokin, R. J. Hemingway, and S. M. Kupchan, *Science* **159,** 1354 (1968).

TABLE I
NATURE OF Na$^+$,K$^+$-ATPase SUBUNITS LABELED WITH DIFFERENT COVALENT
DERIVATIVES OF CARDIAC GLYCOSIDES

Nature of the covalent label	Position of the reactive group	Nature of the labeled subunits	Reference
(Nitroazidophenyl)strophanthidin	Steroid moiety	α	7
Oxidized ouabain	Sugar moiety	α	2
(Haloacetyl)strophanthidin	Sugar moiety	—	3
(Haloacetyl)hellebrigenin	Sugar moiety	—	4
(Diazomalonyl)cymarin	Sugar moiety	α	5
(Nitroazidobenzoyl)ouabain	Sugar moiety	α,γ	8, 9
(Nitroazidophenyl)ouabain	Sugar moiety	α,γ	6, 7
(Diazomalonyl)digitoxin	Sugar moiety	α,β	10
(Iodoazido)cymarin	Sugar moiety	α,β,γ	11

Photoactivatable analogs of pharmacologically active drugs are known to be very useful tools. They are chemically inert in solution when maintained in the dark, but they are activated upon adequate irradiation. Most of the covalent labeling of the digitalis site described thus far has been obtained with cardiac glycosides coupled to arylazido or diazido groups.[5-11] Once irradiated these photosensitive groups are converted into nitrene or carbene radicals which react with the surrounding protein side chains. The use of these digitalis affinity labels has led to the conclusion that the α chain is the most important subunit of Na$^+$,K$^+$-ATPase participating to the formation of the cardiac glycoside-binding site. The involvement of other subunits is still controversial. Data summarized in Table I show that the γ proteolipid[6,8,9,11] and even the β subunit[10,11] can be labeled to a lesser extent depending on the position of the reactive group on the digitalis molecule.

The yield of covalent incorporation into purified Na$^+$,K$^+$-ATPase obtained with most of the derivatives of this generation of affinity labels is near 1%. This level of labeling is too low for experiments aiming to locate the binding site more precisely within the α subunit structure of Na$^+$,K$^+$-ATPase. In addition, this low level of covalent labeling creates problems of specificity when using aryl- or diazo-containing derivatives with mem-

[5] A. E. Ruoho and J. Kyte, this series, Vol. 46, p. 523.
[6] T. B. Rogers and M. Lazdunski, *FEBS Lett.* **98**, 373 (1979).
[7] T. B. Rogers and M. Lazdunski, *Biochemistry* **18**, 135 (1979).
[8] B. Forbush, J. H. Kaplan, and J. F. Hoffman, *Biochemistry* **17**, 3667 (1978).
[9] B. Forbush and J. F. Hoffman, *Biochim. Biophys. Acta* **555**, 299 (1979).
[10] C. Hall and A. Ruoho, *Proc. Natl. Acad. Sci. U.S.A.* **77**, 4529 (1980).
[11] J. Lowndes, M. Hokin-Neaverson, and A. E. Ruoho, *J. Biol. Chem.* **259**, 10533 (1984).

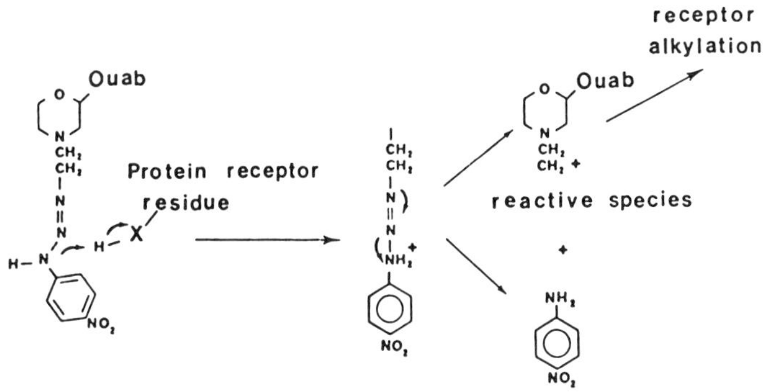

FIG. 1. Probable course for the reaction of NPT-ouabain with the digitalis site of Na^+,K^+-ATPase.

branes containing relatively low amounts of Na^+,K^+-ATPase. Problems of specificity are also due to the fact that irradiation indiscriminately activates specifically and nonspecifically bound ligand as well as free molecules. The hydrophobic character of arylazido or diazo derivatives favors their nonspecific binding to the membrane, contributing to a possible increase in labeling of proteins not directly involved in the formation of the receptor-binding site.

A New Generation of Digitalis Affinity Labels

We describe here the synthesis of two covalent derivatives of ouabain that have a much higher efficiency and specificity than the previous generation of compounds. The first one (p-nitrophenyltriazenylethylenediamine)ouabain (NPT-ouabain), belongs to the category of alkylating agents; the second one (p-aminobenzyldiazonio)ouabain (ABD-ouabain), is a photoactivatable derivative.

Sinnott and Smith[12] have demonstrated that triazene derivative ligands behave as suicide reagents being activated only during their interaction with the receptor protein. At the ligand-binding site a transfer of proton occurs from a side chain of the protein receptor to the triazene moiety of the ligand as shown in Fig. 1; the activated molecule then generates a very reactive carbonium ion. The yield of covalent incorporation (18%) of NPT-ouabain observed on purified *Electrophorus electricus* Na^+,K^+-ATPase[13] is by far higher than that obtained with other digitalis alkylating agents which hardly label the receptor site.

[12] M. L. Sinnott and P. L. Smith, *Biochem. J.* **175**, 525 (1978).
[13] B. Rossi, P. Vuilleumier, C. Gache, M. Balerna, and M. Lazdunski, *J. Biol. Chem.* **255**, 9936 (1980).

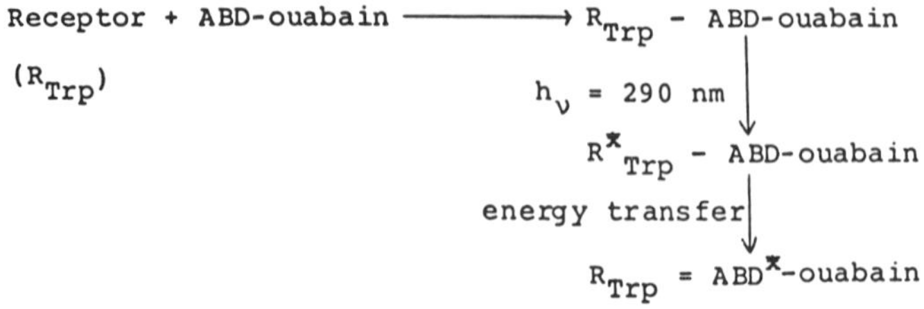

Receptor + ABD-ouabain ⟶ R_{Trp} - ABD-ouabain
(R_{Trp})

h_ν = 290 nm ↓

R^x_{Trp} - ABD-ouabain

energy transfer ↓

R_{Trp} = ABD^x-ouabain

FIG. 2. Strategy of activation of ABD-ouabain by energy transfer.

The idea of the synthesis of ABD-ouabain was linked to the observation by Fortes[14] that energy transfer occurs between a tryptophan residue of the digitalis receptor in the Na^+,K^+-ATPase and anthroylouabain. The strategy of photosuicide affinity labeling has been described by Goeldner and Hirth.[15] Its application for covalent incorporation of ABD-ouabain at the digitalis-binding site is presented in Fig. 2.

Irradiation of the enzyme–ABD-ouabain complex is carried out at a wavelength (λ 295 nm) where little or no direct activation of ABD-ouabain molecules occurs. Tryptophan residues in the receptor structure absorb energy at 295 nm and reemit photons in a range of wavelength near 350 nm corresponding to nearly maximum activation of ABD-ouabain (λ_{max} 375 nm). Therefore, irradiation at 295 nm will indirectly photodecompose by energy transfer via tryptophan residues only the ABD-ouabain molecules which are specifically bound to the digitalis receptor. Nonspecifically bound and free ABD-ouabain molecules will not be activated. This procedure largely eliminates nonspecific covalent labeling. In addition, the ionic character of diazonium salts renders them less hydrophobic than the corresponding azido derivatives, thereby decreasing their tendency for nonspecific binding to the membrane outside the specific receptor site.

Synthesis of NPT-Ouabain

Synthesis of p-*Nitrophenyltriazenylethylenediamine [1-Nitrophenyl-3-(2-aminoethyl)triazene]*

Ethylenediamine (14.65 mg, 0.244 mmol) is dissolved in 3 ml of ice-cold water and the pH adjusted to 7.0 with HCl; p-nitrophenyldiazonium tetrafluoroborate (28.9 mg, 0.122 mmol) is suspended in 2 ml of ice-cold water and then added to the ethylenediamine solution. The suspension is maintained at pH 7.0 by addition of 5 N NaOH. After 5–10 min, when the

[14] P. A. Fortes, *Biochemistry* **16**, 531 (1977).
[15] M. Goeldner and C. Hirth, *Proc. Natl. Acad. Sci. U.S.A.* **77**, 6439 (1980).

pH is stabilized, the solid phase is filtered through a fritted suction funnel and washed with 2 ml of ice-cold water. The pooled filtrates are extracted with 5 ml ice-cold diethyl ether which was previously saturated with water. The aqueous phase is then extracted four times with 5 ml of ice-cold 1-butanol. The combined butanol extracts are evaporated under vacuum at room temperature (15.8 mg, 62% yield). The calculated molecular absorption coefficient ($\varepsilon_{352}^{CH_3OH} = 22{,}000~M^{-1}~cm^{-1}$) is in accordance with that reported by Sinnott and Smith[12] for p-nitrophenylpropyltriazene. The product is either immediately used or stored at $-60°$ and repurified before use on a Sephadex LH-20 column eluted with methanol.

Preparation of Oxidized Ouabain

Ouabain (10 mM) and sodium periodate (11 mM) are incubated in an aqueous solution for 1.5 hr at room temperature. There is a quantitative conversion of ouabain to the oxidized product as determined by (1) the decrease in optical density at 220 nm due to the destruction of periodate [$\varepsilon_{220}^{H_2O}(IO_4^-) = 9500;~\varepsilon_{220}^{H_2O}(IO_3^-) = 1100$]; (2) thin-layer chromatography analysis on silica gel plates of the reaction using [^{3}H]ouabain as a tracer in the reaction with chloroform–methanol (2:1) as the developing solvent, R_f(ouabain) 0.27, R_f(oxidized ouabain) 0.57. At the end of the reaction the solution is passed through a QAE-Sephadex (Pharmacia) column eluted with water in order to remove IO_4^- and IO_3^- ions. The oxidized ouabain peak detected by UV absorption (λ 220 nm) is recovered and lyophilized.

Synthesis of (p-Nitrophenyltriazenylethylenediamine)ouabain (NPT-Ouabain)

Oxidized ouabain (10 mM) is incubated with 20 mM NPT-ethylenediamine in absolute methanol at room temperature. The reaction mixture is adjusted to an apparent pH between 5 and 6 by addition of acetic acid or triethylamine. The Schiff base between the carbonyl function of oxidized ouabain and the amino group of NPT-ethylenediamine is reduced by addition of NaCNBH$_3$ to a final concentration of 20 mM. The reaction (Scheme 1) is followed by analytical thin-layer chromatography on silica gel. Three spots appeared with R_f values of 0.40 (NPT-ouabain I), 0.49 (NPT-ouabain II), and 0.55 (NPT-ouabain III). After 4 hr, the reaction products are separated by preparative thin-layer chromatography on a silica gel 60 plate using chloroform:methanol (4:1 by volume) as the developing system. Cardiac steroids are assayed according to Kedde[16] and primary amines according to Inman and Dintzis.[17] Products con-

---
[16] D. L. Kedde, *Pharm. Weekbl.* **82**, 741 (1947).
[17] J. K. Inman and H. M. Dintzis, *Biochemistry* **8**, 4074 (1969).

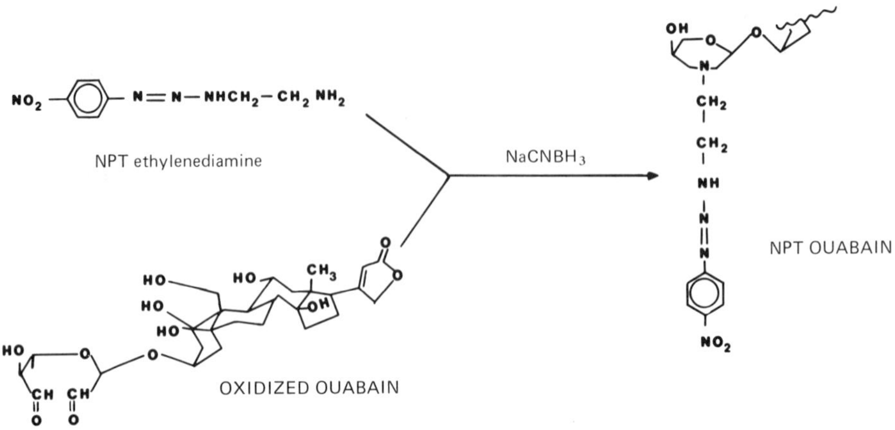

SCHEME 1.

taining a triazene moiety are detected on the plate by long-wave UV absorption.

The three products, each representing approximately 5% of the starting material, are eluted with methanol. After removal of silica gel over glass wool and centrifugation, the methanolic solutions are dried under a nitrogen stream and used immediately or stored in liquid nitrogen for a period not exceeding 2 weeks. The three compounds have identical UV spectra, each consisting of two peaks. The first peak (λ_{max} 362 nm) corresponds to the absorption of the triazenyl group, a shift of λ_{max} occurs during the coupling of NPT-ethylenediamine (λ_{max} 352 nm) to ouabain ($\varepsilon_{362}^{CH_3OH} = 22,000\ M^{-1}\ cm^{-1}$). The second peak ($\lambda_{max}$ 220 nm) corresponds to the absorption of the lactone ring of ouabain and the secondary peak of the p-nitrophenyltriazene moiety.

The order of efficiency of the three derivatives is as follows: NPT-ouabain I > NPT-ouabain II > NPT-ouabain III.[13]

Synthesis of Radiolabeled
(p-Nitrophenyltriazenylethylenediamine)ouabain

Method A. [^{3}H]NPT-ethylenediamine (25.8 nmol, specific radioactivity 30 Ci/mmol) and 200 nmol of oxidized ouabain are incubated in 100 μl of absolute methanol at pH between 6.0 and 7.0. After 15 min NaCNBH$_3$ is added to a final concentration of 10 mM. The reaction products are separated after a 5-hr incubation on an analytical thin-layer silica gel plate as indicated above. Detection of the different products is obtained by counting the radioactivity present in a 0.5-cm-wide lateral strip. The bands containing the desired compounds (R_f 0.4, 0.49, 0.55) are extracted from silica gel with methanol. Analysis by UV absorption and radioactivity counting indicates a total product yield relative to ouabain of 23% (approximately 8% for each of the three derivatives).

Method B. Oxidized [^{3}H]ouabain (2 μmol, specific activity 6 Ci/mmol) and NPT-ethylenediamine (4 μmol) were incubated in 620 μl of absolute ethanol at pH 6.0–7.0. After 15 min NaCNCH$_3$ is added to a final concentration of 10 mM and the reaction allowed to proceed for 4 hr before separation on thin-layer chromatography as indicated above. The final specific radioactivity of oxidized ouabain is about 60% of that of the initial sample of [^{3}H]ouabain. The total yield of NPT-ouabain derivatives is again approximately 20% shared on the three compounds.

Radiolabeled NPT-ouabain fractions are dried and stored as indicated for the unlabeled compound.

Synthesis of ABD-Ouabain

Synthesis of tert-Butyloxycarbonyl(Boc)-p-phenylenediamine. A 0.3 M solution of di-*tert*-butyl dicarbonate in tetrahydrofuran is added with stirring to a 0.3 M solution of freshly sublimated *p*-phenylenediamine (110° at 0.1 mmHg) in tetrahydrofuran. The reaction mixture is stirred for an additional 4 hr. After evaporation of the solvent under vacuum, the residue is taken up in 50 ml of water and extracted twice with an equivalent volume of ethyl acetate. The organic phase is dried over sodium sulfate, and after evaporation it is chromatographed by using a preparative thin-layer silica gel plate [R_f 0.5 with ethyl acetate–hexane (1 : 1) as developing solvent]. The compound is recrystallized from ethyl acetate and a minimum of hexane; the yield is 60%.

Synthesis of ABD-Ouabain. Oxidized ouabain (1 mmol, prepared as described above) in absolute methanol (20 ml) is incubated with 1 mmol of *tert*-butyloxycarbonyl-*p*-phenylenediamine. Acetic acid is added to bring the apparent pH between 5 and 6. After addition of sodium cyanoborohydride to a final concentration of 75 mM, the reaction mixture is stirred for 4 hr at room temperature. The solvent is removed under vacuum and the residue taken up in 50 ml of water. The aqueous phase is then extracted 5-fold with 20 ml of chloroform and the pooled organic phases are dried over sodium sulfate. Removal of the solvent under vacuum leads to a syrupy residue (about 800 mg) which is a mixture of three main compounds on analytical thin-layer silical gel chromatography: R_f 0.76 corresponds to the starting amine; R_f 0.46 corresponds to the coupled aromatic amino ouabain derivative (the spot is detected with a UV lamp at 254 nm); R_f 0.06 corresponds to the starting oxidized ouabain developed in a chloroform–methanol (9:1) solvent. A minor unknown compound appears at R_f 0.53. Chromatography on a silica gel column (50 g of silica gel) with the same eluent separates the desired pure compound (180 mg; yield 20%) in addition to a comparable amount which is contaminated with the unknown compound (see Scheme 2).

Diazotization of the Aromatic Amine. The amino-protecting *tert*-butyloxycarbonyl (Boc) group is removed by stirring in a 0.2 *M* trifluoroacetic acid solution (0.5 ml) at $-5°$ for 45 min under nitrogen. The free aromatic amine was then directly diazotized by treatment with sodium nitrite (10% excess) at $-10°$ using a salt–ice bath. Solid nitrite was added during 20 min in the absence of light. After stirring for another 20 min under the same conditions, trifluoroacetic acid is eliminated by lyophilization. The yellow residue is taken up in 0.3 *M* tetrafluoroboric methanolic solution (0.6 ml) and immediately chromatographed on a BioGel P-2 column (100 g) and eluted with water in the absence of light. After a 80-ml head fraction, the diazonium salt was eluted in four fractions of 15 ml; estimated yield by UV analysis: 60%.

Synthesis of Radioactive ABD-Ouabain. A 33 m*M* methanolic solution of oxidized [^{3}H]ouabain (1 mCi, specific radioactivity 1 Ci/mmol) is incubated in an ice bath with an equivalent amount of *tert*-butyloxycarbonyl-*p*-phenylenediamine and sodium cyanoborohydride at a final concentration of 66 m*M*. The pH is adjusted to 6.2 with a 2% acetic acid solution in methanol, and the reaction mixture kept in the cold overnight. The crude reaction mixture is chromatographed directly on an analytical thin-layer silica gel plate (about 6 cm wide and 10 cm high) with chloroform–methanol (8.5:1.5) as the developing solvent. A 0.5-cm-wide lateral strip is removed and analyzed for radioactivity. The band containing the desired compound (R_f 0.54 detectable with an UV lamp at 254 nm) represents about 40% of the total radioactivity. Extraction of the compound from silica gel with methanol results in a 17% yield. After removal of the silica gel over glass wool and centrifugation, the methanolic solution is

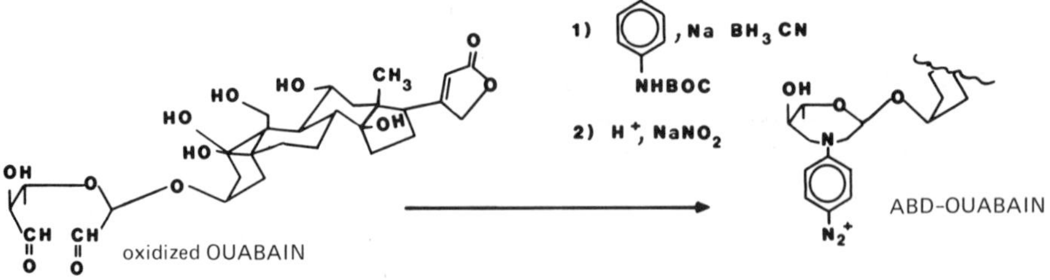

SCHEME 2.

dried under a nitrogen stream and used directly for the diazotization. The procedure of diazotization is the one described for the unlabeled compound (see above); sodium nitrite is added in a water solution (13 mM). The photochemical decomposition of an aliquot of the radiolabeled ABD-ouabain gives the same mixture of compounds as the photodecomposition of the unlabeled sample.

The overall yield was 12% for a compound having a specific radioactivity of 1 Ci/mmol.

Covalent Labeling of the Digitalis Site

NPT-Ouabain. Equilibrium dissociation constants of NPT-ouabain at the digitalis-binding site of purified Na$^+$,K$^+$-ATPases from *Electrophorus electricus* and rabbit kidney were 88 and 30 nM, respectively.[13]

A typical covalent labeling experiment was carried out as follows. The fraction containing Na$^+$,K$^+$-ATPase (50–500 µg/ml) was incubated in 1 ml of triethanolamine buffer (50 mM, pH 7.4) containing 2 mM ATP, 2 mM MgCl$_2$, 100 mM NaCl, 1 mM dithiothreitol, and 1 to 5 mM [^{3}H]NPT-ouabain (0.2–4 Ci/mmol) at 25°. Nonspecific covalent labeling was estimated in experiments where enzyme preparations were first incubated with 1 mM unlabeled ouabain for 10 min. The extent of covalent incorporation was followed by addition of an aliquot (2–20 µg of protein) of the reaction medium to 1 ml of a 5% trichloroacetic solution. After 15 min at 0° the precipitated protein was filtered through glass fiber filters (Whatman) GF/C and rinsed twice with 5 ml of a 50 mM triethanolamine buffer (pH 7.5). The filters were then dispersed in the scintillation liquid and counted. After 1 to 4 hr, depending on the protein concentration used, the extent of covalent incorporation reached a plateau value of 18% with purified Na$^+$,K$^+$-ATPase from *Electrophorus electricus* and 12% with the purified enzyme from rabbit kidney. Under these conditions, only the α subunit was specifically labeled with this affinity probe even when crude microsomal fractions were used.[13]

ABD-Ouabain. The dissociation constant of the complex formed between dog kidney Na$^+$,K$^+$-ATPase and ABD-ouabain was 50 nM, compared to 30 nM for ouabain.[18]

A typical labeling with ABD-ouabain was carried out as follows. Purified Na$^+$,K$^+$-ATPase or a crude microsomal fraction of the electric organ from *Electrophorus electricus* (0.25–0.75 mg/ml) was incubated for 45 min in the absence of light in 1 ml of 50 mM TEA–HCl buffer (pH 6.8)

---

[18] M. Goeldner, C. Hirth, B. Rossi, G. Ponzio, and M. Lazdunski, *Biochemistry* **22**, 4685 (1983).

containing 2 mM MgCl$_2$, 100 mM NaCl, and 1 mM [^{3}H]ABD-ouabain (1 Ci/mmol). The samples were irradiated in a quartz cell with a 1-cm light pathway by using a monochromatic light from a 1000-W Xe/Hg lamp at 290 nm (incident energy 20 mV). During irradiation the reaction solution was continuously maintained under magnetic stirring and kept at 4°. Aliquots containing 4 µg of protein were withdrawn at different times and added to 1 ml of 5% trichloroacetic acid solution. The precipitated protein was filtered through glass fiber filters (Whatman GF/C) and rinsed twice with 5 ml of 50 mM triethanolamine buffer (pH 7.5). The extent of covalent incorporation was estimated by counting the radioactivity trapped on filters. The covalent incorporation reached a plateau value corresponding to about 8% of the total ouabain-binding sites available in the incubation medium after 15 min of irradiation. Nonspecific covalent labeling was estimated in experiments in which enzyme preparations were first incubated with 1 mM unlabeled ouabain for 10 min. Like NPT-ouabain, ABD-ouabain labeled only the α subunit of the Na$^+$,K$^+$-ATPase even when a crude microsomal fraction of electric organ from *Electrophorus electricus* was used as the source of Na$^+$,K$^+$-ATPase.[18]

Determination of Location of the Digitalis-Binding Domain on the α Subunit

In the presence of KCl the catalytic α subunit of dog kidney Na$^+$,K$^+$-ATPase is cleaved by trypsin in two main fragments: an N-terminal peptide of M_r 41,000 and a C-terminal peptide of M_r 58,000.[19]

Purified Na$^+$,K$^+$-ATPase from rabbit kidney or dog kidney was labeled as described above and centrifuged at 45,000 rpm for 45 min. The pellet was resuspended in 25 mM imidazole–HCl (pH 7.4), 1 mM EDTA, and centrifuged again. The final pellet was resuspended in a medium containing 25 mM imidazole, 1 mM EDTA, and 100 mM KCl (pH 7.4) to a final protein concentration of 1 mg/ml. The labeled enzyme was then submitted to a mild proteolysis by addition of diphenylcarbamoyl chloride-treated trypsin (60 µg/ml) for 7 min at 37°. The reaction was stopped by addition of a 2-fold excess of soybean trypsin inhibitor. After separation of the tryptic polypeptides by sodium dodecyl sulfate-polyacrylamide gel electrophoresis, most of the radioactivity initially incorporated in the α subunit was recovered in a polypeptide of apparent M_r 41,000. No significant incorporation of radioactivity was visible in the C-terminal polypeptide of M_r 58,000. The mapping of the digitalis-binding domain on the α subunit is presented in Fig. 3, along with other important sites of the

[19] P. L. Jørgensen, *Biochim. Biophys. Acta* **466**, 97 (1977).

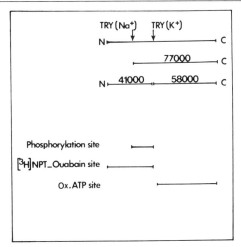

FIG. 3. Mapping of different sites of the α subunit of Na$^+$,K$^+$-ATPase.

enzyme for comparison.[20,21] The same N-terminal peptide was labeled with [^{3}H]NPT-ouabain[22] and [^{3}H]ABD-ouabain.[18]

Acknowledgments

This work was supported by the Centre National de la Recherche Scientifique, the Institut National de la Santé et de la Recherche Médicale (PRC 13106), the Délégation Générale à la Recherche Scientifique et Technique (Grant 81.E.1204), and the Fondation pour la Recherche Médicale.

[20] J. Castro and R. A. Farley, *J. Biol. Chem.* **254**, 2221 (1979).
[21] G. Ponzio, B. Rossi, and M. Lazdunski, *J. Biol. Chem.* **258**, 8799 (1983).
[22] B. Rossi, G. Ponzio, and M. Lazdunski, *EMBO J* **7**, 859 (1982).

[31] Determination of Quaternary Structure of an Active Enzyme Using Chemical Cross-Linking with Glutaraldehyde

By WILLIAM S. CRAIG

Na$^+$,K$^+$-ATPase is the membrane-bound enzyme responsible for catalyzing the active transport of Na$^+$ and K$^+$ across the plasma membrane. While it has been shown that the two polypeptides, designated α and β, that constitute the enzyme exist in an equimolar ratio within the same molecular complex,[1] uncertainty has existed over the definition of the smallest unit of the enzyme necessary to catalyze active transport.[2] A resolution of this conflict might have a profound influence on arguments supporting currently popular theories that suggest quaternary structures for enzymes that catalyze active transport.[3,4]

The determination of the oligomeric structure of Na^+,K^+-ATPase, and most other membrane-bound proteins, usually involves one of three approaches: measurement of the hydrodynamic properties of a complex of lipid, detergent, and protein, measurement of the stoichiometry of ligand binding, or chemical cross-linking. The first approach suffers from the substantial corrections that are necessary to account for bound lipid, detergent, and (if the protein is glycosylated) for covalently attached carbohydrate.[5,6] Interpretation of results from ligand-binding experiments can be confused by the problems of impure preparations or inadequate analytical techniques.[7–9] Finally, substantial uncertainty resides in conclusions drawn from the results of chemical cross-linking when a quantitative reaction is not observed.[10–12] These shortcomings, particularly as they apply to membrane-bound proteins, have been discussed in detail elsewhere.[1,2]

This chapter presents methods for the preparation of a soluble, catalytically competent form of Na^+,K^+-ATPase through the use of the homogeneous, nonionic detergent octaethylene glycol dodecyl ether ($C_{12}E_8$). A quantitative chemical cross-linking assay that utilizes the reagent, glutaraldehyde, also is described. This assay is shown to be capable of an unambiguous assessment of the distribution of Na^+,K^+-ATPase among the various quaternary states present in a solution of detergent.

Reagents. The glutaraldehyde used in these methods was an electron microscopy grade supplied as an 8% solution by Polysciences, Inc. However, in a comparison experiment, glutaraldehyde from several other

---

[1] W. S. Craig and J. Kyte, *J. Biol. Chem.* **255**, 6262 (1980).
[2] J. Kyte, *Nature (London)* **292**, 201 (1981).
[3] S. J. Singer, *Annu. Rev. Biochem.* **43**, 805 (1974).
[4] J. Kyte, *J. Biol. Chem.* **250**, 7443 (1975).
[5] D. F. Hastings and J. A. Reynolds, *Biochemistry* **18**, 817 (1979).
[6] J. R. Brotherus, J. V. Møller, and P. L. Jørgensen, *Biochem. Biophys. Res. Commun.* **100**, 146 (1981).
[7] C. Monteilhet and D. M. Blow, *J. Mol. Biol.* **122**, 407 (1978).
[8] P. L. Jørgensen, *Biochim. Biophys. Acta* **466**, 97 (1977).
[9] G. L. Petersen, R. D. Ewing, S. R. Hootman, and F. P. Conte, *J. Biol. Chem.* **253**, 4762 (1978).
[10] J. Kyte, *J. Biol. Chem.* **250**, 7443 (1975).
[11] G. Giotta, *J. Biol. Chem.* **251**, 1247 (1976).
[12] S. M. Periyasamy, W.-H. Huang, and A. Askari, *J. Biol. Chem.* **258**, 9878 (1983).

sources (J. T. Baker, Sigma Chemical Co., and Eastman Kodak) proved equally effective at providing quantitative cross-linking of the polypeptides. $C_{12}E_8$ was purchased from Nikkol Chemical Company.

Enzyme Assay. The assays for strophanthidin-sensitive Na^+,K^+-ATPase activity, determined either by dilution into detergent-free medium or by direct addition of assay substrates (0.1 vol), have been described.[13]

Preparations of Detergent-Dispersed Na^+,K^+-ATPase

Supernatant–$C_{12}E_8$ Enzyme. Membrane-bound enzyme, purified as previously described,[14] is diluted at 22° into 30 mM imidazolium chloride, pH 7.1, containing KCl and glycerol. A solution of $C_{12}E_8$ in water is added a drop at a time while vortexing the mixture, such that the final concentration of $C_{12}E_8$ is 3 mg (mg of protein)$^{-1}$ and the final concentrations of KCl and glycerol are 0.1 M and 10%, respectively. Without further incubation, the mixture is centrifuged at 100,000 g for 90 min either in a Beckman airfuge, if the volume is less than 1 ml, or, with larger volumes, in a type Ti-60 Beckman rotor. Because the pellet is very flocculent, when the larger rotor is used the sample is layered onto a cushion of 30% sucrose and submitted to centrifugation at 12–15°. The cloudy interface is broad and care must be taken to exclude it when removing the clear supernatant above. Usually, 80% of the supernatant layer can be conveniently retrieved.

In order to detect Na^+,K^+-ATPase activity, this preparation must be assayed by dilution into a solution free of detergent, and the final concentration of the detergent that is carried over must be below its critical micelle concentration.[14] When this is accomplished, the preparation should have a specific activity at 37° of 1000–1500 μmol of P_i (mg of protein)$^{-1}$ hr^{-1}.

Soluble Na^+,K^+-ATPase. This preparation, capable of enzymatic turnover without dilution, is produced by a modification of the above procedure. To a suspension of purified membrane-bound enzyme at a concentration of 10 mg ml^{-1} are added small volumes of concentrated solutions of NaCl, glycerol, and imidazolium chloride, pH 7.1, such that the final concentration of protein in the reaction mixture is not less than 4 mg ml^{-1}.

The following tabulation presents a typical preparation and gives the concentrations of solutions, the volumes added, and the order of their addition.

[13] W. S. Craig, *Biochemistry* **21**, 5707 (1982).
[14] W. S. Craig, *Biochemistry* **21**, 2667 (1982).

Step	Volume (μl)
Purified membrane-bound Na$^+$,K$^+$-ATPase (10 mg ml^{-1}) in 40 mM imidazolium chloride, pH 7.1	175
50% (w/v) glycerol	65
0.5 M NaCl	60
100 mM C$_{12}$E$_8$	45
20 mM imidazolium chloride, pH 7.1	90

The detergent is added slowly while the mixture is rapidly vortexed. The imidazolium chloride, pH 7.1 (90 μl), is added last, with vortexing, and the mixture is immediately subjected to centrifugation at 100,000 g for 60 min. The clear supernatant is removed, dialyzed at 3° against two changes (500 ml each) of 20 mM imidazolium chloride, pH 7.1, containing 5% glycerol, and stored at 3°. The preparation may be diluted up to 10-fold with 40 μM C$_{12}$E$_8$ without an effect on either specific activity or oligomer content. Na$^+$,K$^+$-ATPase does not turn over in the presence of micellar C$_{12}$E$_8$.[14] Therefore, in order to measure enzymatic activity initiated by the addition of substrates (0.1 vol), it is necessary to lower the concentration of free detergent in the sample to a value below the critical micelle concentration (90 μM).[15] This is accomplished in the above protocol by the final addition of buffer after the protein has been dispersed from the membrane. Since it is to be expected that when different sources and preparations of membrane-bound Na$^+$,K$^+$-ATPase are employed, different amounts of C$_{12}$E$_8$ will be lost to the pellet during the subsequent centrifugation, it may be necessary to adjust the volume of the buffer added. The proper amount is the minimum volume of buffer needed for the enzyme to display maximum Na$^+$,K$^+$-ATPase activity upon direct addition of assay substrates (0.1 vol). It should also be kept in mind that dialysis has little effect on the concentration of free detergent (<2% loss) because of the large size of C$_{12}$E$_8$ micelles (65,000).[16]

Nature of the Glutaraldehyde Reaction

Experiments using glutaraldehyde as a fixative, a reagent for the immobilization of proteins, or a coupling reagent are plentiful. A large number of references within volumes of this series alone can be cited (see, for example, Vols. 32, 44, and 46). The uncertainty of the chemistry of its

---

[15] A. Helenius, D. R. McCaslin, E. Fries, and C. Tanford, this series, Vol. 56 [63].
[16] C. Tanford, Y. Nozaki, and M. F. Rohde, *J. Phys. Chem.* **81,** 1555 (1977).

reaction with primary amines[17-20] should not inhibit its use in studies of proteins. Its particularly promiscuous nature and the relative abundance of the ε-amino groups of lysine residues on the surface of proteins make it useful and valuable in a protein chemist's arsenal of modification reagents. Indeed, in recent studies on a soluble, cytoplasmic protein, namely lactate dehydrogenase, chemical cross-linking with glutaraldehyde was shown to accurately assess the distribution of noncovalent oligomers present in a given solution.[21] In that case, and as will be demonstrated here, various tests were performed to prove the cross-linking results accurately assessed the state of oligomerization, and in no case is knowledge of the mechanism of the cross-linking reaction necessary.

Glutaraldehyde Cross-Linking Assay

After a short incubation to establish the desired temperature, cross-linking in all samples is initiated by the addition of 0.1 vol of concentrated glutaraldehyde. Typically, a reaction mixture containing a final concentration of 8 mM glutaraldehyde is incubated for 45 min at 22°, after which the reaction is quenched with an 8–10 M excess of glycine, pH 9.5. Often the sample becomes discolored during the quenching reaction. The pale yellow color, however, does not interfere with subsequent electrophoretic analysis. After a 5-min incubation at 22°, 20% SDS is added [>5 mg of SDS (mg of protein)$^{-1}$] to denature the polypeptides and prepare them for electrophoresis. Placing the sample in a boiling water bath, a procedure commonly employed after the denaturation of polypeptides with SDS, should be avoided unless proteolytic activity is a severe problem. Spurious aggregation of the cross-linked complexes has occasionally been observed when the cross-linked mixtures are heated prior to electrophoresis.[22]

The distribution of the products of a typical cross-linking reaction, visualized by staining with Coomassie brilliant blue, is shown in the scan presented in Fig. 1A. The smallest product of the reaction, component 1, has been shown by a number of criteria to be a covalent heterodimer of one α and one β polypeptide.[1,14] The apparent polymer lengths of the remaining components, 2–4, are approximately integral multiples of the apparent polymer length of the heterodimer, when compared with the

---

[17] F. A. Quiocho, this series, Vol. 44 [38].
[18] P. Monson, P. Germain, and H. Mazarguil, *Biochimie* **57**, 1281 (1975).
[19] P. M. Hardy, G. J. Hughes, and H. N. Rydon, *J. Chem. Soc., Perkin Trans. 1* 2282 (1976).
[20] D. T. Cheung and M. E. Nimni, *Connect. Tissue Res.* **10**, 187 (1982).
[21] R. Hermann, R. Jaenicke, and R. Randolph, *Biochemistry* **20**, 5195 (1981).
[22] W. S. Craig, unpublished observations.

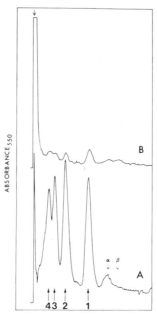

FIG. 1. Scans of SDS–polyacrylamide gels of Na^+,K^+-ATPase cross-linked with glutaraldehyde in $C_{12}E_8$ solution (A) and in the membrane-bound form (B). A sample of supernatant-$C_{12}E_8$ enzyme was cross-linked with 8 mM glutaraldehyde for 45 min at 22°. After the addition of glycine to quench the reaction, the enzyme was denatured with SDS and submitted to electrophoresis on a 3.6% polyacrylamide gel. The gel was stained and scanned at 550 nm. An equal amount of enzyme in membrane-bound form was cross-linked identically and also submitted to electrophoresis. The normal migratory positions of the α polypeptide and β polypeptide are indicated. The positions of the new products observed on the gel are indicated by the arrows labeled 1, 2, 3, and 4. The direction of electrophoresis is from left to right with the upper left arrow marking the top of the gels. From Craig,[14] with permission.

mobilities of the consecutive, noncovalent polymers of bovine serum albumin as standards.[23] Taken together with the observation that all the components demonstrate positive staining for carbohydrate,[14] and therefore contain the β polypeptide, these results indicate that the components 1, 2, 3, and 4, the electrophoretically identified products of the cross-linking reaction, represent the covalent complexes α–β, $(\alpha$–$\beta)_2$, $(\alpha$–$\beta)_3$, and $(\alpha$–$\beta)_4$, respectively, produced from the cross-linking of the noncovalent complexes $\alpha\beta$, $(\alpha\beta)_2$, $(\alpha\beta)_3$, and $(\alpha\beta)_4$ existing in the initial solu-

[23] Noncovalent complexes of serum albumin are most conveniently prepared by denaturing 2–3 μl of a 100 mg ml^{-1} solution of the protein with an equal volume of 20% SDS. The volume is raised to 50 μl by the addition of 0.1% SDS and the sample is submitted to electrophoresis.

tion. These complexes will hereafter be referred to as monomer, dimer, trimer, and tetramer, respectively. Furthermore, when the amount of protein applied to the gel is maintained within the linear range of dye binding, the areas of the individual components, determined from gel scans (Fig. 1A), are directly proportional to the relative weight concentrations of those components that were present in the initial solution.

The results of several experiments support this conclusion that the cross-linking assay provides an accurate method for cataloging macromolecular complexes present in a given detergent solution prior to the addition of the glutaraldehyde. The last five lines in Table I show the effects, within a cross-linking assay, of a variation in the concentration of protein. No changes can be observed in the relative amounts of the covalent oligomers when the concentration of protein is varied by a factor of 10. Therefore, no intermolecular collisions are participating in the cross-link-

TABLE I
EFFECT OF EXPERIMENTAL CONDITIONS ON THE RELATIVE AMOUNTS
OF COVALENT OLIGOMERS OBSERVED WITH
GLUTARALDEHYDE CROSS-LINKING

Experimental conditions[b]	Integrated area[a]			
	$\alpha-\beta$	$(\alpha-\beta)_2$	$(\alpha-\beta)_3$	$(\alpha-\beta)_4$
1.5 mM glutaraldehyde[c]	21	13	3	—[d]
6 mM glutaraldehyde[c]	30	22	13	7
8 mM glutaraldehyde[c]	34	27	18	14
12 mM glutaraldehyde[c]	30	27	19	16
16 mM glutaraldehyde[c]	31	28	19	15
10° (90 min)[e]	40	30	15	9
15° (60 min)[e]	43	29	16	10
22° (45 min)[e]	39	30	16	10
30° (45 min)[e]	37	28	17	8
0.5 mg ml^{-1} protein[f]	41	33	25	—[d]
0.25 mg ml^{-1} protein[f]	46	33	21	—[d]
0.125 mg ml^{-1} protein[f]	43	32	14	—[d]
0.071 mg ml^{-1} protein[f]	51	35	14	—[d]
0.05 mg ml^{-1} protein[f]	47	35	18	—[d]

[a] Percentage of total staining on the gel.
[b] Three separate experiments (c–e) are reported.
[c] Reaction time was 60 min at 22°.
[d] No measurable tetramer.
[e] Parentheses indicate the time of reaction with 8 mM glutaraldehyde.
[f] Reaction time was 60 min at 22° and with 8 mM glutaraldehyde. From Craig,[14] with permission.

ing reaction, and there is no evidence of concentration-dependent interconversion of complexes during the time of the assay. Furthermore, when membrane-bound enzyme is cross-linked with 8 mM glutaraldehyde for 45 min at 22°, none of the covalent products is capable of entering the SDS gel (Fig. 1B). This indicates that the glutaraldehyde is sufficiently reactive to cross-link all of the neighboring proteins in a hydrodynamic complex, in this case the whole membrane fragment. Finally, when a sample of uncross-linked supernatant–$C_{12}E_8$ enzyme is submitted to centrifugation on a sucrose gradient containing $C_{12}E_8$, all of the oligomers can be separated on the basis of their sedimentation coefficients and identified with the cross-linking assay (Fig. 2). The separation of the uncross-linked oligomers on the sucrose gradient provides independent evidence for the existence of the complexes that the cross-linking assay stated were in the original solution of supernatant–$C_{12}E_8$ enzyme. In addition, when a sample of supernatant–$C_{12}E_8$ enzyme is cross-linked prior to centrifugation on a sucrose gradient, the distribution and the relative amounts of the various complexes is very similar.[22] Taken together, these results indicate

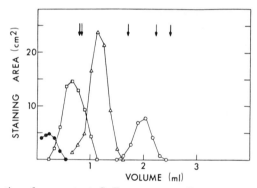

FIG. 2. Separation of supernatant–$C_{12}E_8$ enzyme on a linear sucrose gradient containing $C_{12}E_8$. A sample (100 μl) of Na$^+$,K$^+$-ATPase dissolved in a solution of $C_{12}E_8$ was clarified by centrifugation and submitted to a second centrifugation through a 5–20% linear sucrose gradient containing $C_{12}E_8$ (16 mM). Fractions collected from the gradient were analyzed for the presence of the various complexes, monomer, dimer, trimer, and tetramer, with the glutaraldehyde cross-linking assay. Stained SDS–polyacrylamide gels of each of the cross-linked fractions were scanned in order to quantify the amount of each complex present in the individual fractions. Indicated in the graphs is the absolute staining area of the covalent complexes α–β (-○-), $(\alpha$–$\beta)_2$ (-△-), $(\alpha$–$\beta)_3$ (-□-), and $(\alpha$–$\beta)_4$ (-●-) present in each fraction assayed. The recovered volume from the gradient was 3.7 ml and the bottom is to the left. The positions of sedimentation standards from a companion gradient are indicated on the graph by the upper arrows. The standards used were bovine serum albumin (4.7 S), human transferrin (6.1 S), lactate dehydrogenase (7.2 S), bovine catalase (11.3 S), and aspartate carbamoyltransferase (11.7 S), in that order, right to left. From Craig,[14] with permission.

that the complexes present in the supernatant–$C_{12}E_8$ enzyme preparations are discrete and stable entities, and the distribution of the enzyme among them can be quantitatively determined with the glutaraldehyde cross-linking assay.

The quantitative cross-linking assay has advantages over other methods used to determine the distribution of oligomers of Na^+,K^+-ATPase and other multimeric proteins. First, it does not require uncommon equipment. Thus, it can be performed in any laboratory with standard equipment for electrophoresis. Second, the assay can be performed under a variety of experimental conditions. Once the lowest concentration of glutaraldehyde that produces an unchanging pattern of covalent complexes has been determined (first five lines of Table I), higher concentrations of the reagent can be used to perform a more rapid assay (Table I and Figs. 3 and 4). In addition, as the results in Table I indicate, the temperature of the reaction can be varied without affecting the pattern of covalent complexes. Finally, the cross-linking reaction can be performed in the presence of substrates or other effectors, provided none is a strong nucleophile. These properties of the assay provide a flexibility unshared by other methods.

Applications

Analysis of Diverse Preparations. In several laboratories Na^+,K^+-ATPase has been dispersed from the membrane with nonionic detergents as an enzymatically stable, soluble preparation.[5,6,24] In each case, a claim, based on measurements of hydrodynamic properties, was made that the preparation in question was a monodisperse solution of dimers[5,24] or monomers.[6] When samples prepared in an identical manner are submitted to the cross-linking assay and the relative proportions of all the complexes present are assessed (Table II), it is clear that each of these solutions contains a rather complex mixture of different oligomers. These experiments demonstrate that the cross-linking assay is capable of detecting oligomers that are overlooked in interpretations of gel filtration and ultracentrifugation data.[5,6,24,25] In addition, when Na^+,K^+-ATPase was dissolved in solutions of deoxycholate, 3-[(3-cholamidopropyl)dimethylammonio]-1-propane sulfonate, Triton X-100, or detergents of the Brij series, the solutions all contained a mixture of oligomers.[14] Although the relative

---

[24] M. Esmann, C. Christiansen, K. Karlsson, G. C. Hansson, and J. C. Skou, *Biochim. Biophys. Acta* **603**, 1 (1980).

[25] J. R. Brotherus, J. Jacobsen, and P. L. Jørgensen, *Biochim. Biophys. Acta* **731**, 290 (1983).

TABLE II
OLIGOMER CONTENT OF PREVIOUSLY REPORTED PREPARATIONS OF Na^+,K^+-ATPase IN DETERGENT SOLUTIONS[a]

Detergent	Mg detergent mg protein	Integrated area[b]				
		$\alpha-\beta$	$(\alpha-\beta)_2$	$(\alpha-\beta)_3$	$(\alpha-\beta)_4$	$(\alpha-\beta)_5$
Lubrol WX[c]	15	12	24	14	22	22
$C_{12}E_8$[d]	2	25	28	23	20	—[e]
$C_{12}E_8$[f]	2.7	54	24	17	4	—[e]

[a] From Craig,[14] with permission. Conditions reported earlier were duplicated.
[b] Percentage of total staining on the gel.
[c] Hastings and Reynolds.[5]
[d] Esmann et al.[24]
[e] No measurable pentamer.
[f] Brotherus et al.[6]

amounts of the complexes changed as the type and concentration of the detergent was changed, the cross-linking assay was invariant in its ability to detect and quantify each component.

Cross-Linking during Enzymatic Turnover. As stated above, the quantitative cross-linking assay can be used in the presence of a full complement of enzymatic substrates, provided none is a potent nucleophile. Other methods employed to determine oligomeric structure, based on measurements of hydrodynamic properties, have failed to address the question: What is the quaternary structure of Na^+,K^+-ATPase when it is turning over in an enzymatic assay? More to the point, techniques such as radiation inactivation[26,27] and the stoichiometry of ligand binding,[8,9] although representing a more direct approach by measuring protein mass (active site)$^{-1}$, still, in general, suffer from inherent difficulties and inadequacies.[1,2] Only recently have more reliable values for concentrations of active sites become available.[28,29] On the other hand, a sample of soluble Na^+,K^+-ATPase can be cross-linked in the solution used for its enzymatic assay and the covalent products of the reaction separated by gel electrophoresis. Since the integrated areas from a scan of a gel taken at 280 nm yield a quantitative measure of protein distribution,[1] a direct comparison can be made between levels of enzymatic activity and oligomer content. It

[26] G. R. Kepner and R. I. Macey, *Biochim. Biophys. Acta* **163**, 188 (1968).
[27] J. C. Ellory, J. R. Green, S. M. Jarvis, and J. D. Young, *J. Physiol. (London)* **295**, 10P (1979).
[28] E. G. Moczydlowski and P. A. G. Fortes, *J. Biol. Chem.* **256**, 2346 (1981).
[29] J. Kyte, *J. Biol. Chem.* **247**, 7634 (1972).

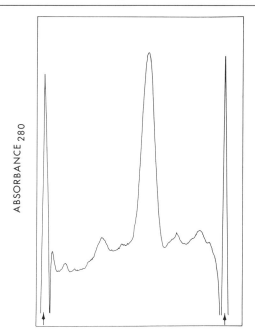

FIG. 3. Scan of an SDS–polyacrylamide gel of cross-linked soluble Na^+,K^+-ATPase. A sample of soluble Na^+,K^+-ATPase was cross-linked during turnover with 80 mM glutaraldehyde at 11°. After denaturation of the enzyme with SDS, the sample was submitted to electrophoresis on a 3.6% polyacrylamide gel. After electrophoresis, the gel was scanned at 280 nm without staining. The direction of electrophoresis is from left to right. From Craig,[13] with permission.

is clear from the gel scan in Fig. 3 that the majority of the enzyme present when it is actually turning over is monomeric.[30] These results provide compelling evidence that Na^+,K^+-ATPase functions as a monomeric unit. In addition, by utilizing the cross-linking assay in the presence of substrates, monomers that possess Na^+,K^+-ATPase activity can be distinguished from inactivated monomers.[13]

Trapping of the Phosphorylated Intermediate. When soluble Na^+,K^+-ATPase is subjected to a rapid cross-linking assay (2 min with 80 mM glutaraldehyde) after a brief incubation with 170 μM [γ-^{32}P]MgATP in the presence of 13 mM NaCl, greater than 80% of the radioactivity incorporated into protein has an electrophoretic mobility coincident with that of

[30] In a companion assay, at 11° and less than saturating concentrations of cations, the soluble Na^+,K^+-ATPase exhibits the same high specific activity as the purified membrane-bound enzyme (18–20 μmol P_i mg^{-1} hr^{-1}).

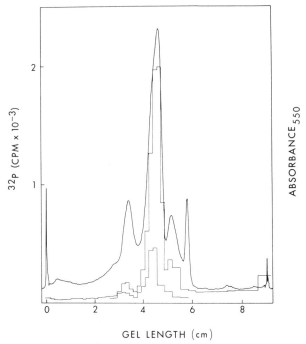

FIG. 4. Scan of and distribution of radioactivity over an SDS–polyacrylamide gel of soluble Na^+,K^+-ATPase phosphorylated with $[\gamma$-$^{32}P]$MgATP. A sample of soluble enzyme was phosphorylated with 170 μM $[^{32}P]$MgATP in the presence of either 13 mM NaCl or 4 mM KCl. Slices of the gels were assayed for ^{32}P and the distribution over the gel of the radioactivity incorporated in the presence of NaCl (upper profile) or KCl (lower profile) is shown superimposed upon a scan of a stained gel of a duplicate, nonradioactive sample. The direction of electrophoresis is from left to right. From Craig,[13] with permission.

the covalent monomer (Fig. 4). Although it might be expected that phosphorylated, soluble enzyme would not maintain its phosphorylated intermediate during the additional 4 min of the cross-linking and quenching reactions it experiences prior to denaturation with SDS, at least 70% of the specific activity (cpm/area of a 550-nm scan) relative to a sample of purified membrane-bound enzyme phosphorylated in an identical manner was recovered.[22] This observation may be related to the enzyme's reported ability to retain Na^+-dependent phosphorylating activity after extensive modification with glutaraldehyde.[31]

[31] D. M. Chipman and A. Levi, *Biochemistry* **22**, 4450 (1983).

Concluding Comments

Some confusion has arisen concerning interpretations of quantitative cross-linking results with Na^+,K^+-ATPase dissolved in solutions of $C_{12}E_8$ because of a recent report that a time-dependent aggregation of α and β polypeptides occurs in these solutions in the absence of a cross-linking agent.[12] In the many control experiments performed by this author, in several different laboratories, cross-linking of the polypeptides of either supernatant-$C_{12}E_8$ enzyme or soluble enzyme has never occurred in the absence of a cross-linking reagent. More to the point, the partial cross-linking reportedly observed in the absence of a cross-linking reagent is irrelevant to the claim that the cross-linking assay, as described here, provides a quantitative catalog of all the macromolecular complexes in a given solution. Any cross-links present beside the one needed to tie covalently two neighboring polypeptides together are irrelevant. Furthermore, since the time-dependent patterns of these aggregates were not affected by dilutions of the sample as large as 20-fold,[12] it can be concluded that, whatever caused it, the partial cross-linking occurred within complexes already present in the solution and was not, therefore, the result of intermolecular collisions. These same complexes would be the ones that the cross-linking assay identifies and quantifies, unambiguously.

Acknowledgment

I thank Jack Kyte (University of California, San Diego), in whose laboratory this work was performed, for his continuing support.

[32] Use of Cross-Linking Reagents for Detection of Subunit Interactions of Membrane-Bound Na^+,K^+-ATPase

By Wu-Hsiung Huang, Sham S. Kakar, Sankaridrug M. Periyasamy, and Amir Askari

The oligomeric structure of Na^+,K^+-ATPase and the existence of subunit interactions accounting for the cooperative kinetics of the enzyme were suggested more than a decade ago.[1] Since then, a great deal of effort has been devoted to the determination of the association state of the

---

[1] A. Askari, *Mol. Cell. Biochem.* **43**, 129 (1982).

enzyme. The highly purified enzyme consists of an α subunit (about 100 kDa) and a β subunit (about 40 kDa). Hence, the question has been whether the functional enzyme of the plasma membrane is the α,β-heterodimer or an oligomer of this dimer. Among the variety of approaches that have been used to resolve this question are chemical cross-linking studies. Although the early experiments showing the covalent cross-linking of the enzyme subunits were readily interpreted to indicate an $(\alpha,\beta)_2$ structure, the belated realization of the limitations of cross-linking experiments with membrane proteins led to the suspicion that such observations could be explained by the random collisions of the enzyme monomers that are highly concentrated within the membrane phase of the purified enzyme.[2] These uncertainties, however, have been resolved by studies of our laboratory.[3-5] The ligand-induced and ligand-modified cross-linkings of the subunits in the presence of various cross-linking reagents, and the kinetics and stoichiometries of these reactions, have clearly indicated that the membrane-bound enzyme is indeed an oligomer of α,β-dimers which exhibits strong interprotomer site–site interactions.[5] What remains to be determined is whether or not the oligomeric structure is essential to the transport function of the enzyme; and if not, what role the protomer interactions play in the regulation of the enzyme function. Further cross-linking studies may be of value in the resolution of these issues. Here, we describe procedures for the covalent cross-linking of the subunits of a purified enzyme preparation from kidney, and show the results of several typical experiments.

Materials. Frozen canine kidneys were obtained from Pel-Freeze Biologicals, Rogers, Arkansas. The membrane-bound enzyme of the kidney medulla was purified according to procedure of Jørgensen,[6] washed repeatedly with 0.25 *M* sucrose, 30 m*M* histidine, 1 m*M* EDTA (pH 6.8), suspended in the same solution, and stored at −30°. The preparations made routinely, and assayed as described,[6] had specific activities of 1000–1500 μmol of ATP hydrolyzed/mg of protein/hr and retained this activity upon storage up to 3 months. Vanadium-free ATP, *o*-phenanthroline, 1,5-difluoro-2,4-dinitrobenzene, and ouabain were obtained from Sigma. [γ-^{32}P]ATP and ^{32}P$_i$ were purchased from New England Nuclear. The former was used as supplied, but the latter was purified[7] before use.

[2] J. Kyte, *Nature (London)* **292,** 201 (1981).
[3] W.-H. Huang and A. Askari, *FEBS Lett.* **101,** 67 (1979).
[4] A. Askari, W.-H. Huang, and J. M. Antieau, *Biochemistry* **19,** 1132 (1980).
[5] S. M. Periyasamy, W.-H. Huang, and A. Askari, *J. Biol. Chem.* **258,** 9878 (1983).
[6] P. L. Jørgensen, *Biochim. Biophys. Acta* **356,** 36 (1974).
[7] A. Askari, W.-H. Huang, and P. W. McCormick, *J. Biol. Chem.* **258,** 3453 (1983).

General Procedure for the Reaction of the Enzyme with Various Cross-Linking Reagents

The enzyme that was stored in a solution containing EDTA and histidine (see above) was freed of these reagents by centrifugation at 100,000 g for 1 hr, and washing in 10 mM Tris–HCl (pH 7.4). The enzyme (0.1 mg of protein) was then added to 0.2 ml of a reaction mixture containing 50 mM Tris–HCl (pH 7.4) and the appropriate concentrations of the cross-linking reagents and other indicated ligands. After incubation at 24° for a desired period, the reaction was stopped in one of two ways: (1) If the cross-linking reaction involved the ^{32}P-labeled α subunit (see below), 1 ml of ice-cold 8% perchloric acid was added, the mixture was centrifuged, and the sediment was washed with cold 0.1% trichloroacetic acid containing 1 mM unlabeled ATP and 10 mM unlabeled P$_i$. The pellet was then dissolved in 5% SDS (without heating), and aliquots were subjected to SDS–polyacrylamide gel electrophoresis at pH 2.4 as indicated below. (2) When the cross-linking reaction involved the unlabeled enzyme, the reaction was terminated by the addition of 0.1 ml of 0.1 M Tris–EDTA (pH 7.2), followed immediately by the addition of 0.1 ml of 20% SDS to solubilize the enzyme. Aliquots of this solution were then subjected to SDS–polyacrylamide gel electrophoresis at pH 7.2.

Gel Electrophoresis

Solubilized samples (5–50 μg of protein), obtained as indicated above, were subjected to disc gel electrophoresis using 5% acrylamide SDS–phosphate gels either at pH 2.4 (for ^{32}P-labeled samples) according to Avruch and Fairbanks,[8] or at pH 7.2 according to Weber and Osborn.[9] Sulfhydryl reagents were omitted from the buffers used for electrophoresis, since some of the cross-linked products are formed through intersubunit disulfide bonds.[4,5] The gels were either stained, or sliced and counted. When necessary, quantitative measurements of stained bands were done with a densitometer.

Ligand-Modified and Ligand-Induced Cross-Linking of the α and β Subunits

Figure 1 shows the results of typical cross-linking experiments in the presence of Cu^{2+} and o-phenanthroline. Depending on the relative con-

---

[8] J. Avruch and G. Fairbanks, *Proc. Natl. Acad. Sci. U.S.A.* **69**, 1216 (1972).

[9] K. Weber and M. Osborn, in "The Proteins" (H. Neurath and R. L. Hill, eds.), Vol. I, p. 179. Academic Press, New York, 1975.

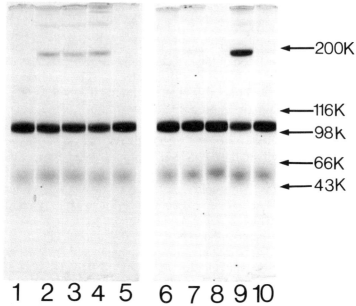

FIG. 1. Ligand-modified and ligand-induced formations of cross-linked α,α-dimer. The enzyme was incubated for 15 min with Cu^{2+} and o-phenanthroline under the standard conditions described in the text. Gel 1 represents the control enzyme not exposed to the crosslinking reagents. Samples represented by gels 2–5 were exposed to 0.25 mM Cu^{2+}, 0.5 mM o-phenanthroline, and the following ligands: gel 2, 100 mM Na^+; gel 3, 25 mM K^+; gel 4, 100 mM Na^+ plus 2 mM ATP; gel 5, 25 mM K^+ plus 2 mM ATP. Samples represented by gels 6–10 were exposed to 0.25 mM Cu^{2+}, 1.25 mM o-phenanthroline, 2 mM Mg^{2+}, and the following ligands: gel 6, no additional ligands; gel 7, 100 mM Na^+; gel 8, 25 mM K^+; gel 9, 100 mM Na^+ plus 2 mM ATP; gel 10, 25 mM K^+ plus 2 mM ATP. Electrophoresis was done at pH 7.2.

centrations of the reagents, dramatically different patterns of cross-linking are obtained. With 0.25 mM Cu^{2+} and 0.5 mM o-phenanthroline (gels 1–5), a cross-linked α,α-dimer is obtained regardless of the presence or absence of any of the physiological ligands of the enzyme (gels 2–4); only when K^+ and millimolar concentrations of ATP are present is the formation of this α,α-dimer prevented (gel 5). On the other hand, when 0.25 mM Cu^{2+} and 1.25 mM o-phenanthroline are used (gels 6–10) no cross-linking occurs unless ligands are present that cause phosphorylation of the α subunit. In experiments illustrated in Fig. 1, phosphoenzyme is formed in the presence of Na^+ and ATP (gel 9), but not in the presence of K^+ and ATP (gel 10). Phosphorylation-induced formation of the α,α-dimer in the presence of Cu^{2+} and o-phenanthroline also occurs when the enzyme is phosphorylated with P_i (Fig. 2); and participation of the phosphoenzyme

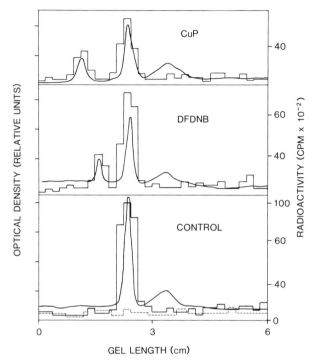

FIG. 2. Phosphorylation-induced formations of cross-linked α,β-dimer and cross-linked α,α-dimer. The enzyme was incubated for 10 min under the standard conditions described in the text in the presence of 0.12 mM $^{32}P_i$, 2 mM Mg^{2+}, and 1 mM ouabain. The control sample contained no cross-linking reagent. The other two samples contained 0.1 mM 1,5-difluoro-2,4-dinitrobenzene (DFDNB), or 0.25 mM Cu^{2+} and 1.25 mM o-phenanthroline (CuP). Electrophoresis was done at pH 2.4. Bars represent the radioactivities of the gel slices, and the curves are the densitometer traces of the stained gels. Direction of electrophoresis was from left to right. The four stained bands shown are, from left to right, the α,α-dimer, the α,β-dimer, the α-monomer, and the β-monomer. Taken from Periyasamy et al.[5]

in the cross-linking reaction may be demonstrated directly if $^{32}P_i$ is used, and gel electrophoresis is conducted at pH 2.4 (Fig. 2).

Ligand-modified and ligand-induced cross-linking may be observed with the use of reagents other than o-phenanthroline–Cu^{2+} complex.[5] For example, when 0.1 mM 1,5-difluoro-2,4-dinitrobenzene is used, there is again no cross-linking unless the enzyme is phosphorylated. With this reagent, however, the product of the phosphorylation-induced cross-linking is an α,β-dimer (Fig. 2). The fact that phosphorylation induces the formation of α,β-dimer in the presence of one reagent and that of α,α-dimer in the presence of another (Fig. 2) is one of several lines of evidence indicating that these cross-linking reactions reflect the existence of a noncovalent oligomer of α,β-dimers within the membrane.[5]

Comments. Experiments similar to those shown here for the purified enzyme may be done with crude membrane preparations. Phosphorylation-induced formations of α,α-dimer and α,β-dimer have been shown in kidney microsomes and in human red cell membranes.[5] Such experiments are possible only with ^{32}P-labeling of the α subunit and the use of acid gels. Studies on the kinetics of cross-linking are also done more accurately with ^{32}P labeling of the enzyme rather than through the measurements of stained bands.[5] It is also appropriate to point out that cross-linking reactions of the membrane-bound enzyme as described here are quite different from those of the detergent-solubilized preparations of the enzyme. For example, while in the membrane-bound enzyme o-phenanthroline–Cu^{2+} complex causes the formation of cross-linked α,α-dimer (Fig. 1), in a deoxycholate-solubilized preparation the same complex produces an α,β-dimer.[10] Another active, but highly unstable, detergent-solubilized preparation undergoes spontaneous cross-linking resulting in number of products, the largest of which is a dimer of α,β-dimer.[5]

Acknowledgments

This work was supported by NIH Research Grant P01HL-36573 awarded by the National Heart, Lung, and Blood Institute, USPHS/DHHS.

---

[10] W. S. Craig and J. Kyte, *J. Biol. Chem.* **255**, 6262 (1980).

Section VIII

Magnetic Resonance Studies of Na^+,K^+-ATPase

[33] Nuclear Magnetic Resonance Investigations of Na^+,K^+-ATPase

By CHARLES M. GRISHAM

Introduction

Na^+,K^+-ATPase is the membrane transport enzyme responsible for the active transport of sodium and potassium across mammalian plasma membranes. The central importance of this enzyme to cell function has stimulated many studies of structure and mechanism. However, the very nature of this system has frequently imposed severe constraints on physical and structural studies. Na^+,K^+-ATPase is an integral membrane protein and as such has been found to be (1) resistant to crystallization, (2) inactive in the absence of lipid[1] or (3) unstable in the presence of detergents,[2] and (4) difficult (at best) to study by optical methods due to its particulate nature. Moreover, the particulate nature of purified preparations of ATPase limits the concentrations that can be attained for spectroscopic studies of this enzyme. In the past few years, creative efforts in several laboratories have begun to overcome these limitations. Thus two-dimensional crystalline arrays of Na^+,K^+-ATPase have been characterized by Jørgensen[3,4] and the enzyme has been solubilized in the detergent $C_{12}E_8$ such that it retains activity for a few hours at 2°.

Nuclear magnetic resonance methods can be an alternative to X-ray diffraction or optical spectroscopy for structural studies with systems such as Na^+,K^+-ATPase. The development over the past 20 years of NMR techniques which rely on rapid chemical exchange of an observed nucleus between sites on macromolecules and the bulk solvent now permits a variety of structural investigations which can be performed on low concentrations of the enzyme. In addition, the development of high-field superconducting NMR spectrometers with wide-bore probes has made possible the observation of very low concentrations of bound substrate nuclei. In this chapter, the details of such investigations with Na^+,K^+-ATPase will be described. No attempt will be made to comprehensively review the theory of NMR, but references to the general theory will be provided.

[1] R. W. Albers, *Annu. Rev. Biochem.* **36**, 727 (1967).
[2] K. Wheeler and R. Whittam, *J. Physiol. (London)* **207**, 303 (1970).
[3] P. L. Jørgensen, *Biochim. Biophys. Acta* **356**, 36 (1974).
[4] P. L. Jørgensen, *Biochim. Biophys. Acta* **694**, 27 (1982).

Methods and Special Considerations

The success of magnetic resonance studies with an enzyme such as Na^+,K^+-ATPase depends intimately on such factors as the purity and state of the enzyme, the concentrations which can be obtained of the enzyme and observed nucleus, the sensitivity of the nucleus, and, in exchanging systems, the exchange rates and affinities for the enzyme of ions, substrates, and inhibitors.

The crucial requirement with respect to enzyme purity in magnetic resonance studies is that contaminants, if any, should not participate in the magnetic interactions themselves, or at least that the behavior of contaminating proteins should be distinguishable from the interactions of interest. In the case of Na^+,K^+-ATPase, the preparation of Jørgensen[3-5] consists of two polypeptides of M_r 90,000 and 35,000[5] with no indications that the contaminating proteins (usually less than 5% of the total protein in the preparation) bind significant amounts of ATP, ouabain, or cations. On the other hand, cation binding by the lipid membrane is a potential problem with this enzyme. The sites for divalent cations on the membrane have been characterized,[6,7] but for the most part the binding of metals to the membrane occurs with low affinity and low specificity.[6] For most purposes, it is of little consequence in NMR and EPR studies of this enzyme.

The concentration limits of Na^+,K^+-ATPase are a somewhat more difficult matter. Owing to the high molecular weight of the functional unit of the ATPase (approximately 250,000), and also to the large amount of lipid associated with the enzyme in even highly purified preparations, the maximum enzyme concentration which can be obtained in solution (i.e., suspension) is approximately 0.3–0.35 mM. This is the approximate concentration of ATPase in a loose centrifuge pellet of ATPase. Higher concentrations might be obtainable if delipidated preparations were used, but the enzyme would probably be unstable in such a state. This concentration limit makes NMR studies of the resonances of the ATPase protein itself unfeasible in all but the most specialized spectrometers. Wide-bore spectrometers accepting 20-mm sample tubes and operating at high field (greater than 300 MHz, for example) might be capable of producing spectra of ATPase protein resonances. However, even in this case, the large number of amino acid residues in this protein (approximately 750 in the α subunit alone) would result in very complex spectra, and it would be unlikely that more than a few resonances could be resolved. On the other

[5] P. L. Jørgensen, *Ann. N.Y. Acad. Sci.* **402**, 207 (1982).
[6] C. M. Grisham and A. S. Mildvan, *J. Biol. Chem.* **249**, 3187 (1974).
[7] C. M. Grisham and A. S. Mildvan, *J. Supramol. Struct.* **3**, 304 (1974).

hand, Fossel et al.[8] have reported the observation of a phosphorus resonance from the ATPase which they ascribe to the phosphorylated intermediate, E–P, which occurs when the terminal phosphate of ATP is transferred to an aspartyl residue of the enzyme during catalysis. This resonance is fortunately shifted with respect to the phosphorus resonance arising from endogenous phospholipids in the preparation.

A far more versatile approach to spectroscopic studies of Na^+,K^+-ATPase lies in the exploitation of rapid chemical exchange of substrates, activators, and inhibitors between solution and sites on ATPase. If exchange is fast compared to the nuclear relaxation rates, $1/T_1$ and $1/T_2$, of the exchanging species, chemical information from the sites on the enzyme (which is in a sense "stored" in the relaxation rates themselves) is effectively transferred to solution. To the extent that the concentration of the exchanging substrate or activator exceeds that of the enzyme, the result is a chemical amplification of the relevant information. One can then extract enzyme site information from the relaxation behavior of the free substrate or activator in solution. There are several NMR experiments which take advantage of this phenomenon. We shall discuss two in detail here. Nuclear relaxation studies with paramagnetic probes permit the determination of site–site distances and detailed geometries of substrates and activators bound to the enzyme. Transferred nuclear Overhauser enhancement (TRNOE) measurements, on the other hand, provide information on the conformation of bound substrates or activators on the basis of dipole–dipole interactions between pairs of bound substrate nuclei, and thus do not depend on interactions with a paramagnetic probe.

Nuclear Relaxation with Paramagnetic Probes

Theory

Unpaired electrons, with magnetic moments which are three orders of magnitude greater than protons, are much stronger magnets. Hence paramagnetic ions and radicals are exceedingly effective in increasing the relaxation rates of ligands. Thus the longitudinal relaxation rate for the protons of water coordinated to the manganese aquocation is 10^5 times greater than that of solvent water. The experimentally determined paramagnetic contributions to the relaxation rates ($1/T_{1p}$, $1/T_{2p}$) are defined as

$$1/T_{1p} = (1/T_1) - (1/T_1)_0 \qquad (1)$$
$$1/T_{2p} = (1/T_2) - (1/T_2)_0 \qquad (2)$$

---

[8] E. T. Fossel, R. L. Post, D. S. O'Hara, and T. W. Smith, *Biochemistry* **20**, 7215 (1981).

where $(1/T_1)$ and $(1/T_2)$ are the measured relaxation rates in the presence of the paramagnetic species and $(1/T_1)_0$ and $(1/T_2)_0$ are the measured relaxation rates in the absence of the paramagnetic species.

The equations relating the experimentally measured parameters $(1/T_{1P}, 1/T_{2p})$ to theoretical parameters were derived from the Bloch equations by Swift and Connick[9] and by Luz and Meiboom[10] to be

$$1/T_{1p} = pq/(T_{1M} + \tau_M) \qquad (3)$$
$$1/T_{2p} = pq/(T_{2M} + \tau_M) \qquad (4)$$

where $1/T_{1p}$ is the paramagnetic contribution to the spin-lattice relaxation rate, $1/T_{2M}$ is the longitudinal relaxation rate in the first coordination sphere of the paramagnet, $1/T_{2p}$ and $1/T_{2M}$ are similarly related to transverse relaxation, pq is the mole fraction of nuclei in the coordination sphere, and τ_M is the mean residence time for nuclei in the first coordination sphere.

When $\tau_M \ll T_{1M}$, distances between nuclei and paramagnetic probes can be calculated from the dipolar effect of the paramagnet on the nucleus. The relationship between T_{1M} and r, the relevant distance, is given by[11]:

$$1/T_{1M} = [qB^6 f(\tau_c)]/r^6 \qquad (5)$$

where q is the coordination number and τ_c is the correlation time defined by

$$(\tau_c)^{-1} = (\tau_r)^{-1} + (\tau_s)^{-1} + (\tau_M)^{-1} \qquad (6)$$

In Eq. (6), τ_r is the time constant for rotation of the paramagnetic complex and τ_s is the electron spin lattice relaxation time. The constant B is given by

$$B = [(2/15)S(S + 1)\gamma_I^2 g^2 \beta^2]^{1/6} \qquad (7)$$

and $f(\tau_c)$ is defined as

$$f(\tau_c) = 3\tau_c/(1 + \omega_I^2 \tau_c^2) + 7\tau_c/(1 + \omega_s^2 \tau_c^2) \qquad (8)$$

From Eqs. (3) and (6), in suitable cases, it is possible to calculate the values of q, τ_M, and r from a study of $1/T_{1p}$ at several frequencies.

[9] T. J. Swift and R. E. Connick, *J. Chem. Phys.* **37**, 307 (1962).
[10] Z. Luz and R. Meiboom, *J. Chem. Phys.* **40**, 2686 (1964).
[11] A. S. Mildvan and M. Cohn, *Adv. Enzymol.* **33**, 1 (1970).

Paramagnetic Probes for Na^+, K^+-ATPase

There are several potentially useful paramagnetic probes for studies with Na^+, K^+-ATPase. These include metal ions bound to the enzyme itself, metal ions coordinated to the ATP substrate, nitroxide spin labels covalently bound to the enzyme itself, or spin label analogs of substrates, inhibitors, or lipids. Studies of this enzyme with spin-labeled lipids are discussed elsewhere in this volume.

One of the most useful paramagnetic probes for NMR studies of Na^+, K^+-ATPase has proved to be the Mn^{2+} ion bound to a single site on the ATPase.[12] Mn^{2+} EPR studies are consistent with the binding of one Mn^{2+} ion per ATP-binding site, per phosphorylation site, or per ouabain-binding site on ATPase, with a dissociation constant of $0.2-0.9 \times 10^{-6} M$. Manganese is a particularly potent paramagnetic probe, due to the five unpaired electrons in its $3d$ orbitals and due also to its characteristically long electron spin relaxation time (typically 2 nsec).

The trivalent metal ion, Cr^{3+}, has also been shown to be a useful paramagnetic probe in magnetic resonance studies of Na^+, K^+-ATPase. The β, γ-bidentate metal–nucleotide complex, $Cr(H_2O)_4ATP$, binds in a competitive manner to the ATP substrate site of Na^+, K^+-ATPase and provides a convenient paramagnetic probe at the site normally occupied by Mg in the MgATP–substrate complex at the active site.[12]

Δ-$Cr(H_2O)_4ATP$ Λ-$Cr(H_2O)_4ATP$

Nitroxide spin labels have been used extensively as paramagnetic probes with a variety of enzymes,[13] and several groups have reported studies of Na^+, K^+-ATPase using spin-labeled probes,[14–16] but to date none of these has employed spin labels as paramagnetic probes in structural studies of ATPase. A spin-labeled analog of ouabain with the struc-

---

[12] S. E. O'Connor and C. M. Grisham, *Biochemistry* **18**, 2315 (1979).
[13] L. J. Berliner, ed., "Spin Labelling: Theory and Applications," Vols. 1 and 2. Academic Press, New York, 1976 and 1979.
[14] C. Grisham and R. Barnett, *Biochemistry* **12**, 2635 (1973).
[15] H. Simpkins and L. Hokin, *Arch. Biochem. Biophys.* **159**, 897 (1973).
[16] M. Esmann, this volume [34].

ture shown below has recently been synthesized.[17,18] This analog provides for the first time a specific paramagnetic probe for the extracellular surface of Na^+,K^+-ATPase.

^{31}P and ^1H NMR Studies of the ATP Site

One of the primary considerations when planning nuclear relaxation studies of an enzyme such as Na^+,K^+-ATPase is to ensure that the complex to be observed is stable and static over the course of the NMR experiments. This is difficult to do with the ATP site of Na^+,K^+-ATPase. Normal ATP is rapidly hydrolyzed by the enzyme, and even in the absence of added Mg^{2+}, Na^+, and K^+ the rate of ATP hydrolysis is substantial. How then to characterize the ATP site of the enzyme?

An additional problem arises when Mn^{2+} is to be used as the paramagnetic probe. In the ATPase–ATP complex, there are two potential high-affinity binding sites for Mn^{2+}. One of these of course is the aforementioned high-affinity site on the ATPase itself, and the other is the triphosphate moiety of ATP. If Mn^{2+} is added to this complex, the metal will distribute between the two sites in a manner dictated by the dissociation constants at the two sites. This is a situation to be avoided, since the integrity of a paramagnetic probe experiment depends on having the probe at one site only.

A solution to both the above problems lies in the design of an ATP analog which is not hydrolyzed by ATPase and which has a metal tightly coordinated to the triphosphate chain. Such a probe exists in the form of $Co(NH_3)_4ATP$, a β,γ-bidentate complex of diamagnetic Co^{3+} with ATP. We have shown that this analog is not hydrolyzed by ATPase under the conditions of the NMR experiment, and further that this analog is a competitive replacement for MgATP at the ATP-binding site. It is well estab-

---

[17] L. P. Solomonson and M. J. Barber, *Biochem. Biophys. Res. Commun.* **124,** 210 (1984).
[18] J. P. Girard and C. M. Grisham, unpublished observations (1984).

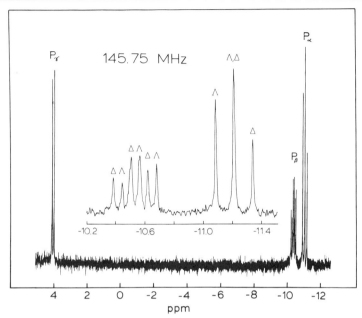

FIG. 1. Phosphorus-31 NMR spectrum of CoATP at 145 MHz at pH 7.5 and 4°. The inset shows the assignments of the resonances of the diastereomers.

lished that CoATP is substitution inert, with a half-time for dissociation of the metal on the order of days.[19]

The 145-MHz ^{31}P NMR spectrum of CoATP is shown in Fig. 1. At this relatively high frequency, the α-P and β-P multiplets are nicely resolved. The spectrum consists of superimposed resonances for both the Δ and Λ diastereomers of CoATP. As shown in Fig. 2, the addition of Mn^{2+} (which binds to its high-affinity site on the enzyme) to a solution of ATPase and CoATP results in an increase in T_1 relaxation rate for the phosphorus nuclei of CoATP. This increase, which is due to a dipolar Mn–P interaction, is not observed in the absence of enzyme and is not observed when diamagnetic Mg^{2+} is added to the enzyme. The inescapable conclusion from these data is that Mn^{2+} and CoATP are bound simultaneously to ATPase. The observed increase in $1/T_1$ could not occur in any other way.

Several points should be made here regarding experiments of this type. The paramagnetic effect of Mn^{2+} on the nuclei of CoATP at the active site is manifested in the slope of the plot in Fig. 2. As can be seen from Eq. (3), the effect is dependent on p, the ratio of the concentration of

[19] D. Dunaway-Mariano and W. W. Cleland, *Biochemistry* **19**, 1496 (1980).

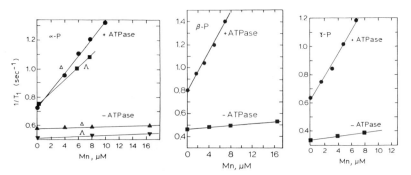

FIG. 2. Effect of Mn^{2+} on the longitudinal relaxation rates of the phosphorus nuclei of CoATP in the presence and absence of Na^+,K^+-ATPase. The solution contained 20 mM CoATP, 82 mM Tris, pH 7.5, 100 mM NaCl, 10 mM KCl, and 10 μM ATPase.

enzyme-bound Mn^{2+} to the concentration of CoATP. This is a consequence of the fast exchange condition, and the measured paramagnetic effect (the slope of the plot in Fig. 2) can be increased by either increasing the enzyme concentration or lowering the concentration of CoATP. There are practical limits here. Lowering the CoATP level will degrade the signal-to-noise in the spectrum, lengthening the time required for the experiment. On the other hand, enzyme supplies are usually limited, and the final choice normally becomes a compromise between maximizing the paramagnetic effect, conserving enzyme, and achieving the maximum signal-to-noise. The value of p in the present case would be 2000 if the enzyme site was saturated with Mn^{2+} and values of 500–2000 are typical.

The levels of Mn^{2+} in the present study were chosen to fall in the range of 10–70% of the enzyme concentration itself. If the affinity of the enzyme for Mn^{2+} is sufficiently high (as is the case here), the plot of relaxation rate vs Mn^{2+} will be linear over this range. Adding Mn^{2+} beyond this point would result in curvature and leveling off of the plot at the point where enzyme site concentration equalled that of Mn^{2+}. If, on the other hand, the affinity of the enzyme for Mn^{2+} were not high, the plot would be curved across the entire titration range. In such a case, the fraction of the added Mn^{2+} which is actually enzyme bound (the site occupancy) would have to be calculated at each point using the known dissociation constants, and the value of p in Eq. (3) would be corrected accordingly.

One final caution should be offered here. Site occupancy may also be a problem for the substrate. In the present case, the level of CoATP employed was sufficient to saturate the substrate site on the ATPase.[20] When

[20] M. L. Gantzer, C. Klevickis, and C. M. Grisham, *Biochemistry* **21**, 4083 (1982).

TABLE I
PARAMAGNETIC CONTRIBUTION TO RELAXATION RATES AT 145.75 MHz FOR
PHOSPHORUS NUCLEI OF Co(NH$_3$)$_4$ATP AND Mn^{2+}–P DISTANCES IN THE ATPase-BOUND
Mn^{2+}–Co(NH$_3$)$_4$ATP COMPLEX

P atom	$T_{1M}^{-1} \times 10^{-3}$ (sec^{-1})	$1/fT_{2p} \times 10^{-4}$ (sec^{-1})	r^a (Å)
P$_\alpha$ (Δ)	0.82 ± 0.06	5.0 ± 0.08	6.6
P$_\alpha$ (Λ)	0.50 ± 0.04	4.8 ± 0.07	7.2
P$_\beta$ (Δ, Λ)	1.67 ± 0.10	23.4 ± 3.0	5.9
P$_\gamma$ (Δ, Λ)	1.67 ± 0.08	7.0 ± 0.5	5.9

^{a} Calculated using a correlation time (τ_c = 1.9 × 10^{-9} sec) obtained from the frequency dependence of H$_2$O proton relaxation.[6]

such is not the case, an additional occupancy correction must be made for the substrate. This also modifies the value of p in Eq. (3).

The data of Fig. 2 can be used along with Eqs. (3) and (5) to calculate Mn-P distances if a suitable value for τ_c, the dipolar correlation time, can be determined. While methods for determining correlation times will not be described here, they have been treated in detail elsewhere.[21] Using a value for τ_c of 1.9 × 10^{-9} sec, Eq. (5) yields the distances in Table I. Also compared in Table I are the paramagnetic contributions to the longitudinal and transverse relaxation rates of the phosphorus nuclei. The largest value of the transverse relaxation rate in Table I (23.4 × 10^4 sec^{-1}) sets a lower limit on $1/\tau_M$ that is much greater than the longitudinal relaxation rates. Thus the fast exchange assumption and the calculation of distances from the values $1/T_{1M}$ are justified.

It is also instructive to consider the effect of Eq. (5) on the error limits for the calculated distances. The sixth root relationship in Eq. (5) means that a 10% error in the measurement of $1/T_{1p}$ or $f(\tau_c)$ will result in only an error of 1.6% in the distances, while an error of 100% in the measured relaxation parameters yields an error of 12% in the distances.

The distances in Table I and Fig. 3 are too large for an inner sphere complex between Mn^{2+} and the phosphates of CoATP, but are typical of second coordination sphere distances. This observation is interesting by itself, but it raises the question of the precise role of Mn^{2+} at the active site. An answer to this question, and additional information on the conformation of ATP at the active site of the ATPase, can be approached by means of ^{1}H NMR studies of CoATP on the enzyme. We shall see that ^{1}H NMR studies offer both advantages and disadvantages compared to the phosphorus studies above. The proton is a more sensitive NMR nucleus

[21] F. Raushel and J. Villafranca, *Biochemistry* **19**, 5481 (1980).

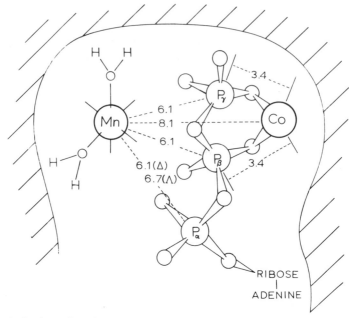

FIG. 3. Conformation of the triphosphate chain of CoATP with respect to ATPase-bound Mn^{2+} ion.

than phosphorus, permitting shorter acquisition times and improved signal-to-noise. On the other hand, proton studies in water face a fundamental dynamic range problem, owing to the high concentration of protons in water (110 M for protons). The use of D_2O (typically 99.8% D or greater) as a solvent in place of H_2O is a necessity in such cases, but this raises the matter of contamination of the NMR solutions by divalent cations, particularly paramagnetic ions, since commercial preparations of D_2O are often heavily contaminated with paramagnetic ions, particularly iron. Ion-chelating resins, such as Chelex 100, can and in fact should be used to remove divalent cations from all solutions prepared for such NMR studies, but D_2O presents a special problem. Chelex treatment of D_2O solutions will remove metals, but will also introduce intolerable amounts of H_2O into the D_2O. We have found that Chelex can be stirred in D_2O and then filtered several times to remove essentially all of the H_2O, while retaining sufficient chelating capacity to then remove metals from fresh D_2O. Alternatively, Chelex 100 can be lyophilized and then rewetted with D_2O with no adverse effects.

The ^1H NMR spectrum of CoATP at 361 MHz is shown in Fig. 4. Seven resonances from the protons of CoATP are observed. It should be

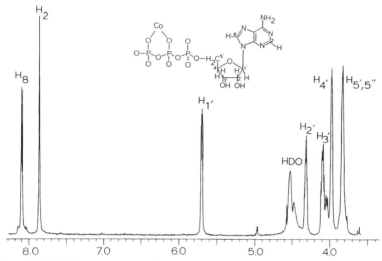

FIG. 4. ^{1}H NMR spectrum (361 MHz) of CoATP at 4°. Solutions contained D_2O (99+%) and the residual water signal was irradiated to reduce its intensity.

pointed out that the H-2' proton is only resolved at low temperatures (4° in the present case), and is "buried" under the HDO resonance at room temperature, due to temperature-dependent chemical shifts. As shown in Fig. 5, addition of Mn^{2+} to solutions of ATPase and CoATP results in linear increases in $1/T_1$ of all seven protons of the nucleotide analog. It is important to point out that the relative magnitudes of these numbers reflect the degree of interaction of each of the protons of CoATP with the bound Mn^{2+} ion. Thus, it is clear from the data of Fig. 5 that the H-8 proton of the nucleotide experiences the greatest degree of interaction with the Mn^{2+} at its site on the enzyme.

We have performed studies of this type in solutions of varying cation composition in an attempt to distinguish between the so-called E_1 and E_2 conformations of ATPase. It is by now well established that in the absence of phosphorylation the E_1 conformation of ATPase predominates in high-sodium solutions, whereas E_2 is the predominant conformation in the presence of large amounts of potassium ion. However, we find no significant differences in our nuclear relaxation studies when sodium ion is replaced with potassium ion. The normalized paramagnetic effect of Mn^{2+} on the relaxation rates of the protons of CoATP is indistinguishable under these two different sets of conditions. Whatever the conformational difference between the E_1 and E_2 forms of the ATPase, it does not seem to be reflected in the conformation of ATP bound at the active site.

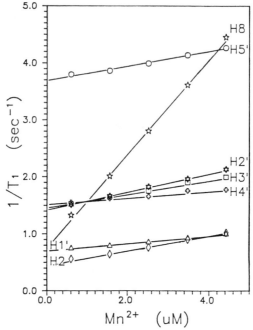

FIG. 5. Effect of Mn^{2+} on the longitudinal relaxation rates of the protons of CoATP in the presence of ATPase. Solution contained 15 mM CoATP, 15 mM Tris, pH 7.5, 150 mM NaCl, and 8 μM ATPase.

The relaxation data of Fig. 5 can be used to calculate Mn–H distances at the active site for the seven unique resonances of CoATP. These calculations again require knowledge of a dipolar correlation time for the Mn–H interactions. Analysis of the frequency dependence of the relaxation rates (data not shown) establishes a correlation time of 5.5×10^{-10} sec. Using this value for τ_c, the Mn–H distances shown in Table II were calculated.

The data of Table II provide unique information on the conformation of ATP at the active site of Na^+,K^+-ATPase. The distances are consistent with a model for the active site in which the bound Mn^{2+} ion is proximal to the N-7 and NH_2 groups of the adenine ring of bound CoATP, as shown in Fig. 6. Model-building studies based on the data of Table II place the Mn^{2+} ion in the plane of the adenine ring. In such a conformation, it might be expected to experience optimum interactions with the adenine ring. All of the distances of Table II can be reconciled in such a model, which yields an antitype conformation for the glycosidic bond linking the adenine base to the ribose sugar. The torsion angle about the glycosyl C-1′–

TABLE II
SUMMARY OF PARAMAGNETIC ^{1}H-NMR DATA

	H-8	H-2	H-1'	H-2'	H-3'	H-4'	H-5',5"
$1/(fT_{1p})^{300}$	8522 ± 104	1323 ± 20	708 ± 11	1530 ± 33	1198 ± 35	623 ± 27	715 ± 29
$1/(fT_{1p})^{361}$	8060 ± 62	1103 ± 13	565 ± 19	1170 ± 36	929 ± 11	517 ± 15	615 ± 18
$\dfrac{(1/T_{1p})^{300}}{(1/T_{1p})^{361}}$	1.092 ± 0.036	1.202 ± 0.023	1.249 ± 0.053		1.207 ± 0.048	1.252 ± 0.083	1.235 ± 0.068
$(1/T_{1M}{}^a$	17000–37600	1410–1530	730–770	1650–1810	1250–1390	620–690	710–800
rMn–H (Å)[b]	4.3–4.9	7.4 ± 0.1	8.2 ± 0.1	7.2 ± 0.1	7.5 ± 0.1	8.4 ± 0.1	8.2[c]

[a] Calculated using $1/\tau_M = 1.395 \ (\pm 0.275) \times 10^4$, reported by range of values.
[b] Calculated using $\tau_c = 5.47 \ (\pm 0.40) \times 10^{-10}$ sec.
[c] This is only an approximate distance, due to orientational uncertainties.

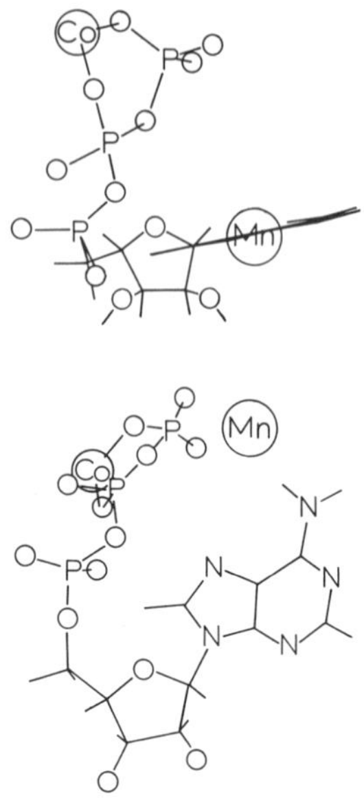

FIG. 6. Conformation of CoATP bound at the active site of Na^+,K^+-ATPase, based on paramagnetic relaxation studies (Table II) and two-dimensional TRNOE measurements (Table III). The structure was modeled on a Silicon Graphics workstation using the Molecular Modelling System software.

N-9 bond that links base to sugar is denoted by χ. The term "syn" refers to values of χ in the range of 180 ± 90°, while "anti" refers to angles of 0 ± 90°. The anticonformation is much more generally found in purine nucleotides and their protein complexes in the crystalline state as well as in solution.

The data and the model in Fig. 6 raise the possibility of bonding interactions between the enzyme-bound Mn^{2+} and the N-7 and N-6 nitrogens. Precedent for such interactions exists in the literature. Chiang et al.[22] have described several potential bonding schemes for metals interacting with nucleotides at these two nitrogens.

It is important to point out that the distances calculated in Table II depend on the choice of a correlation time, τ_c, for the Mn–H interactions. The value used in the present case was that determined from the frequency dependence of the paramagnetic relaxation rates (5.5×10^{-10} sec). Due to possible errors in this quantity and other quantities associated with the distance calculations, it is rarely possible to know these distances in

[22] C. Chiang, L. Epps, L. Marzilli, and T. Kistenmacher, *Inorg. Chem.* **18**, 791 (1979).

an absolute sense to better than ±10%. However, the relative error for the values in Table II will be much smaller, since the same assumptions are used throughout. Thus the differences in the relaxation parameters for the sodium and potassium forms of the enzyme clearly reflect different conformations for bound CoATP in the two forms of the enzyme.

Transferred Nuclear Overhauser Enhancement Measurements

It would be particularly desirable to have a means of corroborating the results of the paramagnetic probe experiments described above. One means of doing this involves the nuclear Overhauser enhancement (NOE), and, in systems undergoing chemical exchange, the transferred nuclear Overhauser enhancement (TRNOE) experiment.[23,24] This measurement, which does not rely on interactions of nuclei with paramagnetic probes, depends instead on dipole–dipole interactions between adjacent nuclei, and is thus a sensitive measure of conformation even in complex molecules.

In the extreme narrowing limit ($\omega\tau_c \ll 1$), proton–proton NOE's are positive with a maximum value of $+0.5$. This is the case for small ligand molecules which are characterized by very short correlation times ($\tau_c < 10^{-10}$ sec). In the spin diffusion limit ($\omega\tau_c \gg 1$), NOE's are negative with a maximum value of -1.0. This is the case for large proteins which are characterized by long correlation items ($\tau_c > 10^{-8}$ sec). The basis of the TRNOE involves making use of chemical exchange between two bound nuclei from the bound to the free state. Thus the aim of the TRNOE is to measure negative NOE's on the easily detectable free or observed ligand resonances following irradiation of other ligand resonances (free, bound, or observed) in order to obtain conformational information on the bound ligand.

Recently, we have explored the possibility of performing this experiment in the two-dimensional mode. This is essentially a two-dimensional NOESY experiment in an exchanging system. There are several advantages to this approach, including a saving of time and also the elimination of the problems associated with selective irradiation of closely spaced resonances. Figure 7 shows a two-dimensional TRNOE contour plot of CoATP bound to Na$^+$,K$^+$-ATPase. The cross-peaks which connect the H-8 proton with the H-2′, H-3′, and H-5′ protons indicate proximity of the H-8 proton with the 2′, 3′, and 5′ protons of the adenine ring. This is essentially the same result obtained in the paramagnetic probe experi-

[23] G. M. Clore and A. M. Gronenborn, *J. Magn. Reson.* **48**, 402 (1982).
[24] G. M. Clore and A. M. Gronenborn, *J. Magnetic Reson.* **53**, 423 (1982).

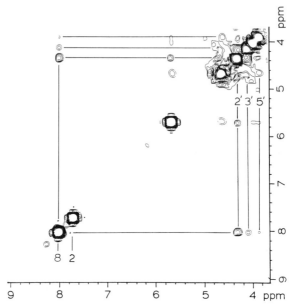

FIG. 7. Contour plot of two-dimensional TRNOE data for the CoATP–Na$^+$,K$^+$-ATPase complex. The cross-peaks indicate spin–spin interactions between protons of the bound CoATP.

ments and supports the notion of an anticonformation for the adenosine moiety of bound CoATP. Consistent with this interpretation is the absence of cross-peaks between the H-2 proton and any of the ribose ring protons. This would be expected (i.e., required) for an anticonformation

TABLE III
CALCULATED PROTON–PROTON DISTANCES:
H-1'–H-2' REFERENCE ($d = 2.9$ Å)

H-8–H-1'	3.6 ± 0.5
H-8–H-2'	3.1 ± 0.5
H-8–H-3'	3.6 ± 0.5
H-8–H-4'	3.9 ± 0.5
H-8–H-5',5"	3.7 ± 1.0[a]
H-1'–H-2'	2.9
H-1'–H-3'	3.8 ± 0.5
H-1'–H-4'	3.5 ± 0.5
H-1'–H-5',5"	4.0 ± 1.0[a]

[a] H-5',5" are equivalent protons and therefore uncertainties are greater.

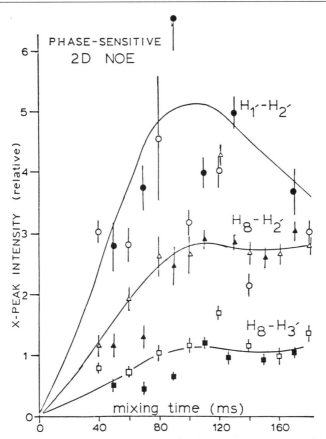

FIG. 8. Cross-peak intensity from two-dimensional TRNOE maps versus mixing time for CoATP bound to Na^+,K^+-ATPase. The initial build-up rates measured from such plots can be used to calculate the distances given in Table III.

of the bound CoATP. Since NOE effects frequently involve three-spin systems, it is difficult to assess the relevance of interactions between any two spins. On the other hand, three-spin interactions generally exhibit a time lag in their buildup, and direct, two-spin interactions can be distinguished from indirect, three-spin interactions by measuring the time evolution of the cross-peaks in two-dimensional TRNOE experiments. Initial build-up rates for the cross-peaks are determined from plots of cross-peak intensity versus the mixing time of the two-dimensional experiment (Fig. 8). From these initial build-up rates, proton–proton distances for the bound CoATP can be calculated. The distances calculated in this fashion are shown in Table III. These distances fit a conformation for the bound

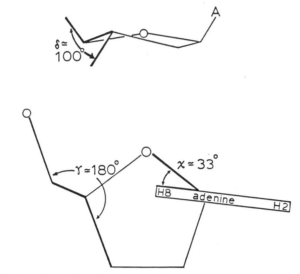

FIG. 9. Conformation of the ribose ring and the glycosidic torsion angle of CoATP bound to Na$^+$,K$^+$-ATPase as determined from two-dimensional TRNOE measurements.

CoATP which is identical to that determined from the paramagnetic relaxation studies (Fig. 6). Thus the distances determined from these two different and completely independent NMR techniques are consistent with each other. In addition to this corroboration of the paramagnetic probe measurements, it is possible to determine from the two-dimensional TRNOE measurements the conformation of the ribose ring of the bound CoATP, as well as a more accurate value for the glycosidic torsion angle. As shown in Fig. 9, the conformation consistent with the two-dimensional TRNOE distances is a C-3′-endo conformation and the glycosidic torsion angle is indeed 33°.[25]

It is tempting to speculate about the future of NMR measurements with systems such as Na$^+$,K$^+$-ATPase. One of the most obvious and exciting possibilities will surely be the use of NMR to determine three-dimensional conformation information for ATPase. This is presently possible only for very small peptides whose primary sequence is known.[26] We can anticipate that primary sequence information for Na$^+$,K$^+$-ATPase will soon be available, and furthermore it is reasonable to expect that small peptides will eventually be isolated from hydrolysates of the ATPase α and β chains. When this has been achieved, NMR methods

[25] J. Stewart and C. M. Grisham, manuscript in preparation.
[26] K. Wuthrich, G. Wider, G. Wagner, and W. Braun, *J. Mol. Biol.* **155**, 311 (1982).

may provide one of the most useful and potent tools for conformational analysis of Na^+,K^+-ATPase and other similar membrane proteins.

Acknowledgment

Work in this laboratory is supported by The Muscular Dystrophy Association of America, the National Institutes of Health (AM19419), and the National Science Foundation (INT-8317417). C.M.G. is a Research Career Development Awardee of the U.S. Public Health Service (AM00613).

[34] Electron Spin Resonance Investigations of Na^+,K^+-ATPase

By MIKAEL ESMANN

Introduction

The electron spin resonance (ESR) technique has been used in several ways to characterize Na^+,K^+-ATPase. Interaction of paramagnetic ions (such as Mn^{2+}) with the enzyme will be reviewed elsewhere in this volume.[1] A second method is to investigate the interaction of spin-labeled phospholipids with the membrane-bound enzyme. Since the technique for introduction of lipid spin labels has been described earlier in this series[2] only reference to recent work on lipid spin labeling of Na^+,K^+-ATPase shall be given.[3-5] A third area where the ESR technique has been used is by introducing covalently bound paramagnetic molecules, usually nitroxide radicals, into the sulfhydryl groups (SH) of Na^+,K^+-ATPase, and this will be dealt with in this chapter.

The α chain of Na^+,K^+-ATPase (M_r 104,000) contains about 14 mol SH groups/mol α chain. Part of these (six SH groups/α chain) can be labeled with maleimide spin labels (MAL-SL) under nondenaturing conditions.[6] This chapter will describe a method for prelabeling SH groups under conditions where about two SH groups are labeled per α chain and no loss of enzymatic activity occurs. This allows for a subsequent selective labeling of SH groups of importance for the Na^+,K^+-ATPase activity.

[1] C. M. Grisham, this volume [33].
[2] D. Marsh and A. Watts, this series, Vol. 88, p. 762.
[3] C. M. Grisham and R. E. Barnett, *Biochim. Biophys. Acta* **266**, 613 (1972).
[4] A. F. Almeida and J. S. Charnock, *Biochim. Biophys. Acta* **467**, 19 (1977).
[5] M. Esmann, A. Watts, and D. Marsh, *Biochemistry* **24**, 1386 (1985).
[6] M. Esmann, *Biochim. Biophys. Acta* **688**, 251 (1982).

Methods

Reagents

Spin labels were obtained from Aldrich Chemicals. Concentrated stock solutions in ethanol were prepared fresh daily. Na$^+$,K$^+$-ATPase was purified from rectal glands of *Squalus acanthias*.[7] All other chemicals were analytical grade.

Prelabeling in the Presence of Glycerol

Na$^+$,K$^+$-ATPase (1 mg/ml) from shark rectal glands[7] in 30 mM histidine (pH 7.4 at 37°) is incubated at 37° in the presence of 150 mM KCl, 5 mM CDTA, and 50 μM MAL-SL (2-maleimido-2,2,5,5-tetramethyl-1-pyrrolidinyloxyl). The reaction is quenched with 1 mM mercaptoethanol (MSH) and the residual activity is subsequently measured as described.[8]

As shown in Fig. 1 (curve A) about 80% of the activity is lost within 45 min. If the incubation medium also contained 40% glycerol no appreciable loss of activity occurs within 45 min (Fig. 1, curve B). Titration of the SH groups shows that about six SH groups are labeled per α chain in the absence of glycerol and about two SH groups per α chain in the presence of glycerol.[6] The same pattern of inactivation is obtained with MAL-SL and N-ethylmaleimide (NEM).

This allows for a differential labeling of the six SH groups/α chain. Class I (the two SH groups labeled in the presence of glycerol; Fig. 1, curve B) can be labeled directly with MAL-SL. The remaining four groups (class II) can be labeled with MAL-SL after prelabeling of class I with NEM, and class I and II can be labeled directly with MAL-SL as shown in Fig. 1, curve A.

ESR Measurement

After labeling with MAL-SL for a given period of time the reaction is quenched with 1 mM MSH, and the nonincorporated label is removed by repeated centrifugations (usually three) of the membranes at 200,000 g for 60 min in 20 mM histidine and 25% glycerol (pH 7.0 at 20°). The labeled membranes can be stored in this buffer at $-20°$. Removal of unbound label can be checked easily by measuring the ESR signal in the final pellets (see below) of Na$^+$,K$^+$-ATPase membranes that have been incubated with MAL-SL which has been prereacted with MSH (and therefore does not incorporate into the protein).

[7] J. C. Skou and M. Esmann, this volume [3].
[8] M. Esmann, this volume [10].

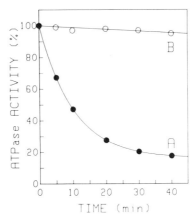

FIG. 1. Time course of inactivation of Na^+,K^+-ATPase activity by MAL-SL. Na^+,K^+-ATPase (1 mg/ml) was incubated at 37° with 100 μM MAL-SL in a 30 mM histidine buffer (pH 7.4 at 37°) containing 150 mM KCl, 5 mM CDTA, and 2.5% (filled circles, A) or 40% glycerol (open circles, B). At the time points indicated an aliquot was withdrawn and mixed with a cold 1 mM MSH solution. The residual Na^+,K^+-ATPase activity is given in percentage of activity at time zero.

Samples for ESR are prepared by suspending the labeled membranes in the appropriate buffer [for example, 100 mM NaCl, 1 mM CDTA, 30 mM histidine (pH 7.4, 37°)]. The ESR measurements are performed at 23° in 100 μl Blaubrand micropipet (Brand) in which the membranes have been pelleted by centrifugation at 10,000 g for 10 min to a volume of about 5 μl, equal to about 0.5 mg protein.

Examples of Spin Labeling

Differential Labeling of SH Groups

Figure 2A shows an example of the spectra obtained when only the class I groups (labeled as in Fig. 1, curve B) are labeled with MAL-SL. Spectrum B in Fig. 2 shows the labeling obtained when the SH groups of both class I and class II are labeled with MAL-SL. The SH groups of the class II can be selectively labeled with MAL-SL as described above, and this spectrum (C) is also shown in Fig. 2. All three spectra are typical two-component spectra[2] with the motionally restricted components in the wings indicated by an arrow and a more mobile component (dashed arrow in Fig. 2) in the center part. The method allows an assignment of the class I groups as "mobile" (spectrum A) and the class II groups as "immobile"

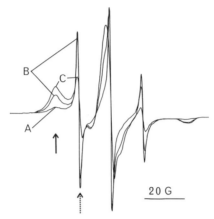

FIG. 2. ESR spectra of MAL-SL-labeled Na$^+$,K$^+$-ATPase. Na$^+$,K$^+$-ATPase was incubated with 100 μM MAL-SL as described in the legend to Fig. 1 in the absence (A) or presence (B) of 40% glycerol, leading to labeling of class I and II groups (spectrum A) or class I groups only (spectrum B). Class II groups (spectrum C) were labeled exclusively with MAL-SL after prelabeling of class I groups with 100 μM NEM in the presence of 40% glycerol, as described in text. ESR spectra were recorded on a Bruker ESR spectrometer at 23° a 9.5 GHz with a scanwidth of 100. The time constant was 0.064 sec, and a power of 20 mW was used.

(spectrum C). This is in agreement with the higher reactivity of the class I groups.

Labeling of Class II Groups with Different Maleimide Spin Labels

Figure 3 shows an example of selective labeling of the immobile class II groups. Na$^+$,K$^+$-ATPase is prelabeled with NEM as described in Fig. 1 (curve B), and subsequently labeled with maleimide spin labels with varying distances between the maleimide functional groups and the spin label (see Fig. 3). The inactivation of the enzyme was to about 30% residual Na$^+$,K$^+$-ATPase activity for all five labels. The motionally restricted component (arrow in Fig. 3) becomes less and less predominant as the spin label is moved away from the SH group that is labeled. This is what should be expected if the class II SH groups are "buried" in the protein (see inset in Fig. 3) and the rotation of the spin label increases in amplitude as the length of the label increases.

Information that Can Be Obtained Using Covalent Modification of Na$^+$,K$^+$-ATPase

Reliable structural and dynamic information from ESR experiments can be obtained only if a specific group in the enzyme is modified with a

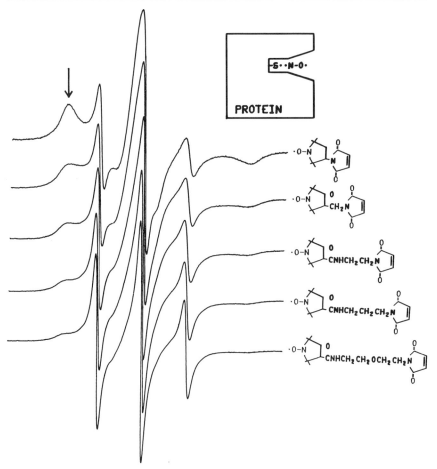

FIG. 3. Labeling of class II groups. After pretreatment with NEM (see text) in order to block class I groups of Na^+,K^+-ATPase, the class II groups were labeled by incubation with maleimide spin labels for 45 min as described in the legend to Fig. 1 (curve A in Fig. 1). The time course of inactivation was roughly the same for all five labels. The structure of the different maleimide spin labels is shown in Fig. 3 together with the ESR spectra obtained. Experimental conditions for ESR spectroscopy as given in legend to Fig. 2.

spin label. This limits the application of covalent modification seriously, especially with large molecules such as Na^+,K^+-ATPase with its many reactive groups.

New techniques such as saturation transfer ESR[9] (ST-ESR) can give

---

[9] D. Marsh, in "Membrane Spectroscopy" (E. Grell, ed.), pp. 51–142. Springer-Verlag, Berlin, 1981.

information about protein rotation in the membrane, provided the spin label is immobilized on the conventional ESR time scale. A future application of the spin label technique is therefore to label the protein in such a way that the spin label is fully immobilized. This will probably require more sophisticated spin labels than those listed in Fig. 3 (see, for example, Ref. 10).

Acknowledgment

The ESR data collection was kindly performed by Dr. J. A. Pedersen.

---

[10] J. F. W. Keana, K. Hideg, G. B. Birrell, O. H. Hankovszky, G. Ferguson, and M. Parvez, *Can. J. Chem.* **60**, 1439 (1982).

Section IX

Biogenesis and Membrane Assembly

[35] Molecular Cloning of Na^+,K^+-ATPase α Subunit Gene Using Antibody Probes

By JAY W. SCHNEIDER, ROBERT W. MERCER, EDWARD J. BENZ, JR., and ROBERT LEVENSON

Introduction

Most cells maintain a high internal potassium (K^+) concentration and a low internal sodium (Na^+) concentration relative to the extracellular fluid. The asymmetric distribution of Na^+ and K^+ across the cell membrane is generated by the energy-dependent exchange of internal Na^+ with external K^+. The transport of Na^+ and K^+ against their respective electrochemical gradients is promoted by an integral membrane enzyme called Na^+,K^+-ATPase. This enzyme, also known as the sodium pump, is fundamental to many physiological processes such as the regulation of cell volume, cotransport of nutrients, and membrane excitability. The known functional capacity of the sodium pump appears to reside solely in the α subunit, which is one of the two polypeptide subunits of the pump. The α subunit is a polypeptide of $M_r \sim 100,000$ that contains the binding sites for Na^+, K^+, ATP, and the cardiac glycoside ouabain, which is a potent inhibitor of the pump. The β subunit is a glycosylated polypeptide of $M_r \sim 55,000$ which has no known physiologic function.

The sodium pump has been extensively characterized at the enzymatic and biochemical levels. Detailed summaries of the enzymatic structure and function of Na^+,K^+-ATPase are presented in Refs. 1–3. The sodium pump has been purified to near homogeneity from cell membranes of a wide variety of tissues, including the mammalian brain and kidney. The α subunit of the pump is largely hydrophobic, a feature of the polypeptide which has made the task of determining its amino acid sequence very difficult. Only fragmentary amino acid sequence information for the α subunit is presently available. The molecular cloning and sequencing of a full-length cDNA would allow the entire amino acid sequence of the α subunit to be deduced. This would open the possibility of designing molecular genetic experiments, such as site-directed mutagenesis, to define the structural and functional domains of Na^+,K^+-ATPase. The availability of a nucleic acid probe for Na^+,K^+-ATPase would also represent a

[1] L. C. Cantley, *Curr. Top. Bioenerg.* **11**, 201 (1981).
[2] J. Kyte, *Nature (London)* **292**, 201 (1981).
[3] P. L. Jørgensen, *Biochim. Biophys. Acta* **694**, 27 (1982).

valuable tool with which to study the biogenesis and assembly of this important enzyme.

Cloning Strategy for Na^+, K^+-ATPase cDNA

Recent experiments carried out in several laboratories[4,5] have demonstrated the utility of employing antibody probes to isolate cDNA representations of rare mRNA's from libraries of recombinant phage. This system, originally developed by Young and Davis[6,7] has been used successfully in our laboratory for isolating a cDNA clone of the rat brain Na^+, K^+-ATPase α subunit. Given the availability of appropriate recombinant DNA libraries and antibody probes, the procedure we describe here should be applicable to the isolation of cDNA's for other ion-transport proteins, even when the mRNA's coding for these transport systems are of low abundance in a recombinant cDNA library.

λgt11 *Expression Libraries*. The basis of the λgt11 cloning system is outlined in Fig. 1. The utility of the λgt11 expression system derives from the fact that antisera directed against a specific polypeptide can be used to isolate a recombinant phage clone harboring a cDNA insert which codes for that polypeptide. A λgt11 library could potentially express portions of all proteins found in a complex tissue, such as the rat brain. λgt11 has been genetically engineered to permit the insertion of cDNA molecules into the unique *Eco*RI site within the coding sequence of the β-galactosidase gene (*lacZ*). The resulting gene construct will express a hybrid protein, consisting of all but the 16 carboxy-terminal amino acids of β-galactosidase fused to the inserted cDNA polypeptide. λgt11 recombinant plaques may be detected by inducing expression of the *lac* operon with isopropylthiogalactosidase (IPTG) and assaying for the insertional inactivation of β-galactosidase by cDNA.

Normally, the accumulation of eukaryotic protein in a bacterial cell is limited by the cytotoxicity of the protein or by its rapid degradation by intracellular proteases. A host strain of *Escherichia coli*, called Y1090, has been developed that allows for growth of recombinant λgt11 and the inducible expression of stable fusion protein. The antigenic signal produced by λgt11 plaques is often very strong. Driven by the efficient *lac* gene promoter, a significant percentage of the intracellular protein of a fully induced λgt11-infected Y1090 bacterium may be fusion protein.

[4] N. R. Landau, T. P. St. John, I. L. Weissman, S. C. Wolf, A. E. Silverstone, and D. Baltimore, *Proc. Natl. Acad. Sci. U.S.A.* **81**, 5836 (1984).
[5] J. E. Schwarzbauer, J. W. Tamkun, I. R. Lemischka, and R. O. Hynes, *Cell* **35**, 421 (1983).
[6] R. A. Young and R. W. Davis, *Proc. Natl. Acad. Sci. U.S.A.* **80**, 1194 (1983).
[7] R. A. Young and R. W. Davis, *Science* **222**, 778 (1983).

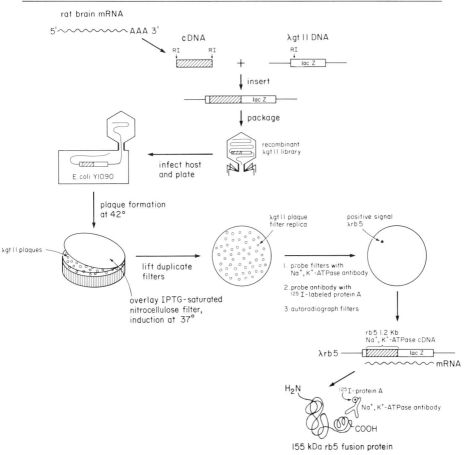

FIG. 1. Scheme for cloning and detection of Na^+,K^+-ATPase α subunit cDNA using the λgt11 expression system. For explanation see text.

The procedures involved in the synthesis of cDNA and its insertion into the λgt11 vector system are described in detail in Refs. 5 and 8. When constructing a λgt11 library several key points should be borne in mind. If the cDNA and antiserum are prepared from different cells, tissues, or species, their compatability should be assessed by spotting antigen from the library source onto nitrocellulose and screening with the antibody. This will ensure that the antibody to be used in screening is immunoreactive with the polypeptide from a heterologous source. Another consider-

[8] T. Maniatis, E. F. Fritsch, and J. Sambrook, "Molecular Cloning: A Laboratory Manual." Cold Spring Harbor Lab., Cold Spring Harbor, New York, 1982.

ation of λgt11 cloning is that the antiserum used to screen the library be directed at core polypeptide. Since bacteria do not have the capacity to posttranslationally modify eukaryotic polypeptides, antiserum directed at carbohydrate moieties, for example, will fail to detect phage clones expressing unmodified core peptides.

Materials and Methods

Screening of a Rat Brain λgt11 Library with Na^+,K^+-ATPase Antiserum

Materials

NZY broth: 10 gm/liter NZ amine (Sheffield), 5 g/liter Bacto-yeast extract (Difco), 5 gm/liter NaCl, 2 g/liter $MgCl_2 \cdot 6H_2O$, 1.5 ml/liter 3 N NaOH

NZY agar: NZY broth + 15 g/liter Bacto-agar (Difco) + 75 µg/ml ampicillin (Sigma)

NZY top agarose: NZY broth + 7.5 g/liter agarose

Overnight: NZY broth + 75 µg/ml ampicillin + 0.2% maltose

IPTG (Sigma): 10 mM, store at $-20°$

TBS: 50 mM Tris–HCl, pH 8.0, 150 mM NaCl

Staphylococcus aureus, ^{125}I-labeled protein A (Amersham)

Methods. The λgt11 cDNA library used to isolate a phage clone containing Na^+,K^+-ATPase α subunit cDNA was constructed from poly(A)$^+$ RNA isolated from 1- and 2-week-old rat brains.[9] Individual phage plaques (500,000) were plated onto *E. coli* strain Y1090 and screened with an antiserum directed against rat Na^+,K^+-ATPase[10] by the method of Young and Davis.[6,7] A fresh overnight culture of Y1090 (300 µl), grown in NZY broth in the presence of maltose and ampicillin, was infected with 25,000 plaque-forming units of phage and plated with 6 ml NZY top agarose onto a 150-mm NZY plus ampicillin agar plate. After a 5-hr incubation at 42°, plates were overlaid with nitrocellular filter disks. The filters had been previously soaked in 10 mM IPTG for 1 hr, then air dried for 1 hr. The IPTG induction inactivates *lac* repressor and allows for fusion protein expression in unlysed host cells at the periphery of a plaque. Accumulation of fusion protein was allowed to proceed for 2 hr at 37°. Duplicate filters were then lifted from each plate and screened with anti-Na^+,K^+-ATPase antiserum. Briefly, the filters were washed for 10 min at

---

[9] The λgt11 library was provided to us by Dr. N. Davidson, California Institute of Technology, Pasadena.

[10] K. J. Sweadner and R. C. Gilkeson, *J. Biol. Chem.* **260,** 9016 (1985).

room temperature in TBS, then incubated overnight at 4° in TBS plus 3% bovine serum albumin (BSA). The filters were reacted with the anti-Na$^+$,K$^+$-ATPase antibody probe for 2 hr in TBS plus 3% BSA at room temperature. (Prior to its use in screening, the antiserum was diluted 1:100 in TBS + 3% BSA, then preabsorbed with 0.1 vol of boiled Y1090 bacterial lysate.[11] After a 2-hr preabsorption of antibody with lysate at 4°, the bacterial debris was removed by centrifugation.) The filters were then washed for 10 min at room temperature first in TBS, then in TBS plus 0.1% Nonidet P-40 (NP-40), then again in TBS. Filters were incubated for 1 hr at room temperature with 1 × 10^6 cpm fresh ^{125}I-labeled protein A per filter and the previous washes repeated. Duplicate filters were keyed with radioactive ink, air dried, and autoradiographed side by side overnight. Positive signals on both filters were selected for subsequent rounds of antibody screening.[12] A useful control experiment can be performed by comparing the signal obtained with an anti-β-galactosidase primary antibody (Cappel) on IPTG-induced and uninduced filters to ensure that the antibody detection system is working. Once a phage clone exhibiting antibody reactivity has been identified, the phage clone should be purified. This usually involves at least four successive rounds of antibody screening and plaque purification.

Characterization of Na$^+$, K$^+$-ATPase α Subunit Fusion Protein

After purification of an antibody reactive phage clone, it is necessary to verify that the antibody reacts specifically with the fusion protein of that clone. This can be accomplished by immunoblotting a bacterial lysate produced by a lysogen which harbors the λgt11 phage clone of interest. A small culture of the λgt11 lysogen will produce large amounts of fusion protein after the inactivation of the thermosensitive *cI* repressor and *lacZ* induction with IPTG. Since the fusion protein represents a molecular hybrid of bacterial β-galactosidase and cDNA polypeptide, it should be immunoreactive with the antiserum used to screen the library as well as with anti-β-galactosidase antiserum. The steps involved in carrying out such an experiment include (1) construction of a recombinant λgt11 lysogen, (2) induction and preparation of a bacterial lysate, and (3) SDS–polyacrylamide gel electrophoresis and Western blotting of the lysates.

A lysogen of the bacteriophage clone containing Na$^+$,K$^+$-ATPase α subunit cDNA (λrb5) was prepared by infecting a high-frequency lyso-

---

[11] D. M. Helfman, J. R. Feramisco, J. C. Fiddes, G. P. Thomas, and S. H. Hughes, *Focus* (*Bethesda Res. Lab.*) **6**, 1 (1984).
[12] R. W. Davis, D. Botstein, and J. R. Roth, "A Manual for Generic Engineering: Advanced Bacterial Genetics." Cold Spring Harbor Lab., Cold Spring Harbor, New York, 1980.

genization (hfl) *E. coli* strain, Y1089, with λrb5 phage. Y1089 bacteria were grown to stationary phase in NZY medium containing maltose and ampicillin, centrifuged, and resuspended in 10 mM MgSO$_4$ to a final concentration of 2×10^9 cells/ml (OD$_{600}$ 2.5/ml). Cells (2×10^6 cells in 0.1 ml) were infected with purified λrb5 phage at a multiplicity of infection (MOI) of 5 and grown for 15 min at 30°. The suspension was then diluted with NZY broth and 2×10^3 cells were plated onto a 90-mm NZY plus ampicillin plate. The plate was incubated for 12 hr at 30°. Fifty colonies were transferred to each of two NZY plates; one plate was incubated at the permissive temperature, 30°, and the other was incubated at the nonpermissive temperature, 42°, for 12 hr. Approximately 30% of the colonies demonstrated growth only at 30° and were selected as λgt11 lysogens. It should be noted that successful lysogenization greatly depends on an accurate MOI, and therefore on a careful titering of the bacteriophage stock.

Bacterial lysates from λrb5 lysogens were prepared essentially as described by Young and Davis.[6] Briefly, lysogens were grown at 30° to a density of 4×10^{10} cells/ml, induced for 30 min by temperature shift to 42°, and then incubated either with or without 5 mM IPTG at 38° for 2 hr. The bacterial suspension was centrifuged at 1000 g and the pellet resuspended in 5 ml of 100 mM NaCl, 10 mM Tris–HCl, pH 8.0, 1 mM EDTA. The suspension was frozen, thawed, and then passed five times through a 19-gauge needle. The lysates were analyzed for the presence of IPTG-inducible fusion protein by electrophoresis through a 7.5% SDS–polyacrylamide gel (Laemmli) followed by Coomassie blue staining.

The results of the experiment designed to analyze λrb5 fusion protein for immunoreactivity with Na$^+$,K$^+$-ATPase antiserum are shown in Fig. 2. Proteins from bacterial lysates were separated by SDS–polyacrylamide gel electrophoresis, electroblotted onto nitrocellulose paper, and probed with either the Na$^+$,K$^+$-ATPase antibody (Fig. 2C) or β-galactosidase antibody (Fig. 2B). A duplicate gel stained with Coomassie blue is depicted in Fig. 2A. As shown in (A), the addition of IPTG (+ lane) induces the expression of β-galactosidase (lower arrow) in wild-type λgt11 lysogens, while in λrb5 lysogens, synthesis of a protein with a molecular weight of approximately 155,000 is stimulated (upper arrow). Immunoblotting of the lysogen proteins demonstrated that β-galactosidase was immunoreactive only with the anti-β-galactosidase antibody, while fusion protein from the λrb5 lysogen reacted with both Na$^+$,K$^+$-ATPase (Fig. 2C) and β-galactosidase (Fig. 2B) antibodies. The reactivity of β-galactosidase and Na$^+$,K$^+$-ATPase antisera with rat brain plasma membranes is shown in Fig. 2B and C. The β-galactosidase antibody failed to react with rat brain membranes while the Na$^+$,K$^+$-ATPase antibody was

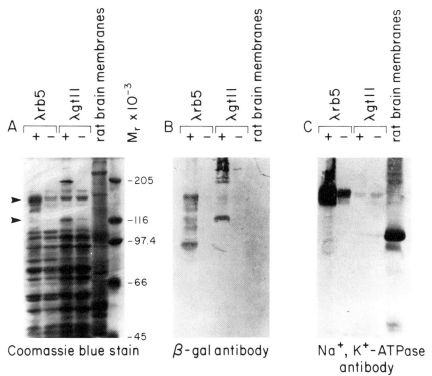

FIG. 2. Western blot analysis of λrb5 fusion protein. The phage clones λrb5 and wild-type λgt11 were used to lysogenize bacterial strain Y1089. Lysogens were grown at 32°, shifted to 42°, and then induced (+) or not induced (−) with IPTG. Lysogen proteins were fractionated by electrophoresis through a 7.5% SDS–polyacrylamide gel and transferred to nitrocellulose. (A) Coomassie blue stain of lysogen proteins. (B) Reactivity with anti-β galactosidase antibody followed by treatment with alkaline phosphatase-conjugated protein A. (C) Reactivity with Na$^+$,K$^+$-ATPase antiserum followed by treatment as in (B). Molecular weight markers are shown at the right of (A). The position of β-galactosidase (lower arrow) was determined from the mobility of purified β-galactosidase. The upper arrow denotes the position of the λrb5 fusion protein.

reactive with the 100-kDa α subunit of Na$^+$,K$^+$-ATPase. These results demonstrate that Na$^+$,K$^+$-ATPase antibody is reactive with the cDNA product of the λrb5 phage clone and not with a bacterial or phage protein.

Subcloning of cDNA into a Plasmid Vector

Following the purification of a phage clone containing a cDNA insert which directs the synthesis of an immunoreactive fusion protein, it is often convenient to subclone the cDNA fragment into a plasmid cloning vector. This procedure eliminates the necessity of working with large, fragile λ phage DNA molecules and increases the proportion of insert DNA sequence relative to vector DNA sequence.

DNA was prepared from phage clone λrb5 by standard procedures.[12] The cDNA insert was liberated from the vector by digestion with the restriction endonuclease *Eco*RI and the DNA fragments resolved by electrophoresis through a 1% agarose gel. The gel band containing the 1200-bp cDNA fragment was excised and recast in a second 1% agarose gel. Purification of the cDNA fragment by electrophoresis through a second gel significantly reduces the possibility that the DNA fragment will contain contaminating λ phage DNA fragments. Following electrophoresis, the cDNA insert was purified from the gel by the glass powder method.[13] The gel-purified 1200-bp cDNA molecule was then ligated into the *Eco*RI site of the plasmid cloning vector pAT153 and amplified in the bacterial host DH-1. The methods used for preparation of vector DNA molecules and large-scale plasmid DNA isolation are described in detail in the handbook *Molecular Cloning*.[8]

Northern Blot Hybridization

Materials

10× Gel buffer: 200 mM MOPS, 50 mM NaOAc, 10 mM EDTA, pH 7.0
Sample buffer: 0.5 ml formamide (Merck Biochemical) + 0.169 ml formaldehyde (J. T. Baker) + 0.10 ml 10× gel buffer. Store at −20°
5× Loading buffer: 5 ml 10× gel buffer + 0.01 g Bromphenol blue + 1.50 g Ficoll 400 + water to 10 ml final volume
20× SSCPE: 175.32 g/liter NaCl, 88.2 g/liter sodium citrate, 35.4 g/liter KH_2PO_4, 7.44 g/liter Na_2 EDTA; pH 7.2
Agarose (Seakem LE grade, FMC Marine Colloids Bioproducts)
Prehybridization buffer: 50% (v/v) formamide, 5× SSCPE, 250 μg/ml sonicated denatured salmon sperm DNA, 1× Denhardt's solution (0.02% each BSA, Ficoll, and polyvinylpyrrolidone)
Hybridization buffer: 4 Parts prehybridization buffer + 1 part 50% (w/v) dextran sulfate + 100 μg/ml poly(A), -(I), -(C) ribonucleotides (Boehringer Mannheim) + 2 × 10^6 cpm/ml probe

Methods. RNA's prepared from various cells or tissues can be analyzed using the cloned cDNA molecule as a hybridization probe. The purpose of this analysis is to determine (1) whether the nucleic acid probe hybridizes to an mRNA of a size sufficient to encode the polypeptide of

---

[13] V. R. Racaniello and D. Baltimore, *Science* **214**, 916 (1981).

interest, and (2) whether the mRNA exhibits an expected pattern of tissue- or cell-specific abundance.

The cDNA insert isolated from λrb5 was found to hybridize to a 27 S mRNA (~5000 bp) in several tissues of the rat as well as the dog kidney. The size of this mRNA is sufficient to code for the α subunit of Na$^+$,K$^+$-ATPase (3 kb of coding sequence required for a 100-kDa polypeptide) and to contain additional 5' or 3' flanking sequences. The 27 S mRNA exhibited a tissue-specific pattern of abundance in rat kidney, brain, and liver which corresponds with the abundance of the Na,K-pump in these tissues (kidney > brain > liver). The results of such a Northern blotting experiment are shown in Fig. 3. For this experiment, RNA was prepared from fresh rat brain, rat liver, rat kidney, and dog kidney by the guanidinium isothiocyanate (Fluka) procedure of Chirgwin et al.[14] The RNA yield was highest when the explanted tissues (or cells) were placed immediately in guanidinium isothiocyanate and thoroughly homogenized with a Tekmar Tissumizer. We find that α subunit mRNA sequences can be detected by hybridization of total cellular RNA without enrichment for poly(A)$^+$ RNA sequences prior to blotting. However, for very rare mRNA's it may be advisable to isolate poly(A)$^+$ RNA prior to blotting.

The following procedure was used for Northern blotting of RNA: 25 μg of total RNA in 4.6 μl water was denatured by the addition of 15.4 μl of sample buffer and incubation for 10 min at 65°. The RNA was then placed on ice and 4 μl of 5× loading buffer was added. The RNA was fractionated by electrophoresis through a 1% formaldehyde agarose gel. A 125-ml gel was prepared by the addition of 21.5 ml formaldehyde, 12.5 ml 10× running buffer, and 12.5 μl of 10 mg/ml ethidium bromide to 1.25 g melted agarose in 91 ml water. The gel was poured in a fume hood and polymerized for at least 1 hr. The RNA was electrophoresed at 25 V overnight in 1× recirculated running buffer, rinsed with water, photographed, and the positions of the 28 S and 18 S rRNA bands noted. The RNA was transferred to nitrocellulose, prehybridized and hybridized, as described by Thomas,[15] with the addition of single-stranded poly(A), poly(I), and poly(C) ribonucleotides to the hybridization buffer. The addition of these polyribonucleotides reduces the nonspecific binding of radiolabeled probe to ribosomal RNA. After hybridization, RNA blots were washed twice for 5 min in 2× SSCPE plus 0.01% SDS at room temperature and then twice for 20 min in 0.2× SSCPE plus 0.01% SDS at 65°. The filters were then air dried and autoradiographed overnight.

[14] J. M. Chirgwin, A. E. Przybyla, R. J. MacDonald, and W. J. Rutter, *Biochemistry* **18**, 5194 (1979).
[15] P. S. Thomas, *Proc. Natl. Acad. Sci. U.S.A.* **77**, 5201 (1980).

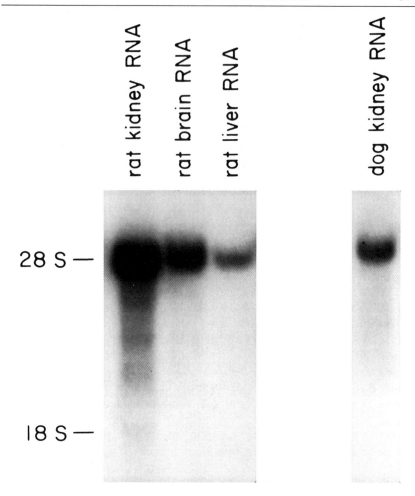

FIG. 3. Northern blot analysis of Na$^+$,K$^+$-ATPase α subunit gene transcription in rat and dog tissues. Total RNA was prepared from various tissues of the rat and dog as shown. Total cellular RNA (25 μg) was run in each lane. RNA was electrophoresed through a 1% formaldehyde-agarose gel, then transferred onto a nitrocellulose filter. The filter was reacted with 2 × 10^7 cpm of radiolabeled rb5 probe. The positions of 28 and 18 S RNA markers are shown on the left.

Southern Blot Hybridization

The cloned cDNA molecule can also be used as a hybridization probe to screen panels of DNA's prepared from a variety of cell and tissue sources. This procedure can provide the investigator with information regarding the number of copies of a gene within a particular genome and how the gene is organized. The Southern blotting procedure can also be used to determine whether potential restriction endonuclease fragment polymorphisms are characteristic features of the gene in different cells, tissues, or species.

The methods involved in carrying out the transfer of DNA fragments from agarose gels to nitrocellulose filters and the conditions for sequence hybridization analysis have been thoroughly described.[8] We prefer to transfer DNA fragments to Zetabind (AMF-Cumo division) because the efficiency of transfer and binding of DNA fragments to Zetabind is quantitatively greater than to nitrocellulose and Zetabind filters can be reprobed several times. Conditions for using Zetabind for DNA transfer are supplied by the manufacturer.

We have used the rat brain Na^+,K^+-ATPase α subunit cDNA probe, rb5, to screen genomic DNA's prepared from a variety of cell lines and tissues. These Southern blotting experiments demonstrated that the gene coding for the α subunit of Na^+,K^+-ATPase exists in only one or very few copies in the rat, dog, and human genomes. On the other hand, we found that in a ouabain-resistant HeLa cell line, C^+, which acquired ouabain resistance in concert with minute chromosomes, the DNA sequences representing the α subunit of the sodium pump were amplified ~40-fold over the level in parental HeLa cells. The correlation between increased numbers of sodium pump genes and sodium pump polypeptides in C^+ cells suggested that gene amplification was the basis for ouabain resistance in this cell line. When C^+ cells were cultured in the absence of ouabain (a procedure that results in the concomitant loss of drug resistance, minute chromosomes, and sodium pump polypeptides), the number of α subunit genes were found to decrease in number. Interestingly, a restriction fragment length polymorphism in the Na^+,K^+-ATPase α subunit gene was observed when C^+ cells were cultured in ouabain-free medium. Thus loss of ouabain resistance, while clearly related to the loss of sodium pump genes, may also involve alterations in the DNA sequence of the sodium pump.

Synthetic Oligonucleotides and DNA Sequence Analysis

In order to establish unequivocally that a cDNA clone is the DNA sequence coding for a specific polypeptide, it is necessary to obtain the nucleotide sequence of the cDNA molecule. The amino acid sequence which can be deduced from the nucleotide sequence of the cDNA should correspond to the known amino acid sequence of the polypeptide. This often requires sequencing of the entire cDNA molecule. However, if specific landmarks in the amino acid sequence of the protein can be used as a guide to direct the investigator to a specific restriction fragment of the cDNA clone for sequence analysis, only a limited amount of DNA sequencing may be necessary.

Only fragmentary amino acid sequence information is currently available for the α subunit of the sodium pump. The sequence of 10 amino acids at the amino terminus of the α subunit has been reported, as has the sequence of a characteristic tryptic digestion fragment of the α subunit.[1] This fragment consists of an 11-amino acid sequence which is specifically labeled by the reagent fluorescein 5'-isothiocyanate. In order to ascertain whether the rb5 cDNA clone contained the nucleotide sequence coding for the α subunit FITC site, we constructed a synthetic oligonucleotide probe based on the FITC amino acid sequence.

The synthetic oligonucleotide probe was chemically synthesized using an Applied Biosystems DNA synthesizer. The amino acid sequence of the FITC-binding fragment and the synthetic oligonucleotide probe we constructed are shown below. It should be noted that the synthetic oligonucleotide actually consisted of a mixture of eight different 17-mers in order to account for third base position degeneracy in codon usage for a particular amino acids. A codon preference table[16] was used to select the most likely codons for Gly and Ala.

NH_2-His-Leu-Leu-Val-Met-Lys-Gly-Ala-Pro-Glu-Arg-COOH   Protein

$$5'\text{-AUG-AA}^G_A\text{-GGC-GCC-CC}^{C\,A}_{\,\,U\,G}\text{-GA-}3'\quad\text{mRNA}$$

$$3'\text{-TAC-TT}^C_T\text{-CCG-CGG-GG}^{G\,T}_{\,A\,C}\text{-CT-}5'\quad\text{Oligo}$$

The synthetic oligonucleotide was end labeled with polynucleotide kinase (Boehringer Mannheim) and [γ-^{32}P]ATP (NEN) and purified by electrophoresis through a 20% polyacrylamide–urea gel. The probe was blotted against various restriction fragments of rb5 cDNA. We found that the probe hybridized exclusively to a BamHI–EcoRI restriction fragment of the clone, suggesting that the FITC-binding sequence was contained

---

[16] R. Grantham, C. Gautier, M. Gouy, M. Jacobzone, and R. Mercier, Nucleic Acids Res. 9, r43 (1981).

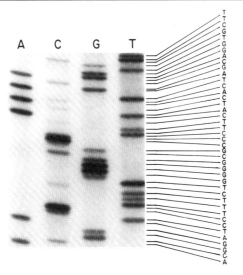

Nucleotide sequence of rb5	TTCGTGGACGATCACTACTTCCCGCGGGGTCTTT**CCTAGG** AAGCACCTGCTAGTGATGAAGGGCGCCCCAGAAA**GGATCC**
Amino acid sequence of rb5	Lys His Leu Leu Val Met Lys Gly Ala Pro Glu Arg

Fluorescein-labeled tryptic peptide from Na⁺, K⁺-ATPase
FTC = fluorescein carbamyl
	His–Leu–Leu–Val–Met–Lys–Gly–Ala–Pro–Glu–Arg
	                              |
	                           FTC

FIG. 4. DNA sequence analysis of α subunit cDNA. rb5 cDNA was digested with BamHI and the EcoRI–BamHI fragments subcloned in the phage vector M13mp8. DNA sequence analysis was carried out by the dideoxy chain termination method.[17] The nucleotide sequence of a portion of rb5 cDNA is shown to the right of the sequencing gel. A comparison of the rbcDNA nucleotide sequence and the FITC peptide amino acid sequence is shown at the bottom of the figure.

within this restriction fragment. The BamHI–EcoRI restriction fragment was subcloned into the phage cloning vector M13mp8 and sequenced by the dideoxy sequencing procedures.[17] A portion of a DNA sequencing gel and the sequence of bases that can be read from the gel are shown in Fig. 4. The nucleotide sequence of rb5 cDNA corresponding to the FITC site begins at the BamHI (highlighted bases in Fig. 4) terminus of the clone. The nucleotide sequence of rb5 cDNA corresponds exactly for all 11 amino acids of the FITC peptide. Thus direct DNA sequence analysis of rb5 cDNA demonstrated that this cDNA clone is a portion of the gene

[17] F. Sanger, S. Nicklen, and A. R. Coulson, *Anal. Biochem.* **132**, 6 (1983).

coding for the Na^+,K^+-ATPase α subunit. rb5 cDNA can now be used as a hybridization probe to isolate a full-length cDNA representation of the sodium pump α subunit. The availability of a full-length α subunit cDNA clone will make it possible to derive the complete amino acid sequence of the α subunit.

Acknowledgments

Work in the authors' laboratories was supported by grants from the NIH (HL-24385 and AM-28076) to E.J.B., Jr., and from the NCI (CA-38992), American Heart Association, March of Dimes Foundation, and Cystic Fibrosis Foundation to R.L. E.J.B., Jr. is the recipient of a Research Career Development Award (HL-01098) from the N.I.H. R.L. is a recipient of a Swebilius Foundation Cancer Research Award.

[36] Preparation and Use of Monoclonal Antibodies to Na^+,K^+-ATPase

By MICHAEL KASHGARIAN and DANIEL BIEMESDERFER

Introduction

Epithelial transport functions have generally been studied using electrical or chemical methods to trace ion movements across intact epithelium or isolated membrane vesicles. Chemical studies of the structure of plasma membranes have demonstrated that specific membrane-associated proteins can be identified as either specific ion channels or enzymatic ion pumps. The utilization of antibody technology and, in particular, the advent of monoclonal antibodies have provided a unique opportunity to unify these varied approaches in the study of transport proteins and have expanded the investigators' capability to use the same reagent or probe to study localization, distribution, and function at the epithelial level as well as molecular structure, synthesis, assembly, and insertion at the cellular level.

While they were studying the genetics of immunoglobulin synthesis, Kohler and Milstein[1] developed a method that involved the fusion or hybridization of antibody-secreting plasma cells with myeloma cells in culture. In the fused product or hybridoma the plasma cell contributes the capacity to produce a specific antibody and the myeloma cells immortalizes that capacity in hybrid cells so that they now form a long-lived cell

[1] J. Kohler and C. Milstein, *Nature* (*London*) **256**, 495 (1975).

line secreting antibodies of one specificity. The basic process of making monoclonal antibodies is a simple one. A mouse is immunized with an antigen and spleen cells from the immunized animal, many of which are producing antibodies, are fused with a myeloma cell line using polyethylene glycol to induce the hybridization. The hybridized cells are cloned and screened for specific reactivity and then are propagated by large-volume culture or tumor ascites. The hybridoma thus is formed by the fusion of a short-lived B cell which secretes antibodies of a single specificity with a long-lived myeloma cell. This facilitates production of a large quantity of pure reagent with a single defined specificity. While the basic outline is a simple one, there are specific problems which must be addressed which are unique to the antigen being studied. The case of the tetrameric membrane transport protein Na^+,K^+-ATPase will be examined here.

Advantages and Limitations of Monoclonal Antibody Techniques

Polyclonal antisera are a mixture of different antibody molecules that react with a spectrum of antigenic determinants. For high specificity they must be prepared with a highly purified antigen. If the antigen is not pure, the resultant antisera must be purified by absorption with unrelated antigens. While polyvalency of antisera offers the advantage of recognizing multiple antigenic determinants on the same molecule and thus of potentially increasing the sensitivity of the reagent there is a greater hazard for cross-reactivity with unrelated proteins. Furthermore, antisera have the capability to activate complement and bind protein A, making detection and analysis relatively simple. Heterogeneity of polyvalent reagents is complicated by the varying ability of immunized animals to respond to different antigenic determinants and the varying amounts of different types of antibodies which make up an antiserum from one bleeding to another. In contrast, monoclonal antibodies which come from a clone of a single hybrid cell generated from an individual antibody forming cell produce only a single antibody with a single specificity for a single antigenic determinant. This provides a well-defined chemical reagent in which a specific domain of a functional protein can be reproducibly and positively identified. This high degree of specificity is especially useful in the study of membrane associated proteins since they frequently cannot be isolated in high purity (see Table I).

Another advantage of monoclonal antibody technology is that it is suited for the preparation of specific antibodies from incompletely purified antigen. Using screening methods, the multiple antibodies produced to such an impure antigen preparation can be separated and the one specific for the protein of interest can be selected.

TABLE I
ADVANTAGES AND LIMITATIONS OF MONOCLONAL ANTIBODIES

Advantages	Disadvantages
Single antibody of a single class with a single specificity	May recognize small relatively ubiquitous domains resulting in cross-reactivity
Can be produced with impure antigens	Requires multiple screening to be certain of specificity
Can be selected by screening to be a defined reagent with specific functional or immunochemical activity	May tend to have low affinity and do not form large complexes requiring secondary antibodies for precipitation and identification. Also may not fix complement or react with protein A

Inherent limitations of monoclonal antibodies also exist. Since monoclonal antibodies generally form dimers with their antigens rather than large antigen–antibody complexes, precipitation of antigens requires use of a secondary antibody. Furthermore, they generally do not fix complement and frequently do not react with protein A, making detection of the antibody more difficult. The high degree of specificity of the monoclonal antibody may itself be one of its limitations. If the antibody is directed against a small antigenic determinant, perhaps in the range of 8 or 10 amino acid residues, the resultant antibody, although specific for that determinant, may be polyspecific if the sequence is common to a number of different proteins. This disadvantage may be overcome by utilization of a library of monoclonal antibodies directed at different determinants of the same molecule rather than relying on a single monoclonal antibody. Despite these caveats there is no doubt that the careful and rigorously controlled utilization of monoclonal antibodies will yield information not readily obtainable by other methods relative to the structure and function of Na^+,K^+-ATPase.

Preparation of Monoclonal Antibodies

The preparation of monoclonal antibodies (Table II) requires a series of steps, each of which requires special considerations unique to Na^+,K^+-ATPase. Success depends on careful evaluation of the procedures chosen at each point. Combined with a necessary degree of serendipity, a library of monoclonal antibodies can be developed for application to a wide variety of investigations. Each of the steps will be discussed in detail with particular reference to the specific application of Na^+,K^+-ATPase.

Antigen Preparation. Although a crude microsomal fraction prepared from renal outer medullary tissue is very rich in Na^+,K^+-ATPase, addi-

TABLE II
PRODUCTION OF MONOCLONAL ANTIBODIES

1. Preparation of antigen
2. Immunization of mice
3. Screen mouse serum for activity (ELISA)
4. Harvest and fuse spleen cells
5. Screen hybridoma supernatants for activity (ELISA)
6. Expand positive cultures
7. Screen supernatants (ELISA)
8. Clone cultures
9. Screen supernatants of clones (ELISA and immunofluorescence)
10. Produce ascites
11. Characterize monoclonals (functional screening, immunoprecipitation, immunoblotting, electron microscopy, immunocytochemistry)
12. Propagate and freeze clones

tional enrichment is relatively easily achieved by the method of Jørgensen[2] as modified by Forbush et al.[3] Outer medulla is dissected from coronally bisected kidneys and homogenized in phosphate-buffered saline with a Teflon–glass homogenizer with a very loose serrated pestle. The homogenate is filtered through cheesecloth and again homogenized in a Teflon–glass homogenizer with a standard pestle. The homogenate is centrifuged at 7500 g for 15 min and the resultant supernatant again centrifuged at 48,000 g for 30 min. The light upper portion of the pellet is resuspended in PBS and subjected to gradient centrifugation. Linear or continuous sucrose, Hypaque, or Ficoll gradients may be used. With sucrose gradients, activity is detected in a broad peak between 37 and 44% sucrose after centrifugation for 15 hr at 102,000 g. Purification is determined by assay of the enzyme activity as related to the total protein content of the preparation.

Enzyme Assay. Na^+,K^+-ATPase activity is assayed by incubating microsomal samples for 10 min at 22° with 1% bovine serum albumin (BSA) in 25 mM imidazole buffer with 20 mM EDTA (control) or with 100 μl of 0.65 mg/ml of sodium dodecyl sulfate in 1% BSA in 25 mM imidazole buffer, pH 7.0 (detergent-activated membranes).[4] After preincubation, 500 μl of 0.3% BSA in 25 mM imidazole buffer is added and 100-μl aliquots are assayed for Na^+,K^+-ATPase in an assay medium containing 125 mM NaCl, 25 mM KCl, 4 mM Na_2ATP, 4 mM $MgCl_2$, 1% BSA, 60 mM Tris–HCl, 0.5 mM EDTA, pH 7.5 (37°), and approximately 1 μg of

[2] P. L. Jørgensen, *Biochim. Biophys. Acta* **356**, 36 (1974).
[3] B. Forbush III, J. H. Kaplan, and J. F. Hoffman, *Biochemistry* **17**, 3667 (1978).
[4] B. Forbush III, *Anal. Biochem.* **128**, 159 (1983).

purified Na^+,K^+-ATPase. After 10 min of incubation at 37°, the reaction is stopped by addition of 1 ml of freshly prepared 1.5% ammonium molybdate, 2% silicotungstic acid in 1.4 N H_2SO_4, and 2 ml of isobutyl alcohol/hexane (1 : 1) and immediately vortexed for 10 sec. An aliquot of the upper phase is diluted with 1.6 ml of 2% H_2SO_4 and 95% ethanol and reduced with 0.1 ml of freshly diluted 1.2 $SnCl_2$ in acidic ethanol. The optical density of the reduced phosphomolybdate is linear at 670 nm with P_i in the assay medium to beyond 1 mM.

Antigen Selection. The use of whole enzyme preparations as an antigen has the advantage of stimulating the production of antibodies to both α and β subunits of the enzyme and to epitopes which are external to the lipid domain of the enzyme. The disadvantage is that the enriched preparation contains approximately 5–15% protein impurities which may also act as antigens. The individual subunits may be used as antigens after purification by SDS–polyacrylamide gel electrophoresis (PAGE). On a uniform 7.5% polyacrylamide gel, the α subunit migrates as a narrow band at approximately 95–100 kDa and the β subunit is a broad glycoprotein band of approximately 50–55 kDa. The subunits can be removed from the gel by electroelution. The disadvantage of using SDS–Page-purified subunits is that the proteins are denatured by the ionic detergent used in purification and antibodies produced to denatured proteins may not react with the native forms. In addition, antibodies which are directed against intramembranous domains of the enzyme may be produced and the utility of these may be limited.

Immunization. Since the most established myeloma cell lines are derived from BALB/c mice and ascites fluid from hybridomas are grown in mice which share BALB/c histocompatibility antigens, BALB/c mice are the most convenient strain to immunize. Genetic factors can influence antigen recognition and the production of antibodies, so different strains will occasionally yield a greater immune response but hybrids must then be maintained by the more difficult method of long-term large-volume cell culture. In our experience, BALB/c mice were satisfactory generators of an immune response to Na^+,K^+-ATPase and long-term generation was easily accomplished by ascites tumor propagation.

Antigen Presentation. The method of immunization has also been found to be of importance in determining the success of the hybridization procedure. The most commonly used method is to emulsify 5–100 μg antigen in 0.1 ml of phosphate-buffered saline with 0.1 ml of Freund's complete adjuvant and to inject the 0.2 ml into the hind footpads and intraperitoneally. After 3 weeks the animals are boosted with an intravenous injection with 5–100 μg of antigen intravenously if the antigen is soluble (i.e., purified subunit) or intraperitoneally if it is particulate (i.e., enriched membrane preparation). While others have found this method to be satisfactory, we have used an alternate method. BALB/c mice were primed with 2×10^9 *Bordetella pertussis* organisms injected intraperitoneally 2 to 6 hr prior to the intraperitoneal injection with 100–150 μg of antigen which was aggregated by adsorption with aluminum potassium sulfate with the antigen in 1 ml of 0.9% NaCl, 25 μl of 1 N NaHCO$_3$, and 50 μl of aluminum potassium sulfate are added. After 1 hr at 4°, the precipitate is spun at 2000 rpm for 5 min and the pellet is resuspended in the desired volume of 0.9% NaCl and used for intraperitoneal immunization. After 3 weeks the mice were bled and their serum tested against the original antigen for antibody activity by an enzyme-linked immunosorbent assay (ELISA). Mice with a high antibody titer were boosted intravenously with 100 μg of antigen and spleens were harvested 3 days after boosting. We have found this method to be superior to the more commonly utilized technique of emulsification in Freund's adjuvant.

Hybridoma Production. The SP2/0 myeloma cell line is probably best suited for hybridoma production since this cell line is unlike other myeloma cell lines in that it does not synthesize any immunoglobulin components. The resultant antibodies are therefore free of any nonspecific, nonimmune heavy and light immunoglobulin chains. Detailed practical descriptions of the fusion procedure are generally available and they will be described only briefly here.[5] Spleen cells are teased from the harvested spleen and in general a spleen will yield approximately 1×10^8 cells. These cells are mixed with myeloma cells in a spleen cell to myeloma cell ratio of approximately 3:1. Fifty percent polyethylene glycol mixed with a minimal essential tissue culture medium is then added to the cells and they are held at 37° for 1 min. The polyethylene glycol is diluted with additional culture medium and the cells are collected by centrifugation and resuspended in HAT (hypoxanthine, aminopterin, and thymidine) medium. The SP2/0 myeloma line is defective in hypoxanthine-guanine phosphoribosyltransferase (HGPRT), an enzyme necessary in the salvage pathway for purine synthesis. Unfused myeloma cells are eliminated in HAT medium since aminopterin effectively inhibits *de novo* purine nucleotide synthesis. Only hybrid cells which have HGPRT activity donated by the genes of the spleen cells of the immunized mice can convert hypoxanthine through the salvage pathway into purines and therefore survive in the HAT medium.

The cells are distributed into the wells of microtiter plates. Since unfused cells will not survive in the HAT medium for more than a few days, only hybrid cells will be viable after day 7, at which time aminop-

---

[5] E. A. Lerner, *Yale J. Biol. Med.* **54**, 387 (1981).

terin may be omitted from the culture medium. The surviving clones will appear macroscopically between days 10 and 20. The supernatants of clones at day 20 should be screened for desired antibody activity.

Once initial screening has determined which of the clones contain hybridomas of potential interest, it is imperative that cells from these wells be expanded to larger volume cultures and subcloned immediately. This is necessary since the hybrids are very unstable at this point and since not all the cells within a colony may necessarily be producing antibodies and non-antibody-producing cells could easily outgrow the desired cells or several different antibody-producing cells may be present within a single well and cloning is necessary to separate them.

Cloning of Hybridomas. The original 0.2-ml microtiter colony is transferred to 1 ml of culture medium to which BALB/c "feeder" thymocytes have been added to stimulate growth. Feeder thymocytes are a necessary adjunct during early expansion and initial cloning. Later subcloning can usually be performed without a feeder layer. Three to 6 days later, the cells should cover the 1-ml culture plate and these should be transferred to a 10-ml culture flask. At this point, the cells should be cloned either on hard agar or by limiting dilution. Culturing on hard agar allows for visually selecting macroscopic clones and expanding them to larger volume cultures. Cloning by limiting dilution attempts to derive a culture of cells derived from one hybrid clone per well. This goal is generally achieved by plating a large number of microtiter wells with varying dilutions of hybrid cells per well and testing the supernatants from the populations of the wells that have growth.

Cloning on hard agar is generally thought to be simpler than cloning by limiting dilution. Tissue culture dishes are coated with agar and hardened in an 8% CO_2 incubator to maintain proper pH. Feeder thymocytes and hybrid cells are resuspended in Beggar's medium and added to the dishes and incubated. Between 1 and 2 weeks visible clones are removed from plates by cutting out a small agar plug. The cloned plus is then incubated with medium in a microtiter well. The supernatant of the mature culture is screened and if positive expanded to first 1-ml and then to 10-ml cultures.

Cloning by limiting dilutions is accomplished by mixing hybrid cells and feeder thymocytes in Beggar's medium in dilutions which approximate 5, 1, and 0.5 hybrid cells/well. After cell growth, the supernatant medium is screened and if positive, the cultures are expanded first to 1-ml and then to 10-ml cultures.

After having confirmed that the desired antibody activity continues to persist in the cloned cultures by using various screening procedures, the cultures may be expanded to large-volume culture or to the generation of tumor ascites.

Tumor Ascites Production. To generate antibody-producing tumor ascites BALB/c or BALB/c F_1 hybrid mice which match the MHC type of the SP2/0 myeloma cell line are used. They are primed with pristane 5 days before injecting with hybrid cells. Hybrid cells ($1-5 \times 10^6$) are injected intraperitoneally in 0.5 ml of PBS. Ascites generally develops in 1–3 weeks and the ascites fluid can be tapped and passed on to additional mice or frozen for storage.

After removal from the peritoneum, immunoglobulin is precipitated from ascites fluid by precipitation with ammonium sulfate. These precipitated immunoglobulins are redissolved in PBS and are purified by immunoaffinity chromatography over a column of rabbit anti-mouse IgG or IgM (depending on the class of the monoclonal antibody to the purified) coupled to CNBr-activated Sepharose 4B agarose gel. Bound antibody is eluted with 1 M acetic acid in 0.15 M sodium chloride neutralized with 1 M Tris base. The antibody is dialyzed overnight with PBS and concentrated with aquacide II-A. The antibody solution may be either frozen or lyophilized for storage.

Screening Procedures

Screening procedures (Table III) are necessary initially to identify the presence of antibody and to determine the specificity of its reactivity during the immunization and hybridization protocols. Later they are used

TABLE III
SCREENING PROCEDURES

Technique	Advantages	Limitations
ELISA	Rapid; simple; sensitive	Specificity depends on purity of antigen
Immunoprecipitation	Accuracy of SDS–PAGE; native antigen is reacted with antibody; Ab may coprecipitate α and β subunits	Antigens must be radiolabeled; proteolysis and aggregation may occur; secondary antibodies or high avidity is necessary
Immunoblotting	Specific to subunit; does not require radiolabeling of antigen; transferred proteins may be stored on nitrocellulose	Antibody may not react with detergent-treated antigen; configurational antibodies may not react with dissociated antigen
Functional	Most specific method	Low yield
Morphologic	Rapid; simple; access to all cell compartments in intact tissues	Antigen may be altered by processing

to characterize the antibody in greater detail in relation to the antigenic determinant to which it is directed and to assess any functional activity.

Screening by ELISA. The simplest initial screening procedure utilizes enzyme-linked immunosorbent assay (ELISA). A stock preparation of Na^+,K^+-ATPase (Jørgensen membranes) with a protein content of approximately 3 mg/ml should be diluted 1:1000. The diluted membranes (100 μl) should be placed in each well of a flat-bottomed microtiter plate. After 1 hr in the refrigerator, the wells are washed with 0.05% Triton X-100 in phosphate-buffered saline (PBS) five times. A 1:10 dilution (100 μl) of culture supernatant or ascites containing the antibody is placed in the well and incubated for 1 hr at room temperature. After washing three times with Triton X-100 in PBS, 100 μl of a 1:100 dilution of 4 mg/ml stock solution of rabbit anti-mouse immunoglobulin is added to the wells and incubated for 1 hr. After again washing multiple times with Triton X-100/PBS, horseradish peroxidase conjugated to protein A is added to the wells and incubated for 1 hr. After washing an additional four times, the final wash fluid is left in the wells and 50 ml o-phenylenediamine substrate solution (20 mg o-phenylenediamine in 50 ml of 0.05 M citric acid and 0.1 M Na_2HPO_4) containing 20 μl of 30% hydrogen peroxide is prepared and 200 μl is placed in each of the wells of two plates for 0.5 hr. Fifty microliters of 6 N HCl is added to each well to stop the reaction and intensify the change. Positive antibody activity is seen by the development of the color by the reagent.

While the ELISA technique is particularly well suited for screening large numbers of cultures and has the advantage of using minute amounts of antibody and avoiding radiochemicals, it depends on the purity of the antigen bound to the plates. Since even the best zonal centrifugation methods will yield only a 95% purification on Na^+,K^+-ATPase in microsomes, the possibility of strongly antigenic minor contaminants exists and more specific screening procedures must be used to define the antibody.

Functional Screening. Functional screening allows for the most specific identification of antibody by attempting to demonstrate that the addition of the antibody will inhibit either enzyme activity or ouabain binding. The screening of antibodies for functional effects on the Na^+,K^+-ATPase is complicated by the presence in culture media of substances which are known to affect various aspects of the Na^+,K^+-ATPase reaction. Two procedures were developed to immobilize Na^+,K^+-ATPase which permitted removal of the antibody-containing medium prior to the actual Na^+,K^+-ATPase assay.[6]

[6] M. Kashgarian, D. Biemesderfer, M. Caplan, and B. Forbush III, *Kidney Int.* **28,** 899 (1985).

In the first, Na^+,K^+-ATPase was diluted in phosphate-buffered saline and 200-μl aliquots were placed into the wells of the microtiter plate and allowed to stand overnight at 4°. The cells were emptied and rinsed with PBS containing 1% BSA and 30 mM EDTA. The 50- to 200-μl aliquots of supernatant solutions from the hybridomas are added to the wells and the plates are shaken for 45 min at room temperature. The wells are emptied and rinsed with 25 mM imidazole buffer containing 1% BSA and 1 mM EDTA. [^{3}H]Ouabain (0.01 mM) in 60 ml of a solution containing 125 mM sodium chloride, 3 mM Na$_2$ATP, 3 mM MgCl$_2$, 0.1% BSA, 25 mM Tris, pH 7.5, or 5 mM MgCl$_2$, 3 mM inorganic phosphate, 2 mM EDTA, 0.2% BSA, 30 mM Tris buffer at pH 7.25 with or without additional unlabeled ouabain in a concentration of 1–4 M. After a 40-min incubation at room temperature, the wells are emptied, rinsed with 100 mM KCl, 25 mM imidazole, 1 mM EDTA, pH 7.5, and the bound [^{3}H]ouabain is removed from the wells with 200 μl of 2% SDS, transferred to scintillation vials, and radioactivity is counted in a liquid scintillation counter. Alternatively, it is possible to screen for antibody inhibition of Na^+,K^+-ATPase by substituting 200 ml of Na^+,K^+-ATPase assay medium (see above) for the [^{3}H]ouabain-binding medium, incubating for 30 min at 30° and analyzing for inorganic phosphate. Note that only approximately 10% of the original Na^+,K^+-ATPase activity is retained in control wells on the microtiter plates used in this method.

A second method employs filtration with cellulose ester filters as a means to immobilize Na^+,K^+-ATPase. This method provides considerably greater reproducibility than the use of microtiter plates but is not suited to screening large numbers of samples. Na^+,K^+-ATPase in the form of SDS-activated membrane (see above) is a concentration of 0.4 mg of protein/ml is preincubated with 1 : 10 dilutions of supernatant or ascites fluid in 0.25 ml of 25 mM imidazole buffer containing 1% BSA and 20 mM EDTA for 45 min at room temperature. Aliquots of 50 ml are filtered onto Gelman GN-6 filters on a filtration manifold. After rinsing briefly with 1 ml of 25 mM imidazole, 200 μl of [^{3}H]ouabain-binding medium (see above) is applied to the surface of the filters. Following a 45-min incubation at room temperature with shaking, vacuum is reapplied to remove the binding medium and the filters are rinsed with 100 mM KCl in buffer and are transferred to scintillation vials and counted in a liquid scintillation counter. In control experiments, maximal ouabain binding as determined by this method is approximately 70% that obtained when incubation is carried out in large-volume test tube analysis. Na^+,K^+-ATPase assay can be carried out in a similar protocol with 1 ml of Na^+,K^+-ATPase assay medium applied to the filters in place of the [^{3}H]ouabain binding medium and after 30 min shaking at room temperature the medium is collected through the filter along with 0.3 ml distilled water rinse and assayed for inorganic phosphate.

In order to study the enzymatic behavior of the antibody Na^+,K^+-ATPase complex, SDS-activated microsomes in imidazole buffer (see above) are incubated with either ascites fluid or culture supernatant in a ratio of approximately equal amounts of enzyme and immunoglobulin protein. The incubation is held at 22° for 40 min, pelleted, and resuspended in 0.1% BSA in imidazole buffer containing 1 mM EDTA. Assays for Na^+,K^+-ATPase activity, ouabain binding, E–P formation,[7] and rubidium-[8] deocclusion can be carried out by standard methodology. Na^+-ATPase activity is assayed the same as Na^+,K^+-ATPase activity but in a medium containing 100 mM Na, 1.6 mM Mg, 0.1 mM ATP, 0.2 mM EDTA, 30 mM Tris buffer, pH 7.2. Potassium-dependent nitrophenylphosphatase activity (KNPPase) is measured by incubation of the enzyme in 1 ml of 10 mM KCl, 5 mM $MgCl_2$, 10 mM p-nitrophenyl phosphate, 25 mM imidazole for 30 min at 37° followed by the addition of 2 ml of 0.2 N NaOH, 20 mM EDTA, 0.2% SDS and absorbance is determined at 410 nm.

Screening by Immunoprecipitation. Immunoprecipitation methodologies further characterize the antibody and its specificity to the specific protein subunits of Na^+,K^+-ATPase and to specific sites within the subunits. A variety of methods have been developed for the detection of specific proteins from crude extracts using gel electrophoresis combined with the antigen–antibody reaction. These methods combine the high resolution of gel electrophoresis with the specificity of the antibody reaction. With the simplest method, either membrane fragments or solubilized microsomes are incubated with an excess amount of antibody to form immune complexes. An alternate method involves the transfer of proteins for SDS–PAGE electrophoretogram to nitrocellulose or similar membrane filter paper. Overlaying the paper containing the transferred proteins with antibody allows correlation of the binding activity of the antibody with the specific polypeptide bands.

For immunoprecipitation of membranes or solubilized proteins, the best results are obtained if the membrane proteins are labeled with a radioactive label. Membranes can be iodinated using the lactoperoxidase technique, but this method has several problems which relate to the labeling procedure. The specific protein in question may not be iodinated or iodination may alter the antigenicity of the molecule. Biosynthetic label-

---

[7] P. L. Jørgensen, *Biochim. Biophys. Acta* **466**, 97 (1977).
[8] I. M. Glynn and D. W. Richards, *J. Physiol. (London)* **330**, 17 (1982).

ing is therefore preferable. Sources of Na^+,K^+-ATPase which may be biosynthetically labeled include rat kidneys, or in the case of dog antigen, cell culture from the established tissue culture line of Madin–Darby canine kidney (MDCK). Isolated rat kidneys are labeled by perfusion with [^{35}S]methionine as described by Kerjaschki et al.[9] A kidney is removed from a rat with a segment of aorta intact and is perfused with minimal essential medium containing 2 mCi [^{35}S]methionine for 4 hr in a perfusion apparatus which isolates and recirculates the radioactive medium. The cortex is homogenized and extracted in 0.1% SDS, 1% sodium deoxycholate, 1% Triton X-100, 150 mM NaCl, 1 mM EDTA, 50 mM Tris–HCl buffer, pH 7.2, with the protease inhibitors; 1 mM diisopropyl fluorophosphate, 1 μg/ml antipain, 1 μg/ml pepstatin A, 1 mM benzamidine. The mixture is centrifuged and the supernatant is used for immunoprecipitation. Supernatants are then incubated with excess antibody and immune complexes are precipitated with agarose beads linked to rabbit anti-mouse Ig antibody (see below). The immune complexes are eluted from the beads with buffer, treated with buffered detergent, and subjected to SDS–polyacrylamide gel electrophoresis. The gels are stained with Coomassie blue, dried, and a contact radioautogram is prepared. The band with radioactivity is compared to the original gel for identification of the specific proteins. Tissue culture cells such as MDCK cells can be metabolically labeled with [^{35}S]methionine by trypsinizing the cells and washing in methionine-free Eagle's minimal essential medium (EMEM). Cells are resuspended in medium containing [^{35}S]methionine and incubated for 2 hr prior to sonic disruption of the cells. The disrupted cells are treated similar to cortical homogenates (see below).

An alternate method of labeling which is more specific for Na^+,K^+-ATPase is to label cortical cell suspensions or MDCK cells with the photoaffinity derivative of ouabain.[3] Cells are suspended in buffered saline and incubated with 1 mM ^{3}H-labeled 2-nitro-5-azidobenzoyl(NAB)-ouabain. The mixture is diluted with buffer and exposed to ultraviolet light for 5 min in order to achieve complete photolysis of the NAB-ouabain. Following labeling with either [^{35}S]methionine or [^{3}H]NAB-ouabain, the cells are disrupted by sonication for 30 sec on ice, washed, and the pellets resuspended in precipitation buffer containing 1% NP-40.

Aliquots of cortical homogenates or resuspended cell pellets (350 μl) are mixed with 6 mg of nonimmune mouse monoclonal Ig and incubated for 30 min at 4°. One of the swollen washed protein A beads which have been preconjugated with 20 mg of affinity-purified rabbit anti-mouse IgG

---

[9] D. Kerjaschki, L. Noronha-Blob, B. Sacktor, and M. G. Farquhar, *J. Cell Biol.* **98**, 1505 (1984).

is then added and incubated for another 30 min. The beads are pelleted and 2 µl of monoclonal antibody or, as a control, 2 µl of nonimmune monoclonal IgG, is added and incubated for 30 min at 4°. Following the incubation, another milligram of protein A rabbit anti-mouse IgG conjugate is added. After 1 hr at 4° the beads are pelleted and washed and the protein is eluted from the beads with buffer and subjected to SDS–PAGE.

Immunoprecipitation techniques can also be combined with enzyme assay techniques to identify antibodies which do not directly alter enzyme function. Incubation of hybridoma supernatants with a standard amount of Na^+,K^+-ATPase followed by addition of secondary antibody and protein A beads may precipitate the enzyme, resulting in a reduction of activity in the microsomal fraction. Elution of the antigen from the beads and reconstitution of Na^+,K^+-ATPase activity from the precipitate will confirm the specificity of the antibody. While this technique may be useful to identify some clones, it is cumbersome and requires rigid controls to eliminate the possibility of nonspecific binding.

Immunoblotting Techniques

Immunoblotting techniques (Western blots)[10] can complement or substitute for immunoprecipitation procedures. The major advantage with immunoblotting is that radiolabeling of the Na^+,K^+-ATPase is not necessary. Several problems may limit the utility of this method. The enzyme is subjected to SDS–PAGE prior to transfer to nitrocellulose and therefore may not react with an antibody if the reactivity of the antibody is directed only at the native configuration. In addition, monoclonal antibodies with low avidity may not form a stable complex with the transferred protein, making detection difficult. Commercial transfer tanks with built-in heat exchangers are now available which simplify the transfer procedure. Briefly, an SDS–PAGE is placed on top of a sheet of nitrocellulose, sandwiched between blotter papers and porous sponges in a tank containing transfer buffer (25 mM Tris, 190 mM glycine, 20% methanol). A voltage (15 V/1.5-mm gel) is applied for 16–24 hr. Following the transfer, the nitrocellulose is washed to remove the SDS by shaking in 1% Triton X-100 followed by 4 M urea to remove the Triton X-100.

The paper may be dried at this time (but this can reduce antibody binding) or kept wet in transfer buffer. Strips of the nitrocellulose are incubated overnight in 1% BSA/PBS as a general blocking step prior to incubation with the monoclonal antibody for 1–2 hr. Following several washes with PBS–0.5% NP-40 the strip is incubated with peroxidase-

[10] H. Towbin, T. Staehelin, and J. Gordon, *Proc. Natl. Acad. Sci. U.S.A.* **76**, 4350 (1979).

labeled anti-mouse IgG or IgM depending on the class of the monoclonal antibody. After washing with PBS to remove excess secondary antibody, the strip is incubated with 1 mg/ml diaminobenzidine (DAB) in 0.1 M Tris buffer with 30% H_2O_2 added to make a 0.02% solution with the buffer immediately before incubation. After color development the strip is removed and washed in PBS with or without HCl.

While immunoblotting and immunoprecipitation are not interchangeable they are alternate and confirmatory techniques which can be helpful in individual circumstances.

Morphologic Screening

Morphologic screening can be a useful and relatively rapid method of rapid assessment of antibody activity. Screening can be performed on frozen sections of whole tissue, on membrane preparations, or on tissue fixed with periodate–lysate–paraformaldehyde (PLP) fixative[11] prior to sectioning. Use of PLP provides better structural preservation than simple cryostat sections with little alteration of the antigenicity of Na^+,K^+-ATPase. Kidneys are perfused retrograde via the aorta initially with mammalian Ringer's solution and followed by PLP fixative which is composed of 0.01 M $NaIO_4$, 0.75 M lysine, 2% paraformaldehyde in 0.0375 M Na_2HPO_4 buffer at pH 6.2.[6] After 5–10 min of perfusion, the kidney is cut into coronal sections which retain the cortical medullary papillary relationships and then postfixed for 6 hr in a perfusate fixative. The tissue is washed in phosphate buffer and cryoprotection is achieved by incubating the tissue in 10% DMSO in 0.1 M phosphate buffer for 1 hr at 4°. Blocks of tissue are frozen by rapidly plunging them into swirling liquid Freon-22 cooled by liquid nitrogen. Cryosections are cut and placed into 1% bovine serum albumin in PBS. Sections are mounted on glass slides and are preincubated with goat anti-rabbit IgG and PBS with 2% BSA for 1–3 hr as a general blocking step and to block any endogenous FC receptors that may be present in the kidney. After washing, the sections are incubated overnight with 50–100 mg/ml of the monoclonal antibody to be tested. After incubation, the sections are again washed with phosphate-buffered saline and incubated with fluorescein-labeled anti-mouse Fab of IgG. Sections may also be labeled by the peroxidase label using peroxidase-labeled anti-mouse Fab of IgG in place of the fluorescein-labeled antibody. Basolateral localization by immunofluorescence microscopy is a confirmatory screening for the identification of antibodies to Na^+,K^+-ATPase (Fig. 1).

[11] I. N. McLean and P. K. Nakane, *J. Histochem. Cytochem.* **22**, 1077 (1974).

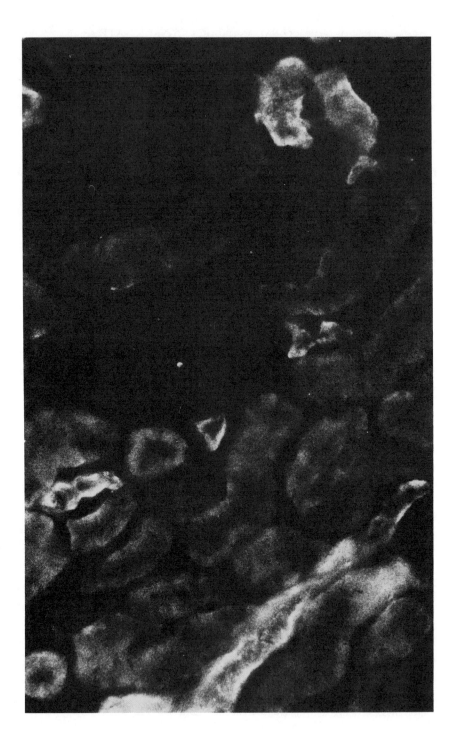

Application of Monoclonal Antibodies to Studies of Na^+,K^+-ATPase

Once having characterized the antibody as being specific for Na^+,K^+-ATPase, it may be used in a variety of different studies to examine many aspects of this transport enzyme. The monoclonal antibody represents a powerful tool in enzyme immunochemistry and can be applied to clarify molecular structure and function, functional characteristics, cellular localization, and distribution in tissues.

Many of the techniques used are essentially identical to those used in the screening procedure but are more specifically oriented to answer a particular question. For example, the ELISA assay can be changed from a screening assay to a specific quantitative immunoassay with sensitivity in the picogram range. Using a constant amount of antibody, a standard curve of intensity of reaction product can be generated with serial dilutions of an Na^+,K^+-ATPase preparation of known activity. Comparison of a test sample to the standard will yield a quantitative assay. This can be of particular value when quantities of the enzyme are small or conditions are not suitable for a standard enzymatic assay. In addition, depending on the specificity of the antibody, ELISA can be used to identify quantities of each specific subunit rather than the holoenzyme in studies of its biosynthesis.

Use of the same procedures used in immunoprecipation or immunoblotting screening can be valuable in examination of the molecular structure of the enzyme. Elucidation of the extramembranous domains or of the three-dimensional structures of the enzyme and their correlations with the functional characteristics can be obtained by immunoprecipitation of proteolytic fragments of each of the subunits. Since the monoclonal antibody is directed against a specific portion of the molecule it also can serve as a probe for peptide mapping and subsequent amino acid sequencing of functionally important domains. Alternatively, the monoclonal antibody can be used as a specific diagnostic tool when combined with molecular biological techniques in cloning cDNA or mRNA. In addition, such techniques allow for assignment of genes encoding the individual subunits to specific chromosomes or regions of chromosomes in somatic cell hybrid studies.

FIG. 1. Immunofluorescence photomicrograph of PLP-fixed section of kidney. The primary antibody is a monoclonal specific for the α subunit of Na^+,K^+-ATPase and the secondary antibody is a fluorescein-labeled rabbit anti-mouse IgG. Basolateral localization is seen in all tubules. Distal nephron segments stain more brightly than proximal tubules. Only a single mesangial cell of the glomerulus is stained. ×500.

The functional screening procedures previously described can also be used in physiologic studies. These studies use essentially the same techniques to examine enzyme activity but are specifically designed to examine the functional characteristics of the molecule and how the transport molecule may behave under different physiologic or ionic conditions. These studies can give information relative to the active configuration of the enzyme and the changes in configuration of the enzyme which occur during transport. They can also help in identification of inhibitory sites and ion-binding sites.

Ball[12] used monoclonal antibodies to Na^+,K^+-ATPase to better characterize the pump molecule. These studies compared monoclonal antibody reagents with polyvalent antisera and revealed that only antibodies directed against the α subunit were effective inhibitors of activities. They also inferred that functionally important molecular sites were unaltered by delipidation and separation from the membrane. In addition, they were able to examine tryptic digests of the enzyme and derive a partial amino acid sequence. Work by McDonough et al.[13] defined antibody-binding regions on the sodium pump subunits that gave information relative to the orientation of the molecule within the membrane. Fambrough and Bayne[14] identified multiple forms of Na^+,K^+-ATPase using monoclonal antibodies. Caplan et al.,[15] using monoclonal antibody and pulse–chase experiments, found that the α subunit was always associated with membrane fractions and that little or no peptide was present in a cytoplasmic pool. These and other studies show the feasibility of using these molecular probes to gain a more complete understanding of the molecule and its function.

Another potentially useful application of monoclonal antibody is in immunoaffinity chromatography. The monoclonal antibody is bound directly to agarose beads and an affinity column is prepared which can then be used for purification of the enzyme subunits. A major advantage of this methodology is again related to the specificity of the monoclonal antibody in identifying a specific protein as opposed to gel electrophoresis where different proteins of the same molecular weight may comigrate. An interesting variation has been utilized by Smith and Garcia-Perez,[16] where monoclonal antibodies are bound to tissue culture dishes to affinity purify specific cell types. In addition, cell populations may be sorted by fluores-

[12] W. J. Ball, Jr., *Biochemistry* **23**, 2275 (1984).
[13] A. A. McDonough, A. Hiatt, and I. S. Edelman, *J. Membr. Biol.* **69**, 13 (1982).
[14] D. M. Fambrough and E. K. Bayne, *J. Biol. Chem.* **258**, 3926 (1983).
[15] M. J. Caplan, M. Kashgarian, D. Biemesderfer, B. Forbush III, and J. D. Jamieson, *J. Cell Biol.* **97**, 115 (1983).
[16] W. L. Smith and A. Garcia-Perez, *Am. J. Physiol.* **248**, F1 (1985).

cence-activated cell-sorting methodology or rosette formation and separation. The immunoaffinity techniques can be applied not only to positively select for Na^+,K^+-ATPase or Na^+,K^+-ATPase-containing cells but also to eliminate Na^+,K^+-ATPase as a contaminating protein, to study the residual proteins in cell membranes, or to separate Na^+,K^+-ATPase-rich cells from proton-transporting cells.

Another major application of monoclonal antibodies is as an immunocytochemical localization to ascertain the distribution of the enzyme to specific cellular domains and to different tissues.[6] Combined with improved immunocytochemical techniques such as cryosectioning and immunolabeling with colloidal gold, accurate and quantitative immunocytochemical studies of the enzyme can be carried out.

Because glutaraldehyde even at concentrations as low as 0.25% alters the antigenicity of Na^+,K^+-ATPase, alternate fixation protocols are necessary. The periodate–lysine–paraformaldehyde fixative (see above) originally described by McLean and Nakane[11] provides adequate structural preservation as well as excellent retention of the antigenicity of the Na^+,K^+-ATPase. The fixative should be prepared just prior to use. Perfusion fixation is preferable to immersion fixation in the study of intact organs. However, immersion fixation provides adequate fixation in isolated tubules or epithelial preparations or with tissue culture cells. After fixation for a period of up to 6 hr, the tissue is washed with 0.1 M Na_2HPO_4 for 1 hr at 4°. Tissue is then frozen rapidly by plunging into swirling liquid Freon-22 cooled by liquid nitrogen. Whole organs are cut into 16-mm-thick cryosections on a cryostat and then placed immediately into 1% BSA in PBS. Isolated tubules or epithelium or tissue culture cells are also thawed in 1% bovine serum albumin in phosphate-buffered saline at 4°. Freeze–thawing appears to permit access of the monoclonal antibody to intracellular domains. For immunoperoxidase labeling the tissue is preincubated in a goat anti-rabbit IgG (5 mg/ml) in PBS with 2% BSA for 1–3 hr as a general blocking step and to block any endogenous FC receptors which may be present. After three washes with PBS plus 1% BSA, the sections are incubated with 50–100 mg/ml of the monoclonal antibody. The sections are again washed three to five times with PBS and incubated for 2–4 hr with peroxidase-conjugated sheep anti-mouse Fab of IgG. Following repeated washing with PBS, the tissue is fixed for 1 hr with 1.5% glutaraldehyde in 0.1 M sodium cacodylate buffer, pH 7.4, with 5% sucrose. This is followed by three washes in 50 mM Tris–HCl, pH 7.4, with 7.5% sucrose. The tissue is placed in 0.2% diaminobenzidine in Tris sucrose buffer for 5 min at which time H_2O_2 is added to produce a final concentration of 0.01%. Incubation is carried out for 5–20 min at room temperature depending on the amount of Na^+,K^+-ATPase present. The

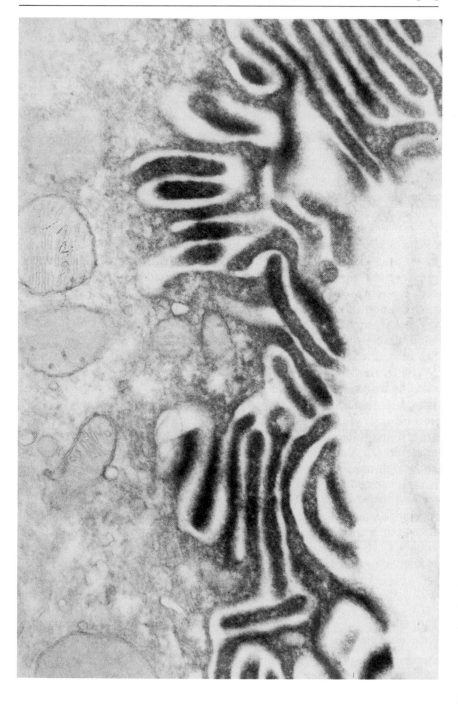

incubation times are varied by visualization of the reaction product by eye. The reaction is stopped by washing the sections in Tris buffer. The tissue is then treated with osmium tetroxide, dehydrated, and embedded in plastic for electron microscopy. Thin sections are cut and may be viewed unstained or stained with 2% aqueous uranyl acetate and lead citrate and then examined with an electron microscope (Fig. 2).

For immunogold labeling either the monoclonal antibody may be bound to gold directly or a secondary antibody bound to gold may be used to localize the primary monoclonal antibody. Colloidal gold solutions are prepared using the method of Faulk and Taylor[17] and bound to either the monoclonal antibody or to a secondary rabbit anti-mouse IgG by the method of Slot and Geuze.[18] With the two-step procedure, tissue that has been fixed with PLP and then frozen, cryosectioned, and thawed is incubated with the monoclonal antibody for 1–2 hr. The tissue is then washed with PBS and embedded in a polar embedding media such as Lowacryl K4M, as has been described by Altman et al.[19] Thin sections are mounted on Formvar-coated grids and then labeled with a secondary rabbit anti-mouse IgG gold conjugate (Fig. 3). For the single-step surface localization technique after fixation and freeze–thawing the tissue is embedded in Lowacryl without prior incubation with the antibody. After thin sectioning, the sections are incubated with the primary antibody labeled with colloidal gold. The advantage of the two-step technique is that a certain degree of amplification occurs over that seen with the direct labeling. An advantage of the direct labeling technique is that it may be used quantitatively to determine the density of antigenic sites. A caveat which must be remembered with monoclonal antibody immunocytomchemistry is that a single antibody may not reveal every place where the molecule is located. Configurational changes or neighboring proteins may hide the antigenic epitope from recognition by the antibody. Pre- or posttransitional microheterogeneity may also affect recognition.

---

[17] W. P. Faulk and G. M. Taylor, *Biochemistry*, **8**, 1081 (1971).
[18] W. J. Slot and H. J. Geuze, *J. Cell Biol.* **90** (1981).
[19] L. H. Altman, B. G. Schneider, and D. S. Papermaster, *J. Cell Biol.* **97**, 309a (1983).

FIG. 2. Electron micrograph of basal infoldings of renal tubular epithelial cell. Tissue was fixed with PLP and incubated with a moncolonal antibody to an intracellular domain of the α subunit of Na^+,K^+-ATPase followed by a secondary antibody to mouse IgG labeled with peroxidase. Reaction product is present along the cytoplasmic surface of the plasmalemma. ×42,000.

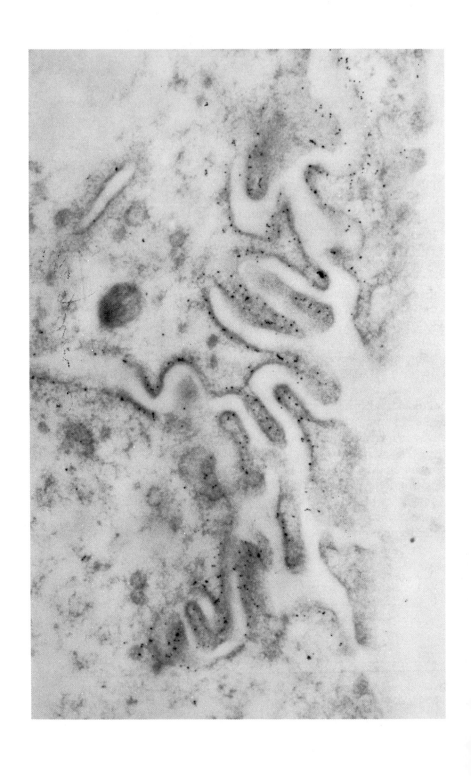

Summary

Application of monoclonal antibody techniques to the study of Na$^+$,K$^+$-ATPase offers a unique opportunity to investigate the molecular basis of sodium potassium transport. As a highly specific tool it can help define the transport process of a molecular, epithelial, and organ level by allowing biochemical, physiological, and morphological approaches to use a single probe. With the use of proper screening procedures, a highly specific reagent can be developed with great utility for a variety of different technical approaches.

FIG. 3. Electron micrograph of basal infoldings of a renal tubular epithelial cell. PLP-fixed tissue was incubated with the same antibody used in Fig. 2. After Lowacryl embedding, thin sections were incubated with a secondary antibody to mouse IgG labeled with 5-nm colloidal gold particles. Punctate colloidal gold particles are seen decorating the internal aspect of the cell membrane. ×42,000.

Section X

Microscopy of Na^+,K^+-ATPase

[37] Histochemical Localization of Na^+,K^+-ATPase

By HIROSHI MAYAHARA and KAZUO OGAWA

There are three techniques with independent rationale for the microscopic localization of Na^+,K^+-ATPase: the enzyme cytochemical method developed by Ernst,[1,2] the autoradiographic method for the localization of ouabain-binding sites by Stirling,[3] and the immunocytochemical method by Kyte.[4]

Enzyme Cytochemistry of Na^+,K^+-ATPase

Enzyme Cytochemistry of Na^+, K^+-ATPase Using ATP as Substrate

In 1957, Wachstein and Meisel[5] established a simple histochemical method for localizing various phosphatase activities including ATPase. In this method, ATP is used as a substrate and lead nitrate as a capture reagent which reacts with phosphate ions liberated by the enzymatic hydrolysis of ATP and forms water-insoluble precipitates on or close to the site of enzyme activity. The species of ATPase activity detected by this method is not that of Na^+,K^+-ATPase because the original Wachstein–Meisel reaction medium contains only Mg^{2+} as activating ions. To localize Na^+,K^+-ATPase, it is necessary to show the enhancement of the activity by Na^+ and K^+ and the inhibition by ouabain. In the past, several authors have claimed that they had succeeded in localizing Na^+,K^+-ATPase activity using Wachstein–Meisel-type incubation media.[6–9] However, none of these methods has been broadly applied because they had very poor reproducibility.[10] After intensive efforts to find the cause of the failure to localize Na^+,K^+-ATPase activity with the Wachstein–Meisel-type methods, it became clear that the enzyme activity is easily lost by weak fixa-

---

[1] S. A. Ernst, *J. Histochem. Cytochem.* **20**, 13 (1972).
[2] S. A. Ernst, *J. Histochem. Cytochem.* **20**, 23 (1972).
[3] C. E. Stirling, *J. Cell Biol.* **53**, 704 (1972).
[4] J. Kyte, *J. Cell Biol.* **68**, 287 (1976).
[5] M. Wachstein and E. Meisel, *Am. J. Clin. Pathol.* **27**, 13 (1957).
[6] I. T. McClurkin, *J. Histochem. Cytochem.* **12**, 654 (1964).
[7] A. Palkama and R. Uusitalo, *Ann. Med. Exp. Biol. Fenn.* **48**, 49 (1970).
[8] T. Tervo and A. Palkama, *Exp. Eye Res.* **21**, 269 (1975).
[9] J. Russo and P. Wells, *J. Histochem. Cytochem.* **25**, 135 (1977).
[10] J. A. Firth, *Histochem. J.* **10**, 253 (1978).

tion or by the presence of heavy metal ions of low concentrations.[11,12] Marchesi and Palade[12] used a modified Wachstein–Meisel medium with a low (0.5 mM) concentration of lead nitrate to localize Na^+,K^+-ATPase activity in unfixed erythrocyte ghost membranes. Although some remaining Na^+,K^+-ATPase activity was shown biochemically under these conditions, these investigators failed to show any cytochemically detectable activation by Na^+ and K^+ or inhibition by ouabain. Moreover, attempts to apply this method to other tissues met with another difficulty; the heavy deposition of nonspecific reaction products on the nuclei and other organelles, a condition which occurs inevitably under such low lead concentrations.[13,14] Thus, it is generally accepted that there is no reliable enzyme cytochemical method to localize Na^+,K^+-ATPase using ATP as substrate at present.[10,15]

Enzyme Cytochemistry of Na^+, K^+-ATPase Using p-Nitrophenyl Phosphate (NPP) as Substrate

Strontium–Lead Method. In 1972, Ernst[1,2] developed a cytochemical method for localizing Na^+,K^+-ATPase without using ATP but using p-nitrophenyl phosphate as a substrate. In this method, NPP is used as a substrate for the K^+-dependent phosphatase component of the Na^+,K^+-ATPase complex,[16–18] and strontium chloride is used as the first capturing reagent for the liberated phosphate ions from NPP. The reaction sequence is two step, in which the first reaction product, SrP_i, is converted into lead phosphate in the second reaction medium because SrP_i is not suitable for electron microscopy owing to the low electron density. The most significant feature of this method is the successful demonstration of the activation by K^+ and the inhibition by ouabain. Thus, the Sr–Pb method has been widely accepted among cytochemists and successfully applied to localize Na^+,K^+-ATPase or the sites of sodium pumps in a variety of tissues (Table I).

Cytochemical Procedure of Sr–Pb Method[19]

Tissue Preparation: Tissues are perfused for 1 min with cold physiological saline followed by a 10-min perfusion with cold 1% paraformalde-

[11] J. M. Tomey, *Nature* (*London*) **210**, 820 (1966).
[12] V. T. Marchesi and G. E. Palade, *J. Cell Biol.* **35**, 385 (1967).
[13] H. L. Moses and A. S. Rosenthal, *J. Histochem. Cytochem.* **16**, 530 (1968).
[14] N. O. Jacobsen and P. L. Jørgensen, *J. Histochem. Cytochem.* **17**, 443 (1969).
[15] S. A. Ernst and S. R. Hootman, *Histochem. J.* **13**, 397 (1981).
[16] J. D. Judah, K. Ahmed, and A. E. M. McLean, *Biochim. Biophys. Acta* **65**, 472 (1962).
[17] M. Fujita, T. Nakao, Y. Tashima, N. Mizuno, K. Nagano, and M. Nakao, *Biochim. Biophys. Acta* **117**, 42 (1966).
[18] H. Bader and A. K. Sen, *Biochim. Biophys. Acta* **118**, 116 (1966).
[19] S. A. Ernst, *J. Cell Biol.* **66**, 586 (1975).

hyde–0.25% glutaraldehyde in 0.1 M cacodylate buffer, pH 7.5. The fixed tissues are quickly excised and cut into small blocks (3 mm^3), rinsed in cacodylate and Tris–HCl buffers, and frozen in a cryostat for immediate use.

Cytochemical Procedures: Cryostat sections (50 μm) of tissues are incubated in the reaction medium containing 20 mM disodium or diTris NPP, 20 mM MgCl$_2$, 30 mM KCl, 20 mM SrCl$_2$, and 250 mM Tris–HCl buffer, pH 9.0. Incubations are performed at room temperature or at 30° for 15–60 min. After incubation, the sections are rinsed with 0.1 M Tris buffer, pH 9.0, and then treated with 2% Pb(NO$_3$)$_2$ to convert precipitated strontium phosphate to lead phosphate for visualization. The sections are then washed and postfixed for 15 min with 1% OsO$_4$. Some sections are treated after osmication with 1% (NH$_4$)$_2$S to convert lead phosphate to lead sulfide for light microscopy. Osmicated sections are dehydrated in ethanol and embedded in the low-viscosity resin of Spurr.[20] Sections for light microscopy (1 μm) are cut with an ultramicrotome and lightly stained with methylene blue. Ultrathin sections for electron microscopy are cut with an ultramicrotome and examined either unstained or doubly stained with uranyl acetate and lead citrate in an electron microscope.

Cytochemical Controls: The following controls are carried out to substantiate the specificity of the cytochemical procedure because NPP is not an intrinsic substrate for Na$^+$,K$^+$-ATPase:

1. Removing either K$^+$ or Mg^{2+} from the incubation medium to demonstrate the dependency of the activity on these ions
2. Inhibition of the reaction by adding ouabain (10 mM) to the reaction medium
3. To demonstrate the localization of nonspecific alkaline phosphatase (ALPase) activity and to differentiate it from that of K$^+$-NPPase, since NPP can be a substrate of ALPase. This is achieved by substituting β-glycerophosphate for NPP in the medium and by showing that the sites of ALPase activity are insensitive to the absence of K$^+$ or the presence of ouabain in the medium, and that they are sensitive to 10 mM cysteine, a potent inhibitor of ALPase
4. To demonstrate the loss of precipitation by removing the substrate. This is necessary to compare the sites of enzymatically produced precipitates from those produced nonenzymatically

Localization of K$^+$-NPPase Activity by the Sr–Pb Method: The final reaction product, lead phosphate, is found along the inner (cytoplasmic) surface of the plasma membrane in a variety of transporting epithelia (see

---

[20] A. R. Spurr, *J. Ultrastruct. Res.* **26**, 31 (1969).

Table I). In some cases, the reaction products are present on the nuclei or basement membranes, perhaps because of the diffusion artifacts.

Advantages and Disadvantages of the Sr–Pb Method: The advantages of this method, when compared with the Wachstein–Meisel-type methods using ATP as substrate, are that it can clearly demonstrate the activation of the phosphatase reaction by K^+ and its inhibition by ouabain. These characteristics of the reaction provide a strong theoretical background to qualify this method as a cytochemical tool for localizing the K^+-NPPase component of the Na^+,K^+-ATPase complex. On the other hand, there are two methodological disadvantages in the Sr–Pb method. First, the reaction sequence to obtain the final reaction products is indirect. As the first reaction product (SrP_i) has a low electron density and is soluble in OsO_4 solutions,[2] it must be converted into lead phosphate for electron microscopy by dipping the specimen in a lead nitrate solution. Such a conversion step may cause redistribution of the reaction products as repeatedly ar-

TABLE I
REPORTS OF K^+-NPPase CYTOCHEMISTRY

Methods	Tissue	References
Sr–Pb	Avian salt gland	1
	Rat and rabbit kidney cortex	19, 41
	Desert iguana nasal gland	42
	Rat cornea	43
	Rat submandibular gland	44
	Frog, rabbit, and rat choroid plexus	45
	Shark rectal gland	46
	Rat cerebral cortex	47, 48
	Marine turtle salt gland	49
	Chick chorioallantois	50
	Rat cerebral cortical capillary	51
	Rat liver	52, 53
	Rat and rabbit colon	54
	Teleost gill chloride cell	24, 55
	Human blood platelet	56
	Rat lens	57
	Rat kidney medulla	25
	Guinea pig inner ear	58
$CoCl_2$	Rat central nervous system	27
	Rat brain synaptosome	28
	Rat, rabbit, and human kidney	29
	Dog submandibular gland	30
Lead citrate	Rat kidney cortex and medulla	31, 33
	Rat choroid plexus	72

TABLE I (continued)

Methods	Tissue	References
	Guinea pig retina	34, 59
	Developing rat kidney	60
	Rat cardiac muscle	61
	Rabbit, guinea pig, rat, and mouse kidney	62
	Mouse central and peripheral nervous system	63
	Rat central nervous system	64, 65
	Rat sweat gland	66
	Guinea pig inner ear	67
	Human thyroid	68
	Rat lacrimal gland	69
	Rat liver	70
	Rat brain capillary endothelium	71
	Mouse gall bladder epithelium	a
	Teleost epidermal ionocytes	b
	Chick epiphyseal growth plate	c
	Rat kidney macula densa	d
	Crustacean calcium transporting epithelium	e
	Rat aortic endothelium and smooth muscle cells	f
	Guinea pig parietal cells	g

[a] T. Fujimoto and K. Ogawa, *J. Ultrastruct. Res.* **79**, 327 (1982).
[b] G. Zaccone, S. Fasulo, and A. Licata, *Histochem.* **81**, 47 (1984).
[c] T. Akisaka and C. V. Gay, *Acta Histochem. Cytochem.* **19**, 21 (1986).
[d] H. Mayahara and K. Ogawa, in "Recent Development of Electron Microscopy 1985, Proceedings of the 3rd Chinese–Japanese Electron Microscopy Seminar" (H. Hashimoto et al., eds.), p. 263. Jap. Soc. Electron Microsc., Kyoto, 1986.
[e] J. C. Meyran and F. Graf, *Histochem.* **85**, 313 (1986).
[f] K. S. Ogawa, K. Fujimoto, and K. Ogawa, *Acta Histochem. Cytochem.* **19**, 601 (1986).
[g] K. S. Ogawa, K. Fujimoto, and K. Ogawa, *Acta Histochem. Cytochem.* **20**, 197 (1987).

gued by cytochemists.[21–23] The nuclear staining and diffusion of the reaction products seen in photographs in reports of studies using the Sr–Pb method[2,19,24,25] may reflect this methodological limitation. Second, the cytochemical reaction in the Sr–Pb method is performed at pH 9.0 although the optimal pH for K^+-NPPase is 7.0.[26] This is because the first reaction product, SrP_i, is soluble at neutral pH. However, the sensitivity of this

[21] A. Mizukami and R. J. Barrnett, *Nature (London)* **206**, 1001 (1965).
[22] H. Mayahara and K. Ogawa, *J. Histochem. Cytochem.* **16**, 721 (1968).
[23] E. Essner, in "Electron Microscopy of Enzymes: Principles and Methods" (M. H. Hayat, ed.), p. 57. Van Nostrand-Reinhold, New York, 1973.
[24] S. R. Hootman and C. W. Philpott, *Anat. Rec.* **193**, 99 (1979).
[25] S. A. Ernst and J. H. Schreiber, *J. Cell Biol.* **91**, 803 (1981).
[26] R. W. Albers and G. J. Koval, *J. Biol. Chem.* **247**, 3088 (1972).

method severely deteriorates with such a pH shift because most of the enzyme activity is lost at pH 9.0.[1,26]

Cobalt Chloride Method. The second disadvantage of the Sr–Pb method was resolved by Guth and Albers,[27] who presented the $CoCl_2$ method to localize K^+-NPPase activity of the Na^+,K^+-ATPase complex. They added 25% (v/v) dimethyl sulfoxide (DMSO) to the incubation medium; this shifts the optimum pH of K^+-NPPase from 7.0 to 9.0, enabling the incubation at the optimum pH, and, at the same time, increasing the specific activity severalfold.[26]

Procedure of the $CoCl_2$ Method. Cryostat sections of nonfixed tissues, 6–24 μm, are cut, mounted on albuminized slides, and allowed to dry at room temperature for 15–20 min. The slides are then placed in the complete incubation medium which contains (final concentrations): KCl (30 mM), $MgCl_2$ (5 mM), p-nitrophenyl phosphate (5 mM), dimethyl sulfoxide (25%, v/v) and 2-amino-2-methyl-1-propanol buffer (70 mM, adjusted to pH 9.0 with HCl). For enzyme inhibition, 1 mM ouabain (final concentration) is added to the incubation mixture. After incubation for 1–3 hr at 37°, the slides are placed directly in 2% $CoCl_2$ for 5 min, rinsed briefly in distilled water, washed in three 30-sec changes of the same buffer (70 mM, pH 9.0), and placed in a 1:50 (v/v) dilution of ammonium sulfide for 3 min. These sections are then washed for 5 min in running tap water, dehydrated in ethanol and xylene, and mounted in Permount.

Advantages and Disadvantages of $CoCl_2$ Method: The advantages of this method are that the cytochemical reaction is performed at the optimum pH of K^+-NPPase, pH 9.0 in the presence of 25% DMSO, and that the incubation medium does not contain heavy metal ions which may inhibit the enzyme activity. Because of these two factors, it may be expected that this method is highly sensitive; however, the sensitivity is far lower than that of the Sr–Pb method or lead citrate method (mentioned later) judging from the unreasonably long incubation time of this method (1–3 hr at 37°). The disadvantage of the $CoCl_2$ method is that the postcoupling system involved allows only light microscopic observation because of the poor localization of the reaction products.[27–30]

One-Step Lead Citrate Method. In 1980, Mayahara et al.[31] presented a new one-step lead citrate method for the light and electron microscopic localization of the K^+-dependent phosphatase activity of Na^+,K^+-ATPase complex. In this method, all the disadvantages in the Sr–Pb or $CoCl_2$

[27] L. Guth and R. W. Albers, *J. Histochem. Cytochem.* **22**, 320 (1974).
[28] A. Daniel and L. Guth, *Exp. Neurol.* **47**, 181 (1975).
[29] R. Beeuwkes III and S. Rosen, *J. Histochem. Cytochem.* **23**, 828 (1975).
[30] I. Nakagaki, T. Goto, S. Sasaki, and Y. Imai, *J. Histochem. Cytochem.* **26**, 835 (1978).
[31] H. Mayahara, K. Fujimoto, T. Ando, and K. Ogawa, *Histochemistry* **67**, 125 (1980).

methods are resolved by using lead citrate for the direct capture of the liberated phosphate ions from NPP, DMSO (25%, v/v) as an activator, and a glycine buffer as a stabilizer of the incubation medium. The use of this one-step method allows a better localization of the reaction products. The use of DMSO not only amplifies the K^+-NPPase activity, but also shifts the optimum pH to around 9.0,[26] where the incubation is performed in this new method, allowing for high sensitivity. The use of a glycine buffer in this method has created extremely fine reaction products caused by the chelation of lead by glycine. Moreover, glycine may be another factor for higher sensitivity in this method because it reportedly enhances Na^+,K^+-ATPase activity by chelating heavy metal ions.[32] With carefully controlled biochemical studies,[31] it was shown that enough K^+-NPPase activity remains in the properly fixed specimens in the presence of 4 mM lead citrate in the present incubation medium, and the cytochemical localization of K^+-NPPase was successfully made by this method[31,33] (see Table I).

Cytochemical Procedures of the Lead Citrate Method

Tissue Preparation: Tissues are fixed either by immersion or perfusion. In immersion fixation, tissue slices, less than 1 mm thick, are fixed in a cold mixture of 2% paraformaldehyde and 0.5% glutaraldehyde in 0.1 M cacodylate or 0.05 M PIPES buffer, pH 7.2, for 30–50 min.[31] In perfusion fixation, tissues are perfused with a cold mixture of 2% paraformaldehyde and 0.25% glutaraldehyde in the same buffer, pH 7.2, for 5 min.[34] After fixation, the tissues are washed overnight in a 0.01 M cacodylate or 0.05 M PIPES buffer, pH 7.2, containing 8% sucrose and 10% (v/v) DMSO. After they are washed, frozen sections for light microscopy, 10–14 µm thick, are cut with an electrofreezing microtome. The sections for electron microscopy, 30–50 µm thick, are cut with a vibratome[35] or a microslicer[36] without freezing.

Cytochemical Procedures: Tissue sections are incubated in a reaction medium whose composition is shown on the next page.

The appropriate incubation time depends on the intensity of the enzyme activity of the specimen: 5–15 min at room temperature for kidney distal tubules,[33] 10–25 min for retinal pigment epithelium.[34] After incubation, frozen sections are washed in distilled water, treated with 1% ammo-

[32] S. C. Spechet and J. D. Robinson, *Arch. Biochem. Biophys.* **154**, 314 (1973).
[33] H. Mayahara and K. Ogawa, *Acta Histochem. Cytochem.* **13**, 90 (1980).
[34] S. Ueno, H. Mayahara, I. Tsukahara, and K. Ogawa, *Acta Histochem. Cytochem.* **13**, 679 (1980).
[35] R. E. Smith, *J. Histochem. Cytochem.* **18**, 590 (1970).
[36] H. Mayahara, K. Fujimoto, T. Noda, I. Tamura, and K. Ogawa, *Acta Histochem. Cytochem.* **14**, 211 (1981).

Reagents[a]	Final concentrations
1.0 M Glycine–KOH buffer, pH 9.0	2.5 ml (glycine, 250 mM; K$^+$, 25 mM)
1% Lead citrate[b]	4.0 ml (lead citrate, 4.0 mM; K$^+$, 20 mM)
Dimethyl sulfoxide (DMSO)	2.5 ml (25%, v/v)
0.1 M p-nitrophenyl phosphate (NPP, Mg^{2+} salt)[c]	1.0 ml (NPP, 10 mM; Mg^{2+}, 10 mM)
Levamisole[d]	6.0 mg (2.5 mM)
Final solution	10.0 ml, pH 8.8

[a] All the reagents must be added in the top-to-bottom order indicated dropwise under stirring. The completed medium is a crystal-clear, light yellow solution stable for hours.

[b] Dissolved in 50 mM KOH. Product of K and K Lab., New York, is recommended. Must be freshly prepared before use.

[c] Magnesium salt of NPP (Sigma) has a slightly low solubility. It takes about 30 min to dissolve under rigorous shaking. Sodium salt of NPP is usable but not recommended because Na$^+$ inhibits activation by K$^+$.[37] When sodium salt is used, 10 mM MgCl$_2$ must be added to the incubation medium. Di-Tris salt of NPP is not usable because it forms a sediment in the medium.

[d] Tetramisole (5.0 mM) or bromotetramisole (0.5 mM) can also be used. These are added as specific inhibitors of nonspecific alkaline phosphatase (ALPase).[38] Be careful, however, that these inhibitors do not inhibit the intestinal ALPase activity.[38]

nium sulfide for 2 min for visualization, mounted in glycerin jelly, and observed with a light microscope. For electron microscopy, nonfrozen sections are washed in 0.1 M cacodylate or 0.05 M PIPES buffer, pH 7.2, containing 8% sucrose but no DMSO, dehydrated with alcohol, embedded in the low-viscosity resin of Spurr.[20] Ultrathin sections are cut with an ultrotome, poststained with uranyl acetate, and observed with an electron microscope.

Cytochemical Controls: The following controls should be carried out to substantiate the specificity of the cytochemical procedures:

1. To demonstrate the K$^+$ dependency of the NPPase activity, K$^+$ is removed from the original incubation medium. This is achieved by substituting glycine–NaOH buffer for glycine–KOH, and dissolving lead citrate in 50 mM NaOH

2. To demonstrate the inhibition of the reaction by ouabain, add ouabain (10 mM) to the incubation medium. Use a glass rod to grind the ouabain crystal to dissolve quicker

[37] K. Nagai, F. Izumi, and H. Yoshida, *J. Biochem.* (*Tokyo*) **59,** 259 (1966).
[38] M. Borgers, *J. Histochem. Cytochem.* **21,** 812 (1973).

3. To compare the localization of nonspecific alkaline phosphatase (ALPase) activity with that of K^+-NPPase, demonstrate the ALPase activity by removing levamisole and K^+ from the incubation medium or by the lead citrate method for ALPase by Mayahara et al.[39] with sodium β-glycerophosphate as a substrate

Advantages and Disadvantages of the Lead Citrate Method: The advantages of the lead citrate method are as follows:

1. The localization of the reaction products at the electron microscopic level seems to be the best among the three enzyme cytochemical methods, perhaps because the direct capturing system eliminates possible diffusion artifacts which occur during the conversion step in the earlier methods (refer to the papers listed in Table I).

2. The highest sensitivity among the three methods was obtained in the lead citrate method by using DMSO, which amplifies the K^+-NPPase severalfold and at the same time enables incubation to occur at the optimum pH of the enzyme.[26]

3. Moreover, the use of glycine buffer, a metal chelator,[40] is another factor that increases the sensitivity of the lead citrate method by protecting the enzyme activity by chelating the heavy metal ions used as a capture.[32]

4. The size of the reaction products seems to be the finest in the lead citrate method, perhaps because of the lead-chelating effect of glycine and citrate in the reaction medium. This means that the highest resolution at the electron microscopic level is available by this method.

The disadvantage of this method is that it does not allow for quantitative study because the reaction product is amplified at the sites of K^+-NPPase activity as it is in all other cytochemical methods.

Localization of K^+-NPPase Activity with the Lead Citrate Method: The reported localizations of the reaction product, lead phosphate, obtained by the lead citrate method are similar to those by the Sr–Pb method and are found along the inner (cytoplasmic) surface of the plasma membrane in a variety of cells and tissues (Figs. 1–3, and Table I[41–72]). In the

---

[39] H. Mayahara, H. Hirano, T. Saito, and K. Ogawa, *Histochemie* **11**, 88 (1967).
[40] S. Chaberek and A. E. Martell, "Organic Sequesting Agents," p. 540. Wiley, New York, 1959.
[41] J. A. Firth, *J. Histochem. Cytochem.* **22**, 1163 (1974).
[42] R. A. Ellis and C. C. Goertemiller, Jr., *Anat. Rec.* **180**, 285 (1974).
[43] P. M. Leuenberger and A. B. Novikoff, *J. Cell Biol.* **60**, 721 (1974).
[44] B. I. Bogard, *J. Ultrastruct. Res.* **52**, 139 (1975).
[45] T. H. Milhorat, D. A. Davis, and M. K. Hommock, *Brain Res.* **99**, 170 (1975).
[46] C. C. Goertemiller, Jr., and R. A. Ellis, *Cell Tissue Res.* **175**, 101 (1976).

(footnotes continued on page 427)

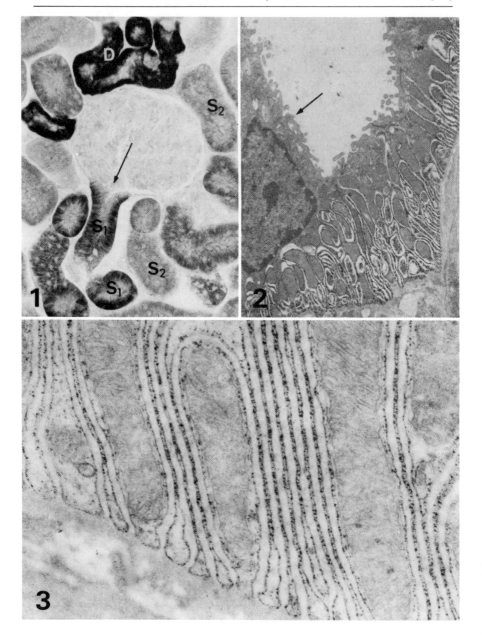

[47] W. L. Stahl and S. H. Broderson, *J. Histochem. Cytochem.* **24**, 731 (1976).
[48] S. H. Broderson, D. L. Patton, and W. L. Stahl, *J. Cell Biol.* **77**, R13 (1978).
[49] R. A. Ellis and C. C. Goertemiller, Jr., *Cytobiology* **13**, 1 (1976).
[50] A. S. M. Salenddin, C. P. M. Kyriakides, A. Peacock, and K. Simkies, *Comp. Biochem. Physiol.* **54A**, 7 (1976).
[51] J. A. Firth, *Experientia* **33**, 1093 (1977).
[52] B. L. Blitzer and J. L. Boyer, *J. Clin. Invest.* **62**, 1104 (1978).
[53] D. S. Latham and M. Kashgarian, *Gastroenterology* **76**, 988 (1979).
[54] P. B. Vengasa and V. Hopfer, *J. Histochem. Cytochem.* **27**, 1231 (1979).
[55] S. R. Hootman and C. W. Philpott, *Am. J. Physiol.* **238**, R199 (1980).
[56] L. S. Cutler, M. B. Feinstein, and C. P. Christian, *J. Histochem. Cytochem.* **28**, 1183 (1980).
[57] N. J. Unakar and J. Y. Tsui, *Invest. Ophthalmol. Visual Sci.* **19**, 630 (1980).
[58] K. Mees, *Otolaryngology (Rochester, Minn.)* **95**, 277 (1983).
[59] S. Ueno, H. Mayahara, I. Tsukahara, and K. Ogawa, *Acta Histochem. Cytochem.* **14**, 186 (1981).
[60] H. Mayahara, T. Ando, Y. Ishikawa, and K. Ogawa, *Acta Histochem. Cytochem.* **15**, 421 (1982).
[61] K. Fujimoto and K. Ogawa, *Acta Histochem. Cytochem.* **15**, 388 (1982).
[62] T. Kuwahara, H. Mayahara, C. Kawai, and K. Ogawa, *Acta Histochem. Cytochem.* **15**, 717 (1982).
[63] A. W. Vorbrodt, A. S. Lossinsky, and H. M. Wisniewsky, *Brain Res.* **243**, 225 (1982).
[64] K. Inomata, H. Mayahara, K. Fujimoto, and K. Ogawa, *Acta Histochem. Cytochem.* **16**, 277 (1983).
[65] F. Nasu, *Acta Histochem. Cytochem.* **16**, 368 (1983).
[66] H. Mayahara, T. Ando, T. Fujimoto, and K. Ogawa, *J. Histochem. Cytochem.* **31**, 224 (1983).
[67] H. Seguchi and T. Kobayashi, *Acta Histochem. Cytochem.* **16**, 610 (1983).
[68] Y. Mizukami, F. Matsubara, and S. Matsukawa, *Lab. Invest.* **48**, 411 (1983).
[69] S. Ueno, H. Mayahara, M. Ueck, I. Tsukahara, and K. Ogawa, *Cell Tissue Res.* **234**, 497 (1983).
[70] K. Yamamoto, H. Mayahara, and K. Ogawa, *Acta Histochem. Cytochem.* **17**, 23 (1984).
[71] K. Inomata, T. Yoshioka, F. Nasu, and H. Mayahara, *Acta Anat.* **118**, 243 (1984).
[72] T. Masuzawa, T. Saito, and F. Sato, *Acta Histochem. Cytochem.* **13**, 394 (1980).

FIG. 1. A light micrograph of K^+-NPPase activity in rat kidney cortex. The strongest activity is found in the distal convoluted tubules (D), while a weaker activity is found in the first segments (S_1) of the proximal convoluted tubules, and the weakest activity in its second segments (S_2). The glomerulus is negative. The arrow shows a urinary pole which enables the identification of S_1. ×280.

FIG. 2. An electron micrograph of K^+-NPPase activity in a cortical distal tubule (rat kidney). The reaction products are found on the plasma membrane of basal infoldings. Note that no reaction products are found on the luminal plasma membrane (arrow). ×6600.

FIG. 3. A high-magnification electron micrograph of K^+-NPPase activity in a cell of the ascending thick segment of the loop of Henle (rat kidney). Note that extremely fine reaction products are found on the inner (cytoplasmic) surface of the plasma membrane of basal infoldings. ×40,000.

transporting epithelia, K^+-NPPase reaction products indicating the localization of Na^+,K^+-ATPase are localized on the basolateral plasma membranes except in the case of the choroid plexus, in which the activity is localized in the microvilli of the apical surface.[72]

Autoradiographic Method for Localizing Ouabain-Binding Sites

Ouabain, a cardiac glycoside, is a specific inhibitor of Na^+,K^+-ATPase and binds with a high affinity to Na^+,K^+-ATPase.[73] Therefore, ouabain is an ideal probe to localize Na^+,K^+-ATPase complexes in intact cells.

Autoradiographic Localization of [^{3}H]Ouabain-Binding Sites[3]

Cells or tissues are preincubated for 10 min at 37° in Ringer's solution buffered with bicarbonate, and then incubated in Ringer's solution containing 25 Ci/ml [^{3}H]ouabain and sufficient unlabeled inhibitor to reach 10^{-6} M. The incubation time is 30 min at 37°. For microscopy, the conventional methods of dehydration are inadequate since ouabain is highly soluble in most organic solvents. Therefore, the tissues are cut into small blocks and quickly frozen in liquid propane cooled to $-175°$ with liquid nitrogen. The frozen tissues are then freeze dried at a low temperature, fixed in osmium tetraoxide vapor, and vacuum embedded in a mixture of silicon and epoxy resin. Sections, 1–2 μm thick, cut with glass knives and collected over water, are then placed on microscope slides, coated with liquid photographic emulsion (Kodak NTB-2), and exposed for 4–25 days. After development, the sections are stained with basic fuchsin and examined by bright-field microscopy. The localizations of Na^+,K^+-ATPase obtained by this method are similar to those obtained by enzyme cytochemical methods for K^+-NPPase.[3,74,75,76] In some cases, however, the apical membranes exhibit weak binding of [^{3}H]ouabain,[75,76] indicating nonspecific binding.

The advantage of this method is that it allows the enzyme sites to be studied quantitatively by radioautographic grain counting and liquid scintillation counting on the paired samples. The major disadvantage is that the use of this method is limited to the light microscopic level because of the poor resolution of the radioautography at the electron microscopic level. Moreover, ouabain binds to the enzyme in a relatively reversible

---

[73] A. Schwartz, G. E. Lindenmayer, and J. C. Allen, *Pharmacol. Rev.* **27**, 3 (1975).
[74] J. W. Milles, S. A. Ernst, and D. R. Dibona, *J. Cell Biol.* **73**, 88 (1977).
[75] S. A. Ernst and J. W. Milles, *J. Cell Biol.* **75**, 74 (1977).
[76] J. L. F. Shaver and C. Stirling, *J. Cell Biol.* **76**, 278 (1978).

manner in some tissues, for example toad urinary bladder[77] and rat kidney,[78] resulting in a low density of the developed silver grains and possible diffusion artifacts. Further, there is a lower limit to the sensitivity of the technique, which has been indicated by the failure to localize Na^+,K^+-ATPase in cells with low levels of the enzyme (for example, erythrocytes and fibroblasts) with this method.[74] Detailed discussion of the validity of this technique and of its application for a variety of epithelial cells is not presented here because space is limited; however, excellent reviews are available.[79,80]

Immunocytochemical Method

In 1976, Kyte[4,81] used an immunocytochemical method to demonstrate the ultrastructural localization of Na^+,K^+-ATPase over the plasma membrane of the renal convoluted tubule from canine kidney. Rabbit antibodies, specific for either the holoenzyme or the catalytic subunit, were prepared. Ultrathin frozen sections were cut from fragments of renal cortex and specifically stained with antibodies, and then with ferritin-conjugated goat anti-rabbit γ-globulins. It was demonstrated that Na^+,K^+-ATPase is distributed uniformly and at a high concentration along the basolateral plasma membranes. When antiholoenzyme was used, the enzyme is distributed along the luminal surface of the tubules as well, but at a much lower concentration. However, no staining was shown when anti-large-chain antibody was used.[81]

The advantages of this method are that it allows a quantitative evaluation of the enzyme and a high resolution by electron microscopy. The disadvantage is that it is technically demanding to prepare purified antigens, excellent antibodies, and ultrathin frozen sections. Moreover, the positive reaction does not indicate the presence of active enzyme since this technique depends only on the antigenicity. For example, the positive reaction of the luminal surface of the renal proximal and distal tubules observed with the immunoferritin technique[4,81] may indicate the presence of the antigenic determinants, probably in the sialoglycoprotein, but not the presence of the active Na^+,K^+-ATPase.

An alternative immunocytochemical technique at the electron microscopic level was presented by Wood et al.[82] using horseradish peroxidase

[77] J. W. Milles and S. A. Ernst, *Biochim. Biophys. Acta* **375**, 268 (1975).
[78] J. C. Allen and A. Schwartz, *J. Pharmacol. Exp. Ther.* **168**, 42 (1969).
[79] S. A. Ernst and J. W. Milles, *J. Histochem. Cytochem.* **28**, 72 (1980).
[80] S. A. Ernst and S. A. Hootman, *Histochem. J.* **13**, 398 (1981).
[81] J. Kyte, *J. Cell Biol.* **68**, 304 (1976).
[82] J. G. Wood, D. H. Jean, J. N. Whitacker, B. J. McLaughlin, and R. W. Alvers, *J. Neurocytol.* **6**, 571 (1977).

(HRP) substituted for ferritin in the goat anti-rabbit IgG conjugate. The advantage of this method is that antigenic sites are localized with an enzyme marker which amplifies the sensitivity of the technique. However, it cannot be used for quantitative studies for the same reason.

As an immunocytochemical method for localizing Na^+,K^+-ATPase, the immunofluorescence method is also available although its use is limited to the light microscopic level. Details of this technique are not treated here as it is the subject of the following section. Detailed methods for preparing purified antigens and specific antibodies is also treated in the next section.

In conclusion, each method for localizing Na^+,K^+-ATPase has its advantages and disadvantages. It is advisable to select the best method matched with the purpose of the study or to use different methods in combination to reveal the real localization of the active Na^+,K^+-ATPase.

Addendum

A new cytochemical method[83] for detecting K^+-NPPase activity was published recently. This method is a modification of the method of Robinson and Karnovsky,[84,85] which uses cerium salt for a capture reagent and allows good electron microscopic localization of the reaction products. This method has a disadvantage of no applicability or poor applicability for light microscopy, but has an advantage of being useful at physiological pH.

[83] T. Kobayashi, T. Okada, and H. Seguchi, *J. Histochem. Cytochem.* **35**, 601 (1987).
[84] J. M. Robinson and M. J. Karnovsky, *J. Histochem. Cytochem.* **31**, 1190 (1986).
[85] J. M. Robinson and M. J. Karnovsky, *J. Histochem. Cytochem.* **31**, 1197 (1986).

[38] Analysis of Na^+,K^+-ATPase by Electron Microscopy

By ARVID B. MAUNSBACH, ELISABETH SKRIVER, and PETER L. JØRGENSEN

Introduction

Examination of isolated membranes by electron microscopy after negative staining and freeze–fracture can provide information about the structure of the protein molecules and their organization in the membranes. Such information may include frequency and dimensions of the molecules and the interrelationships of the molecules and their subunits. Negative staining primarily visualizes those parts of the membrane proteins that project above the surface of the lipid bilayer while freeze–fracture discloses intramembrane aspects of the protein units and, if the membranes have been etched, some surface characteristics of the membranes.

Na^+,K^+-ATPase can be isolated from outer renal medulla in pure, membrane-bound form by selective extraction of a crude membrane fraction with SDS (sodium dodecyl sulfate) in the presence of ATP followed by isopycnic centrifugation.[1] In ultrathin sections (Fig. 1) the preparations exhibit exclusively flat or cup-shaped membrane fragments which have diameters of 0.1–0.6 μm.[2] The uniformity and chemical purity of this preparation make it suitable for structural analysis of the enzyme after negative staining and freeze–fracture and the observed structural features of the membranes can be related to the enzymatic and chemical composition of the membranes.[3,4] In this chapter, we describe the procedures for analyses of membrane-bound Na^+,K^+-ATPase by electron microscopy after thin sectioning, negative staining, and freeze–fracture as well as the main steps in ultrastructural studies of Na^+,K^+-transport vesicles reconstituted with purified Na^+,K^+-ATPase.

Preparation of Membrane-Bound Na^+,K^+-ATPase for Electron Microscopy

Thin Sectioning

Fixation. The sample of purified Na^+,K^+-ATPase containing 50–100 μg of enzyme protein is diluted to 0.8 ml with 25 mM imidazole buffer, pH 7.2. A pellet is prepared by centrifugation, e.g., for 100 min at 40,000 rpm in a Beckman type 65 rotor using the adaptor for 0.8-ml tubes (Beckman Instruments, Spinco Div., Palo Alto, CA). After centrifugation the supernatant is removed with a Pasteur pipet and the pellet overlayered with 2% glutaraldehyde in 0.1 M cacodylate buffer, pH 7.2. After 2 hr the pellet is rinsed twice with 0.1 M cacodylate buffer, pH 7.2, and postfixed for 1 hr with 1% osmium tetroxide in 0.029 M veronal acetate buffer, pH 7.2.

---

[1] P. L. Jørgensen, *Biochim. Biophys. Acta* **356,** 53 (1974).
[2] A. B. Maunsbach and P. L. Jørgensen, *Proc. Int. Congr. Electron Microsc., 8th* **2,** 214 (1984).
[3] N. Deguchi, P. L. Jørgensen, and A. B. Maunsbach, *J. Cell Biol.* **75,** 619 (1977).
[4] A. B. Maunsbach, E. Skriver, and P. L. Jørgensen, in "International Cell Biology 1980–1981" (H. G. Schweiger, ed.), p. 711. Springer-Verlag, Berlin and Heidelberg, 1981.

An alternative fixation method, which gives similar results, is to fix the membranes in suspension by mixing the sample (50–100 μg enzyme protein in 0.4 ml imidazole buffer) with an equal amount of 4% glutaraldehyde in 0.1 M sodium cacodylate buffer, pH 7.2. After 2 hr the fixed membranes are sedimented, rinsed, and postfixed as described above.

After osmium tetroxide fixation the pellet is rinsed twice with veronal acetate buffer. To enhance contrast the pellet can be stained en bloc for 1 hr with 0.5% uranyl acetate in 0.05 M sodium maleate buffer at pH 5.2.[5] Before and after staining the pellet is rinsed twice in 0.05 M maleate buffer, pH 5.2.

Embedding, Sectioning, and Staining. The bottom of the tube with the pellet is cut off and the pellet dehydrated in a graded series of acetone or ethanol (70, 90, 95, 100%). If acetone is used care is taken that the pellet is not damaged when the cellulose tube dissolves toward the end of dehydration. Pellets are embedded either in Epon 812 or Vestopal W. During embedding the pellet is oriented in such a way that thin sections can be cut at a right angle to the surface of the pellet. In order to objectively judge the content of the pellet the ultrathin sections should reach from top to bottom of the pellet. The sections are stained either with uranyl acetate and lead citrate or lead citrate alone.[6] Additional structural information can be obtained if tannic acid is included in the glutaraldehyde fixative.[7]

Negative Staining

The sample of purified membrane-bound Na^+,K^+-ATPase is diluted with 25 mM imidazole buffer, pH 7.2, to a concentration of 0.1–0.2 mg of enzyme protein/ml. The enzyme suspension is kept at 0–4° until use. Negative staining can be carried out on carbon-reinforced Formvar films but carbon films supported by 300- or 400-mesh copper grids are preferable. The carbon films are prepared on freshly cleaved mica. Supporting films are made hydrophilic by glow discharge.[8] An alternative way to make the supporting film hydrophilic is to mix the enzyme sample with an equal volume of a solution containing 150 μg/ml bacitracin.[9]

Staining is initiated by placing approximately 5 μl of the enzyme suspension on the grid. The membrane fragments are allowed to settle on the grid for 1 min and excess liquid is removed with filter paper which has been moistened with distilled water in order to retard the removal of the

---

[5] M. J. Karnovsky, *J. Cell Biol.* **35**, 213 (1967).
[6] E. Reynolds, *J. Cell Biol.* **17**, 208 (1963).
[7] A. Saito, C.-T. Wang, and S. Fleischer, *J. Cell Biol.* **79**, 601 (1978).
[8] E. Namork and B. V. Johansen, *Ultramicroscopy* **7**, 321 (1982).
[9] D. W. Gregory and B. J. S. Pirie, *J. Microsc. (Oxford)* **99**, 261 (1973).

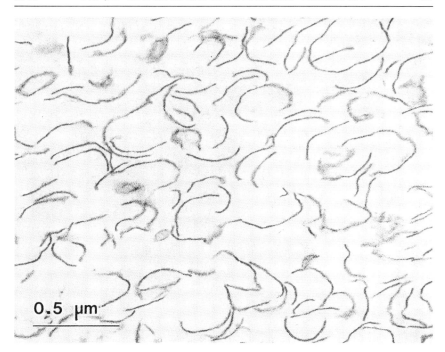

FIG. 1. Electron micrograph of a thin section through a pellet of purified, membrane-bound Na^+,K^+-ATPase from pig kidney. The pellet contains a uniform population of small membrane fragments, which are open and do not form closed vesicles. ×46,000. From Deguchi et al.[3]

enzyme suspension. If the preparation is left longer on the grid, e.g., 4 min, more membrane fragments become attached to the supporting film. Immediately after removal of the enzyme suspension a droplet of stain is placed on the grid with a clean Pasteur pipet. The stain may be either 1% uranyl acetate in distilled water (freshly prepared and filtered) or 2% phosphotungstic acid, which is adjusted to pH 7.2 with KOH or NaOH.[10] After 1 min the droplet is drained off with filter paper and the grid allowed to dry. It is examined in the electron microscope immediately or within 1–2 days (Figs. 2 and 3). If a thinner layer of uranyl acetate is desired staining is carried out with 0.5% uranyl acetate. However, two-dimensional crystals of Na^+,K^+-ATPase should be stained with uranyl acetate[11,12] since phosphostungstic acid often disturbs the regularity of the membrane crystals.

[10] E. A. Munn, this series, Vol. 32, p. 20.
[11] E. Skriver, A. B. Maunsbach, and P. L. Jørgensen, FEBS Lett. **131**, 219 (1981).
[12] E. Skriver, A. B. Maunsbach, H. Hebert, and P. L. Jørgensen, this volume [8].

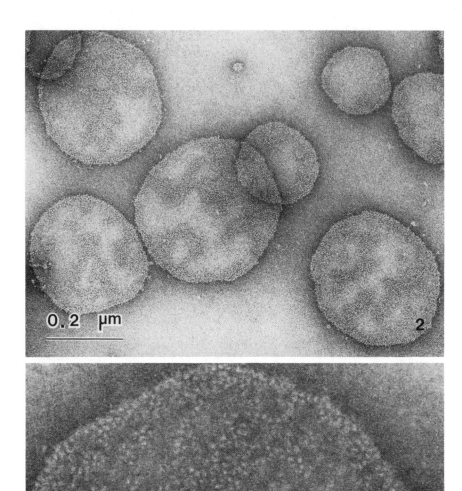

Freeze–Fracture and Freeze–Etching

Freeze–fracture[13] is carried out on frozen samples prepared from concentrated suspensions of membrane-bound Na^+,K^+-ATPase containing 10–20 mg enzyme protein/ml. The frozen enzyme may be fractured either unfixed or subsequent to fixation in suspension with glutaraldehyde (see above, Thin Sectioning) and with or without glycerol as a cryoprotectant.

Freezing. If a cryoprotectant is used a pellet containing about 200 µg enzyme protein is first prepared by centrifugation from a dilute sample of purified Na^+,K^+-ATPase. The supernatant liquid is carefully removed and the pellet is resuspended in the centrifuge tube in one drop of 50% glycerol in 25 mM imidazole buffer, pH 7.2. The suspension is allowed to stand in the cold for 1–2 hr, which improves the uniformity of the suspension. Small droplets are then placed on specimen support disks and quickly frozen by immersion in Freon-22 cooled with liquid nitrogen.

A device for rapid freezing is necessary to preserve fine structure without the use of cryoprotectants, e.g., in connection with etching of membranes before shadowing. Among several cryofixation methods[14] the procedure of sandwich cryogen jet freezing is suitable for membranes and other small objects in suspension. Small samples (2–4 µl) of purified Na^+,K^+-ATPase containing 10–20 mg enzyme protein/ml are placed between thin copper foils and frozen by double-sided jets of propane cooled by liquid nitrogen (e.g., Balzers Cry-Jet, Balzers AG, Lichtenstein).

Fracture and Replication. The frozen specimen is fractured in a freeze–fracture apparatus (BAF 300, Balzers AG, Lichtenstein) at -100 to $-120°$ in a vacuum of 1×10^{-6} torr or better. Shadowing is performed with platinum–carbon. The last cut with the knife is carried out immediately *after* initiation of platinum shadowing in order to minimize the time from fracture to replication and thus avoid contamination of the fracture

[13] K. Fisher and D. Branton, this series, Vol. 32, p. 35.
[14] H. Plattner and L. Bachmann, *Int. Rev. Cytol.* **79**, 237 (1982).

FIG. 2. Membrane fragments from a preparation of purified Na^+,K^+-ATPase following negative staining with phosphotungstic acid. The membranes have the appearance of rounded disks and show clusters of small surface particles. ×105,000. From A. B. Maunsbach, E. Skriver, H. Hebert, and P. L. Jørgensen, *Verh. Anat. Ges.* **78**, 55 (1984).

FIG. 3. Higher magnification of a single membrane fragment from a preparation similar to that in Fig. 2. The membrane shows surface particles with a diameter of 30–50 Å arranged in clusters and strands which are separated by empty areas. The surface particles are interpreted as $\alpha\beta$ units of the enzyme protein and the surrounding empty areas as the lipid domains of the membrane. ×400,000. From A. B. Maunsbach, E. Skriver, H. Hebert, and P. L. Jørgensen, *Verh. Anat. Ges.* **78**, 55 (1984).

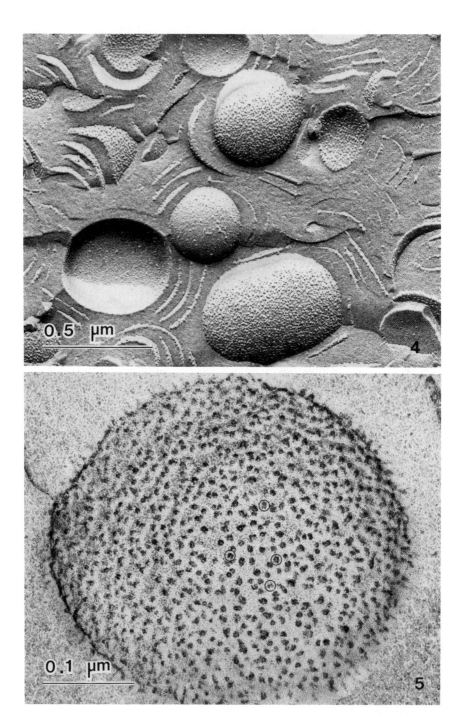

face. The angle of unidirectional shadowing is usually 45° and the thickness of the shadowing material is adjusted to 20–30 Å by a quartz crystal thin-film monitor. After shadowing with platinum the replica is reinforced by evaporation of carbon. To further strengthen the replica an additional brief evaporation of carbon is carried out later when the replica has reached a temperature of approximately −40°. The replica is cleaned with 15% sodium hypochlorite for 2 hr, rinsed with distilled water, and mounted on carbon-coated Formvar films supported by 200-mesh copper grids (Fig. 4).

Rotary Shadowing. Rotary shadowing of fractured membranes may reveal structural details that are not demonstrable with unidirectional shadowing[15] and is carried out by placing the specimen on a rotating cooling table in the freeze–fracture apparatus. The angle of shadowing is usually 25° in order to get large areas with symmetrical shadowing. If lower angles are used areas with optimal shadowing are smaller. Rotation of the specimen is initiated immediately after the last fracture and at the same time as shadowing with platinum–carbon is initiated. The speed of rotation is set in advance to about 30 rpm. Rotary shadowing of freeze-fractured membrane-bound Na$^+$,K$^+$-ATPase (Fig. 5) reveals that some intramembrane particles are resolved into dimeric structures.[16]

Freeze–Etching. Enzyme samples which have been frozen without glycerol, e.g., by cryojet, can be etched before replication. In order to avoid contamination of the fracture surface during etching it is essential that the pressure is as low as possible. The evaporation of platinum–carbon is started 2–10 min after the last cut. During the period of etching the knife, cooled by liquid nitrogen, is placed over the specimen as a cool trap. The knife is moved to the side immediately after the evaporation of platinum is initiated.

[15] L. H. Margaritis, A. Elgsaeter, and D. Branton, *J. Cell Biol.* **72**, 47 (1977).
[16] A. B. Maunsbach, E. Skriver, and P. L. Jørgensen, in "Na,K-ATPase: Structure and Kinetics" (J. C. Skou and J. G. Nørby, eds.), p. 3. Academic Press, London, 1979.

Fig. 4. Freeze–fracture replica of purified Na$^+$,K$^+$-ATPase membranes from pig kidney. Tangentially fractured membranes are either convex or concave. The fracture faces show either many or very few intramembrane particles, suggesting a specific orientation of the protein molecules in the membrane. The direction of shadowing is from below. ×60,000. From A. B. Maunsbach, E. Skriver, H. Hebert, and P. L. Jørgensen, *Verh. Anat. Ges.* **78**, 55 (1984).

Fig. 5. Membrane fragment of purified Na$^+$,K$^+$-ATPase following freeze–fracture and rotary shadowing at an angle of 15°. The shadowing is symmetrical in the center of the membrane fragment and many intramembrane particles here show an asymmetric structure, some with a dimeric appearance (circles). ×250,000. From A. B. Maunsbach, E. Skriver, H. Hebert, and P. L. Jørgensen, *Verh. Anat. Ges.* **78**, 55 (1984).

Microscopy and Measurements

Samples prepared by negative staining are usually analyzed in the electron microscope at direct magnifications ranging from ×20,000 to ×80,000. Care is taken to minimize radiation damage of the specimen in the electron microscope. In order to record available information in the specimen a direct magnification of at least ×50,000 is usually required. Freeze–fracture preparations are analyzed at magnifications ranging from ×15,000 to ×30,000, but ×50,000 is useful for the detailed structural analysis of intramembrane particles, in particular in rotary shadowed preparations.

After negative staining of the Na^+,K^+-ATPase membranes the enzyme protein is visualized as a uniform population of surface particles (Figs. 2 and 3). In freeze–fracture preparations tangentially fractured membrane fragments show intramembrane particles with different dimensions and frequencies (Fig. 4). Measurements of particle diameters, distances between centers of particles, and particle frequencies are made on enlargements of electron micrographs with final magnifications of ×200,000 both for negative staining and for freeze–fracture preparations. Particle diameters are measured to the nearest 0.1 mm through a magnifier. In freeze–fracture preparations the diameter of a particle is determined as the width of the metal deposit at right angle to the direction of shadowing. Areas are measured by point counting.

Electron Microscopy of Na^+,K^+-Transport Vesicles

Phospholipid vesicles (liposomes) reconstituted with different amounts of purified Na^+,K^+-ATPase[17] can be analyzed by thin sectioning and freeze–fracture electron microscopy to correlate the ultrastructure of the vesicles with their cation transport capacity.[18]

Thin Sectioning

Freshly prepared vesicles are concentrated, e.g., by centrifugation at 100,000 rpm for 30 min at 20° in a Beckman airfuge. The pellet is fixed with 1% glutaraldehyde in 0.1 M cacodylate buffer, pH 7.2, for 2 hr and postfixed in 1% osmium tetroxide in the same buffer for 0.5 hr. It is then thoroughly rinsed in cacodylate buffer and stained with 1% tannic acid in

---

[17] B. M. Anner, L. K. Lane, A. Schwartz, and B. J. R. Pitts, *Biochim. Biophys. Acta* **467**, 340 (1977).

[18] E. Skriver, A. B. Maunsbach, and P. L. Jørgensen, *J. Cell Biol.* **86**, 746 (1980).

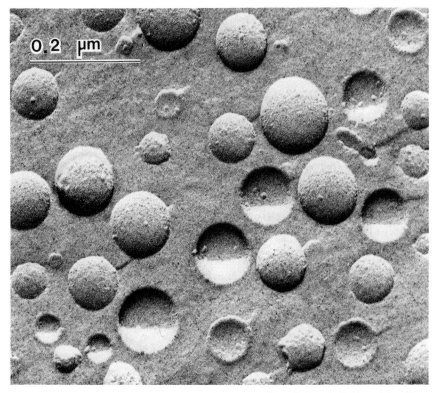

FIG. 6. Electron micrograph of freeze–fracture replica of phospholipid vesicles (liposomes) reconstituted with Na^+,K^+-ATPase (protein:lipid ratio, 669×10^{-4}). Occasional intramembrane particles are present on both convex and concave fracture faces. A quantitative analysis of many fracture faces reveals that the particles are equally frequent on the two types of fracture faces, suggesting that the Na^+,K^+-ATPase molecules are inserted randomly into the vesicle membrane. ×150,000. From Skriver et al.[18]

the same buffer for 0.5 hr.[19] The pellet is then stained en bloc in 0.5% uranyl acetate in veronal acetate buffer, pH 5.5, dehydrated in acetone, and embedded in Vestopal. Ultrathin sections are stained with lead citrate.

Freeze–Fracture

Pellets of vesicles are resuspended and equilibrated for 2–3 hr at 0–2° with glycerol to a final concentration of 20% (v/v) glycerol in the buffer in

[19] N. Simionescu and M. Simionescu, *J. Cell Biol.* **70**, 608 (1976).

which the vesicles are prepared. Aliquots of vesicles are placed on gold disks and quickly frozen in Freon-22 cooled by liquid nitrogen. The specimens are fractured at $-100°$ and shadowed with platinum–carbon as described above for membrane-bound Na^+,K^+-ATPase.

Comments and Interpretations

Following negative staining with phosphotungstic acid, the surfaces of purified Na^+,K^+-ATPase membranes exhibit a uniform population of surface particles (Figs. 2 and 3). The particles have an estimated diameter of 30–50 Å, which is close to the estimated resolution of 20 Å after negative staining with uranyl acetate. In all membranes the particles are arranged in clusters and strands, which are separated by empty areas interpreted as lipid regions. In glutaraldehyde-fixed membranes the clustering is more extensive than in unfixed preparations. The frequency of surface particles is $12,500/\mu m^2$ and both membrane surfaces show a similar distribution of particles. On the basis of the chemical composition of the preparation we have calculated that an $\alpha\beta$ unit (protomer, monomer) should have a diameter of about 50 Å and a frequency of $13,700/\mu m^2$. Since these calculated parameters are very close to the observed values for the surface particles we have concluded that one surface particle represents one protomer ($\alpha\beta$ unit) of the enzyme protein.[3]

Freeze–fracture electron microscopy reveals the organization of the Na^+,K^+-ATPase protein within the membrane (Fig. 4). The frequency of intramembrane particles, determined as the combined particle frequency of the two fracture faces of the membranes, is 6100 particles/μm^2. The diameter of unidirectionally shadowed particles is approximately 90 Å. These quantitative results demonstrate that negative staining and freeze–fracture electron microscopy reveal different aspects of the Na^+,K^+-ATPase protein. Since the ratio between the frequencies of surface particles and intramembrane particles is close to 2:1 it is likely that the intramembrane particles are oligomers, most likely dimers, of the protein units that constitute the surface particles.[3,16]

Freeze–fracture followed by rotary replication provides further evidence for an oligomeric organization of the intramembrane particles. Symmetric shadowing of fracture faces with a thin layer of platinum at an angle of 10–25° reveals that many particles with a total diameter of about 90 Å appear divided into two approximately equal subunits (Fig. 5). These observations are consistent with the interpretation that the intramembrane particles observed by freeze–fracture represent Na^+,K^+-ATPase units with a dimeric composition. Direct evidence that intramembrnae particles represent the protein of the Na,K-pump has been obtained by correlating the ultrastructure (Fig. 6) and the transport characteristics of phospholipid vesicles reconstituted with different amounts of pure Na^+,K^+-ATPase.[18]

Acknowledgments

Supported by the Danish Medical Research Council and the Research Foundation at the University of Aarhus.

Author Index

Numbers in parentheses are footnote reference numbers and indicate that an author's work is referred to although the name is not cited in the text.

A

Acs, G., 273
Adams, R. J., 203
Ahmed, K., 71, 238, 418
Ahrens, E. H., 129
Aisen, P., 251
Akera, T., 9, 207, 213(14), 223
Albers, R. W., 6, 10(26), 21(26), 18, 19(72), 29, 116, 215, 302, 353, 421, 422, 423(26), 425(26)
Aldanova, N. A., 15
Allen, J. C., 9, 203, 206, 428, 429
Allison, R. D., 238
Almeida, A. F., 371
Altbauer, A., 85
Altman, L. H., 411
Altweger, L., 273
Alvers, R. W., 429
Ames, B. N., 130, 139(16)
Amory, A., 123, 124(13)
Amos, L. A., 85
Anderson, M. G., 68
Ando, T., 422, 423(31), 427
Anner, B. M., 127, 141, 142, 153(8), 154, 155, 156, 232, 438
Antieau, J. M., 346, 347(4)
Apell, H.-J., 154
Arndt-Jovin, D. J., 279
Aronson, I. K., 10
Arzamazova, N. M., 15
Ash, J. F., 121
Ashbrook, J. D., 201
Askari, A., 8, 238, 244, 252, 258(12), 302, 334, 345, 346, 347(4, 5), 349(5), 350(5)
Ausiello, D. A., 215
Avruch, J., 297, 347

B

Bachmann, L., 435
Bader, H., 418
Baginski, E. S., 107, 159
Bagshaw, C. R., 275
Baker, P. F., 215
Balaban, R. S., 214
Balerna, M., 325, 328(13), 331(13)
Ball, W. J., 408
Ball, W., 97(24), 100, 101
Baltimore, D., 380, 386
Banerjee, S. P., 24, 302
Bangher, B. W., 116
Barber, M. J., 358
Barnakov, A. N., 16, 85
Barnett, R. E., 237, 371
Barnett, R. J., 421
Barnett, R., 357
Barrett, P. Q., 226
Bartlett, G. R., 52
Bashford, C. L., 248
Baskin, S. I., 223
Baumeister, W., 85
Baxter-Lowe, L. A., 48, 61(3), 64(3)
Bayley, H., 300
Bayne, 408
Beaugé, L. A., 251, 252, 253(16), 254, 255(10, 15, 20), 257(10, 20), 259, 260, 261(10), 262(10), 263(10, 20), 264(10), 265, 266(22, 28), 267(11, 15, 22), 272, 274(8), 275(8), 282
Becker, R. R., 64
Beeuwkes, R., III., 422
Behringer, L., 315
Beissner, R. S., 116
Berberian, G., 254, 255(20), 258(20), 259, 263(20), 265(20)
Bergmeyer, H. U., 314
Berliner, L. J., 357
Biemesderfer, D., 400, 405(6), 408, 409(6)
Bilezekian, J. P., 214, 223
Birdsall, N. J. M., 116
Birrell, G. B., 376
Blackshear, P. J., 124
Blitzer, B. L., 427

Blostein, R., 7, 8(32), 10, 168, 173, 174(12), 176, 178, 245, 261, 265(25)
Blow, D. M., 334
Bobis, G., 319, 322(33)
Bogard, B. I., 425
Boldyrev, A., 110
Bond, G. H., 251, 265(9)
Bonting, S. L., 13, 14(59), 120, 121(3), 241, 302
Boon, N. A., 10
Borgers, M., 424
Botstein, D., 383
Boyer, J. L., 427
Brangon, D., 435, 437
Braun, W., 370
Bredenbröcker, B., 318
Brinley, F. J., 252, 254(14)
Broderson, S. H., 427
Brody, T. M., 223
Bron, C., 213
Brotherus, J. B., 30, 31(27), 33(14, 15), 42(15), 43(14, 15)
Brotherus, J. R., 13, 334, 341, 342(6)
Broude, N. E., 15
Brown, L., 201
Brown, T. A., 15
Brunner, J., 300
Burg, M. B., 213
Butcher, G., 96

C

Cantley, L. C., Jr., 9, 64, 86, 119, 163, 233, 236, 237, 242, 251, 252(5, 8), 254, 257(5), 258(5), 265(5, 8), 266(5, 8), 299, 379
Cantley, L. G., 9, 86, 233, 251, 252(7), 265(8), 266(8)
Cantley, L., 300
Caplan, M. J., 400, 405(6), 408, 409(6)
Carilli, C. T., 92, 299
Castro, J., 291, 298(3), 299(3), 313, 322(8), 333
Cavieres, J. S., 252, 255(10), 257(10), 261(10), 262(10), 263(10), 264(10), 265(10)
Cerami, A., 273
Chaberek, S., 425
Chan, S. S., 279
Chapman, C., 203
Chappell, B., 314
Chappell, J. B., 135, 307
Charnock, J. S., 248, 371
Cheng, V.-J. K., 207, 213(14)
Cheung, D. T., 337
Chiang, C., 366
Chihara, G., 303, 306(26)
Chin, G., 127, 128(8), 139(10), 154, 300
Chipman, D. M., 344
Chirgwin, J. M., 387
Christian, C. P., 427
Christiansen, C., 11, 12, 13(53), 30, 33(17), 72, 75(6), 114, 342(24)
Christiansen, N. O., 6, 18(25), 19(25), 21(25), 22(25), 24(25)
Chu, L., 7, 8(32)
Churchill, L., 61, 62(21), 63(22), 64(22)
Clark, A. W., 110
Clarke, D. A., 303, 305(22)
Clausen, T., 123
Cleland, W. W., 243, 317, 359
Clore, G. M., 367
Cohn, M., 356
Collins, J. H., 64, 87, 90, 92(4)
Collins, J., 100
Colowick, S. P., 191
Connick, R. E., 356
Conte, F. P., 48, 62(5), 334
Cooper, J. R., 120, 121(5)
Copenhaver, J. H., Jr., 49, 72, 87, 90(2), 110
Cornelius, F., 12, 17(55), 21(26), 72, 73(4), 157, 160(8), 162(8), 164(8), 167(8)
Coulson, A. R., 390
Craig, W. S., 14, 334, 335, 336(14), 337, 338(14), 339(14), 340(14, 22), 341(14), 342(14), 343(13), 344(13, 22), 350
Croft, D. N., 221
Cu, V. M. S., 10
Cutler, L. S., 427
Czarnecki, J. J., 319
Czarnecki, J., 314

D

Dahl, J. L., 11, 43, 49, 72, 141, 143(5), 144(5), 149(5)
Dall-Larsen, T., 312
Daniel, A., 422

Datta, A., 116
Davis, D. A., 425
Davis, R. A., 380, 382(6, 7), 384(6)
Davis, R. W., 383
Dayer, J. M., 215
De Weer, P., 10
Deguchi, N., 32, 38(16), 80, 431, 433(3), 440(3)
Demin, V. V., 16, 85
DePamphilis, M. L., 317, 319(22)
DePont, J. J. H. H. M., 13, 14(59), 110, 120, 121(3), 241, 302
Deth, R. C., 201
Deupree, J. D., 11, 43, 49, 72, 141, 143(5), 144(5), 149(5)
Deutch, H., 98
DeVries, G. H., 65, 68
Dibona, D. R., 428, 429(74)
Dintzis, H. M., 327
DiPolo, R., 252, 254, 255(15), 267(16)
Dissing, S., 22, 173
Dixon, J. F., 11, 29, 30, 38(16), 43, 72, 110, 139, 141, 142, 143, 144(5, 10, 11), 146(20), 147(20), 148(20), 149(5, 20), 150(10, 20), 151(20), 152(20), 153(20), 154(20, 23), 155(20, 23)
Doerr, J. L., 303, 305(22)
Dorvall, G., 99
Dose, K., 312, 313, 314(10)
Doucet, A., 34
Dowd, F., 206
Drapeau, P., 7, 8(32), 173, 174(12)
Duke, J. A., 303
Dunaway-Mariano, D., 359
Dunham, P. B., 174
Dunn, M. J., 223
Dux, L., 81
Dzhandzhugazyan, K. D., 30, 42(25)
Dzhandzhugazyan, K. N., 16, 85
Dzhandzhugazyan, K., 100, 291

E

Eccleston, J. F., 272, 275
Eckstein, F., 310
Edelman, I. S., 13, 213, 408
Eisner, D. A., 242
Elgsaeter, A., 437
Ellis, R. A., 29, 425, 427

Ellory, J. C., 13, 248, 342
Engvall, E., 99
Epps, L., 366
Erdmann, E., 10, 191, 201, 202, 203(7), 211(7), 302, 306(10), 309, 310(10), 311(10)
Erickson, H. P., 80
Ernst, S. A., 29, 214, 218, 219(6), 222(6, 8), 227(6, 8), 417, 418, 420(2), 421, 422(1), 428, 429
Esmann, M., 4, 5, 8, 10(34), 11, 12, 13(53), 14, 21(26), 29, 30, 33(17), 43, 46, 72, 73, 74, 75(6), 76(7), 77, 78(11), 79(10), 110(3), 114, 115, 119, 157, 193, 239, 245(17), 272, 274(9), 278, 279(1, 3), 281, 341, 342(24), 357, 371, 372
Ewing, R. D., 48, 61(5), 62(5), 334

F

Fahn, S., 6, 10(26), 21(26), 302
Fairbanks, G., 114, 297, 347
Fambrough, D. M., 120, 408
Farley, R. A., 15, 92, 291, 298(3), 299, 300, 313, 318, 322, 333
Farquhar, M. G., 403
Farr, A. L., 37, 47, 52, 111, 158, 177
Farrants, G., 85
Fasold, H., 303, 306(24), 310, 312
Faulk, W. P., 411
Faust, U., 310
Fehlman, M., 213
Feinstein, M. B., 427
Feramisco, J. R., 383
Ferguson, G., 376
Fiddes, J. C., 383
Field, M., 227
Fineman, R. M., 121
Firth, J. A., 417, 425, 427
Fischer, L. W., 156
Fisher, J. A., 48, 61, 63(22), 64(3, 22)
Fisher, K., 435
Fisher, L. W., 141
Fiske, C. H., 107
Flashner, M. S., 236, 244
Fleischer, S., 30, 35(24), 432
Fletcher, J. E., 201
Foa, P. P., 107, 159
Fogac, M., 154

Forbush, B., III., 12, 30, 31(22), 34(22), 90, 92, 239, 282, 324, 395, 400, 403(3), 405(6), 408
Forgac, M., 127, 128(8), 139(10), 300
Fortes, P. A. G., 13, 14(58), 110, 112(15), 242, 302, 342
Fortes, P. A., 326
Fossel, E. T., 47, 48(4), 355
Foury, F., 123, 124(13)
Fox, J. J., 303, 305(22)
Frantz, C., 215
Freychet, P., 213
Freytag, W., 32, 85
Fries, E., 30, 336
Fritsch, E. F., 381, 386(8)
Fritzsch, G., 302, 310(17), 311(17), 312
Frizzell, R. A., 227
Frye, J. S., 303, 305(23)
Fujimoto, K., 422, 423, 427
Fujita, M., 418
Fukushima, Y., 244
Fullerton, D. S., 71
Furukawa, H., 213, 223

G

Gabriel, S., 315
Gache, C., 325, 328(13), 331(13)
Galfré, G., 96
Gallacher, D. W., 226
Gantzer, M. L., 360
Garay, R. P., 248
Garcia-Perez, A., 408
Gardner, J. D., 215
Garg, L. C., 213
Garrahan, P. J., 4, 25(6, 7), 127, 248
Garray, R. D., 4, 25(6)
Gasko, O., 181
Gathiram, P., 203
Gautier, C., 390
Geahlen, R., 314
Germain, P., 337
Gerring, K., 213
Geuze, H. J., 411
Gilekson, R., 203
Gilkeson, R. C., 65, 69(2), 382
Giotta, G., 300, 334
Girard, J. P., 358
Girardet, M., 213
Gitler, C., 291, 299(2, 4), 300(2, 4)

Glynn, I. M., 1, 4, 5, 6, 10, 18(13, 18, 27), 19(13), 21(1), 22(42), 24(20, 42), 127, 135, 153, 173, 179, 181(1, 2), 246, 248, 251, 252, 254, 255, 257(10), 261, 262(10), 263(10), 264(10), 265(10), 266, 267, 271, 272, 273(3), 274(8), 275(8), 276(2), 277(2, 3), 279, 282, 286(2), 314, 402
Glynn, J. M., 307
Goeldner, M., 326, 331, 332(18)
Goertemiller, C. C., Jr., 29, 425, 427
Goffeau, A., 123, 124(13)
Goldin, S. M., 72, 127, 128(6), 131(6), 135(3), 138(6), 141, 150(2), 153(9), 156, 164(4)
Goldman, D. W., 300
Goody, T., 275
Gordon, J., 404
Goto, T., 422
Gottikh, B. P., 315
Gouy, M., 390
Grahame-Smith, D. G., 10
Grant, R., 223
Grantham, J. J., 251, 252, 254(2), 255(10), 257(10), 261(10), 262(10), 263(10), 264(10), 265(10)
Grantham, R., 390
Greeb, J., 16, 65
Green, J. R., 13, 342
Green, N. M., 92
Gregory, D. W., 432
Griffith, O. H., 30, 31(27)
Grisham, C. M., 240, 244, 354, 357, 358, 360, 361, 371
Grishin, A. V., 15
Gronenborn, A. M., 367
Gruber, W., 314
Grupp, G., 203, 206
Grupp, I. L., 203, 206
Guidotti, G., 9, 120, 121(6), 122(6), 123(6), 163, 254
Guillory, J. R., 315, 316(17)
Guth, L., 422

H

Hackney, J. F., 11, 29, 30(5), 43, 49, 72, 110, 141, 143(5), 144(5), 149(5)
Haley, B. E., 312, 314, 318(2)
Hall, C., 324
Halloway, P. W., 158

Hampton, A., 303
Hankovszky, O. H., 376
Hansen, O., 113, 123
Hansson, G. C., 12, 13(53), 342(24)
Hara, Y., 6, 18(27), 234, 266, 282
Hardy, P. M., 337
Hart, W. M., 302
Hasse, W., 10
Hasselberg, M., 302, 307, 310(16, 30), 311(30), 312(30)
Hastings, D. F., 13, 30, 33(18), 72, 230, 233, 334, 341(5), 342(5)
Hawke, D., 92
Hayahara, H., 427
Hayashi, K., 95
Hayashi, Y., 79, 230, 232, 235, 236(17)
Hayashida, H., 15
Hayat, M. A., 421
Hebert, H., 29, 42(11)
Hegerl, R., 85, 87(8, 16)
Hegyvary, C., 4, 5(11), 7, 21(26), 191, 240, 242(19), 275, 279, 310, 319, 320(28), 323
Helenius, A., 336
Helenius, E., 30
Helfman, D. M., 383
Hemingway, R. J., 323
Henderson, G. R., 302
Henderson, R., 85
Herbert, H., 16, 17(71), 19(71), 80, 81, 82(8), 83, 85, 86(8), 87(8, 16), 291, 292(5), 293(10), 295(5), 299(5), 300(5), 433
Hermann, R., 337
Hexum, T., 240
Hiatt, A., 408
Hideg, K., 376
Hilberg, C., 46, 109, 302
Hilden, S., 127, 135(4), 141, 144(7), 150(3, 7), 151(3), 152(7), 153(7), 156
Himes, R. H., 240
Hinkle, P. C., 142, 153(21), 179
Hirano, H., 425
Hirose, T., 15
Hirsch, J., 129
Hirth, C., 326, 331, 332(18)
Höckel, M., 312
Hoffman, J. F., 12, 22, 92, 153, 173, 261, 312, 318(2), 324, 395
Hokin, L. E., 11, 29, 30, 31(27), 38(16), 43, 48, 49, 51(12), 53(12), 56(12), 57(12), 59, 60(12, 20), 61, 62(12), 63(22), 64(3, 20, 22), 72, 110, 127, 135(4), 139, 141, 142, 143, 144(5, 7, 10, 11), 146(20), 147(20), 148(20), 149(5, 20), 150(3, 7, 10, 20), 151(3, 20), 152(7, 20), 153, 154(20, 23), 155(20, 23, 24), 156, 323, 357
Hokin-Neaverson, M., 324
Hollis, R. J., 173
Homareda, H., 230, 232, 233(1), 234(11), 235, 236(1, 17)
Hommock, M. K., 425
Hootman, S. R., 48, 61(5), 62(5), 214, 218, 219(6), 220(10), 222(6, 8, 10), 223, 227(6, 8, 10), 229(25), 418, 421, 427, 429, 334
Hopfer, V., 427
Hopkins, B. E., 46, 47(1), 64
Horenstein, A., 4, 25(7)
Horowitz, B., 15
Hovmöller, S., 85
Howard, J., 96
Howe, S., 96
Howland, J. L., 282
Huang, W., 238, 252, 258(12), 302
Huang, W.-H., 334, 345(12), 346, 347(4, 5), 349(5), 350(5)
Hudgins, P. M., 251, 265(9)
Hughes, G. J., 337
Hughes, S. H., 383
Hulla, F. W., 302, 303, 306(24), 310, 311(17), 312
Hulla, F., 312
Hunston, D. L., 201
Hynes, R. O., 380, 381(5)

I

Ikehara, M., 313
Imai, Y., 422
Inayama, S., 15
Inman, J. K., 327
Inomata, K., 427
Ishikawa, Y., 427
Iwasaki, T., 50
Iwata, H., 120, 121(5)
Izumi, F., 424

J

Jacobsen, J., 341
Jacobsen, L., 30, 33(15), 42(15), 43(15)
Jacobsen, N. O., 418
Jacobzone, M., 390

Jaenicke, R., 337
Jamieson, J. D., 408
Jarrell, J. A., 215
Jarvis, S. M., 13, 342
Jaworek, D., 314
Jean, D. H., 429
Jeng, S. J., 315, 316(17)
Jenkin, C., 99
Jensen, J., 4, 5(10), 7, 15, 32, 113, 115, 116, 191, 193(5), 195(3), 201, 240, 279, 309, 310(31), 319, 320(28)
Jesaitis, A. J., 302
Johansen, B., V., 432
Johansen, T., 10
Johnson, D., 68
Johnstone, R. M., 178
Jørgensen, P. L., 4, 11, 13, 14(57), 16, 17(70, 71), 19(71), 29, 30, 31(2, 12, 26), 32, 33(12, 14, 15, 26, 33), 36(2, 4, 26), 38(33), 40(31), 38(16), 42(11, 15, 25, 33), 43, 46, 49, 55(10), 65, 69, 71(8), 80, 81, 82(8), 83, 85(8), 86, 87(8), 110, 127, 130, 155, 179, 191, 232, 234, 245, 272, 274(7), 275, 291, 292, 293(5, 10), 294, 295(5), 296(7), 299(2, 4), 300, 302, 307, 309(28), 312, 317, 318, 322(1), 332, 334, 341, 342(6), 346, 353, 354, 379, 395, 402, 418, 431, 433, 437, 438, 439(18), 440(3, 16), 441(18)
Josephson, L., 9, 86, 233, 237, 251, 252(7), 265(8), 266(8)
Jost, P. C., 30, 31(27)
Jovin, T. M., 279
Judah, J. D., 238, 418

K

Kagawa, M., 180
Kagawa, Y., 127
Kalka, T., 121
Kanazawa, T., 322
Kaniike, K., 191, 230, 233(2), 309
Kant, J. A., 168
Kaplan, J. H., 92, 173, 324, 395
Karlish, S. J. D., 4, 5, 18(13, 14), 19(13), 24(20), 86, 173, 179, 181(1, 3, 5), 182(3), 183(5), 184(3), 185(5, 7), 186(4, 7), 187(4, 6), 188(4), 239, 246, 248, 255, 259, 265, 266(22), 267(22), 271, 272, 273(3), 274(5, 7), 275(4, 5, 7), 276(2, 4, 5), 277(2, 3, 5), 279, 291, 293, 299, 300(2, 4, 8, 13)
Karlsson, K., 342(24)
Karlsson, K.-A., 12, 13(53)
Karnobsky, M. J., 430, 432
Kasahara, M., 142, 153(21), 179
Kashgarian, M., 400, 405(6), 408, 409(6), 427
Kato, H., 50
Kawai, C., 427
Kawamaki, K., 15
Kawamura, M., Jr., 15, 30, 43(19), 48, 48(4), 57(4), 58(4), 59(4), 61(4), 64, 302
Kazazoglou, T., 323
Keana, J. F. W., 376
Kedde, D. L., 327
Kellerman, G. M., 116
Kepner, G. R., 13, 342
Kerjaschki, D., 403
Khorana, H. G., 303, 306(25)
Kimimura, M., 230, 232
Kirley, T. L., 92
Kistenmacher, T., 366
Klevickis, C., 360
Klodos, I., 6, 18, 19(25), 21(25), 22(25), 24(25), 110
Klotz, I. M., 201, 215
Klug, A., 80
Klungsøyr, L., 312
Knepper, M. A., 213
Knowles, A. F., 180
Kobayashi, T., 427, 430
Koepsell, H., 302, 310(17), 311(17), 312
Köhler, G., 96
Kohler, J., 392
Koppers, A., 13, 14(59), 302
Koval, C. J., 29
Koval, G. J., 6, 10(26), 21(26), 116, 215, 302, 421, 422(26), 423(26), 425(26)
Koyal, D., 8
Kraehenbuhl, J.-P., 213
Kramer, R., 116, 307, 308(29), 309(29)
Krawietz, W., 201
Krayevsky, A. A., 315
Kreickmann, H., 319, 320(29)
Kryri, H., 312
Kume, S., 7, 18, 21(26), 302
Kupchan, M., 323
Kuriki, Y., 6
Kuwahara, T., 427

Kuzin, A. P., 16, 85
Kyriakides, C. P. M., 427
Kyte, J., 14, 29, 32, 49, 85, 141, 324, 334, 337(1), 342, 346, 350, 379, 417, 439

L

Laemmli, U. K., 52, 59(18), 93, 124, 296(11), 297, 318
Landau, N. R., 380
Lane, L. K., 15, 49, 72, 87, 90, 92, 100, 110, 127, 141, 153(8), 156, 202, 203, 205(6), 211(6), 230, 233(2), 438
Latham, D. S., 427
Lauf, P. K., 87
Lauger, P., 154
Lazdunski, M., 300, 312, 313(5), 322(5), 323, 324, 325, 328(13), 331, 332(18), 333
Lechene, C., 9
Lee, A. G., 116
Lee, S.-W., 203
Leffert, H., 101
Lemischka, I. R., 380, 381(5)
Leonard, K., 80
Lerner, E. A., 397
Lessard, J., 97(24), 101
Leuenberger, P. M., 425
Levy, H. M., 106, 237
Lieb, W. R., 179, 185(7), 186(7)
Lin, J. C., 120, 121(6), 122(6), 123(6)
Lindenmayer, G. E., 9, 49, 72, 87, 90(2), 110, 202, 203, 230, 233(2), 428
Lindquist, R. N., 252
Ling, E., 90, 92(4)
Lingrel, J. B., 16, 65
Lingrell, J. B., 15
Litman, B. J., 128
Lo, C. S., 213
Loeb, J. N., 214, 223
Lopez, J. A., 237
Lopez, V. J., 252
Lossinsky, A. S., 427
Lowndes, J., 324
Lowry, O. H., 37, 47, 52, 111, 158, 177, 237
Lubran, M., 221
Lunev, A. V., 16, 85
Luz, Z., 356
Lunch, C. J., 201
Lytton, J., 120, 121(6), 122(6), 123(6)

M

MacDonald, R. J., 387
Macey, R. I., 13, 342
Macknight, A. D. C., 215
Maezewa, S., 79
Maguire, M. H., 303
Mandel, L. J., 214
Maniatis, T., 381, 386(8)
Marchesi, V. T., 418
Marcus, M. M., 141, 154
Margaritis, L. H., 437
Marsh, D., 371, 373(2), 375
Martell, A. E., 425
Martin, D. W., 116
Martonosi, A., 81
Maruyama, Y., 226
Marzilli, L., 366
Masuzawa, T., 428
Matsubara, F., 427
Matsuda, T., 120, 121(5)
Matsui, H., 79, 201, 203, 230, 232, 233(1), 234(11), 235, 236(1, 17)
Matsukawa, S., 427
Maunsbach, A. B., 16, 17(70, 71), 19(71), 29, 30, 32, 38(16), 42(11), 80, 81, 82(8), 83, 85, 86(8), 87(8, 16), 127, 291, 292(5), 293(10), 295(5), 299(5), 300(5), 431, 433, 437, 438, 439(18), 440(3, 16), 441(18)
Mayahara, H., 421, 422, 423, 425
Mayrand, R. R., 71
Mazarguil, H., 337
McCaslin, D. R., 30, 336
McClurkin, I. T., 417
McCormick, P. W., 346
McDonough, A. A., 15, 408
McLaughlin, B. J., 429
McLean, A. E. M., 418
McLean, I. N., 405, 409(11)
McParland, R. H., 64
McPherson, G. A., 212
Mees, K., 427
Meiboom, R., 356
Meisel, E., 417
Mercer, R. W., 251, 252(6), 265(6)
Mercer, R., 174
Mercier, R., 390
Mernissi, G. E., 34
Mertens, W., 310
Metcalfe, J. C., 116

Meyer, C., 312
Michael, L. H., 202
Mildvan, A. S., 240, 244(22), 354, 356, 361(6)
Miles, J. W., 428, 429
Milhorat, T. H., 425
Miller, R. P., 15
Mills, J. W., 215
Milstein, C., 96, 392
Misell, D. L., 80
Mitchinson, C., 92
Miyata, T., 15
Mizukami, A., 421
Mizukami, Y., 427
Mizuno, N., 418
Moczydlowski, E. G., 110, 112(15), 242, 342
Modyanov, N. N., 15, 16, 85, 100
Modzydlowski, I. D., 13, 14(58)
Moffat, J. G., 303, 306(25)
Mohraz, M., 85
Mokotoff, M., 323
Molinoff, P. B., 215
Møller, J. D., 13
Møller, J. V., 30, 33(14), 43(14), 334, 341(6), 342(6)
Monastyrskaya, G. S., 15
Monk, B. C., 116
Monson, P., 337
Monteilhet, C., 334
Moosmayer, M., 141
Morgan, M., 121
Morohashi, M., 30, 43(19), 48, 49(4), 48(4), 57(4), 58(4), 59(4), 61(4), 64
Morrison, J. F., 239
Moses, H. L., 418
Mullins, L. J., 252, 254(14)
Mumson, K., 319, 320(31)
Munkries, K. D., 93
Munn, E. A., 433
Munson, K. B., 41, 312, 319(4), 320(4), 322(4)
Munson, P. J., 212
Murphy, A. J., 303

N

Nagai, K., 110, 424
Nagano, K., 15, 203, 302, 418
Nakagaki, I., 422

Nakagawa, Y., 110
Nakane, P. K., 405, 409(11)
Nakanishi, Y. H., 50
Nakao, M., 203, 234, 418
Nakao, T., 418
Namork, E., 432
Nasu, F., 427
Nechay, B. R., 251, 252(7), 265(7)
Nemenoff, R. A., 120
Neufeld, A. H., 106, 237
Nicklen, S., 390
Nicolas, R. A., 300
Nimni, M. E., 337
Noda, M., 15
Noda, T., 423
Noguchi, S., 15
Nojima, H., 15
Nojima, K., 15
Nørby, J. G., 4, 5(10), 6, 7, 13, 18, 19(25), 21(25), 22(25), 24(25), 32, 113, 116, 117(2), 191, 193(5), 195(3), 201, 240, 279, 309, 310(31), 319, 320(27)
Noronha-Blob, L., 403
Novikoff, A. B., 425
Nozaki, Y., 336
Numa, S., 15

O

O'Connor, S. E., 244, 357
O'Hara, D. S. O., 47, 48(4), 355
O'Keefe, K. R., 303, 305(23)
O'Sullivan, W. J., 239
Ochs, D. L., 223, 229(25)
Oetliker, H., 154
Ogawa, K., 421, 422, 423, 425, 427
Ohashi, T., 110
Ohta, T., 15, 302
Okada, T., 430
Okigaki, T., 50
Onur, G., 315
Orcutt, B., 18, 302
Ortanderl, F., 303, 306(24), 310, 312
Osborn, M., 114, 347
Ottolenghi, P., 12, 13, 15, 32, 37, 110, 113, 115, 178, 191, 193(5), 201
Ouchterlony, O., 98
Ovchinnikov, Y. A., 15, 16, 100
Owens, K., 130

P

Pachence, J. M., 13
Palade, G. E., 418
Palkama, A., 417
Papermaster, D. S., 411
Parvez, M., 376
Patton, D. L., 427
Patzelt-Wenczler, R., 302, 305(14), 306(10), 308(15), 309(10, 15), 310, 311(10, 14, 15, 36), 312, 319, 320(29)
Pauls, H., 302, 306(10), 309(10), 310(10), 311(10), 318, 319, 320(29)
Peacock, A., 427
Pedemonte, C. H., 265, 266(28)
Penefsky, H. S., 116, 180
Pennington, J., 153, 155(24)
Perdue, F., 43
Perdue, J. F., 11, 49, 72, 141, 143(5), 144(5), 149(5)
Periyasamy, S. M., 334, 345(12), 346, 347(5), 349(5), 350(5)
Perlinger, H., 315
Perlman, D. M., 299
Perrone, J. R., 29, 30(5), 110, 168
Pershadsingh, H. A., 7, 8(32)
Peters, W. H. M., 13, 14(59), 110, 302
Petersen, O. H., 226
Peterson, G. L., 48, 49, 51(12), 52, 53(12), 56(12), 57(12), 59, 60(12, 20), 61, 62(5, 21), 63(22), 64(20, 22), 74, 77(8), 111, 159, 285, 334
Petrukhin, K. E., 15
Philpott, C. W., 421, 427
Pick, U., 179, 181(3), 182(3), 184(3), 265, 293, 300(8)
Pirie, B. J. S., 432
Pitts, B. J. R., 127, 141, 153(8), 156, 202, 203, 205(6), 211(6), 244, 438
Plattner, H., 435
Plesner, I. W., 108, 265
Plesner, L., 18, 108, 265
Ponzio, G., 300, 312, 313(5), 322(5), 330, 332(18), 333
Post, R. L., 4, 5, 7, 8(28), 18, 21(26), 47, 48(4), 119, 121(2), 153, 177, 178(16), 191, 240, 242(19), 244, 266, 279, 302, 310, 319, 320(28), 355
Potter, J. D., 87
Powell, T., 203

Prowse, P. Ey. S., 99
Przybyla, A. E., 387
Pullman, M. E., 116
Purich, D. L., 238
Purygin, P. P., 315

Q

Quiocho, F. A., 337

R

Racaniello, V. R., 386
Rack, M., 303, 306(24), 310(24), 312
Racker, E., 6, 116, 127, 141, 156, 180
Randall, R. J., 37, 47, 52, 111, 158, 177
Randerath, E., 194
Randerath, K., 194
Randolph, R., 337
Rasmussen, H., 226
Raushel, F., 361
Ray, M. V., 232
Rega, A. F., 4
Reich, E., 273
Reichstein, E., 168
Rempeters, G., 245, 309, 312, 313(3), 316(3), 317(3), 320(3), 322(3), 319, 320(29)
Renaud, J. F., 323
Resh, M. D., 9, 120, 121, 122(10), 123(10), 124(8), 163, 214, 254
Reynolds, E., 432
Reynolds, J. A., 13, 30, 33(18), 72, 116, 334, 341(5), 342(5)
Rhee, H. M., 141, 150(3), 151(3)
Richards, D. E., 4, 6, 18(18, 27), 242, 266, 267, 272, 282, 286(2)
Richards, D. W., 402
Risi, S., 312
Robbard, D., 212
Robinson, J. D., 7, 24, 236, 237, 238(8), 240, 241(26), 242, 243, 244, 245, 246(8, 30, 34, 44), 247(44), 249, 250, 251, 252(6), 265(6), 423, 425(32)
Robinson, J. M., 430
Robinson, L. J., 249
Roelofson, B., 12
Rogers, F. N., 119, 121(2), 244
Rogers, T. B., 324

Rohde, M. F., 336
Rosebrough, N. J., 37, 47, 52, 111, 158, 177
Rosen, S., 422
Rosenthal, A. S., 5, 418
Rossi, B., 300, 312, 313(5), 322(5), 323, 325, 328(13), 331, 332(18), 333
Rossier, B. C., 213
Roth, J. R., 383
Rudolph, F. B., 116
Ruoho, A. E., 323, 324
Russo, J., 417
Rutter, W. J., 387
Rydon, H. N., 337

S

Sachs, J. R., 4, 10(8)
Sacktor, B., 403
Saito, A., 432
Saito, M., 322
Saito, T., 425, 428
Salenddin, A. S. M., 427
Sambrook, J., 381, 386(8)
Samson, F. E., Jr., 240
Saneyoshi, M., 303, 306(26)
Sanger, F., 390
Sasaki, S., 422
Sata, F., 428
Saunders, J. P., 251, 252(7), 265(7)
Saxton, W. O., 85
Scatchard, G., 215
Schacter, E., 237
Schäfer, G., 315
Schäfer, H.-J., 313, 314(10)
Schatzmann, H. J., 9
Scheiner-Bobis, G., 313, 319, 320(9, 30), 322
Schenk, D., 101
Scheurich, P., 313, 314(10)
Schneider, B. G., 411
Schoenenborg, B. P., 13
Schoner, W., 116, 191, 245, 302, 305(14), 306(10), 307, 308(15, 29), 309, 310, 311(10, 14, 15, 36), 312, 313, 316(3), 317(3), 319(9), 320(3, 9, 29, 30), 322
Schreiber, J. H., 421
Schulman, M., 96
Schultz, S. G., 227
Schuurmans Stekhoven, F. M. A. H., 110, 120, 121(3), 241, 302

Schwartz, A., 9, 15, 49, 72, 87, 90, 92(4), 97(24), 100, 101, 110, 127, 141, 153(8), 156, 201, 202, 203, 205(6), 206, 211(6), 230, 232, 233(2), 428, 429, 438
Schwarzbauer, J. E., 380, 381(5)
Segal, I. H., 246, 249(49)
Seguchi, H., 427, 430
Seiler, S., 30, 35(24)
Sen, A. K., 5, 153, 177, 178(16), 302, 418
Serpersu, E. H., 319, 320(29)
Seubert, W., 116, 307, 308(29), 309(29)
Shaver, J. L. F., 428
Shaver, J., 215
Shertzor, H. G., 180
Shively, J. E., 92
Shortman, K., 167
Shull, G. E., 15, 16, 65
Siegel, G. J., 29, 116, 215
Silverstone, A. E., 380
Simkies, K., 427
Simmons, T. J. B., 24
Simpkins, H., 357
Singer, S. J., 334
Sinionescu, M., 439
Sinionescu, N., 439
Sinnott, M. L., 325, 327(12)
Skiver, E., 16, 17(70), 19(71), 29, 42(11), 80, 81, 82(8), 83, 85, 86(8), 87(8, 16), 291, 292(5), 293(10), 295(5), 299(5), 300(5), 431, 433, 437
Skou, J. C., 2, 3, 4, 5, 6(4), 7, 8, 10, 11, 12, 13(53), 14, 17(55), 18(35), 21(26, 43), 25(35), 29, 30, 31(2), 33(17), 36(2), 43, 46, 72, 73, 75(6), 76(7), 77(7), 105, 109, 110(3, 8), 114, 115, 157, 160(8), 162(8), 164(8), 167(8), 193, 233, 239, 245(17), 272, 274(9), 277, 278, 279(1, 3), 281, 302, 342(24), 372
Skriver, E., 30, 438, 439(18), 440(16), 441(18)
Slot, W. J., 411
Smith, P. L., 325, 327(12)
Smith, P. R., 85
Smith, R. E., 423
Smith, R. L., 242, 251, 252(5), 257(5), 258(5), 265(5), 266(5)
Smith, T. W., 46, 47, 48(4), 64, 355
Smith, W. L., 408
Solomonson, L. P., 358
Soulina, E. M., 180

Spechet, S. C., 65, 71(3), 423, 425(32)
Spector, A. A., 201
Spurr, A. R., 419, 424(20)
St. John, T. P., 380
Staehelin, T., 404
Stahl, W. L., 427
Steck, T. L., 114, 168, 174, 239
Stein, W. D., 179, 181(5), 183(5), 185(5), 186(4), 187(4, 6), 188(4), 239
Stevens, T., 252
Stirling, C. E., 215, 417, 428
Stowring, L., 303
Strauss, J. H., 168
Subbarow, Y., 107
Sverlov, E. D., 15
Swank, R. T., 93
Swann, A. C., 239
Swarts, H. G. P., 110, 120, 121(3), 241
Sweadner, K. J., 30, 48, 65, 68, 69, 71(3, 9), 120, 121(4), 127, 203, 382
Swift, T. J., 356

T

Tagaki, T., 79
Taguchi, M., 235, 236(17)
Takahashi, H., 15
Tamkin, J. W., 380, 381(5)
Tamura, I., 423
Tanford, C., 30, 116, 336
Taniguchi, K., 266
Tarussova, N. B., 315
Tashima, Y., 418
Taylor, E. A., 10
Taylor, G. M., 411
Tener, G. M., 303
Tervo, T., 417
Thomas, G. P., 383
Thomas, P. S., 387
Thomas, R. C., 1, 127, 128(8)
Titus, E. O., 302
Tobin, T., 18, 223, 302
Toda, G., 119, 121(2), 244
Tomey, J. M., 418
Tong, S. W., 72, 127, 135(3), 141, 150(2), 156
Tonomura, Y., 230, 234, 322
Toon, P. A., 116
Towbin, H., 404
Tran, C. M., 92, 322

Trentham, D. R., 272, 275
Trinnanman, B. K., 92
Tsilevich, T. L., 315
Tsui, J. Y., 427
Tsukahara, I., 423, 427
Twist, V. W., 203

U

Ueck, M., 427
Ueno, S., 423, 427
Uesugi, S., 313
Ugi, J., 315
Unakar, N. J., 427
Unwin, P. N. T., 85
Uusitalo, R., 417

V

van Deenen, L. L. M., 12
Vanaman, T., 93
Vengasa, P. B., 427
Villafranca, J., 361
von Braun, J., 314
von Ilberg, C., 116, 307, 308(29), 309(29)
Vorbrodt, A. W., 427
Voter, W. A., 80
Vuilleumier, P., 325, 328(13), 331(13)

W

Wachstein, M., 417
Wagenknecht, B., 201
Wagner, G., 370
Wagner, H., Jr., 46, 47(1), 64
Wallach, D. F., 168
Wallach, D. H. F., 114
Wallick, E. T., 92, 202, 203, 205(6), 206, 211(6), 230, 232, 233(2)
Wang, C.-T. 432
Ward, D. C., 273
Warner, R., 9
Warren, G. B., 116
Warren, L., 142
Watts, A., 371, 373(2)
Weber, K., 114, 347
Wehling, M., 206

Weigensberg, A. M., 178
Weiland, G. A., 215
Weinstein, R. S., 168
Weissman, I. L., 380
Wells, P., 417
Wellsmith, N. V., 202
Welsch, K., 99
Wempen, J., 303, 305(22)
Werdan, K., 201
Wetzel, R., 310
Wheeler, K., 353
Whitacker, J. N., 429
Whitmer, K., 206
Whittman, R., 353
Wider, G., 370
Wiedner, H., 310
Wigzell, H., 99
Wilde, C., 96
Wilderspin, A. F., 92
Williams, J. A., 214, 218(8, 10), 220(10), 222(8, 10), 223, 227(8, 10), 229(25)
Willis, J. S., 215
Winslow, J. W., 302
Wire, B., 121
Wisniewsky, H. M., 427
Wolf, S. C., 380
Womack, F. C., 191
Wong, S. M. E., 24, 302
Wood, J. G., 429
Wuthrich, K., 370

Y

Yamaguchi, M., 230, 234
Yamamoto, K., 427
Yanagisawa, M., 9
Yates, D. W., 4, 5(13, 14, 15), 18(13, 14), 19(13), 271, 272(2), 273(3), 275, 276(2, 4), 277(3), 279
Yeats, D. W., 246
Yoda, A., 110, 215
Yoda, S., 110, 215
Yoshida, H., 110, 424
Yoshioka, T., 427
Young, J. D., 13, 342
Young, R. A., 380, 382(6, 7), 384(6)
Yount, R. G., 303, 305(23)

Z

Zahler, W. L., 114
Zak, B., 107, 159
Zamphigi, G., 32, 85
Zimmerman, J., 315
Zinn, K., 242, 251, 252(5), 257(5), 258(5), 265(5), 266(5)
Zot, A. S., 64
Zot, A., 90, 92(4)
Zwibler, B., 201

Subject Index

A

ABD-ouabain, 325
 activation, by energy transfer, 326
 covalent labeling of digitalis-binding site, 331–332
 radioactive, synthesis, 330–331
 synthesis, 329–331
Acidic pH polyacrylamide gel electrophoresis, 124
Acinar cells
 Na,K-pump stimulation in, Ca^+ dependency, 226–227
 ouabain-sensitive oxygen uptake, effect of carbachol and epinephrine on, 228–229
 ouabain-sensitive $^{86}Rb^+$ uptake, effects of hormones and neurotransmitters on, 229
 tritiated ouabain binding, effect of carbachol and epinephrine on, 228–229
Action potential, 1
Active transport, 1, 119. See also Na^+,K^+-coupled active transport
 secondary, 2
Adamantane diazirine, in identification of functional domains of α subunit, 299, 300
Adenosine triphosphate. See ATP
ADP. See Nucleotide binding to Na^+,K^+-ATPase
Albers–Post scheme, 17, 19, 237, 238
Albumin, in protection against detergent inactivation of Na^+,K^+-ATPase, 76–77
(p-Aminobenzyldiazonio)ouabain. See ABD-ouabain
Anas platyrhynchos. See Duck
Antibody probes, for isolating cDNA clones, 380
Artemia salina. See Brine shrimp
ATP. See also Nucleotide binding to Na^+,K^+-ATPase
 analogs
  in inactivation/reactivation studies of Na^+,K^+-ATPase, 302–303, 310–312
 photoaffinity labeling with, 312–322
 preparation, 303–307
 hydrolysis, 2, 178
  as function of Na^+ plus K^+ concentrations, 2–4
  intrinsic protein fluorescence signals accompanying, 275
  turnover rate, 4
 photoaffinity labels, synthesis of, 313–317
8-Azidoadenosine 5'-triphosphate
 chromium(III) complex of, synthesis, 317
 radioactive, preparation, 314
 synthesis, 313–314
3-(4-Azido-2-nitrophenyl)propionic acid
 preparation, 315–316
 synthesis of 3'-ribosyl ester and ATP with, 316–317
3'-O-[3-(4-Azido-2-nitrophenyl)propionyl]-adenosine 5'-triphosphate
 chromium(III) complex of, synthesis, 317
 synthesis, 315

B

Bacteriophage λgt11 expression library, 380–382
 rat brain, screening of, with Na^+,K^+-ATPase antiserum, 382–383
Brain stem, axolemma
 isolation, 66–69
 Na^+,K^+-ATPase, 65–71
  $\alpha(+)$ isozyme, 65–71
  purification, 69–71
Brij 56, 79
Brine shrimp
 development of Na^+,K^+-ATPase activity during embryogenesis, 50, 51
 Na^+,K^+-ATPase
  α subunit
   molecular forms, 48, 61
   NH_3-terminal sequence, 64, 65

amino acid analysis, 61, 62
assay, 51–52
biosynthetic properties, 63–65
$C_{12}E_8$-solubilized, gel filtration of, 58–59
large-scale preparation, 50–51, 55–58
membrane-bound, preparation, 50–53, 55
partially purified, preparation, 50–51, 53–58
preparation, 48–65
purification, 58
purified, properties of, 60–65
small-scale preparation, 50–51, 52–55
subunits
　isolation, 58–60
　molecular properties, 60–62
　molecular weights, 60–62
rearing, 49–50
storage, 49–50
tert-Butyloxycarbonyl(Boc)-p-phenylenediamine, synthesis, 329

C

Calcium
　effect on Na,K-pump, 223
　effect on tritiated ouabain binding stimulated by agonists, 226
Carbachol, effects on Na,K-pump, 224, 228–229
Cardiac glycosides
　alkylating derivatives, 323
　binding to Na^+,K^+-ATPase, 201–213
　　estimation of affinities, from measurement of enzyme activity, 203–206
　　model, 203
　　radioligand binding, 206–213
　covalent derivatives, labeling of subunits of Na^+,K^+-ATPase, 324
　effect on Na^+,K^+-ATPase, 9, 323
$C_{12}E_8$, 11–13, 17, 31, 33, 72, 79, 334, 345, 353
　irreversible inactivation of Na^+,K^+-ATPase, 76
　in preparation of Na^+,K^+-ATPase, 335–336
　solubilization of Na^+,K^+-ATPase, 73–77
　solubilizing efficiency, 73

$C_{12}E_{10}$, 79
　as protective agent against detergent inactivation of Na^+,K^+-ATPase, 77
Cell suspensions, preparation, for tritiated ouabain binding to Na,K-pumps, 217–219
Cholate dialysis, 127, 141, 142
Cross-linking assay, with glutaraldehyde, in determination of quaternary structure of Na^+,K^+-ATPase, 333–345
Cross-linking reagents, for detection of subunit interactions of membrane-bound Na^+,K^+-ATPase, 345–350
Cr^{3+}, as paramagnetic probe for NMR studies of Na^+,K^+-ATPase, 357
Cs^+, occluded ions, in Na^+,K^+-ATPase, measurement, 287
Cyano-6-thioinosine triphosphate, radioactive, synthesis, 307

D

Deoxycholate, 11
　in preparation of Na^+,K^+-ATPase, 43–45
(Diazomalonyl)cymarin, labeling of subunits of Na^+,K^+-ATPase, 324
(Diazomalonyl)digitoxin, labeling of subunits of Na^+,K^+-ATPase, 324
Digitalis, inhibition of Na^+,K^+-ATPase, 323
Digitalis-binding site
　affinity labels
　　first generation of derivatives for, 323–325
　　newer, 325–326
　　types, 323
　covalent labeling of, 331–332
Digoxin, tritiated, binding to Na^+,K^+-ATPase, 201
S-(2,4-Dinitrophenyl)-6-mercaptopurine riboside 5'-monophosphate, preparation, 306
S-(2,4-Dinitrophenyl)-6-mercaptopurine riboside 5'-triphosphate, preparation, 306
Divalent cations. See also specific cation
　effect on Na^+,K^+-ATPase conformational state, fluorescence measurement, 275

Duck, supraorbital salt gland, Na^+,K^+-ATPase activity, 46, 47
  assay, 47
  preparation, 46–48
  stability, 47
  yield, 47, 48

E

Eel. *See Electrophorus electricus*
Egg phosphatidylcholine, preparation, 128–130
Electron microscopy
  analysis of Na^+,K^+-ATPase, 430–441
    embedding technique, 432
    fixation method, 431–432
    freeze–etching for, 437
    freeze–fracture for, 435–437
    results, 438
    rotary shadowing for, 436, 437
    staining for, 432
    thin sectioning for, 431–432
  in histochemical localization of Na^+,K^+-ATPase, 417–430
  of Na^+,K^+-transport vesicles, 438–440
Electron spin resonance study of Na^+,K^+-ATPase, 371–376
  applications, 371
  ESR measurement, 372–373
  information yielded by, 374–376
  methods, 372–373
  prelabeling in presence of glycol, 372
  reagents, 372
Electrophorus electricus, Na^+,K^+-ATPase
  from electroplax of, 32–33
  reconstitution, 141–155
Eosin, 5
  binding site, 278
  binding to Na^+,K^+-ATPase, assay, 278–279
  derivatives, 280–281
  as fluorescence probe for measurement of conformational states of Na^+,K^+-ATPase, 278–281
  limitations, 281
Eosin B, 280
Eosin S, 280
Eosin Y, 278, 280

Epinephrine, effects on Na,K-pump, 223–224, 228–229
Erythrocytes
  inside-out vesicles from
    adjustment of intravesicular ionic composition, 173–175
    Na^+,K^+-ATPase in
      ATP hydrolysis, 178
      enzyme assays, 171, 177–178
      K^+-activated phosphatase, 178
      Na^+–$K^+(Rb^+)$ exchange, 176
      Na^+–O flux and Na^+ exchange, 176–177
      phosphoenzyme intermediate, 177–178
      transport assays, 175–177
    preparation, 171, 173–175
  intact human
    rubidium (potassium) influx, effects of external Na and rubidium concentrations on, 255–259
    sodium efflux, effects of vanadate, 259–261
    potassium efflux from, effects of vanadate, 261–263
Erythrosin B, 280
Erythrosin Y, 280

F

FDP. *See* Formycin nucleotides
Fluorescein isothiocyanate, 5
  labeling of Na^+,K^+-ATPase, 272
Fluorescein isothiocyanate-labeled enzyme, in measurement of Na^+,K^+-ATPase conformational states, 271–277
FMP. *See* Formycin nucleotides
Formycin nucleotides
  in measurement of Na^+,K^+-ATPase conformational states, 271–277
  preparation, 272–273
Formycin triphosphate. *See* FTP
Freeze–thaw sonication, 142, 179–180
FTP (formycin triphosphate), 5. *See also* Formycin nucleotides
  binding and release, 279
  hydrolysis, transient fluorescence changes accompanying, 277

G

Glutaraldehyde
  chemical cross-linking of Na^+,K^+-ATPase with, 333–344
    assay, 337–341
    distribution of products of, on SDS–PAGE, 337–339
    effects of variation of protein concentration, 339–340
    nature of reaction, 336–337
    quantitative cross-linking assay with, advantages, in determining distribution of oligomers of multimeric proteins, 341

H

Hellebrigenin, haloacetyl derivative, 323
  labeling of subunits of Na^+,K^+-ATPase, 324
Hill equation, 246
  and allosteric enzymes, 246–247
Hormones
  effect on electrolyte transport, 213
  effect on Na,K-pump, 214, 223–224
Hybridoma, 392–393

I

Immunogold labeling, using monoclonal antibodies to Na^+,K^+-ATPase, 411–413
Internal dialysis, 252–253
Intrinsic protein fluorescence, in measurement of Na^+,K^+-ATPase conformational states, 271–277
(Iodoazido)cymarin, labeling of subunits of Na^+,K^+-ATPase, 324
Iodonaphthylazide, in identification of membrane-embedded regions of α subunit, 299, 300
$2',3'$-O-Isopropylidene-6-mercaptopurine, synthesis, 303–304

K

Kidney
  crude membranes, preparation by differential centrifugation, 36
  dissection of tissue from, 35
  gross structure, 35
  tissue homogenization, 36
K^+, 2–4
  effect on formation of phosphoenzymes in presence of Na^+, 24
  occluded ions, in Na^+,K^+-ATPase, measurement, 287
K^+–K^+ exchange
  (ATP + P_i)-dependent, uncoupled Na^+ efflux and, 22–23
  in red blood cells, effects of vanadate, 261–263
K^+-phosphatase, 109–110
  assay, 238
  cation requirements, 109
  detergent inactivation, 75–76
  effect of Mg^{2+}, 244
  inactivation by chymotrypsin or trypsin, 292–293
  incubation, 109
  p-nitrophenol product, measurement, 109–110
  reaction blind, 109
  substrate, 109

L

Lamb, kidney, Na^+,K^+-ATPase, preparation, 88–90
Lipid vesicles. *See also* Liposomes
  reconstitution with Na^+,K^+-ATPase, 157–158
  sealed, containing inside-out and right-side-out Na,K-pump, resolution of, 140
  separation of sealed from unsealed, 139–140
Liposomes. *See also* Phospholipid vesicles
  assay of Na^+ and K^+ transport in, 144–145
  electron microscopic study of, 438–441
  incorporation of purified Na^+,K^+-ATPase into, 144
  preparation, 143–145
  reconstitution with Na^+,K^+-ATPase, 157–158
    protein determination after, 158–159
    time course of, 158, 159
  transport studies with, 151, 179

Lowry method, 111–112, 158–159
Lubrol, 11, 13
Lubrol WX, 79

M

Maleimide spin labels
  differential labeling of sulfhydryl groups with, 373–374
  labeling of class II groups with, 374, 375
  labeling of sulfhydryl groups of Na$^+$,K$^+$-ATPase, 371
    method, 372
    results, 372–373
Medullary collecting tubule, 35
Membrane potential, 1
Membrane vesicles, 114
  inside-out, from erythrocytes. See Erythrocytes
  isolation, 34
  Na$^+$,K$^+$-ATPase purification from, 35
  reconstituted
    activation by cytoplasmic sodium, 183–184
    ATPase activity, effect of readdition of cholate, 138–139
    ATPase assays, 135–136
    ATP-dependent Na$^+$–K$^+$ exchange, 181–184
      after treatment with valinomycin plus FCCP, 184
    ATP-dependent Na$^+$–Na$^+$ exchange and uncoupled Na$^+$ flux, 185
    ATP-dependent ^{86}Rb efflux, 184
    double-label determinations of ATP-dependent changes in intravesicular Na$^+$ and K$^+$ content, 138
    by freeze–thaw sonication, 179–180
    inside-out and right-side-out populations, evidence for, 138–139
    net Rb$^+$ flux, 188
    passive Rb$^+$ fluxes in absence of ATP and phosphate, 187
    Rb$^+$–congener exchange, 188
    Rb$^+$–Rb$^+$ exchange
      activated by ATP or phosphate alone, 187
      activated by ATP and P$_i$, 185–188
      Rb$^+$ activation of, 187
    time course of ATP-dependent ^{22}Na uptake, 181–182
    transport assays, 136–139, 180–181
    uptake of externally added ^{22}Na$^+$, 137–138
    separation from open membrane fragments in Metrizamide gradients, 36, 37
6-Mercaptopurine riboside 5′-monophosphate, preparation, 304
6-Mercaptopurine riboside 5′-triphosphate, preparation, 304–305
Mg^{2+}, effect on photoinactivation of Na$^+$,K$^+$-ATPase by ATP analogs, 321, 322
Mn^{2+}, as paramagnetic probe for NMR studies of Na$^+$,K$^+$-ATPase, 357–366
Monoclonal antibodies
  advantages and limitations of, 393–394
  to Na$^+$,K$^+$-ATPase, preparation, 394–399
  production, hybridoma technology for, 392–393
  in study of epithelial transport functions, 392, 413

N

Na$^+$. See Sodium
Na$^+$,K$^+$-ATPase, 2. See also Na,K-pump
  active unit, 14–15
  activity
    coupled assay, 116–119
    definition of, 105
    and incubation time, 106–107
    measurement, 105–110
      in terms of P$_i$ release in colorimetric assays, 237
    and temperature, 106
    test of, 75–76
  affinities for ligands, 202
  α and/or α(+) forms, 120
    front-door phosphorylation from ATP, 121
    identification and quantitation, 123
    immunological identification, 120–121
  αβ units, 87
    electron microscopic analysis, 440
    soluble, with full activity, preparation, 42–43

α chain
 isoforms, 16
 molecular weight, 13
α_{III} isozyme, 65
α subunit, 11–12, 291, 323, 346, 379
 amino acid sequence, 15–16, 379, 390
 concentration in lipid bilayer, 32
 digitalis-binding domain, localization, 332–333
 E_1 form, cleavage by α-chymotrypsin A, 292– 293
 E_2 form, cleavage by chymotrypsin in presence of $MgCl_2$ and ouabain, 294–295
 folding through membrane, 300–301
 functional domains, identification of, 299–301
 labeling with covalent derivatives of cardiac glycosides, 32
 molecular forms of, 48–49
 monoclonal antibodies, 88
 peptides, preparation, 91–94
 proteolytic fragments, orientation, 298–300
 sites of proteolytic cleavage, 291
 tryptic cleavage of E_1 or E_2 form of, 293–294
α subunit cDNA
 cloning and detection of, scheme for, using λgt11 expression system, 380–382
 as probe for Northern blot hybridization, 386–388
 as probe for Southern blot hybridization, 389
 sequence analysis, 389–392
 subcloning into plasmid vector, 385–386
 synthetic oligonucleotide probe for, 390–392
α subunit fusion protein, characterization, 383–385
α subunit gene, molecular cloning, 379–392
 materials, 382
 methods, 382–392
antibodies, preparation, 87–101
antibody binding
 control proteins for, 100
 determination, 99–100

assay, 36–37, 47, 121–122
 in absence of K^+, 237–238
 during inactivation/reactivation, 309
 in presence of ATP, Mg^{2+}, Na^+, and K^+, 237
ATP and ADP binding to, measurement, 191–201
ATPase and phosphate activity, 105–115
ATP-binding site, 302, 312
 affinities of, 309, 310
 high-affinity
  binding capacity of, 309
  photoaffinity labeling at, 313
 NMR studies of, 358–367
back-door phosphorylation, 119–124
 acidic pH PAGE of pellet from, 124
 data analysis, 122–123
 method, 121
β chain, molecular weight, 13
β subunit, 11–12, 291, 323, 346, 379
 amino acid sequence, 15–16
 labeling with covalent derivatives of cardiac glycosides, 324
binding capacities
 for ATP and phosphate, 32
 for ouabain, 40, 41
 for vanadate, 40, 41
biogenesis studies of, 48
biological role, 213
bonding interactions between bound Mn^{2+} and N-7 and N-6 nitrogens, 366
cardiac glycoside binding, effects of other ligands, 202
cation binding, order of efficiency, 235
cation requirements, 105
cDNA, cloning strategy for, 380–382
$C_{12}E_8$-solubilized, incorporation into liposomes, 156–167
conformational states, 4, 271
 effect of K^+ on, measurement, 281
 effects of phosphate, vanadate, ouabain, and divalent cations, 275
 eosin as fluorescence probe for, 278–281
 equilibrium or steady state fluorescence measurements, 273–275
 fluorescence methods for measurement of, 271–277

SUBJECT INDEX 461

intrinsic protein fluorescence, monitoring, 274–275
K$^+$-nucleotide antagonism, 274
NMR studies, 363
photoreactive ATP analogs as probes of, 319–322
and sites of proteolytic cleavage, 291
transition, 4–5
  stopped-flow fluorescence measurement of, 275–277
conformation of CoATP bound at active site of, 365–366
coupled assay, 116–119
  principle, 116–117
  procedure, 117–118
  reagents, 117
covalent modification, information yielded by, 374–376
cross-linking studies
  materials, 346
  in presence of Cu^{2+} and o-phenanthroline, 347–349
  in presence of 1,5-difluoro-2,4-dinitrobenzene, 349
  procedure, 347–350
crystalline arrays
  image analysis of, 85, 86
  two-dimensional, 80–86, 353
  types of, 85, 86
crystallization, 16
  with vanadate/magnesium, 81–83
  with various ligands, 83, 84
detergent-dispersed, preparation, 335–336
detergent effect on, 11, 30, 72, 79
detergent inactivation of, 31
  protection against, 76–77
detergent-solubilized, 353
  cross-linking studies, 350
  reconstituted into phospholipid vesicles, sidedness, 156
digitalis-binding site, covalent labeling, 331–332
digitalis receptor, 323–333
electron microscopic analysis of, 430–441
electron spin resonance studies, 371–376
enzyme cytochemistry, 417–428, 430
  cobalt chloride method, 422
  for detecting K$^+$-NPPase activity, 430

for localization of K$^+$-NPPase activity, 419–421, 425–428
one-step lead citrate method, 422–425
strontium–lead method, 418–422
using ATP as substrate, 417–418
using p-nitrophenyl phosphate as substrate, 418–428
functions, 379
γ protein, 12
γ proteolipid, labeling with covalent derivatives of cardiac glycosides, 324
immunocytochemical method for localization of, 417, 429–430
inactivation and reactivation studies, 307–309
  using analogs of thioinosine triphosphate, 310–312
  using ([γ-^{32}P]s^{6}ITP)$_2$, 307–309
  ([γ-^{32}P]s^{6}ITP)$_2$-inactivated, tryptic map of, 309
inactive, 115
inhibitors, 9–10, 108
integrity, criteria for, 32–33
irreversible inactivation by C$_{12}$E$_8$, 76
isolation, 10–12
isozymes, 65. See also Na$^+$,K$^+$-ATPase, α subunit, molecular forms
K$^+$ binding, characteristics of, 235
kinetic studies, 236–237
  antagonism between K$^+$ and ATP, 242–243
  competition of cations for occupancy of binding sites, 245
  effect of conformation on affinity for ligand, 246
  effects of inhibitors and modifiers, 248–251
  kinetic equation for cation data, 246–248
  measurements of binding at high- and low-affinity sites, 240–242
  MgATP as substrate for, 240–241
  monvalent cation interactions, 244–248
  with multisubstrate reactions, ligand competition for binding sites in, 246
  patterns of product inhibition, 243

with *p*-nitrophenyl phosphate as substrate, 243-244
precautions for, 238-239
selection of kinetic model for, 246-247
substrate interactions, 239-244
labeling, with fluorescein isothiocyanate, 272
lamb kidney, preparation, 88-90
in large membrane fragments, preparation, for crystallization in two dimensions, 42
latency in plasma membrane preparations, 30-31
ligand binding
affinities of different conformational families in, 239
competition between ligands for specific sites, 239
kinetic studies, 236-251
ligand-modified inactivation, kinetic studies, 249-251
lipid requirement, 12
lipid spin labeling of, 371
from mammalian kidney, purity, 32
membrane-bound
cross-linking studies of subunit interactions of, 345-350
crystallization in two dimensions, 80-87
intramembrane particles, electron microscopic analyses, 440-441
isolation, 431
preparation for electron microscopy, 431-437
purification, 29, 81, 292
surface particles, electron microscopic analyses, 431-432, 434-438, 440-441
membrane crystals, electron microscopy, 83-85
microscopic localization, techniques, 417
molar activity, 111-114
molecular weight, 13-14, 33
molecules per cell, 10
monoclonal antibodies, 392-413
applications, 413
in examination of molecular structure of enzyme, 407
in immunoaffinity chromatography, 408-409

in immunocytochemical localization of enzyme, 409-413
in immunogold labeling, 411-413
incubation with, for electron microscopy of periodate-lysate-paraformaldehyde-fixed tissue, 410-413
isotype, identification of, 98
in physiologic studies, 408
in quantitative immunoassay using ELISA assay, 407
monoclonal antibody production, 96-99, 394-399
antigen preparation, 394-395
antigen presentation, for immunization, 396-397
antigen selection, 396
enzyme assay, 395-396
hybridoma cloning, 398
hybridoma production, 397-398
immunization of mice, 396
summary of steps in, 394-395
tumor ascites production, 399
monoclonal antibody screening, 399-406
ELISA method, 399, 400
functional, 399, 400-402
immunoblotting techniques, 399, 404-405
by immunofluorescence photomicrography of periodate-lysate-paraformaldehyde-fixed tissue, 405-407
by immunoprecipitation, 399, 402-404
methods, adaptation for enzyme characterization, 407-412
morphologic methods, 399, 405-407
Na^+ binding, characteristics of, 235-236
Na^+ and K^+ binding
assay, 230-233
and EDTA chelation, 232
effect of buffers, 232
effect of cation concentration, 233
effect of oligomycin, 232
effect of pH, 232
effect of temperature, 232
measurement
by centrifugation methods, 229-236
difficulties of, 229-230
micromodification of assay, 233
ouabain-sensitive, specificity, 234

simultaneous measurement of, 233
Na$^+$-ATPase activity, 108–109
Na$^+$-dependent phosphorylating activity, after modification with glutaraldehyde, 344
negative staining, for electron microscopy, 432–435
NMR studies of, 353–371
and concentration limits of enzyme, 354–355
future of, 370–371
methods, 354–355
nuclear relaxation studies with paramagnetic probes, 355
special considerations, 354–355
transferred nuclear Overhauser enhancement measurements, 355, 366, 367–371
nucleotide requirements, 106
oligomeric structure, 345–346
determination, approaches to, 334
optical density at 280 nm, 112, 113
optimal Na$^+$ plus K$^+$ concentration for, 105
ouabain-binding sites. See Ouabain-binding sites
ouabain-sensitive ATPase activity, reverse K$^+$-free effect on, 255
ouabain-sensitive K$^+$ binding, measurement, 231
from outer renal medulla
preparative procedures, 33–35
selective extraction by SDS, 33
tissue preparation, 33
paramagnetic probes for, 357–358
partial cross-linking, in absence of cross-linking agent, 345
partially active, 115
pH optimum, 105
phosphatase activity, 7–8
phosphopeptides
gradient slab gels for separation of, at low pH, 296–297
phosphorylation and separation of, after proteolysis, 295–299
separation on TSK gel filtration columns in SDS, 297, 298
sequential proteolytic cleavage, procedure for, 295
phosphorylated intermediate, trapping,
by quantitative cross-linking assay, 343–344
phosphorylation-induced formation of α,α-dimer and α,β-dimer, 348–349
in crude membrane preparations, 350
structural significance, 349–350
photoaffinity labeling with ATP analogs
effects of MgCl$_2$, 321, 322
results, 319–322
photoinactivation of, 317–318
photolabeled, 317–318
digestion with TPCK-trypsin, 318
gel chromatographic separation of peptides obtained in trypsinolysis of, 319
photoreactive ATP analogs as affinity labels for, 319–322
physical and structural studies of, difficulties of, 353
polypeptide cross-linking with glutaraldehyde
in analysis of diverse preparations, 341–342
applications, 341–344
reagents, 334–335
preparation, for nucleotide binding experiments, 193
preparations, 29–30
in various detergent solutions, oligomer content of, 341–342
preparative problems, 30–33
preparative procedures, 33–43
protease digestion, 92
protective agents, 108
protein purity, 114, 115
proteolytic cleavage, 291–301
procedures, 292–298
protomer interactions, cross-linking studies of, 346
purification, 10–12, 29–43
centrifugation in airfuge, 42
centrifugation in angle rotor, 41–42
centrifugation in zonal rotor, 39–41
determination of optimum concentration of SDS for incubation, 38
preparation of sample for sucrose gradient centrifugation, 39
results of, 16–17
purified, stability, 43

purity
    criteria of, 31–32, 110–114
    in NMR studies, 354
quaternary structure, determination, using chemical cross-linking with glutaraldehyde, 333–345
rabbit polyclonal antibodies to, preparation, 95–96
Rb^+ binding, measurement, 234
reaction scheme for, 237, 238
reaction with ATP, 5–7
reconstituted
    ATPase activity, 159
    catalytic activity, 17
    electron microscopic analyses, 438–441
    sidedness
        determination of, 160–167
        and protein:lipid ratio of proteoliposomes, 164–167
    transport activity, 17
reconstitution of, methods, 141–142
reincorporation into membranes of lipid vesicles, 17
release of occluded ions, rapid ion-exchange technique for measuring
    apparatus, 282–284
    principle, 281–282
    procedure, 283–287
    sensitivity, 287
reversal K^+-free effect, in vanadate-poisoned cells, 267
from shark, solubilized, purification, 77–79
solubilization, 72–79, 133–135
    goals of, 72
    methods, 73–77
    use of detergents, 72
solubilized
    affinity chromatography on Con A-Sepharose, 77, 78
    analysis, 77–79
    preparation, 157
    purification, 77–79
    size-exclusion chromatography, 77–79
soluble
    activity, 33
    preparation, 335–336
    purification, 29–31
    purity, 33

specific activity, 110–112, 114
    and cation binding, 232
    in crude membrane fragments, determination of, 36–38
specific binding of cations
    criteria for, 233–234
    effect of heat treatment, 233
strophanthidin-sensitive activity, assay, 335
structural analysis, using electron microscopy, 431
structure, electron microscopic analysis, 80–82, 438, 440
subunits
    antibodies, cross-reactivity with holoenzyme, 101
    antigenicity, 101
    isolation, 90–91
    labeling with covalent derivatives of cardiac glycosides, 324
    ligand-modified and ligand-induced cross-linking of, 347–350
    preparation for immunization, 95
sulfhydryl groups
    differential labeling of, with spin labels, 373–374
    electron spin resonance studies of, 371
    labeling with maleimide spin labels, 371–373
    modification of, 302–312
        effect of ATP and ITP, 302–303
        reversible and irreversible, 302–303
    number of, 302
supernatant–$C_{12}E_8$ enzyme, 335
    separation on linear sucrose gradient containing $C_{12}E_8$, 340
temperature, and activity, 73, 75
test medium, composition, 106
three-dimensional model, 16
tissue sources, 29
Tl^+ binding, and EDTA chelation, 232
transport assays, 120
turnover, quantitative cross-linking assay during, 342–343
turnover number for ATP, 123
from whole brain microsomes, purification, 71
Na^+,K^+-coupled active transport
    in dialyzed squid giant axon, 252–255

in red blood cells, 255–259
  effects of vanadate, 259–261
Na$^+$–K$^+$ exchange
  and ADP-dependent Na$^+$–Na$^+$ exchange, 18–21
  by Na$^+$,K$^+$-ATPase
    coupling between chemical reactions and transport, 17–23
    scheme for, 17–19, 24
    simultaneous binding of K$^+$ to extracellular sites and Na$^+$ to cytoplasmic sites, 25
Na,K-pump. See also Na$^+$,K$^+$-ATPase
  abnormal modes of cation transport, 179
  activity
    agents affecting, 214
    changes in, assessment of, 223–228
    corroboration of [^{3}H]ouabain-binding data with other techniques, 228–229
  cholate dialysis reconstitution of, 127–140
    cholate solubilization and hollow fiber dialysis, 130–135
    disadvantages of, 141
    preparation of egg phosphatidylcholine, 128–130
  as energy transducer, 1
  equilibrium binding conditions for ouabain, determination, 219–221
  equilibrium binding constant for ouabain, determination, 219
  incorporated into sealed vesicles, density gradient purification of, 139–140
  in inside-out vesicles
    K$^+$-dependent p-nitrophenylphosphatase, 172, 173
    Na$^+$–K$^+$ exchange, 171–173
    Na$^+$–Na$^+$ exchange, 172, 173
    Na$^+$/O flux, 172, 173
    phosphoenzyme intermediate, 172, 173
  Na,K transport after reconstitution
    coupling ratio of, 153–154
    effect of temperature on, 151–153
    electrogenicity of, 154
  physiological function, 1–2
  quantitation in intact cells, 213–229
  reconstituted. See also Membrane vesicles, reconstituted

    inside-out fraction, tansport activity, 127–128
    Na$^+$ transport, effect of purity of Na$^+$,K$^+$-ATPase on, 149
    sidedness of, 156
  reconstitution by freeze–thaw sonication, 142–155
    effect of asolectin concentration, 146–147
    effect of sodium concentration during reconstitution, 147–149
    effect of sonication time, 145–146
    effect of various parameters on, 145–151
    efficacy, 154–155
    materials, 142–143
    optimal conditions for, 155
    potassium concentration during reconstitution, 149
    stability of Na$^+$,K$^+$-ATPase to sonication after freeze–thaw step, 149–151
  translocation reactions in, inhibition by vanadate, 265–267
  turnover
    information yielded by, 214
    in intact cells, estimation, 213–229
  vanadate inhibition of, 265–267
    effect of ATP, 252
    effect of sodium, 252
    K$^+$ as cofactor in, 252
    Mg^{2+} requirement, 251
Na$^+$–Na$^+$ exchange
  ADP-sensitive, 18–21
  ATP hydrolysis-dependent, 20–22
  in dialyzed squid giant axon, 252–255
  in red blood cells, effects of vanadate, 259–261
Na-pump. See Sodium pump
Nephron, 35
(Nitroazidobenzoyl)ouabain, labeling of subunits of Na$^+$,K$^+$-ATPase, 324
(Nitroazidophenyl)ouabain, labeling of subunits of Na$^+$,K$^+$-ATPase, 324
(Nitroazidophenyl)strophanthidin, labeling of subunits of Na$^+$,K$^+$-ATPase, 324
1-Nitrophenyl-3-(2-aminoethyl)triazene. See p-Nitrophenyltriazenylethylenediamine

p-Nitrophenyltriazenylethylenediamine, synthesis, 326–327
(p-Nitrophenyltriazenyl)ouabain. *See* NPT-ouabain
NPT-ouabain, 325
  covalent labeling of digitalis-binding site, 331
  radiolabeled, synthesis, 328–329
  reaction with digitalis-binding site of Na$^+$,K$^+$-ATPase, 325
  synthesis, 326–329
Nuclear magnetic resonance, in investigations of Na$^+$,K$^+$-ATPase, 353–371
Nuclear relaxation with paramagnetic probes
  applications, 355
  and determination of site–site distances in enzyme, 361–362, 364–365
  probes for studies of Na$^+$,K$^+$-ATPase, 357–358
  ^{31}P and ^{1}H NMR studies of ATP site of Na$^+$,K$^+$-ATPase, 358–367
  theory, 355–356
Nucleotide binding to Na$^+$,K$^+$-ATPase, measurement, 191–201
  binding experiment, 191–192
  binding isotherm, 192
  centrifugation method, 191, 200
  data processing, 200–201
  isotherms, information obtained from, 191
  principle, 191–192
  rate of dialysis method, 191, 193–200

O

Ocatethylene glycol dodecyl monoether. *See* C$_{12}$E$_8$
Oligomycin, 24
  inhibition of Na$^+$,K$^+$-ATPase, 10
Osmotic regulation, 1
Ouabain, 9, 379
  binding to Na$^+$,K$^+$-ATPase, 40, 41, 233
  configuration specificity, 215, 216
  covalent derivatives, affinity labeling of digitalis-binding site of Na$^+$,K$^+$-ATPase, 325–326
  in differentiation of Na$^+$,K$^+$-ATPase incorporated into liposomes and unincorporated enzyme, 150–151
  dissociation from Na$^+$,K$^+$-ATPase, 211–213
  effect on Na$^+$,K$^+$-ATPase conformational state, fluorescence measurement, 275
  and identification of α and $\alpha(+)$ forms of Na$^+$,K$^+$-ATPase, 120
  inhibition of Na$^+$,K$^+$-ATPase, 106, 205–206
  oxidized, 323
    labeling of subunits of Na$^+$,K$^+$-ATPase, 324
    preparation, 327
  spin-labeled analog, as paramagnetic probe for NMR studies of Na$^+$,K$^+$-ATPase, 357–358
  tritiated
    in assessment of changes in Na,K-pump activity in intact cells, 223–228
    binding in intact cells, 213–229
    binding to Na,K-pump
      agonist stimulation of, concentration dependence of, 224–225
      effects of Ca$^+$ on, 223
      effects of K$^+$ on, 222–223
      saturation analysis of, 221–222
      specificity, 215
    binding to Na$^+$,K$^+$-ATPase, 201, 202, 207–213
    to monitor Na,K-pump activity, rationale, 215–217
Ouabain-binding sites, autoradiographic localization, 417, 428–429
Ouabain/ionophore assay, for determination of sidedness of reconstituted Na$^+$,K$^+$-ATPase activity, 160–162

P

Periodate–lysate–paraformaldehyde fixation, for tissue studies of Na$^+$,K$^+$-ATPase using monoclonal antibodies, 406–413
Phloxine, 280, 281
Phloxine B, 280, 281
Phosphate
  effect on Na$^+$,K$^+$-ATPase conformational state, fluorescence measurement, 275

inorganic
  Baginski measurement method, 107–108
  Fiske–Subbarow measurement method, 107
Phospholipid, as protective agent against detergent inactivation of Na$^+$,K$^+$-ATPase, 77
Phospholipid vesicles. *See* Liposomes
Plasma membrane
  latency of Na$^+$,K$^+$-ATPase in preparations of, 30–31
  Na$^+$,K$^+$-ATPase, 10–11
  from outer renal medulla, isolation, 33–35
Potassium. *See also* K$^+$
  efflux from red blood cells, effects of vanadate, 261–263
Proteoliposomes
  hydrolytic activity of, 160–161
  protein:lipid ratios, 164–167
Pump reversal, in red blood cells, effects of vanadate, 262–263

Q

Quantitative amino acid analysis, 111–112

R

Rb$^+$, occluded ions, in Na$^+$,K$^+$-ATPase, measurement, 287
Reabsorption processes, 2
Red blood cells. *See* Erythrocytes
Red cell ghosts, resealed, sodium efflux in absence of external Na$^+$ and K$^+$, 263–265
Reticulocytes, vesicles from, Na$^+$,K$^+$-ATPase activity, 178
Rubidium. *See also* Rb$^+$
  influx into intact human red blood cells, effects of external Na and rubidium concentrations on, 255–259

S

Saponin, in preparation of Na$^+$,K$^+$-ATPase, 44, 45
Saturation transfer electron spin resonance, 375–376

Secretory epithelial cells, tritiated ouabain binding, effect of chloride depletion on agonist stimulation of, 227–228
Sodium deoxycholate, 30–31
Sodium dodecyl sulfate, 11, 31
Sodium, 2–4
  occluded ions, in Na$^+$,K$^+$-ATPase, measurement, 287
  phosphoenzymes formed in presence of, 24
Sodium pump, 119. *See also* Na,K-pump; Na$^+$,K$^+$-ATPase
Spectra/Por hollow fiber bundles, 131–133, 135
Spin-labeled probes, as paramagnetic probes for NMR studies of Na$^+$,K$^+$-ATPase, 357–358
Squalus acanthius
  Na$^+$,K$^+$-ATPase, reconstitution, 141–155
  rectal glands
    microsomes, preparation of, 44
    Na$^+$,K$^+$-ATPase from, 33
      assay, 46
      preparation, 43–46
      purification procedure, 44–46
      purity, 32, 46
      reconstitution, 141–155
Squid giant axon, dialyzed, sodium efflux studies, 252–255
Strophanthidin, 171
  haloacetyl derivative, 323
  labeling of subunits of Na$^+$,K$^+$-ATPase, 324

T

Thick ascending limb of Henle, 33–35
Thin ascending limb of Henle, 35
6-Thiocyanato-9-β-D-ribofuranosylpurine 5′-triphosphate, preparation, 306
Thioinosine monophosphate
  preparation, 304
  tributylammonium salt of, preparation, 305
Thioinosine triphosphate
  derivatives
    as affinity labels of Na$^+$,K$^+$-ATPase, 310–312
    preparation of, 303–307

disulfide, preparation of, 305–306
preparation, 304
[γ-^{32}P]Thioinosine triphosphate, preparation, 307
Tl$^+$, occluded ions, in Na$^+$,K$^+$-ATPase, measurement, 287
Transferred nuclear Overhauser enhancement measurements, in studies of conformation of bound substrates or activators, 355, 366–371
Trifluoromethylphenyldiazirine, in identification of membrane-embedded regions of α subunit, 299, 300

U

Uncoupled Na$^+$ efflux, in resealed red cell ghosts, effects of vanadate, 263–265
Uphill transport, 2

V

Vanadate
binding site, 252
binding to Na$^+$,K$^+$-ATPase, 40, 41, 265–267
crystallization of Na$^+$,K$^+$-ATPase with, 80–81, 85–87
effect on Na$^+$,K$^+$-ATPase conformational state, fluorescence measurement, 275
inhibition of ATPase, 237
inhibition of Na$^+$,K$^+$-ATPase, 9–10
inhibition of translocation reactions, 251–267
Vanadate-binding assay, for determination of sidedness of reconstituted Na$^+$,K$^+$-ATPase activity, 162–167
Vanadate-poisoned cells, reverse K$^+$-free effect in, 255
Vanadium, in regulation of Na,K-pump, 251

231921